Introduction to Social Statistics

This book is dedicated to
Elizabeth and Maximilian

Introduction to Social Statistics

The Logic of Statistical Reasoning

Thomas Dietz and Linda Kalof

A John Wiley & Sons, Ltd., Publication

Blackwell Publishing was acquired by John Wiley & Sons in February 2007. Blackwell's publishing program has been merged with Wiley's global Scientific, Technical, and Medical business to form Wiley-Blackwell.

Registered Office
John Wiley & Sons Ltd, The Atrium, Southern Gate, Chichester, West Sussex, PO19 8SQ, United Kingdom

Editorial Offices
350 Main Street, Malden, MA 02148-5020, USA
9600 Garsington Road, Oxford, OX4 2DQ, UK
The Atrium, Southern Gate, Chichester, West Sussex, PO19 8SQ, UK

For details of our global editorial offices, for customer services, and for information about how to apply for permission to reuse the copyright material in this book please see our website at www.wiley.com/wiley-blackwell.

Library of Congress Cataloging-in-Publication Data

Dietz, Thomas.
Introduction to social statistics / Thomas Dietz and Linda Kalof.
    p. cm.
  Includes bibliographical references and index.
  ISBN 978-1-4051-6902-8 (hardcover : alk. paper)   1. Social sciences–Statistical methods.
2. Statistics. I. Kalof, Linda. II. Title.
  HA29.D469 2009
  519.5–dc22

                                                                    2008030380

A catalogue record for this book is available from the British Library.

Set in 10.5/13pt Minion by Graphicraft Limited, Hong Kong

1   2009

CONTENTS

TABLES

FIGURES

PREFACE: A STRATEGY FOR APPROACHING QUANTITATIVE ANALYSIS

Outline

"Statistics is hard. Statistics is important."

It is rare to have a student who takes introductory statistics as an elective. This seems consistent with the first statement about the difficulty of learning statistics but is not consistent with the second about the importance of statistics. Our experience in teaching statistics has led us to an approach that tries to make statistics less difficult while clarifying why statistics is important. In this preface, we want to sketch the thinking that underlies our approach.

Statistics is Hard

If you approach the introductory statistics course thinking that this is a difficult subject, you are not alone in that perception. The statistician Persi Diaconnis is reported to have said, "Our brains are just not wired to do probability problems very well" (Quoted in Bennett, 1998, p. 1). A variety of interesting evidence reinforces the belief that introductory statistics is a difficult subject to master. Both history and psychological research indicates that statistical thinking does not come easily to humans.

History

As early as 3000 BC, people in ancient civilizations in the Middle East were using dice to add a random element to games. Random processes and chance are mentioned in the *Mahabharata* of India, the *Iliad* of Greece, and the Old Testament of the *Bible*. Random selection of hexagrams is the basis of the *I Ching* of China. So humans have known about the random character of some events, and made use of that randomness, for 5,000 years.

But it was not until the 1600s, when Galileo correctly analyzed chances in games based on dice, that people began to understand the probabilities that underpin random processes. As we will see it's not too hard to figure out the odds when you have just one die. But in the ancient and medieval world it was common to use three dice in many games, rather than the two dice that we typically use. This makes the problem much harder.

It took the keen insights of a scientist like Galileo to understand what the odds of various results were when the game used several dice. So it's not surprising that something that has taken humanity four and a half millennia to understand might be hard to master in a semester.[1]

Psychology

Cognitive and developmental psychology also support the idea that it takes some effort to learn statistics. Studies of young children indicate that statistical reasoning is difficult. Piaget and Inhelder (1975) found that children come to understand the ideas of chance and random processes at a later age than when many other forms of quantitative thinking are mastered. A substantial body of work in cognitive psychology demonstrates the difficulty that even rather sophisticated adults have in working with evidence that is couched in terms of probability and uncertainty.[2] So again, it is not surprising that a course in statistics is viewed with some dread.

The difficulty of learning statistics is compounded for the introductory *social* statistics course. Many social science majors have only limited backgrounds in mathematics. For most students in introductory statistics courses it has been several years since they have had a course that required the use of numbers other than page numbers.

Most social science courses make little if any use of quantitative material, so the social science major is not likely to get much exercise in quantitative thinking. In contrast, those who teach statistics often have substantial and frequently exercised math skills. This can create an unfortunate pedagogical gap – the teacher may have forgotten, or may never have known, just how hard it is for most people to learn statistics. One of our major goals in this book is to get past that block – to present statistics in a way that is both faithful to the way statistics is used in the twenty-first century and to explain statistics in a way that builds on how you can best learn it.

Statistics is Important

Why are social science majors required to take a statistics course? Why are students who want to go into a helping profession such as social work or nursing and who have no intention of conducting their own research or going on to graduate school, required to take the course? We believe there are two very good reasons that undergraduates in the social sciences and related fields should understand statistics.

First, statistics is an important way to look at the world. Life is filled with chance events, and every day, virtually every hour, most of us must make decisions that are based on an assessment of uncertainty and chance. Will it rain? How long will it take to get to work or school? Will my friend be late for our lunch date? Some of the most consequential decisions we make incorporate uncertainty. Medical diagnoses and treatment outcomes are known only in terms of chances – some diagnoses are wrong, and almost any treatment sometimes fails. In US law, we ask juries to make judgments based either on "a preponderance of evidence" or "beyond a reasonable doubt." Both phrases indicate that the jury will not be able to know with absolute certainty the guilt or innocence of the accused.

We have to deal with *probability problems* – problems of making judgments in the face of uncertainty – every day. Statistics provides a set of carefully developed tools for dealing with that uncertainty. The logic of statistics, the insight it gives into the role that randomness plays in our lives, can be very helpful.

In a compelling example we will discuss in more detail later, the evolutionary theorist Stephen Gould has reported how his knowledge of statistical concepts was a great comfort to him when he was diagnosed with a usually fatal cancer (Gould,

1996). Therefore, one reason to study statistics is as an aid to thinking about key aspects of our daily lives.

A second reason for studying statistics moves from the personal to the public. The term statistics is derived from the Latin term for state or government. Statistics is an essential tool for understanding the modern world, and few professionals can conduct their jobs without relying on statistics, either directly or indirectly. The reputation of cities, municipal governments and police departments depend in part on crime and judicial statistics and their changes. Colleges are very concerned with how they rank in the guidebooks published each year, because a move up or down in the rankings is believed to have great impact on applications. Most of the environmental threats to human health and well-being are understood as a result of statistical analysis – whether of the risk of cancer from lead in drinking water or the threat of climate change from greenhouse gas emissions. Whatever one's choice of profession and one's job, statistics convey critically important information.

It is commonplace to be cynical about statistics. Mark Twain reported that the British statesman Benjamin Disraeli once said: "There are three kinds of lies: lies, damn lies and statistics."[3] More recently, the postmodern theorist Jean Baudrillard (2003) has said: "Like dreams, statistics are a form of wish fulfillment."

These criticisms are provocative. But they are the result of an uncritical and naïve belief in the veracity of numeric data, followed by the rage of a jilted suitor who has found the object of affection to be untrue. It is possible to lie and distort with statistics, though not as easily as it is to lie and distort with prose. Statistical data can be misinterpreted by the naïve, or distorted and selected to serve predetermined ends. As Andrew Lang once said: "He uses statistics as a drunken man uses lampposts – for support rather than illumination" (quoted in Gaither and Cavazos-Gaither, 1996, p. 249). But this is not an inherent fault of statistics but rather a fault of analysis whenever public discourse is not careful to examine critically the logic of an argument and the empirical support for it.

One reason to learn more about statistics is to have a sharper ability to think critically about the quantitative data that is so ubiquitous in the modern world, and as a result be able to separate what is reasonable from what is not. Those who reject quantitative data blind themselves to much important information, but those who learn to understand and critically evaluate statistical reports can have their vision much improved.

Quantitative Analysis as Craftwork

Part of the motivation for this book comes from the experiences of one of the authors in learning statistics. In this section, we sketch that process.

"I enrolled in my first course in statistics as an undergraduate. Like most students, I found the introductory course boring and rather difficult. Even when I understood the techniques being presented, it was not clear how they could be applied to problems of theoretical or practical interest, other than the contrived or dull examples in the text. It was three years before I took another course in statistics.

Like the first course, the second course was well taught, and it seemed a little more interesting. But I don't recall any special interest or excitement. Since then I've taken perhaps a dozen other courses in statistics. Some were well taught, while in others the pedagogy left a great deal to be desired. But over time, my interest in the field grew, to the point where I found (and still find) considerable intellectual excitement in good statistical work.

As I learned more statistics, I found it less boring because I could more easily see the ways in which statistical techniques could be applied to important problems. At first I thought this was because I had finally gotten on to the "good stuff," the techniques that were powerful enough to be of use in problems other than contrived examples. To some degree that is true. In the social, environmental and health sciences, the simplest statistical tools cannot cope with the complexity of the world. Our data usually come, not from controlled experiments but from observations of the historical trajectory of the world as it evolved. We share this problem with astronomy, geology, and, to some extent, ecology and evolutionary biology. So we need to cover in the first course the kind of tools that make sense of our data and answer questions that were of practical importance and theoretical interest."

But being able to see uses for statistical techniques is not simply a result of learning more sophisticated methods. Indeed, increased interest and understanding comes not so much from formal knowledge of statistics as from a growing understanding of the craft of statistics and quantitative analysis. When beginning the study of statistics, most students assume that the way to approach the subject matter is as a cookbook, or a paint-by-numbers kit. "For situation A, use technique number 23." Many statistics texts highlight a "decision tree" that asks a series of questions about the character of the data to be analyzed. At each branch in the tree, the student answers a question about the data and is led to another branch, until finally the tree suggests the proper statistical tool to use. Some computer statistical packages also include such aids. By following the path corresponding to the answers, the student selects from the tree an appropriate statistical procedure for the problem under consideration.

An ability to match statistical techniques with the appropriate conditions for their use must be part of education in statistics. But we are skeptical of too much reliance on decision trees that provide a "cookbook" approach to the choice and interpretation of statistical tools. A mechanical approach to data analysis will not provide any insight into the data being analyzed.

Worse, a mechanical approach easily becomes divorced from the substantive and theoretical interests that fascinate and motivate most scientists. The decision-tree approach can easily be automated into a computer program using rather simple elements of artificial intelligence, thus eliminating the need for the thoughts and skills of a quantitative analyst. The result is that users of statistics don't think carefully about what they are doing. They miss important insights that might emerge and they may make egregious errors.

In a series of papers over the past decade, the statistician David Freedman has offered a series of criticisms of statistical practice in the social sciences and many prominent social statisticians have responded to his critique.[4] We recommend the debate to anyone who wishes to do quantitative research in a non-experimental field. In our view, many of the problems that Freedman and others have identified come from a mechanical approach to quantitative analysis. For good results, applications of statistics require understanding basic principles, careful thought about the problem in hand, and craft in using available tools to shape the available materials into an interesting and defensible result. In this regard, statistical analysis is like many other human endeavors.

Cooking provides an analogy. With standard ingredients, a good dish can be prepared by strictly following a cookbook. But a great dish will be rare. The great cooks of the world are sensitive to the character of the materials at hand – the smell, taste, and texture of produce, spices, and so on. With highly variable ingredients, the cookbook result may be less than what could be achieved with more flexibility. Far worse, if you assume you have what the cookbook calls for when you don't, the results can be a disaster. If you try to bake at 350°F, as called for by the recipe, but your oven actually produces a temperature of 450°F when set at 350, your food will likely burn.

Mechanical application of a recipe for data analysis will work if all the ingredients are as described in the cookbook, which usually means that all assumptions underlying the procedure are met. But careful thought about the assumptions, the theory, and the data, rather like an understanding of the actual material available for the recipe, can produce a better result. And if you make assumptions without giving any thought to the actual situation, you may be baking at a temperature 100° too hot.

What is missing in a mechanical approach is the factor critical in making quantitative methods interesting and useful: a sense of *quantitative data analysis as a craft*. Formal statistical techniques are the tools of the craft of quantitative social science. The student must understand those tools, their strengths and weaknesses, where they can be applied with power and subtlety and where they provide only crude force. But that is only the first, and least interesting, step in the process of becoming a quantitative analyst. More important is developing skill, insight, judgment, and ultimately a sense of esthetics regarding good work.

The dynamics of industrial capitalist economies over the past 100 years have led to the deskilling of labor in many realms of human activity.[5] As a result, most students of statistics are not familiar with the traditional process by which an individual learned a craft. Indeed, our particular understanding of the process in its traditional form comes mostly from conversations Dietz had with his father, who was a master machinist. The process by which he learned his craft is exactly that by which most quantitative social scientists develop their skills. Thus we believe it is worth sketching.

In traditional crafts, the student begins as an apprentice. Before the spread of formal education the apprentice might know nothing of the craft he or she would study and spend the first few years learning the absolute rudiments of how to use the tools and materials appropriate to the craft. Recently, high schools and trade schools began to provide some of the basic education required for a craft. This is parallel to learning the fundamental concepts of statistics in an introductory course, including the sampling experiment, the sampling distribution and the use of hypothesis tests and confidence intervals. Whether in school or as an apprentice in the shop, the student's education in a craft would require that considerable time be spent on menial tasks and the repetitive use of a few simple tools and procedures. This is rather like the exercises that students must complete as part of basic statistics courses. Practice is essential for mastering the basics.

After a few years, or a few courses in the case of statistics students, the apprentice had a good command of the basic tools and procedures of the craft, and was promoted to the stage of journeyman.[6] Originally the journeyman would in fact journey from place to place to see a variety of different styles, tools, materials, and approaches to problems, working with several masters and learning from each. In more recent times, it became possible to develop diverse experience without traveling.

In the process of learning, the journeying craftsperson would begin to develop a personal style and approach. The culmination of this stage comes with the production of a masterwork (a master piece), which demonstrates not only superb command of appropriate skills and tools but also the ability to design and execute an original work of high quality.

In learning quantitative social science analysis, the journey phase corresponds to taking additional courses and learning new techniques, as well as to that part of the graduate and professional career during which the student reads the literature, writes papers and reports based on the use of various techniques and works with one or more senior researchers.

The point of this comparison is that quantitative social scientists cannot develop adequate skills simply by learning all that one text or one set of lectures offers. Learning a variety of techniques and approaches is necessary but not sufficient. Real skill comes with repeated application to problems, to broad reading and critique of the applications of others, and to serious effort devoted to producing work of increasing quality, perhaps eventually leading to the masterpiece. Dietz's experience with statistics was one of boredom until he had mastered sufficient tools and developed sufficient skill to be able to do interesting, "journeyman" work. Once he reached that level of skill, each new technique and each new instructor added to his repertoire, and provided new perspectives on old problems. And that in turn made the process of learning quantitative methods quite interesting though it remains hard work.

This implies that we take a somewhat different approach from most statistics texts. The traditional introductory statistics text for social science majors proceeds as if there are a key set of techniques to be mastered – the most basic elements from the decision tree mentioned above. Few methods invented since 1950 are covered in that kind of text. The result is a student ready to read the research literature of the 1950s, but who will be lost in the current literature – a point we will return to below.

What may be worse, the student learns these basic techniques sufficiently well to pass tests and carry out exercises, but we worry that they never learn to think statistically. The specific tools are useful, we believe, only as examples of a more general way of thinking about data and data analysis. Most students forget most of the specific techniques soon after the end of the course. But if they have mastered the key concepts of statistics, they can relearn these tools very quickly, and can learn new tools as well. So our goal in this text is to emphasize those key concepts and ways of thinking. The specific tools are examples of the general ideas and approaches used in current quantitative analysis.

The Discourse of Science

How does science work and what does that imply about quantitative methods? Too often, science is perceived as asocial and absolute. A person posits a law and it is right or wrong. A model does or does not describe reality. (The Freedman debate we recommended above is weaker than it might be because most participants in it seem to hold, at least implicitly, this static, individualistic view of science.)

In fact, science is a process, a discourse of people guided by rules about how to make an argument that are shared in a community. The sharing of rules is imperfect, but good enough that mutually coordinated action goes on. People understand what other people are saying and accept some assertions as correct and others as incorrect, some as legitimate and others as not legitimate.

Consider arguments that might be made about why states in the US differ in their homicide rates. Some might suggest that poverty breeds violence, and therefore states with high levels of poverty will have high rates of homicide. Such a theory, stated in simple and absolute terms, would be something like: "The more poverty in a state, the higher the homicide rate." The graph in Figure P.1 provides some evidence regarding that argument. It shows the relationship between the percentage of people in a state below the federal poverty line in 2002 and the homicide rate in 2003.

Instead of thinking of the theory that poverty causes homicide as either right or wrong, we can think of a process by which we learn about what drives homicide. In looking at Figure P.1 someone might argue that this data seems consistent with the poverty/homicide theory. States with more people in poverty tend to have higher homicide rates. But a key rule in the discourse of science is the need to entertain alternative explanations. Indeed some philosophers of science feel a proposed explanation must be falsifiable if it is to be considered a scientific explanation. In this view, an explanation is a scientific one, only if we can think of ways it could be disproven. If we try to disprove it and fail, then that lends credibility to the explanation. If no evidence could disprove the explanation, then it doesn't fall within the realm of science.

There could be a number of problems with the argument that poverty causes homicide. It may be that both poverty and homicide are concentrated in cities and

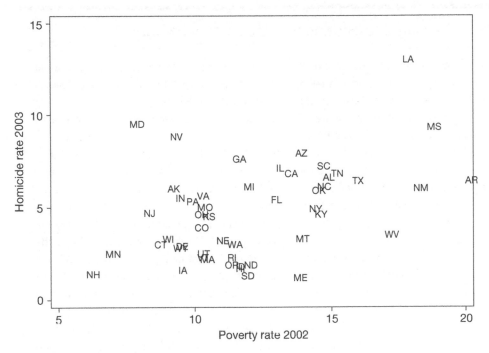

Figure P.1   Scatterplot of homicide rate versus poverty rate, N = 50
Data source: US Census Bureau 2000, 2002, 2003, analyzed with Stata.

so it is really urban life that drives homicide. Many of the high homicide states are in hot climates, so perhaps climate has something to do with homicide – perhaps warm weather allows more of the kind of social interactions that lead to violence. In the discourse of science, we might propose those explanations and find data to see if they provide a better explanation than the argument that poverty is what is driving homicide.

Science proceeds by proposing models, comparing them to data, considering alternative explanations and trying to find ways to choose among the alternatives. The plot in Figure P.1 gives some support to the idea that homicide and poverty are related. It is consistent with the argument that poverty causes homicide. But the plot is also consistent with the two other arguments we have suggested. Figuring out what is really going on involves a discourse in which both data and theory are invoked to assess the plausibility of arguments.

To examine these other two arguments, we would want data on urbanization and on climate. We could examine how well those explanations fit the data. Probably, we would find that there is some truth to more than one theory, that homicide isn't the result of a single factor but that several factors contribute to it. Science is a process in which we hold either an internal dialogue ("What else should I consider in this analysis?") or a public dialogue ("Your argument is incorrect because you didn't take account of the following factors . . .") or both.

Feature P.1 Reading Statistics: "Quants" vs "Quals"

There are two major approaches to empirical research in the social sciences. One is labeled quantitative because it uses numbers to try to understand the social world. In this tradition, data are collected by conducting surveys in which everyone is asked the same set of questions, or by making use of the numbers collected by government or by other organizations or by otherwise gathering information in a form that allows what is observed to be captured by numbers.

In contrast, qualitative research makes use of words and sometimes images rather than numbers. Researchers may do in-depth interviews where each respondent is asked questions as they are in a survey, but as the interview proceeds, questions are tailored to what the respondent has already said. Researchers may do participant observation where they "hang out" in a setting of interest to them and take notes about what is going on. Documents, including texts and photographs, and sound recordings may also be examined.

Over the past few decades, there has been some tension between "quants" (those using numbers) and "quals" (those not using numbers in their research). Until recently, most researchers were trained to use the tools of one approach but not the other. In extreme cases, a kind of intellectual xenophobia developed in which quals denied that quantitative research was valid, and quants denied the utility of qualitative research.

We find these distinctions fuzzy at best and the prejudices that accompany them a real drawback for the progress of research. Contemporary research methods are breaking down the traditional barriers. For example, it is has been customary for qualitative researchers to use a convenience sample (interviewing whoever they could access easily) and quantitative researchers to use a probability sample (one in which everyone of the type of person being studied has the same chance of being interviewed). (We will discuss sampling in more detail later.) But now qualitative researchers often use a probability sample when one is available.

One of us (Kalof, 1993) has shown how content analysis and statistical analysis can be joined to give a more thorough understanding of perceptions of media images than either approach alone could do. And one of our colleagues (McLaughlin, 1996) combines a detailed historical account of the development of co-ops with sophisticated statistical analysis of their foundings and failures.

Some questions seem to call for qualitative methods, some for quantitative. But most research areas benefit from a healthy mix of methods. We strongly advocate "methodological pluralism." And we urge you to be suspicious of claims that one way of doing research is superior to others.

We want to mention that science always involves simplifying the world. The graph in Figure P.1 tells us something about a pattern in the world. But to focus our attention on this pattern of necessity means ignoring other elements of the world. Science is no different than art in this regard. A skilled artist draws our attention to a part of the stream of reality and so does a skilled scientist. There are other traditions of human knowledge that are concerned with awareness of the whole, such as the meditative traditions of Taoism and Zen. But science, like art, is about producing pleasing and useful focus.

A Strategy for Learning

As a field of inquiry, statistics makes substantial use of abstract mathematical reasoning. But that approach is alien to most students in the social sciences and helping professions. We have found that examples help many introductory students understand the principles. But we have also found that examples work best if they are cumulative.

In the chapters that follow we will use a common set of examples across all chapters. Some of the examples will be worked out within the text of the chapter. The examples not considered in the text will be included in an "Applications" section at the end of the chapter. In this way, the examples will become "old friends." Understanding how an analysis makes sense of patterns in the data in one chapter will aid in understanding a different way of analyzing the data in another chapter. But from example to example, the exact format of tables and figures will vary a bit to give you a feel for the different formats and styles you will find in the professional literature.

We have structured the book around basic principles and approaches that recur across many different procedures. As a result we will explain the same idea in several places. Your familiarity with a basic concept from earlier chapters will make a new use of it seem like an old friend.

As you move through the book, learning statistics requires practice in analyzing data, both with paper and pencil and with statistical software. As you work examples and do exercises, you will often compare your work with that of your peers, your teacher and the results from the examples in the text. If you seem to get a different answer, there are several things to check. First, as we mention often in the text, statistical calculations are influenced by "rounding error." If you use a different number of digits in a calculation than someone else, you may get a slightly different answer. Even the two statistical packages we used in preparing the text, SPSS and Stata, give slightly different results in complicated calculations. Second, while we have checked the text many times for mistakes (as have professional proofreaders) we may have missed something. We will post any errors we learn about on the website, so be sure to check before assuming you've done something wrong. Third, if you are using a software package, double check how the package is handling "missing data." Some people don't answer every question in a survey and some countries don't report certain data, so most data sets have data that is missing for some people or countries on some variables. As you use different variables from analysis to analysis, the sample size will shift a bit because of this. So if your results differ from those of others, check on the sample size for the analysis. Finally, it is possible that you are doing something wrong. If so, don't despair, that's why we do exercises. The only way to learn how not to make mistakes is to make mistakes and learn what went wrong.

What Have We Learned?

We began by noting that considerable evidence from history and from psychology demonstrate that understanding statistics does not come easy for most of us. If it were easy, perhaps statistics courses wouldn't be required. But we also argued that it is important to learn statistics for at least two reasons. First, statistics is a view of the world that aids us in dealing with probabilities and events that are, for practical purposes, random. This randomness is a part not only of scientific research but also of daily life, so statistics can offer a perspective that is quite practical in the realm of the personal.

Second, statistics underpins much of what we know in the modern world. Our technologies, governments, and businesses depend on statistical information. So understanding statistics helps us evaluate what these large forces in our lives are doing and saying. In particular, in most professions, the information you will need to make decisions will have a substantial statistical component. Even the expert judgments based on years of professional experience represent a kind of statistical generalization from observation.

We then argued that good quantitative analysis is not achieved by simple recipes but by thinking about analysis as craftwork. Mastery of the tools of the craft – in our case specific statistical methods – is essential. But exceptional work emerges from practice and critical reflection.

Finally, we suggested that science is a discourse and statistics part of the language of that discourse. Theories are abstractions from reality that can help us understand reality. The process by which that understanding emerges is one of dialogue with oneself and with others. Each theory is flawed, but a good theory is one whose flaws suggest further insight into how the world works.

References

Baudrillard, J. 2003. *Cool Memories IV*. London and New York: Verso Books.

Bennett, D. J. 1998. *Randomness*. Cambridge, MA: Harvard University Press.

Berk, R. A. 1991. Toward a methodology for mere mortals. In P. V. Marsden (ed.), *Sociological Methodology 1991*, pp. 315–24. Washington, DC: American Sociological Association.

Blalock, H. M., Jr 1991. Are there really any constructive alternatives to causal modeling? In P. V. Marsden (ed.), *Sociological Methodology 1991*, pp. 325–35. Washington, DC: American Sociological Association.

Braverman, H. 1974. *Labor and Monopoly Capital*. New York: Monthly Review Press.

Freedman, D. A. 1991a. Statistical models and

shoe leather. In P. V. Marsden (ed.), *Sociological Methodology 1991*, pp. 291–313. Washington, DC: American Sociological Association.

Freedman, D. A. 1991b. A rejoinder to Berk, Blalock and Mason. In P. V. Marsden (ed.), *Sociological Methodology 1991*, pp. 353–8. Washington, DC: American Sociological Association.

Gaither, C. C. and Cavazos-Gaither, A. E. 1996. *Statistically Speaking: A Dictionary of Quotations*. Bristol, England: Institute of Physics Publishing.

Gould, S. J. 1996. *The Mismeasure of Man*. New York: W. W. Norton.

Kahneman, D., Slovic, P., and Tversky, A. 1982. *Judgement Under Uncertainty: Heuristics and Biases*. Cambridge, England: Cambridge University Press.

Kalof, L. 1993. Dilemmas of femininity: Gender and the social construction of sexual imagery. *Sociological Quarterly* 34, 639–51.

Mason, W. M. 1991. Freedman is right as far as he goes, but there is more, and it's worse. Statisticians could help. In P. V. Marsden (ed.), *Sociological Methodology 1991*, pp. 337–51.

Washington, DC: American Sociological Association.

McLaughlin, P. 1996. Resource mobilization and density dependence in cooperative purchasing associations in Saskatchewan, Canada. *Rural Sociology* 61, 326–48.

Piaget, J. and Inhelder, B. 1975. *The Origin of the Idea of Chance in Children*. New York: W. W. Norton.

Salsburg, D. 2001. *The Lady Tasting Tea: How Statistics Revolutionized Science in the Twentieth Century*. New York: W. H. Freeman and Company.

US Census Bureau 2000. Table 33. Urban and rural population, and by state: 1990 and 2000. (http://www.census.gov/prod/cen2000/index.html)

US Census Bureau 2002. Historical poverty tables: Table 21. Number of poor and poverty rate, by state: 1980 to 2006. Year 2002 (http://www.census.gov/hhes/www/poverty/histpov/)

US Census Bureau 2003. Table 295. Crime rates by state, 2002 and 2003, and by type, 2003 (http://www.census.gov/prod/2005pubs/06stata "www.census.gov/prod/2005pubs/06statab/law.pdf).

CHAPTER 1
AN INTRODUCTION TO QUANTITATIVE ANALYSIS

Outline

What is Statistics?

Models to Explain Variation

Explaining Variation

The Use of Statistical Methods

Types of Error
   Error in models
   Sampling error
   Randomization error
   Measurement error
   Perceptual error
   Comparison to random numbers

Assumptions

What Have We Learned?

Advanced Topics

Applications

Why do some states have high homicide rates while in other states the occurrence of a homicide is very rare? Why are some countries more likely than others to participate in environmental treaties? Why do some people feel animals have rights while others feel animals can be treated as objects? Why do some people know how the AIDS virus is transmitted and others don't?

In this text we will explore each of these questions using quantitative methods. We will try to answer them by developing models that help us understand why people, states within the US, or countries in the world differ from one another. In attempting to answer these questions we will introduce the standard tools of modern quantitative analysis in the social sciences – statistics. Our answers will always be tentative and never certain. But the scientific method applied by using statistical tools can help us make better decisions about how the world works.

Of course, this is a statistics text, not a book about homicide or the environment. The questions we examine are intended to introduce the tools of statistics. Once you understand the tools, you will be able to see how they can be used to answer many other questions across a range of issues. Perhaps more important, they can help you to think critically about research that is presented to persuade – whether in a paper from a scientific journal, a technical report from a government agency, or in a newspaper story.

Each question we pose in the first paragraph is about variation – why do some people, states or countries differ from others? We attempt to understand that variation by building models. In quantitative analysis we use the term model in much the same way as it is used in everyday life. A model is a representation of something. For example, a model train captures the look of a real train, but in miniature, and a model apartment shows you what the apartment you are thinking of renting might be like. A fashion model shows how you might look wearing the clothes being modeled.[1]

In quantitative analysis we build models with numbers. We want to understand why the things being studied vary – why people are different from one another, why states differ, why countries differ. Like the models of everyday life, quantitative models are useful if they capture the key features of what we are studying. But they also simplify reality and can be deceptive if we don't look at them with a critical eye.

Building models and using them to understand variation is a central theme in quantitative analysis. In this text, we explain how models are developed, show how they can be used to explain variation, and examine how models of variation relate to the ongoing dialogue that is science.

Statistics is intended, in large part, to deal with models that include some error. All of our models will be imperfect descriptions of reality, and the difference between the model and what we observe in the real world is the error in the model. This idea of error is a bit removed from the everyday sense of model trains, model apartments and fashion models. But there is an analogy. As makers of model trains strive to add more and more detail, they are in a sense trying to reduce the error – the difference between the model and reality. If the actual apartment you rent does not much resemble the model you were shown, you have a sense that things are wrong. And taking fashion models with very unusual physiques as a representation of the typical man or woman is clearly an error.

Since all data include error, we will discuss the kinds of error that are most important in social science data. This is the starting point for understanding how statistics allows us to take error into account in our models. As we will see, the error is a critically important part of our models, and we will think as hard about the error as we do about the rest of the model.

What is Statistics?

What does the term statistics mean? There are two definitions of statistics in the typical dictionary (Brown, 1993):

1   the field of study that involves the collection and analysis of numerical facts or data of any kind; and
2   numerical facts or data collected and analyzed.

The first definition refers to the field of study to which this book provides an introduction. The second definition refers to what, in some sense, statisticians study – numerical (or quantitative) data. This is the everyday use of the term statistics – the numbers that are intended to represent some aspect of life. Everyone encounters sports statistics, statistics on cars, statistics on how the economy is doing and so on.

> *A third and more technical definition of statistics* is discussed at the end of this chapter as an Advanced Topic.

The use of such quantitative data goes back at least to the earliest city-states. We know the Babylonians and Egyptians collected numerical data on crops, for example. In fact the term statistics has its roots in the Latin word for "state" indicating the historical linkage between the government and numerical data.

As a field of study, statistics develops tools that allow us to generate better numbers to describe the world. As we noted above, all models of the world involve some error. One of the major concerns in the field of statistics is to understand how error may enter our models and in the numbers that we use to describe the world. By understanding these errors we may be able to reduce them substantially. And even when we can't make them small enough to ignore, statisticians have given us tools to help us understand how large the errors may be. This allows us to guard against making decisions that treat what may be error as if it were fact.

Models to Explain Variation

To understand and use quantitative methods, we must have some sense of the process of proposing models, criticizing them, and learning from the process. Generally, a model has the form:

$$Y = f(X) + E \qquad\qquad (1.1)$$

We say this as "Y equals f of X plus E."

Sometimes the equation is shown with a small subscript i after Y, X, and E. Then the equation would be $Y_i = f(X_i) + E_i$. This is sometimes done to emphasize that every observation – every person in a survey, every country in a cross-national study – can have its own value for Y, X, and E. Of course, two observations may have the same scores on a variable. Two people might have the same level of education or income or two states might have the same homicide rate. But X, Y, and E can vary from person to person even if not every person has a unique value on each of the variables. We won't use the subscripts because they can be confusing when you are first learning statistics.

X and Y can vary across observations, so they are called **variables**. Y is the **dependent variable**, the thing we are trying to explain. If we are trying to understand why states might vary in their homicide rates, then Y would be the homicide rate. If we are trying to understand who knows that the transmission of AIDS is reduced by condom use and who doesn't, Y would be each person's response to a survey question about AIDS transmission.

X is the **independent variable**. The equation suggests that the variation in Y across observations may be the result of variation in X. The equation implies that because X is different from observation to observation then Y will be different. In explaining homicide rates we might think that homicide is a result of poverty, so X would be some measure of poverty in a state. For example, we could use the percent of the state's population below the federal poverty line for X. We might think that a person's knowledge about the AIDS virus depends on their gender, so then X would be gender. One helpful way of remembering the difference between dependent and independent variables is that the dependent variable *depends* on changes in the independent variable.

In some fields, terms other than independent and dependent variables are used to describe the thing we are trying to explain and the thing used to explain it. For example, for obvious reasons the dependent variable is sometimes referred to as the "left hand side variable" and the independent variable is called the "right hand side variable." And sometimes the dependent variable is called the response (because it is responding to the independent variable) while the independent variable is called the "carrier," the "predictor variable," or the "covariate."

The "f" in the equation is just an abbreviation that covers any way X might be linked to Y. You may remember "f" from algebra. It means "is a **function** of." It says that there is a relationship between X and Y, but it doesn't say what that relationship will look like. It might be that as X gets larger Y gets larger. That is what our theory of homicide is saying: states with more poverty will have higher homicide rates. But the "f" by itself is not that specific. It could allow for Y to get smaller as X gets larger, or for some more complicated relationship. For example, using just the "f" allows for the possibility that the states with the highest and lowest poverty rates have the lowest homicide rates, with the highest homicide rates in the states with an intermediate level of poverty. To actually have a model we can apply to data, we have to be more precise about how X and Y are related.

The E term, often called the **residual** or **error term**, suggests that things other than X cause Y to vary from observation to observation. Thus the equation indicates that Y is a function of X, plus an error term. In other words, our model says that poverty rates are not the only cause of homicide rates; other variables may be causes of homicide rates, and these are represented by the "E" in the equation. Just as Y and X can be different from person to person or state to state, E can also vary from person to person and state to state.

Let's pursue the example of homicide rates. Figure 1.1 graphs the homicide rate against the percentage of the population in each state in poverty. This graph is called a scatterplot because we are plotting the "scatter" of Y and X. We will discuss scatterplots in more detail in Chapter 5. The poverty data are for 2002, the homicide data for 2003. The homicide rate is the number of homicides per 100,000 population in the state. The poverty rate is the percentage of families whose income was

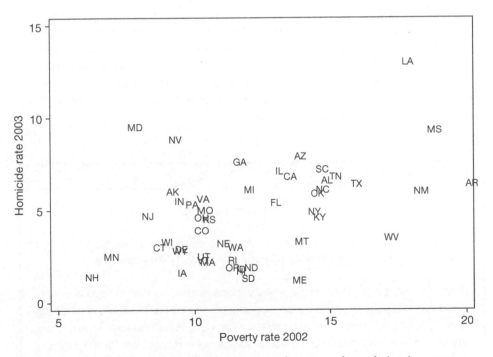

Figure 1.1 Relationship between homicide rate and percent of population in poverty
Data source: US Census Bureau (2002, 2003), analyzed with Stata.

less than the federal poverty line for families. The data are described in more detail in the applications at the end of the chapter. We usually want the independent and dependent variable to be for about the same point in time, and the custom is to have the independent variable be for a slightly earlier point in time than the dependent variable when they are not both for the same time.

Using words instead of letters to form the equation, the model we are proposing is:

$$\text{homicide rate} = f(\text{percent in poverty}) + \text{Error} \tag{1.2}$$

The model says that one reason that states vary in their homicide rates is that they have different levels of poverty. We don't know that the model is correct. Indeed as the quote from George Box (see Box 1.1) suggests, all models, including this one, are wrong. By wrong we mean that no model will predict the data perfectly. But as the quote from Samuel Karlin suggests, we can learn something from imperfect models. We do this by looking at the pattern in the scatterplot.

There does seem to be some tendency for the homicide rate to be higher in the states with the highest poverty levels. Note also that even though there is some pattern, it is far from perfect. The data points don't seem to fit any pattern perfectly even if there is something of a general pattern. States with the most poverty do seem to have the highest homicide rates. The deviations from the pattern are also interesting. States like Mississippi (MS) and Louisiana (LA) have higher levels of homicide than we might expect from their levels of poverty, while Maine (ME) and South Dakota (SD) have lower levels of homicide than we might expect from our model. For example, Arizona and Maine have about the same poverty level (13.5 and 13.4, respectively). But the homicide rate for Maine is only 1.2 while for Arizona the homicide rate is 7.9. So the poverty rate clearly doesn't predict the homicide rate perfectly. In the language of the model, the states that are different from the overall pattern will have large E values. Things in addition to level of poverty are having an influence on the homicide rate.

Box 1.1   Models

All models are wrong. Some models are useful. (George E. P. Box, 1979)

The purpose of models is not to fit the data but to sharpen the questions. (Samuel Karlin, 11th R. A. Fisher Memorial Lectures, Royal Society, 20 April 1983)

Both of these quotes from eminent statisticians remind us that models are tools to aid our understanding. As we develop models we can get lost in the modeling itself. We should always reflect back on the purpose of building and testing models: to help us understand the world. Sometimes a model that doesn't describe data very well tells us as much or more than a model that fits well.

To go beyond looking at a scatterplot, we have to be more precise about how X is related to Y. The shape of the relationship between X and Y is called the "functional form" of the model. In some software it is called the "linking function" because it is what links X to Y. The functional form or linking function is a very important part of the model, one that should be specified, at least tentatively, by theory. In the poverty/homicide rate example, we might suggest that the link is best represented by a straight line that indicates that as poverty rates go up, homicide rates go up. Then f(X) becomes the equation for a straight line:

$$f(X) = A + (B^*X) \tag{1.3}$$

We read this equation as "F of X equals A plus B times X." When working with the equations, pay attention to where the parentheses are. In solving this equation you should multiply X times B, then add A. Here f(X) is the function that links X to Y. The result is a prediction of Y based on X, f(X) is the prediction of Y using X as the predictor. The equation f(X) is not the same as Y itself unless X can predict Y perfectly, which won't be the case with real data. So there has to be an E in the equation for Y itself. That is the equation for Y is now:

$$Y = A + (B^*X) + E \tag{1.4}$$

While this is the simplest and most commonly proposed functional form linking dependent and independent variables, it is not the only possibility. There is nothing except more complicated algebra preventing us from saying that the link between X and Y is a curve rather than a straight line. As we suggested above, we might have the idea that states at a moderate level of poverty have higher homicide rates than those at the high and low extremes. This implies a curve that looks like an upside down letter U. We don't see this pattern in Figure 1.1, but it might occur for other variables. But it's best to start with the simplest models and the straight line is the simplest model we can have for the relationship between two variables.

Remember, E is still the error or residual term. It implies that we don't expect all variation in Y to be predicted by X when we assume that the relationship is a straight line. By including E we acknowledge that factors other than affluence may explain homicide rate. The inclusion of E in the model indicates that we don't expect poverty rates to predict homicide exactly – not all data points will fall exactly on the line.[2]

It is useful to think of even this very simple model as having four parts. First there is the *dependent variable*, Y, the thing whose variation from observation to observation (that is from person to person, state to state, country to country or year to year) we want to explain. In our example, this is the homicide rate. Second is the *independent variable*, X, the thing that we believe can, with its own variation, explain some of the variation in the dependent variable. In the example, this is the percent of the population below the poverty line. Third is the *functional form* (f) that links the independent variable to the dependent variable. For now we'll talk only about straight lines, but more complicated functional forms are often realistic. The key point for now is that we are making a statement that says not only that we believe that the independent variable can predict the dependent variable but also that we will use a straight line to indicate the link between them. The fourth component of the model is the *error term* (E). E describes the difference between the actual value of Y and what we predicted using X. It takes account of the fact that X does not perfectly explain Y. The error term is not usually discussed by sociologists or other theorists, but it is a key issue for statistical theory. Indeed, as will become clear later, the meaning of any statistical procedure rests on the meaning of the error term, and statistical analyses are only as sound as our assumptions about the error term. So our four parts of the model are the dependent variable, the independent variable, the functional form that links them, and the error term.

The functional form, $Y = A + (B*X) + E$, suggests that a straight line describes the link between X and Y. But what values do we pick for A and B – that is, what line would we draw to represent the relationship between A and B? We could pick an A value and a B value by "eyeballing" the data. That is, we could take a ruler and draw a straight line through the graph and then use methods we learned in high school algebra and geometry to find the values of A and B for the equation from the line we drew. But that's not how we find the line in statistical analysis. Chapter 14 will describe in detail how we pick the line. For now, take our word for it that a good value to pick for A is 0.59 and for B a good value is 0.36. Figure 1.2 shows the graph where the line f(X), which for a straight line is $A + (B*X)$, has been drawn in.

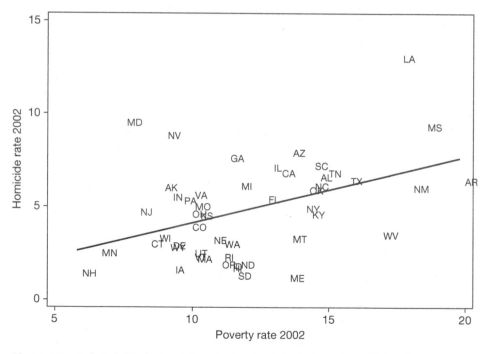

Figure 1.2   Relationship between homicide rate and percent of population in poverty with prediction line
Data source: US Census Bureau (2002, 2003), analyzed with Stata.

A is the value we predict for the homicide rate if there were no families living in poverty in the state – the situation when X equals 0. To see this, look at Equation 1.2. If X = 0 then B*X = 0. Then the prediction for Y is just A, which for this example is 0.59. Often the A term in a prediction doesn't tell us much on its own. Our real interest is in the B value. Remember the B value tells us how Y changes with X – how homicide rates change with levels of poverty. That's the whole reason for doing the analysis. In this case the A term doesn't mean much because there are not states at, or even very close to, a zero poverty rate. So in this example A just gives a reasonable line to describe the data. If the data included values of zero for X, then A would be more meaningful.

B says that for every 1-point difference in the poverty rate, the homicide rate increases by 0.36. (Remember that the poverty rate variable is a percent, so a 1-point increase is a 1% increase.) Let's look at what the model predicts for two states, Vermont and New York. For Vermont, the poverty rate is 9.9% so the model predicts a homicide rate of (0.59 + (0.36*9.9)) = (0.59 + 3.56) = 4.15. For New York, the poverty rate is 14% so the model predicts a homicide rate of (0.59 + (0.36*14)) = 5.63. Remember, these are predictions from the model. The predictions usually won't be exactly right. For example, the actual difference between the homicide rate for Vermont is 2.3 – 4.15 = −1.85. This means the model predicts too high a homicide rate for Vermont by about 1.9 points. For New York the difference is 4.9 − 5.63 = −0.73. Here the model predicts too high by nearly three quarters of a point. Of course, if we had picked other states, we would see that sometimes the model underpredicts.

Let's look at what the model is saying for a few states. Table 1.1 shows the values of Y, X, f(X) and E for a few states. For California, X = 13.1. We multiply this by B, which is 0.36 and get 0.36*13.1 = 4.72. Then we add A, which is 0.59, so (0.59) + 4.72 = 5.31. This is what we predict the homicide rate for California to be based on the poverty rate. The E for California is then the actual homicide rate minus the prediction. Then E = 6.8 – 5.31 = 1.49. We can see that for California the prediction was pretty close, but it underestimated the homicide rate slightly. For Nevada the prediction also underestimated the homicide rate. The actual rate was 8.8 and the predicted rate was 3.79, so the actual rate was about five points higher than what the model predicted. Note that positive values of E mean that the model *under*estimates the value for that state, while negative values indicates that the model *over*estimates the homicide rate.

Table 1.1   Poverty rate, homicide rate, predicted value and error for selected US states

State	X_i (% in poverty)	Y_i (Homicide rate)	$F(X_i) = A + (B^*X_i)$	E_i
California	13.1	6.8	5.31	1.49
New York	14.0	4.9	5.63	−0.73
Nevada	8.9	8.8	3.79	5.01
Vermont	9.9	2.3	4.15	−1.85
Oklahoma	14.1	5.9	5.67	0.23

In Figure 1.2, we see the same thing by noticing that the line, which represents the model's prediction, is very close to the data point for California (CA), but pretty far away from the data point for Nevada (NV) and Vermont (VT). Underestimates mean that the state is above the line and the E value for that state will be positive. Overestimates mean that the state is below the line and the E value will be negative.

Explaining Variation

Variation explained is a central concept in many applications of statistics. Social scientists are concerned with why people, social institutions, communities, cultures, nations and other units of social analysis vary from one to another. We want to know why some people are rich and others are poor, why some nations have a high quality of life and others do not, why some social movements succeed while others fail. In the model just presented, we want to explain why states vary in their homicide rates. In most discussions, we'll actually talk about variance explained rather than variation explained. Both the terms "variance" and "variation" have precise definitions in statistics, as we'll see in Chapter 4. Until we get there and see the definitions, we'll use the term *variation* in its everyday sense – **variation** means that things differ from person to person or country to country.

It is important to remember that the term **explanation** in the context of statistics has a very precise meaning. In its technical usage, it refers to the ability to predict one variable based on another variable or set of variables. So in the example above when we say we can partially "explain" the homicide rate in terms of the amount of poverty in a state we mean the poverty rate can predict the homicide rate, though of course not perfectly. Explaining variation in this technical sense of *predicting variation* across observations can lead to explanation in a broader sense. This happens when the explanation of variation is linked to a practical or theoretical framework. Lacking such a framework, statistics cannot produce much understanding. We may develop models that have great explanatory power in the sense that they predict quite well but provide no theoretical or practical insight. Or we may learn a great deal about the social world from a model that has limited explanatory power but that reveals important patterns. Indeed, the fact that a model doesn't predict very well can lead to important insights, as the quotes in the box above indicate.

Quantitative methods are powerful tools, but they achieve that power only when combined with sound theoretical thinking. In contemporary social sciences, it is often the case that theorists pay little attention to issues of method and sometimes methodologists don't think about theory. Successful research requires making links between theory and method – between developing understanding of phenomena in a theoretical or practical way and explaining variation in a technical way. Indeed, this is the stage at which the craft of research is of paramount importance. Good research makes connections between developing understanding of phenomena in a theoretical or practical way and explaining variation in a technical way.

Consider the relationship between homicide and poverty. Let's suppose we can predict the homicide rate with reasonable accuracy based on the percentage of each state's population in poverty. In the statistical sense we have "explained" a good bit of the variation in homicide. But if we do not have a theoretical model of the relationship between homicide and poverty, we have not really learned much.

We have to link the model to theory. For example, it may be that poor people are alienated and suffer from social disorganization and thus are more likely to engage in crime and violent activities. If we think that's the case, we might want to see if there is a relationship between poverty and other kinds of crime. This would support the idea that poverty leads to disorganization, which leads to crime though there still might be other things to consider. Or it may be that poor people are concentrated in cities and that homicide is more likely in cities simply from the volume of social interactions. Then we would want to look at the relationship between urbanization and both homicide and poverty. As we develop a theoretical explanation to work with our quantitative analysis, the theory often suggests other analyses with other variables. Looking at the data may answer some questions, but it always raises new ones.

Students learning the basic tools of statistical analysis sometimes mistake statistical explanation for theoretical explanation. As we will emphasize, effective use of quantitative methods requires a theoretical context for guidance. Quantitative methods can be a great aid in evaluating the ability of theories to inform us about the social world, but they cannot be used in the absence of theory. This is turn provides a strong challenge to theory. We are convinced that attempts to apply quantitative methods to our theories will greatly improve those theories. Quantitative methods force us to think hard about what we mean and what we assume.

The Use of Statistical Methods

Statistical methods are tools that aid in seeing the empirical world more clearly. Scientists observe phenomena, and from their observations they try to develop better understandings of the world. But many factors cloud and distort our observations, making it difficult to draw conclusions from them. Scientists, like all observers of the empirical world, see "through a glass darkly."

Statistical methods are designed to clarify that glass, to minimize the cloudiness, to help us sort "truth" from "error." Statistical methods cannot eliminate error, nor provide "truth," but they do provide an assessment of the magnitude of error that is there, and thus clarify our perceptions. In the equation presented earlier, E represents all the things in the world that may distort our understanding of the link between X and Y. It is all the things that prevent X from perfectly predicting Y. By understanding E we can improve our understanding of the relationship between X and Y.

Types of Error

There are several ways to think about the role of E – error – in scientific observation. We will discuss five ways to think about the use of statistics in the face of error: error in models, which keep models from predicting perfectly; sampling error; randomization error in experiments; measurement error; and comparisons of independent variables to random error. In this chapter, we introduce these ideas. But they will return again and again in subsequent chapters. We're sure that the more you see them the clearer they will become, so don't worry if the seem a bit difficult to grasp at first. Eventually, with enough understanding of statistics, you'll learn that the different kinds of errors are really different ways of thinking about the same thing.

Error in models

We've already looked at one example of a model – the simple model that uses the poverty level of a state to predict the homicide rate. This seems like a plausible idea, and the scatterplot in Figure 1.2 seems consistent with the argument. But we don't expect perfect prediction with our models. Not all the states have the values for homicide rate that the model predicts, in fact for most states the model is a bit off. This is not surprising – there must be more that causes homicide than just poverty, even if poverty turns out to be part of a good explanation. So our model of homicide rates has some error.

One way to think of the error is that it is everything else that causes the homicide rate. Some of those things are other variables on which we can obtain data, so we could expand the model to include those things, using methods we'll discuss in later chapters. But even if we include all the variables that theory suggests might be important, we still wouldn't expect the homicide rate to be perfectly predicted by the model. If our theories are good descriptions of the world, adding more variables will reduce the errors associated with the prediction for each state, but the error will never completely disappear for all states.

Consider a tragic example. If we had used homicide data for 1995, rather than 2003, the data for Oklahoma would have included the 169 deaths that resulted from the right wing terrorist bombing of the McMurra federal building in Oklahoma City. Such a tragedy cannot be predicted by a model of state homicide rates, though sociological analysis can lend considerable insight into terrorism. We never expect models to predict perfectly, but only to let us understand better how the world works. So our models will always have error.

Sampling error and randomization error are found in most applications of statistics. Before we discuss them, it is useful to introduce a concept central to much of statistical analysis – the arithmetic mean. You already are familiar with the mean but you know it by its everyday name – the average. As you know, the average score of a class on a quiz is calculated by adding up all the scores and dividing by how many people took the quiz. Your personal average on all quizzes in a course is the sum of what you scored on all the quizzes divided by how many you took. A special case of the average is the proportion or percentage of people in a particular category. For example the proportion of people in a class who get an A is just the number who get an A divided by the number of people in the class. We will refer to the average in the next few examples, but you already know enough about it to follow the examples.

Sampling error

Many data sets are based on a sample of objects. We might do a survey of individuals or households. Or we might have a sample of things like states, cities, organizations, or nations. While data are available for a sample, our practical and theoretical concerns usually are with the population from which the sample was drawn. Survey interviews with 1,500 US citizens are of interest to the extent they lead to conclusions about the attitudes and values of all Americans. A researcher may have data on 100 school districts but would like to speak about all school districts. Researchers can analyze the data in hand, the sample, and draw conclusions about it. But in most circumstances we also want to generalize to the population from which the sample was drawn.

Why not collect data on everyone or every organization or every state? When there are a very large number of people or organizations in the population we are studying, the costs in time and money of getting data from everyone can be prohibitive. In fact, statisticians have shown that we often can get a better understanding of a population by being very careful about getting data from a sample rather than having the same resources spread very thin in trying to get data on everyone. Of course, sometimes we do have data on every unit, as in the case of our analysis of state homicide rates.

Suppose, as is often the case, that we are interested in a population average or percentage. For example, in one of our continuing examples, we will examine the percentage of people in a Ugandan national sample of adults who knew condom use can help prevent the transmission of AIDS. (We will discuss this example in more detail in the Applications sections at the end of every chapter.) In the survey 8,310 people answered the question. Of those 8,310, 6,420, or 77% said "yes" that condoms can reduce transmission. That's interesting, but what we really want to know is what percentage of people in the Ugandan adult population, not just the sample, would say "yes" to that question. The percentage in the sample is a guide, but we also expect it not to be exactly the same as the percentage we would get if we interviewed all adults.

We can think of the relationship between the sample percentage and the population percentage in terms of a simple model that looks like the models we have already examined.

Sample percentage = Population percentage + sampling error   (1.5)

The sample percentage is what we calculate from the sample. We know it is 77 percent. We want to know the population percentage, but we shouldn't assume that it is exactly the same as the sample percentage. So we allow for the sample percentage to differ from the population percentage because of sampling error. For the right kinds of samples, statistical procedures allow us to learn a lot about the sampling error and thus about how our sample percentage may differ from the population percentage we would like to know.

The patterns in the sample may not accurately represent the population for several reasons. First, there is the problem of having a sample that was selected using a non-representative process, one that, intentionally or by accident, includes in the sample exceptional rather than typical cases. For example, we might use a convenience sample in which we interview the first 100 students we encounter in the student union. When we have a non-representative sample it is usually not possible to determine the relationship between what is seen in the sample and what is true in the population. We don't know how to generalize from the first 100 students we run into to the entire student body. With non-representative samples, statistics are of little help in going from the sample to the population. But graphical and descriptive statistical techniques can be used to understand the sample itself.

Generally, we would prefer to have a "representative" sample, but what does that mean? One way of thinking about **representative samples** is to have a sample constructed in such a way that every member of the population we're studying has the same chance of appearing in the sample. This is called an *equal probability of selection sample.*

Even with such a representative sample, conclusions drawn from the sample may not be perfectly accurate representations of the population because of chance processes. Sometimes an honest coin may come up tails for 10 flips in a row. In the same way, a random sample may display patterns that aren't typical of the population, purely by bad luck. If we drew names of students at random from the registrar's list, we might by luck get too many women, or too few chemistry majors, or some other non-representative mix. The advantage of a probability sample over a convenience sample is that in the probability sample, statistical procedures let us estimate the likely magnitude of the error produced by sampling. With convenience sampling the magnitude of the error cannot be known.

Ways of drawing probability samples are discussed at the end of this chapter as an Advanced Topic.

When the process by which the sample was drawn is understood, as is the case with simple random samples and other probability samples, then statistical tools make it possible to put probable upper and lower bounds on the errors generated in sampling. They don't eliminate sampling error, but they do indicate how large it may be, and can be used to place appropriate hedges on conclusions. This understanding

Box 1.2   What Are the Chances of Getting a Badly Non-Representative Sample?

What are the chances of drawing a sample composed entirely of men? Suppose we draw a sample of 100 students, from a student body (the population we want to sample) that is half men and half women from a very large university, so that 100 students is a tiny fraction of the study body. If we use a convenience sample by taking the first 100 people walking out of the student union, we can't calculate the chances of getting a sample of all men, but we might imagine things that would make that happen – perhaps we go to the student union right after a fraternity rush event ends. But if we draw the sample at random from the registrar's list of students, we can do the calculation of how likely it is to get all men in the sample. It's the same as the probability of getting 100 heads in a row when tossing a fair coin. The chances are about 0.0000000000000000000000000000001 or one in ten nonillion. Pretty small chances, not something we really need to worry about. In Chapter 7 we'll show you how to do these calculations.

is a fundamental insight in statistics, one that emerged in the late nineteenth and early twentieth centuries. Before that time, scientists tended to work with whatever data were available to them – a convenience sample. They had no clear sense of what data might be representative, and thus useful for generalizations, and what data might be misleading because it was not representative of a larger group. Now in most branches of science, careful attention is given to obtaining a representative sample of the things being studied.

Randomization error

Sociologists and most other social scientists, except psychologists, don't often conduct experiments. But we will discuss randomization error because it has played a central role in the development of statistical thinking and because in some fields like psychology experiments are very important sources of evidence.

When a simple experiment is conducted, the subjects are sorted at random into two groups, one labeled the **experimental group** and the other the **control group**.[3] In experiments with **random assignment**, the two groups are created by a chance process, such as the flip of a coin. "Heads" and you're in the control group, "tails" and you're in the experimental group.

During the experiment, something is done to the experimental group that isn't done to the control group. Then the two groups may differ from one another for two reasons. One is because of the factors manipulated by the experimenter – the thing done to the experimental group but not the control group. The other is the chance process by which subjects were assigned to groups. The power of the experimental approach comes from its relatively unambiguous ability to attribute differences between groups to the experimental manipulation when it is not reasonable to believe that the differences between the groups were due to chance.

Suppose the two groups are created by the flip of a coin. The experimental group might watch a music video that shows stereotypical gender images.[4] The other group watches a music video that has no explicit gender content. After watching the films each group fills out a questionnaire that measures attitudes about gender relations. If the experimental group, on average, scores higher than the control group on items indicating a belief in adversarial gender relations, there are two possible explanations. One explanation is that the gender stereotyped video had an effect, relative to the "neutral" video. The other explanation is that the two groups had different attitudes at the start.

Since people were placed in the groups by the flip of a coin, statistics can assess the likelihood that the two groups differ in gender attitudes as a result of the coin flip that sorted them into groups. If the difference in attitude between the two groups is too large to plausibly attribute to chance then there seems no reasonable explanation except the argument that the film had an effect on the viewers.

Of course, the coin flip could have assigned all those with conservative gender attitudes to one group purely by chance. But statistical analysis that we will learn to do later says the chances of that happening are too small to be believed. So we have more faith in the explanation that the video content had an effect on attitudes.

On the other hand, if the differences between the two groups in gender attitudes were of the kind that flips of a coin could easily create, then the safest explanation may be that the video had no effect on attitudes. In the experiment, the probability that the differences were a result of sorting people into two groups was .021, allowing Kalof to conclude that exposure to the stereotyped video did have an effect on attitudes.

Again, a simple model can serve to explain what might have happened. We could calculate the average score on the gender relations scale for the experimental group and the average for the control group. Then if the video has no effect, the model would say that the two groups differ in their scores on the scale just by luck of the coin flip, which is random error. The model is:

$$\text{Average of experimental group} = \text{Average of control group} + \text{Randomization Error} \qquad (1.6)$$

Remember, all the error we are dealing with can be positive or negative, so the control group could have an average higher or lower that the experimental group. Also, there's an equivalent way of thinking about this model. We can subtract the control group average from both sides of the equation. Then we have:

$$\text{Average of experimental group} - \text{Average of control group} = \text{Randomization Error} \qquad (1.7)$$

Or to put it more clearly,

Difference between experimental and control group
  = Randomization Error   (1.8)

If this model explains our results well – that is if the difference between the two groups is the kind of thing a coin flip could generate – then we would conclude that the video had no effect. If the difference we find is not what we would expect as a result of random error, as was the case in Kalof's actual experiment that we are using as an example, then we would conclude the experimental treatment had an effect. So in a sense we are comparing the difference between the two groups to a random number and saying we have an effect if the difference is bigger than a random number would be.

Measurement error

Most scientific observation involves some mis-measurement of the variables being studied. Survey questions tap individual attitudes and values; official statistics provide information on economic, social and environmental processes in cities, states and countries. However skillful the designer of the survey, however honest the respondent, however careful the statistical office, errors inevitably will creep into the data. These errors in measurement may distort observed patterns.

In the example of homicide rates, we know the homicide rate for each state almost certainly has some error in it. Some homicides are never detected while some deaths that are not homicides might be misclassified as murders. Statistics can provide tools to minimize this error, and under the proper circumstances provide an estimate of how large such errors may be. In fact, many statisticians work on finding ways to reduce measurement errors in studies where the outcomes are important. The US federal government employs many highly trained and dedicated statisticians who are continually developing methods to better measure things like the population or the unemployment rate so that we can make decisions using the most accurate possible information.

Measurement error was central to the origin of statistical analysis (see Bennett, 1998: ch. 6). By the sixteenth century, astronomers and other scientists were noting that if they made the same observation over and over, they got slightly different results each time. In 1632 Galileo noted that:

- the errors were inevitable;
- they were mostly small with relatively few large errors;
- they were symmetric in the sense that overestimation was as common as under-estimation; and
- the true value was in the area where most observations were clustered.

By the eighteenth century, a number of scientists and mathematicians had suggested that a bell-shaped curve described the measurement errors. They began to link the

Figure 1.3   Galileo
Source: http://commons.wikimedia.org/wiki/Image:Galileo-sustermans2.jpg.

idea of errors to random processes. Thus the observation could be thought of as the true value plus (or minus) some measurement error that was generated by a process rather like tossing dice or drawing lottery tickets. With this idea the basis for modern statistics was developed – our observations of the world include random elements and that statistical procedures can help separate out the random elements and provide a better understanding as a result. In statistics, we call processes that generate such random error "stochastic" processes. The term stochastic is derived from the Greek term for "skillful at aiming or guessing." A simple model of this error would be:

$$\text{Measured value} = \text{Actual Value} + \text{Measurement Error} \tag{1.9}$$

The more we know about the measurement error, the better we are able to know what the actual value of the thing measured will be.

In the case of the homicide rate and poverty example, the idea of measurement error would suggest that if we had perfect measurement of homicide rates (that is if the E term for every state were zero) then we would perfectly predict homicide with poverty. This does not seem like the most reasonable way to interpret the E term in the model. We don't think that the model misses badly for Nevada simply because the measurement error for homicide rates is very large there, while it is smaller for California. There are better ways of thinking about the error in our example. But for many other problems, such as those where we are measuring attitudes or values, the measurement error interpretation of the E term is very helpful.

Perceptual error

The social world is complex, and available theories suggest many factors that are important. In addition, the behavior of individuals, groups, institutions, and nations are not rigidly determined, so it can be difficult to see patterns in data, or conversely, patterns that are not really there may seem to leap from the page or screen.

One value of statistical analysis is that it can provide powerful methods for arranging data, including graphs and summary statistics that make it easier to see patterns and to identify particular observations that deviate from the general pattern. Descriptive statistics and graphics help us minimize perceptual error. Over the last 20 years or so, some of the best minds in statistics have devoted much of their time to developing new graphs and summaries that reveal patterns in data.[5]

Comparison to random numbers

One way to think about all the sorts of error we have mentioned is to ask if the independent variable in a model acts any differently than a random number. We introduced this idea in discussing a model in which the difference between experimental and control groups is seen as just random error resulting from the process of assigning people to groups. Suppose we generate a variable by flipping a coin and recording head or tails (0 or 1) or by tossing a die and recording the number that comes up, or by having the computer generate a number picked at random. If an independent variable has as much effect on the dependent variable as a random variable generated by flipping a coin or tossing a die, then it's hard to argue that the independent variable has an important effect on the dependent variable. A study by the statisticians Freedman and Lane (Freedman and Lane, 1983) provides a nice example of this kind of logic.

Freedman and Lane were interested in the fact that at a prestigious graduate university, 28 percent of men applying in a given year were admitted, but only 24 percent of women. All other things being equal (an assumption to keep the example simple), is the difference in admissions rate real or accidental? Should the researcher believe that the admissions process is blind with regard to gender, or is there a reason to be concerned about discrimination?

The model might look like this:

$$Y = f(X) + E \tag{1.10}$$

In this case Y is whether an applicant was admitted. X is the gender of the applicant and E is the error term that suggests that gender may not perfectly predict admissions. Then the question becomes how to interpret E.

Feature 1.1   Tea Tasting and Random Numbers

Freedman and Lane draw on some ideas from R. A. Fisher. Fisher is one of the founders of modern statistics. His name will come up throughout the book. One famous example of this kind of logic – comparing a real variable to a random number – is the story of "The Lady Tasting Tea" which is apparently a true story (Salsburg, 2001, pp. 1–8). It seems that at a formal tea at Cambridge University in England in the 1920s, a woman claimed she could tell the difference between cups of tea depending on whether the milk was poured into the cup before the tea was poured or the tea was poured before the milk. R. A. Fisher was in the room and proposed an experiment.

The woman left the room and several cups of tea were prepared, all in identical cups, some with the tea poured first, some with the milk poured first. The woman was then asked to taste each cup and state whether it was a "tea first" or "milk first" cup. If you guessed randomly by letting a coin flip determine your guess with heads being "milk first" and tails being "tea first" you'd get about half the cups correct. So for the woman to demonstrate that she could really tell the difference, she had to perform considerably better than a random process (a coin flip). Apparently she did, indicating that it's likely that she could tell the difference (or she was very lucky that day).

Freedman and Lane make the following argument. Suppose that, instead of gender, the researcher had assigned each person a score on a random number that has two values. That is, suppose each person seeking admission was assigned a score of "heads" or "tails" based on a coin toss. Then the researcher cross-tabulated that number with admissions, calculating an admissions rate for "heads" people and "tails" people.

Are the results for gender (with two theoretically meaningful categories, female and male) much different than those that come from the random variable with the non-meaningful categories of "heads" and "tails"? To put it more precisely, how often would a difference of 4% result from calculating admissions rate differences between "heads" people and "tails" people?

If the observed gender difference looks like a typical result obtained from the random "coin flip result" variable, it is hard to argue that gender was an important factor in the admissions process. It turns out that many standard statistical methods apply to such problems, including the chi-square method we will learn in Chapter 12. In this case chances are about one in four that a difference of 4% would occur if admissions rates were calculated based on the random variable. Thus we would conclude that nothing important is going on, and the observed difference may well be a fluke.

Comparison of the independent variable to a random number is one way to think about the error in our model of state homicide rates. The data are not from a sample. We have all of the data available. In other words, we have data for the entire population of 50 states. Nor were the values of the independent variable assigned at random as they would be in an experiment – we haven't randomly assigned states to different poverty levels. While there may be measurement error, we are not comfortable saying that measurement error is the only reason (other than poverty) that states vary in the rate of homicide. So what does the E mean? We could ask the question of whether the poverty rate is behaving the same way that a random number would in its predictions of the homicide rate. Unless the poverty rate is a better predictor of the homicide rate than a random number, we wouldn't put give much credence to the theory that says poverty is a cause of homicide.

Another way to think about this situation is to imagine a **superpopulation**. If we have data on all the states in the US or all the nations in the world, we don't really have a sample in the conventional sense, we have a population. But we can imagine the US states having evolved a bit differently than they did, and thus would have different values on the variables we are studying than they actually have.

The same argument could be applied to the nations of the world. We may have the full set of nations, and in that sense we are studying a population – or at least the process that leaves some nations out of the data set cannot be considered random sampling. But we can conceive of a hypothetical population of nations with different mixes of values on the variables of interest. We call this population of all the nations or all the states that "might have been" a superpopulation.

Then we can think of random error as sampling error that gives us a particular set of values for the countries or states in our actual data sets (the ones from the world in which we actually live) just in the way that sampling error in a survey describes the way the mix of people in the sample may differ from the population. We treat the data we actually have as a random sample from the superpopulation.

The statistical procedures are intended to tell us whether what we see in our data are likely to also be true in the "superpopulation," just as statistics tell us about the likely ways a sample may differ from the population from which it was drawn. The difference is that in a survey, the population is real and with enough money and time we could do a survey of everyone, while the superpopulation is just an idea that helps us understand error.

To summarize, we always assume that our models contain error. The error may come from limitations of a model linking an independent to a dependent variable, sampling, from random assignment in an experiment, from measurement flaws or from comparing a real variable to a random number. Statistical methods can help us understand how large the error in our analysis might be. This allows us to make statements that take account of the error, and draw conclusions in the face of that error.

Assumptions

It is common in discussing statistics to mention that all statistical procedures make assumptions about the data, the population and the processes that generated the data. Such discussions then note that statistical techniques are valid only insofar as the assumptions that underpin them are met by the data being analyzed. But rather than talking about assumptions being correct or incorrect, it makes more sense to talk about our models being more or less correct. Remember that the model contains a random element as well as the social variables we are studying. We have to ask if the random model is a pretty good description of the world.

That is, to use statistics we have to postulate a model, and our results depend on that model. If we are ignoring key factors, if the model badly misrepresents reality, then the conclusions we will draw from comparing the model with reality are likely to be wrong. The ability of statistical tools to separate error from truth and to estimate the magnitude of error depends on the random part of the model being roughly correct.

In some cases the random part of the models we use is not hard to justify because the process by which data were collected is well known and matches the conditions under which a technique works (the assumptions flow from these conditions). This is the case when we apply statistical tools to random samples and to data from experiments with random assignment to experimental and control groups. But in the case of measurement error and comparisons of real variables to random numbers, we have to be careful to think through what the results of a statistical analysis really mean.

So again we see there must be a constant and critical interplay between statistics and theory. When we consider various statistical procedures in later chapters, we will indicate the assumptions they presume. We view these assumptions as part of the model to be assessed critically. Sometimes a model may be very implausible because some of the underpinning assumptions seem unreasonable. In other cases, the assumptions might be a bit inaccurate, but we can still get a reasonable description of the world.

Feature 1.2   Diversity in Statistics: Profiles of African American[6] and Mexican American[7] and Women[8] Statisticians

There is a rich history of mathematics in Africa south of the Sahara, a fact that until recently was largely ignored by most historians of mathematics. Indeed even the black African roots of Egyptian mathematics are often denied or otherwise rendered invisible by Eurocentric views of both "history" and "mathematics." But in fact, the Greek civilization that we tend to revere in the history of mathematics was in fact a Mediterranean civilization more than a European one. Ideas were flowing back and forth from Greece, Italy, the Middle East and Africa. During the periods when Greece then Rome dominated the Mediterranean and on to the Middle Ages, European mathematics was "stuck" because there were religious and philosophical objections to the idea of zero and infinity. But during that period, Indian, Arab, and African mathematicians were making great strides because they had no religious objections to zero and infinity (Seife, 2000).

But in more recent times, science has been dominated by men of European origin, and statistics is no exception in this regard. Some of this is because of poverty and lack of opportunity, some of it because of outright discrimination. So as we move through the text and make reference to statisticians who developed the methods we use, it would be easy to get the impression that nearly all statisticians are men of European origin. But that is not true. Many important contributions to statistics have been made by women and people of color. We won't see all those people in the main text because statisticians often worked on topics more advanced than we can cover in the introductory book. To highlight their contributions to statistics, we provide brief profiles of three African American mathematical statisticians, a Mexican-American statistician, and two women statisticians.

David Harold Blackwell was the seventh African American to receive a PhD in Mathematics in the US (University of Illinois, 1941, and he was only 22 years old at the time). Dr Blackwell was also the first African American named to the prestigious National Academy of Sciences – the highest honor an American scientist can win short of the Nobel Prize. Dr Blackwell's most famous work was on game theory, and his book *Theory of Games and Statistical Decisions* (1954) is considered a classic in the field. He started his academic career at Howard University where he was promoted from instructor to Professor in three years. He then moved to the Statistics Department at the University of California at Berkeley where for many

Figure 1.4   David Harold Blackwell
Photograph by Skip Coblyn; © 2008 National Visionary Leadership Project

years he was Professor and Chair of one of the most prestigious statistics department in the US.

Charles Bernard Bell, Jr, received his PhD from the University of Notre Dame in 1953. He has held appointments as Professor of Mathematics and Statistics at Case Western Reserve University, the University of Michigan in Ann Arbor, and Tulane University. Dr Bell has worked with mathematicians in Kenya and India, and has developed courses in mathematics for teachers in Nigeria. He is the author of numerous papers on nonparametric statistics and stochastic processes – that is on statistics that don't make many assumptions about the data and on random processes that are at the heart of statistics.

Albert Turner Bharucha-Reid studied at the University of Chicago (1950–53), after receiving his BS in Mathematics and Biology from the University of Iowa at the age of 19. He left Chicago before finishing his PhD (which he thought was a waste of time). He has held faculty appointments at the University of Oregon and Wayne State University in Detroit (as Full Professor and Dean of the School of Arts and Sciences). Bharucha-Reid has published more than 70 papers and 6 books, including *Probabilistic Analysis and Related Topics* (Academic Press, 1983), *Random Polynomials, Probability and Mathematical Statistics* (Academic Press, 1986), and *Probabilistic Methods in Applied Mathematics* (Academic Press, 1968).

Javier Rojo earned his PhD in Statistics from the University of California at Berkeley, and he is currently Professor of Mathematics at the University of Texas, El Paso. Dr Rojo and his four sisters grew up in Juarez, Mexico. His parents did not finish grade school. He was always good at mathematics, and after high school he earned a Bachelor's degree in Mathematics at the University of Texas at El Paso and then a Master's degree in Statistics at Stanford University before going on to Berkeley for his doctorate. Dr Rojo views statistics as a critical tool for better understanding the problems in society. Dr Rojo has examined the impact of the 20-year-old Clean Air Act on pollution in the national parks, and he has studied the mapping of genes in the human genome to determine whether a particular gene has an impact on one's chances for getting certain diseases.

Florence Nightingale David earned her PhD in Statistics at the University College, London, in 1938, after five years working under the guidance of Karl Pearson and Jerzy Neyman, two influential modern statisticians. During World War II she developed models of the effects of the German Blitz on London that were very important in allowing the government to save lives and provide essential services during the bombing. Dr David held faculty appointments at the University of California at Berkeley and the University of California at Riverside, where she was Professor and Chair of the Department of Biostatistics. Dr David authored more than 100 scientific papers and 9 books, including the classic book on the history of probability theory, *Games, Gods, and Gambling.*

Figure 1.5 Florence Nightingale David
Source: http://mathdl.maa.org/images/
upload_library/1/Portraits

Gertrude Mary Cox earned Bachelor's and Master's degrees at Iowa State University with the support of George Snedecor, one of the pioneers in using statistics to analyze experiments. She then went to the University of California at Berkeley to earn her PhD in statistics. Snedecor encouraged her to return to Iowa State, and together they developed tools for analysis of experiments that are still in use. Her classic book with Snedecor, *Experimental Designs* (1950), is cited in the research literature hundreds of times each year. When Snedecor was asked to recommend someone to start the statistics department at North Carolina State University, he listed ten men and then added a line: "These are the ten best men I can think of. But, if you want the best person, I would recommend

Figure 1.6 Gertrude Mary Cox
Source: North Carolina State University Libraries.

Gertrude Cox." Cox went on to become a Professor of Statistics at North Carolina State University at Raleigh and the founder of one of the strongest statistics departments in the US. The "Research Triangle" area of North Carolina, where North Carolina State, the University of North Carolina and Duke University are located is a major center for statistical work, and much of this can be traced to the leadership of Cox. She was the first woman elected into the International Statistical Institute in 1949 and was named to the prestigious National Academy of Sciences three years before her death in 1978.

What Have We Learned?

Quantitative analysis proceeds by developing models to explain the variation in one variable, the dependent variable. We do this by finding one or more independent variables whose variation we believe causes variation in the dependent variable. Thus quantitative analysis builds models to explain variation.

Statistics provides the tools for building and assessing models. Of course, we don't expect that any model will explain the world perfectly, so all quantitative models include error. This error is as important as the rest of the model. There are a number of things that can generate error in our models.

One is sampling error, the fact that data that are a sample from a population may not perfectly represent the population. Another is randomization error in experiments where people are sorted into experimental and control groups through a random process like the flip of a coin. For both sampling and randomization error, if we understand the process by which the sample was drawn or by which people were sorted into experimental and control groups, statistics can tell us how large the error is likely to be and give us a sense of when we are seeing valid results and when what we see is probably just error.

In some cases we have data on whole populations, and there is no experimental assignment. Then the interpretation of random error in our models is more subtle. Sometimes it's helpful to think of the error we encounter as measurement error. At other times the error takes the form of comparing a variable of interest to a random number. But whatever we are studying, in order to use statistical tools, we have to understand what may have caused random error to influence our data. Once we have thought through the origins of error in our data, we can use statistical tools of the kinds we'll learn in the following chapters to understand our data despite those random errors.

Advanced Topic 1.1   A Third and More Technical Definition of Statistics

There is a third definition of statistics that is more technical than the other two. As noted in the first definition, one of the primary uses of statistical methods is to draw conclusions ("make inferences" in the language of statistics) about a population based on data from a sample of that population. For example, we often have a survey of the population of the US and want to use the survey to estimate (guess) at what people in the whole population think. It is common to refer to the numbers we calculate using the sample data as the sample statistics.

We can think through this process by considering the Demographic and Health Surveys (DHS) (www.measuredhs.com). The DHS coordinates nationally-representative household surveys with over 75 countries on health, nutrition, and HIV-related topics. In 2000, a sample of Ugandan citizens were interviewed about their knowledge of AIDS, including whether they think condoms can help prevent the transmission of AIDS.

In the survey 6,420 people said "Yes" (correctly answered the question), 819 people said "No," and 1,071 people said "Don't Know." Since 8,310 people answered the question (6,420 + 819 + 1,071), the percentage saying yes in the sample is 6,420 divided by 8,310, which equals about 77 percent. We know this number is true for the sample. In the technical language of statistics, this percentage is called a "sample statistic." A *sample statistic* is just a number that is calculated to describe the sample.

The parallel numbers in the population (the percent in the whole population who would correctly answer the question) are then called *population parameters*. So if we had data on the whole Ugandan population and calculated the percentage saying "Yes" we would have the population parameter. We make this distinction because we know, as a result of simple arithmetic, the sample statistics. But we don't have data on the whole population so we don't know the population parameters. An important part of statistics is learning how to use data in the sample to make good estimates (guesses) of the population parameters. Statistical methods can tell us how to use the information that 77 percent of the sample answered correctly the question about how AIDS is transmitted.

Advanced Topic 1.2   Ways of Drawing Probability Samples

We need a "probability sample" to use statistical techniques to understand the size of sampling error and be able to use sample information to construct good guesses of what is true in the population. A probability sample is one in which we know the probability (the chances) that each and every member of the population ended up in the sample. The simplest kind of probability sample is the "simple random sample" in which every member of the population has the same chance of being in the sample. The statistical formulas for handling sampling error in simple random samples are simpler than those for other kinds of probability samples. We will only present the formulas that go with simple random samples in this book.

But sometimes it isn't practical to do a simple random sample. Suppose we want to compare two groups of very unequal size – say a majority group that is 90 percent of the population and minority group that is 10 percent of the population. If we draw a simple random sample of 500 people we'll have about 450 members of the majority group and only about 50 members of the minority group. We have a representative sample of the population, but our comparisons between groups will be limited by the small number of minority group members. We might be better off by intentionally *oversampling* the minority group. We might design the sample so that we get a simple random sample of 250 people from the majority group and a simple random sample of 250 people from within the minority group. We can do this by drawing a simple random sample of each group. This gives us a much better ability to draw conclusions about the differences between groups than would a simple random sample of the whole population. And as long as we know how the groups are split in the population (90–10 in this example) then we still have a probability sample. This approach is called a **stratified sample**. We can still handle the sampling error involved but the formulas are a bit more complicated than for a simple random sample.

Another practical constraint leads to what is termed a **cluster sample**. It is often a good idea to collect data with face-to-face interviews. Sometimes we have to visit the offices of an organization to get data from their files. For example, many studies of the criminal justice system draw samples from police or court documents. But if we draw a simple random sample of people or organizations from a large geographic area such as the whole country, then interviews and site visits are scattered about, literally at random, all over the country. The time and travel costs involved gets very high.

As an alternative, some study designs draw the sample in stages. For a survey we might first pick counties within the US. We could set the sampling up so that each county has a chance of being picked that is proportional to its population. Thus counties with very large populations are almost certain to be picked in the sample while counties with very small populations have very little chance of being in the sample. Then we might pick blocks within the county, again with the probability that a block is picked proportional to the population of that block.

When we have finished this process, we will have a sample in which the probability that any member of the population is selected is known but where interviews will be clustered in a reasonable number of areas. Again, there are statistical procedures that allow us to generalize from a sample drawn in this way to the population, but the formulas for those procedures are a bit more complicated than those we use with a simple random sample. We will discuss the ways of drawing samples again in Feature 7.4.

Applications

Some students learn best by focusing on the theory first, others do best by beginning with examples. Even for the students who prefer the theory, examples provide a check on their understanding. In the text, we will use several extended examples that will come up in each chapter. This will allow you to build on a base of prior understanding rather than encountering each example without a starting point. Each of these examples is based on a research question that appears in the literature and a data set that can be used to try to answer that question. Here we introduce the examples.

Example 1:   Why do some US states have higher than average rates of homicide?

We have begun to explore this question in this chapter. As you can see from the graphs we've used in the chapter, homicide rates are higher in states with high levels of poverty than in other states. By looking more carefully at the homicide rates, we can see that high homicide states are clustered in the southern part of the United States. This might be because that's also where the high poverty states are located. But the literature on homicide offers other explanations.

One theoretical explanation for the high homicide rates in the south is that there is a culture of violence in that region of the country that promotes homicide, driven in part by its history of slavery and lynching and a widespread use of guns (Baron and Straus, 1988). Another theory argues that the high rate of homicide in southern states is the result of the high rate of poverty in the region. Thus, poverty and economic inequality and their links with social disorganization cause homicide.

There is, however, a third line of reasoning that might explain why there are more homicides in the south than in other regions of the country: environmental conditions. For example, research has documented the connection between aggressive behavior and increases in temperature (Anderson and Anderson, 1984; Harries and Stadler, 1983). The high homicide rates in the south might be due to the hot climate of the region.

Example 2:   Why do people differ in their concern for animals?

Animals are of substantial importance in human society. It has been theorized that research on how humans regard other animals, or their degree of concern for other animals, provides important information about how we organize our social worlds and how we see our connection to other living things (Arluke and Sanders, 1996). Thus, understanding variation in animal concern may provide us with insights into human character.

We could first look at differences between women and men in their concern for animals. Examining variation in animal concern by gender is based on the theoretical argument that women are more likely than men to be caring and to make moral decisions based on an ethic of care (Gilligan, 1982). Thus we would expect that women would be more likely than men to be concerned about animals.

Another argument suggests that an individual's concern for animals is rooted in a connection the individual makes with the oppression and exploitation of other living organisms. Thus, the experience of oppression produces empathy for other oppressed individuals, human and nonhuman. According to this line of thinking, women would have higher levels of concern for animals than men because of their experiences with oppression and exploitation. Minority groups that have been subject to discrimination would also have higher levels of animal concern.

Example 3:   Why are some countries more likely to participate in environmental treaties than other countries?

It's widely accepted that the world faces severe environmental problems. Burning fossil fuels, like gasoline and coal, has led to a build-up in the atmosphere of "greenhouse" gases that are changing the earth's climate. Species of plants and animals are going extinct at one of the fastest rates in the history of life on earth. Humans use about half the freshwater that flows through the earth's rivers and lakes.

In seeking solutions to environmental problems, treaties have been an important way for countries to make international promises to address global environmental problems. Nations differ quite a bit in their responses to environmental treaties. Some nations are quick to ratify most environmental treaties, other nations ignore them completely, and some nations are selective in their participation in treaties. Numerous factors, besides the merits of particular treaties, have been proposed to explain differences in environmental treaty participation among nations. (Dietz and Kalof, 1992; Frank, 1999; Roberts, Parks, and Vasquez, 2004)

One potential explanation is that international relationships play a role in treaty participation, with nations being encouraged to share common global values, including environmental protection, by other nations, especially in the context of membership in international organizations. Pressure by citizens, environmental movements, and other organizations to make environmental commitments may also be important. It also has been argued that nations with strong democratic institutions will be more likely to ratify treaties, especially in nations in which political accountability to citizens is high (e.g., politicians need to worry about being reelected and therefore want to be responsive to the public's interests). Additionally, countries that are environmentally vulnerable may have more incentives to participate in environmental treaties.

Example 4: Why do people differ in their knowledge of how the AIDS virus is transmitted?

It has been said that since the bubonic plague in the fourteenth century, which killed about one-third of the population in Europe alone, no epidemic has had as strong an impact on population growth in the world as HIV/AIDS. According to the World Health Organization (WHO) and the Joint United Nations Programme on HIV/AIDS (UNAIDS), over 34 million people are living with HIV today (2000 statistic). And an estimated 18.8 million people in the world have died from AIDS since the beginning of the epidemic. Knowledge about HIV/AIDS prevention has been and will continue to be the key to reducing the spread of the disease (Population Reference Bureau, 2001; UNAIDS, 2000; World Health Organization and UNAIDS, 2006).

In 2000, 72 percent of the people in the world with AIDS were living in Africa. Uganda, located in East Africa, was one of the first countries to experience the HIV/AIDS epidemic. While rates of HIV/AIDS have been rising in most African countries in the past few decades, Uganda is one country where rates have been significantly declining. This success has been attributed to the nationwide effort – on the part of the government, non-governmental organizations, religious leaders, and community groups – to increase HIV/AIDS knowledge among its citizens. Efforts at improving education have been concentrated in schools via sex education programs and in radio programs. Since the 1990s, there has been a large push to increase citizens' use of condoms, to educate individuals who have other sexually transmitted diseases in particular, and to increase the availability of HIV/AIDS tests.

In Uganda and worldwide, the AIDS pandemic has become feminized since the 1990s, meaning the HIV/AIDS virus is increasingly found among women (in sub-Saharan Africa today, around 60 percent of those testing positive for HIV are women). Furthermore, especially in developing countries, rates of HIV/AIDS infection have been increasing among married women. One reason is that many married women are at risk because they are not using any contraception during sex. Since knowledge is a key to reducing the transmission of HIV/AIDS, it may be that women and married people (married women in particular) have less knowledge about how HIV/AIDS is transmitted. Furthermore, it is reasonable to expect that people with higher levels of education are more likely to be knowledgeable about how HIV/AIDS is transmitted. Finally, the extent of HIV/AIDS knowledge may vary among those living in urban and rural areas. Urban dwellers have greater access to sources of education, including schools, radio programs, and health care facilities. While this example will draw on Ugandan data, it is applicable to other countries, including the United States and Europe, where rates of HIV/AIDS infection have been growing.

References

Anderson, C. A. and Anderson, D. C. 1984. Ambient temperature and violent crime: Test of the linear and curvilinear hypotheses. *Journal of Personality and Social Psychology* 46, 91–7.

Arluke, A. and Sanders, C. R. 1996. *Regarding Animals*. Philadelphia: Temple University Press.

Baron, L. and Straus, M. A. 1988. Cultural and economic sources of homicide in the United States. *The Sociological Quarterly* 29, 371–90.

Bennett, D. J. 1998. *Randomness*. Cambridge, MA: Harvard University Press.

Blackwell, D. and Girshick, M. A. 1954. *Theory of Games and Statistical Decisions*. New York: John Wiley & Sons.

Box, G. E. P. 1979. Robustness in the strategy of scientific model building. In R. L. Launer and G. N. Wilkinson (eds), *Robustness in Statistics*, New York: Academic Press.

Brown, L. 1993. *The New Shorter Oxford English Dictionary on Historical Principles*. Oxford: Clarendon Press.

Cleveland, W. S. 1993. *Visualizing Data*. Summit, NJ: Hobart Press.

Cleveland, W. S. 1994. *The Elements of Graphing Data*. Summit, NJ: Hobart Press.

Cox, G. M. 1950. *Experimental Designs*. New York: John Wiley and Sons.

David, F. N. 1962. *Games, Gods and Gambling: The Origins and History of Probability and Statistical Ideas from the Earliest Times to the Newtonian Era*. New York: Hafner Publishing Company.

Dietz, T. and Kalof, L. 1992. Environmentalism among nation-states. *Social Indicators Research* 26, 353–66.

Frank, D. J. 1999. The social bases of environmental treaty ratification, 1900–1990. *Sociological Inquiry* 69 (Fall), 523–50.

Freedman, D. A. and Lane, D. 1983. Significance testing in a nonstochastic setting. In P. J. Bickel, K. A. Doksum and J. L. J. Hodges (eds), *A Festschrift for Erich L. Lehmann*, pp. 185–208, Belmont, CA: Wadsworth.

Gilligan, C. 1982. *In a Different Voice: Psychological Theory and Women's Development*. Cambridge: Harvard University Press.

Harries, K. D. and Stadler, S. J. 1983. Determinism revisited: Assault and heat stress in Dallas, 1980. *Environment & Behavior* 15, 235–56.

Kalof, L. 1999. The effects of gender and music video imagery on sexual attitudes. *Journal of Social Psychology* 139, 378–85.

Population Reference Bureau. 2001. *2000 World Population Data Sheet* (www.prb.org).

Roberts, J. T., Parks, B. C., and Vasquez, A. A. 2004. Who ratifies environmental treaties and why? Institutionalism, structuralism and participation of 192 nations in 22 treaties. *Global Environmental Politics* 4(3), 22–64.

Salsburg, D. 2001. *The Lady Tasting Tea: How Statistics Revolutionized Science in the Twentieth Century*. New York: W. H. Freeman and Company.

Seife, C. 2000. *Zero: The Biography of a Dangerous Idea*. New York: Penguin Books.

Tufte, E. R. 1982. *The Visual Display of Quantitative Information*. Chesire, CT: Graphics Press.

Tufte, E. R. 1990. *Envisioning Information*. Chesire, CT: Graphics Press.

Tufte, E. R. 1997. *Visual Explanations*. Chesire, CT: Graphics Press.

UNAIDS. 2000. *Report on Global HIV/AIDS Epidemic, 2000* (www.unaids.org).

US Census Bureau. 2000. Table 33. Urban and rural population, and by state: 1990 and 2000 (http://www.census.gov/prod/cen2000/index.html).

US Census Bureau. 2002. Historical poverty tables: Table 21. Number of poor and poverty rate, by state: 1980 to 2006. Year 2002 (http://www.census.gov/hhes/www/poverty/histpov/hstpov21.html).

US Census Bureau. 2003. Table 295. Crime rates by state, 2002 and 2003, and by type, 2003 (http://www.census.gov/prod/2005pubs/06stata"www.census.gov/prod/2005pubs/06statab/law.pdf).

Williams, S. W. no date. Mathematicians of the African Diaspora (www.math.buffalo.edu/mad).

World Health Organization and UNAIDS. 2006, December. AIDS Epidemic Update (www.who.int/hiv/mediacentre/2006_EpiUpdate_en.pdf).

CHAPTER 2
SOME BASIC CONCEPTS

Outline

Key Terms
 Variables
 Levels of measurement
 Tools for working with measurement levels
 Scaling
 Other terms
 Labeling variables
 Constants
 Functions

Units of Analysis

Data Structure

Sample Size and Sample Selection

What Have We Learned?

Applications

Exercises

In this chapter we will define a few basic concepts that are necessary before we go on to examine data using models. You may already have a common sense understanding of these concepts or have encountered them in other courses. But we want to clarify what these concepts mean as they are used in quantitative analysis. We will reinforce the discussions of chapter 1 by paying special attention to the concepts that are needed to understand models and how we apply data to evaluating models.

Key Terms

We have to define a number of terms before we can proceed. While at first this seems a dull enterprise, we can enliven it a bit by keeping in mind that in defining terms we are constructing a way to look at the world. Our terms generate a conceptual framework, and in the case of statistics and quantitative methods it is a framework that has substantial power and subtlety.

Variables

Variables are in some fundamental sense what we are trying to understand. They are the properties or characteristics of people or organizations or nations that we want to explain. In particular, we want to know why people, organizations, or nations are different from one another – we want to explain variation. We want to know why the characteristics (the variables) vary. So variables are characteristics, such as gender or years of formal education for an individual or level of affluence for a nation.

A variable must vary – across individuals or countries or organizations it must take on different values. One can distinguish disciplines by what varies in what they study and what does not. For example, the species under study does not vary for most social scientists – we usually study only human beings. So species is not a variable for most social scientists. But zoologists and some biological anthropologists and psychologists are concerned primarily with differences across species, so for these researchers, species is a variable.

Variables have attributes that describe the variation in the characteristics they measure. For the variable "gender" the attributes are male and female. For the variable "years of formal education" the attributes are 0, 1, 2, 3, and so on. For the poverty rate for a state, the rate can range from 0 percent, when no one is poor, to 100 percent, if everyone is poor. Of course, we may not have any states at 0 percent or at 100 percent or at any other particular value. But the variable *might* take on those values.

Levels of measurement

The different ways we collect data lead to measurements with different characteristics. In quantitative work we always assign numbers to what we observe in the

Box 2.1 Keeping Percentages and Proportions Straight

The poverty rate is calculated by counting the number of people in poverty. Or, if we can't count the number of poor people, we can use surveys and the statistical tools that apply to them to estimate the number of people in poverty. Then the number in poverty is divided by the total population. This yields a number between 0 and 1.00. This is referred to as the *proportion*. For convenience we often change this into a *percentage*. A percentage runs between 0 percent and 100 percent. Remember that "percent" is Latin for "divided by 100." So 50 percent is not 50, it is 50/100 or 0.50. For most people it's easer to read 50 percent than 0.50, but remember that 50 percent is not the same as 50. It's the same as 0.50. The proportion is calculated by dividing the size of the group of special interest, the number in poverty in this case, by the total population. The percentage is just the proportion multiplied by 100. While this distinction may seem minor, it becomes important when we do arithmetic with either proportions or percentages, as we will in later chapters. If we use percentages when we should use proportions we will get the wrong answers.

Sometimes we use percentages and proportions to examine how much something has changed. For example, we might have data on how much the homicide rate changed over 10 years. When the proportion or percentage is calculated based on how many people (or countries or states) have a certain characteristic compared to the population, then the proportion must be between 0 and 1.0 and the percentage between 0 and 100. But if we look at change these restrictions don't apply. Suppose the number of homicides went from 50 to 150 over a few years. The change would be the new number of homicides (150) minus the old number (50), which is an increase in homicides of 100. We then divide by the original number (50) to get the *proportional change* (100/50 = 2.0). This means that the proportional change is 2. If we want the percentage change we would multiply by 100 and get a 200% increase in homicides. So when we look at change, there are no limits to the numerical values we can get for proportions and percentages.

By the way, when looking at proportional or percentage changes, be sure to keep in mind the base. A change from 50 to 150 seems like an important change. But a change from 5 to 15 is also a 200 percent increase. You have to decide what changes are important. Percentages and proportions help but you must use judgment too.

world, following the idea expressed by the physicist William Thomson in the quote in Box 2.2. We can have different kinds of information expressed by the numbers we assign: in some cases they are just labels, like numbers on an athletic jersey. In other cases they are the "ordinary" numbers of arithmetic. The amount of information in a number we use as a measurement is called the **level of measurement**. For some kinds of numbers, the standard tools of arithmetic apply, for others they do not. This in turn determines what statistical techniques can be applied to variables.

Most researchers refer to four levels of measurement: **nominal, ordinal, interval** and **ratio**. The idea of four levels of measurement comes from a famous paper by the psychologist S. S. Stevens (Stevens, 1946). His typology is used in most statistics books, and many researchers feel we should be very cautious about using only statistical methods that match the level of measurement. But many other researchers, including us, are less strict about this. A good review of the controversy can be found in Velleman and Wilkinson (Velleman and Wilkinson, 1993). We will explain the controversy after we describe the levels of measurement.

Nominal

In *nominal* measurement we simply assign observations to categories, and there is no ordering among the categories. One example would be student ID numbers – each person in a study gets a different number so their information can be identified. Another example is gender, which we can usually think of as having two attributes: male and female. These attributes have no special numerical value and cannot be rank ordered in any meaningful way. Thus, gender is a nominal level variable.

Sometimes numbers are given to the nominal categories of a variable, for example 1 = female and 2 = male, because numbers are more convenient in computer programs used to do statistical analysis. The numbers do not imply rank ordering, but are arbitrary. We could just as well make male equal to 1 and female equal to 2. The numbers assigned don't matter (as long as we keep track of them, so we don't confuse categories). Thus we might assign numbers for marital status as follows:

1   Married;
2   Widowed;
3   Divorced;
4   Separated but married; and
5   Never married.

Or for continents we might assign them as:

1 Africa;
2 Antartica;
3 Asia;
4 Australia;
5 Europe;
6 North America; and
7 South America.

A nominal variable should assign every person (or country or state) to a category, and to only one category. We think the marital status variable does this – everyone could be assigned into a category and no one would fit into more than one category. The continent variable works only if we leave out the many island nations that are not traditionally "assigned" to a category. And we also have to make a

---

Box 2.3 Historical Point about Assigning Categories to Variables

Developing procedures for assigning people to categories in a categorical variable sometimes has important policy implications. In the US, every Census since the first in 1790 has included a racial classification variable, but the categories used in this nominal variable have changed over time, and the racial categories were always related to public policy (Hattam, 2005; Prewitt, 2005; Snipp, 2003). In the first Census in 1790 there were three categories: "European," "African," and "Indians not taxed." As the phrase indicates, "Indians not taxed" were not counted for the purposes of representation in Congress, while for the purposes of allocating seats in Congress, a slave counted as "three fifths" of a freeman. The federal marshals who conducted the Census made the determination of who fitted into each category. The assumption of the time was that for people with parents, grandparents, and other ancestors of African heritage, the "African" group "dominates" in assigning an individual to one category in the nominal variable. This is called the "hypodescendent presumption" or the "one drop rule" in which any African ancestry led an individual to be classified as of "African" descent (Hollinger, 2005). Such a rule obviously does not reflect the reality of race and ethnicity in the US or elsewhere. Starting in 2000, the US Census used five primary categories: "American Indian/Native Alaskan," "Asian," "Black/African-American," "Native Hawaiian/Pacific Islander," and "White" plus "Some Other." To get past the hypodescendent presumption, respondents to were allowed to "select one or more." This leads to a lot of categories once one looks at all the possibilities (63 categories in fact) but gets closer to the reality of race/ethnicity than the earlier, cruder distinctions.

decision about what to do with, for example, Russia, since Russia spans both Europe and Asia. We should also note that Iceland, Ireland and Great Britain are usually assigned to Europe even though they are islands. We could add a category to account for Oceania and one for the Caribbean but there are still other island nations.

With nominal levels of measurement it is not legitimate to perform arithmetic on the variables since the numbers are simply shorthand versions of words, and we cannot perform arithmetic on words. But there is a special trick that is worth mentioning. Suppose we think of a nominal variable as indicating whether an observation has some property. This works best with just two categories, like gender. We can call these binary nominal variables or just **binary variables**. Then we can create a variable that is coded 1 for women and 0 for men. These are sometimes called "dummy" variables because when they were first developed there was a sense that they were not "real" variables. The term has stuck but as we will see, binary variables are very useful and just as "real" as other variables.

The new binary variable indicates whether an individual has the property "female." It captures the same information as the 1, 2 gender variable, but it has the advantage that certain kinds of arithmetic are legitimate with the binary variable that would not make sense with nominal variables that don't have values of only 0 and 1. For example, if we take the average of the new female variable, we would add up all the 1s and 0s and then divide by the number of people in the data set. In doing the addition, each woman counts 1, and each man counts 0. So when we take the average of a binary nominal variable we are really adding up the number of women and dividing by the number of people. But if we took the average of "continent" or "marital status" variables with numbers assigned as above, the results would not make sense.

Taking the average of a binary variable is how we calculate the proportion of people in the category labeled 1 – the number of people in the category of interest divided by the total number of people. So taking the average of a 0–1 variable is the same as calculating the proportion of people in the category coded 1. For example, in the 2000 International Social Survey Programme (ISSP) data set, there were 13,964 men and 17,064 women. If we add those together we get 31,028 respondents. Then if we want the proportion of women, we would divide 17,064 by 31,028 to get 0.550 or 55.0 percent.

But if we score the men 0 and the women 1 and take the average we will add together 17,064 ones and 13,964 zeros for a total of 17,064. Then, in taking the average, we take the sum and divide by the number of observations, so we divide 17,064 by 31,028 to get 0.550, or 55.0 percent. So whether we take the proportion or the average of a "zero, one" variable, we get the same thing. But remember this *only* works for a "zero, one" variable and only gives us the proportion in the category scored one. When we have more than two categories, things get a bit more complicated, and we will save the explanation of how to deal with that situation until later.

Ordinal

In ordinal variables numbers represent rank ordering, like the finishing order in a race. But we do not know from them how far apart the ranks are. Many attitude scale items take this form. For example in a survey we might ask:

> Tell us how you feel about the following statement. Do you strongly agree, agree, neither agree nor disagree, disagree or strongly disagree: It is right to use animals for medical testing if it might save human lives.

We then code the responses as:[1]

1   Strongly agree;
2   Agree;
3   Neither agree nor disagree;
4   Disagree; and
5   Strongly disagree.

Here the numbers mean more than they did in a nominal scale. Given how the numbers have been assigned, we know that people coded 1 (strongly agree) feel more strongly that animal testing is alright than those coded 2 (agree) who feel more strongly than the people coded 3 (neither agree nor disagree) who feel more strongly than those coded 4 (disagree) who feel more strongly than the people coded 5 (strongly disagree). But we do not know the distance between a "strongly agree" and an "agree," nor do we know if that is the same as between any other two points on the scale. We know the rank ordering but we can't assume that the difference in feeling between a "strongly agree" and an "agree" is the same as between a "disagree" and a "strongly disagree" (even though if we consider the 1, 2, 3, 4 and 5 values as regular numbers, we would think the distances are the same). As with nominal data, we have to be careful how we use the numbers we've assigned.

Why is that? What happens if we take the average? (This is in fact something we really shouldn't do with ordinal data.) In the 2000 ISSP data set, the distribution of valid responses was as shown in Table 2.1.

We could add up a 1 for each "strongly agree," a 2 for each "agree," a 3 for each "neither agree or disagree," a 4 for each "disagree" and a 5 for each "strongly disagree." Then we would divide by 29,486, the number of respondents, to get the average of these scores.

Doing the arithmetic, we will find that the average is 2.49. But since this is ordinal data, it's not clear what that number means. We don't know that the difference in strength of feeling about animal testing, say between "strongly agree" and "agree," is exactly one point, even though that is how it's scored. Nor do we know that the difference between "agree" and "neither agree nor disagree" is also exactly one point. There has been a long debate in the statistical literature on whether we should apply techniques that require arithmetic to interval data. We will discuss this issue below.

Table 2.1 Distribution of responses to question on animal testing

Response	Number	Percent
1-Strongly agree	5,520	18.7
2-Agree	12,803	43.4
3-Neither agree nor disagree	4,617	15.7
4-Disagree	4,161	14.1
5-Strongly disagree	2,385	8.1
Total	29,486	100.0

Data source: 2000 ISSP data set, analyzed with SPSS. Details of sampling strategy and related information for continuing examples are provided in the Applications section.

Interval

Here we have "regular" numbers in which the number indicates how far apart observations are. Such things as homicide rate for a state or income, and years of formal education for individuals are interval observations. For all practical purposes we can perform conventional arithmetic with these numbers and get meaningful results. For example, the sum of the incomes of everyone in a city is a meaningful summation – it is the total income for everyone.

Ratio

Ratio numbers have all the properties of interval numbers, but also have a natural zero point. The natural zero point (having none, or zero, of the quality being measured) is important for performing division. But in practice we will not worry about this distinction and just refer to interval variables whether they have a meaningful zero point or not.

Tools for working with measurement levels

We can always move from higher levels of measurement to lower. For example, when we have an interval number, like age, we also know the rank order of people in terms of age, so we can treat age as an ordinal number. We can label people with a particular age, so we can create a nominal variable. But we have to be careful about treating variables at lower orders of measurement as if they had the properties of higher orders of measurement. Why? Many methodologists and statisticians argue that converting ordinal variables into interval variables is not a good idea because we only know the rank orders, not the distance between scores. Standard arithmetic assumes that we know the distance between numbers. For example, the difference in age between a person who is 20 and one who is 21 is one year, and the difference in age between someone who is 42 and one who is 43 is also one year. But as we noted above, the difference on a survey response between a "strongly agree" and an "agree" may not be the same as the distance between an "agree" and a "neither agree nor disagree." So the arithmetic may lead us astray. And with nominal variables we don't even have a rank ordering so any arithmetic is almost certain to lead us astray.

Most statistical tools have been developed for either interval or nominal variables, taking account of their different properties and the kinds of mathematics that can be applied to them. We have extensive tools for interval variables. We have a number of good tools for nominal variables and a mixture of nominal and interval variables. But we have only very limited tools for ordinal variables and mixtures of ordinal and nominal or interval variables. This is unfortunate because in the social sciences we collect a lot of data through surveys, often using attitude questions like the one above that give us interval data.

There are three ways to deal with ordinal data. One is to use the rather restricted set of tools that take account of the fact that we have only rank orderings, and we don't know the actual distances between values and thus can't do normal arithmetic. Another is to convert ordinal measures to nominal measures and use the extensive set of tools that exist for nominal measures. This is a reasonable strategy but it has some costs. First when we do this we throw away the information about the order of responses, treating them just as categories with no order. Second, for technical reasons you'll see later the techniques for analyzing nominal measures often require large sample sizes.

The third strategy is more controversial. We can pretend that ordinal measures are actually interval (this is sometimes called "scaling up"). We can proceed to do the kinds of arithmetic that aren't really justified for ordinal numbers, like adding the numbers together and dividing by the number of observations to get the average. There is some research that indicates treating ordinal measurements as if they were really interval will not lead us far astray.[2] But there are no guarantees that in any particular analysis we will get the right answers. A good strategy for this and other problems in statistical analysis is to try to do the analysis several ways and see if they all lead to the same conclusion.

Scaling

Ordinal levels of measurement are not easy to analyze, but are very common because that's the kind of data we get from most survey questions. To get past this problem, many researchers combine several nominal or ordinal variables into a new variable called a "scale." There are many statistical tools that help us develop good scales, and scale development is a specialty for many bright researchers.

We create scales for two reasons. First, we can't do ordinary arithmetic on the individual ordinal items, but we are usually comfortable doing arithmetic with a scale, allowing us to use many techniques in analyzing the scale that we can't use on a single ordinal item.

Second, we build scales because we believe that several items in a survey get at the same value or belief or attitude of our respondents. If we combine them into a single variable, a scale, we hope we'll get a better measure of the value, belief, or attitude. This is because we think that the errors in one question will balance out the errors in the other questions. This is a common way of dealing with measurement error. If we have several measurements and we average them, the average is likely to be a more accurate measure than any single measurement.

For example, say we had data on the following question:

> Tell us how you feel about the following statement. Do you strongly agree, agree, neither agree nor disagree, disagree or strongly disagree: Animals should have the same moral rights that human beings do.

We could combine this with the animal testing item, since we feel that both these items reflect a person's attitude regarding the position of animals and humans in the world. We hope combining the two items into a single scale would more accurately reflect people's attitudes than either item alone. Unfortunately though, this question was only asked in the American survey of the ISSP and was not available for respondents in other countries, so in this book we will have to use only the animal testing question to measure animal concern.

Other terms

While nominal, ordinal, and interval are the classical terms used to distinguish levels of measurement, other terms are used as well. We sometimes refer to **qualitative** and **quantitative variables**. By qualitative we mean nominal variables, and quantitative refers to ordinal and interval levels of measurement.[3] Nominal variables are also called **discrete variables** (because they consist of separate, distinct categories). Interval variables are called **continuous variables** (because they can take on any possible value).[4] Ordinal variables are somewhere in between, although most authors call them discrete. If we have only a few possible scores on a variable, an ordinal variable looks discrete. This is the case for the five possibilities on the response scale for the animal rights question: "Strongly agree, agree, neither agree nor disagree, disagree, strongly disagree."

Labeling variables

In the standard notation we use in this book, letters from the end of the alphabet will indicate variables. Recall from chapter 1 that the letters X and Y refer to variables. Thus we might label gender X, level of education Z, and attitude about animal testing Y. For states we labeled the poverty rate X and the homicide rate Y. Or sometimes we use computer abbreviations, which are just short sets of letters to stand for the variable. So gender could be "gender" or "sex" and years of education could be "education," "ed," or "educ." What approach we use doesn't matter as long as we keep track of the variables.

Remember that each variable takes on some value for every unit of analysis included in our data. If X is gender, then every person in the data set will have a score of 0 or 1 (if that is how we coded gender), and there will be as many values of gender as there are people (of course many people will share the same value).

We use numbers to label people 1, 2, 3, . . . , and so on. These are just arbitrarily assigned "ID" numbers that help the computer keep track of the data. For example, we might assign ID numbers to states after listing them in alphabetical order so that Alabama gets a 1, Alaska a 2, and Wyoming a 50. In a survey, people are assigned ID numbers either at random or as their data are entered into the computer. But however the IDs are assigned, they don't mean anything; they are just a way of keeping track of the data. For states, we could use the state names but it's somewhat harder for the computer to work with words than with numbers. In surveys, we almost always guarantee respondents confidentiality, so we don't want their names in the data set.

We also distinguish between *independent* and *dependent variables*, as noted in chapter 1. Recall that dependent variables are what we are trying to explain. We want to know why they vary from person to person, state to state or country to country. Independent variables are variables we are using to try to understand why the dependent variables vary. We might think that homicide rate is a function of poverty. Then the independent variable is the poverty level and the dependent variable is the homicide rate. We might think that attitudes towards animals are a function of gender. Then gender is the independent variable and may be one of the questions in a survey measuring *animal concern*, which is the dependent variable. Theory might suggest that differences across people, states or countries in the independent variable produce variation in the dependent variable.

Constants

Variables vary across observations. But many statistical procedures also produce constants, which are numbers that are the same for everyone in a data set. A simple example is the average score of all members of a class on a test. Each student has her or his own score, but there is one average for the class. In that case, the average is a constant. But we might compare test scores across classes. Now our unit of analysis has shifted from individuals in a class to classes, and the average score for a class becomes a variable since it will vary across classes.

Functions

As noted in chapter 1, a function links the independent to the dependent variable. We are going to expand our discussion of functions here because we will use the idea of a function throughout the rest of the book. Please don't worry about the algebra. Here and in the chapters that follow we'll take things slowly, one step and a time, so you won't get overwhelmed even if you find the equations daunting. Remember that in reading equations, look inside each group of parentheses first, then out to the next set, and so on.

To start, remember that we usually call X the independent variable (for example the poverty rate of a state) and Y is the dependent variable (for example the homicide rate). In this model, we would be thinking that perhaps the variability across states in the homicide rate is caused by states differing in their poverty rates. To show that idea we would link the homicide rate Y to the poverty rate X with a function. It is useful at this point to reproduce the equation from chapter 1 that illustrates the relationships between X, Y, E and f:

$$Y = f(X) + E \tag{2.1}$$

As we discussed in chapter 1, E is an error term that indicates we do not expect the poverty rate to exactly predict the homicide rate. E is a variable because the amount by which the prediction for the dependent variable misses will differ from person to person or country to country. So this simple function has three variables: Y, X, and E.

The function, f, represents how Y is related to X. For the simplest models (which are often very useful), we assume the function is a simple equation that maps values of X to values of Y in a straight line. A straight line has the form $Y = A + (B*X)$, which we read "Y equals A plus B times X." That equation defines a straight line. But when we analyze data we need to include an error term E that indicates that we don't expect the data to fall exactly on the line. So the equation that is our little model of the homicide rate for states is:

$$Y = (A + (B*X)) + E \tag{2.2}$$

Now that we have actually written down a specific function, rather than letting f(X) stand for any possible function, we have two constants A and B in addition to the three variables, Y, X, and E. There will be one A and one B for all the states. A and B describe the link between X and Y for the group of states studied. The equation says that a state's homicide rate is predicted to be:

$$(A + (B*X)) + E \tag{2.3}$$

(Remember that the parentheses tell you the order in which to do the arithmetic: you should multiple B times X first, then add A.)

In algebra our equations do not have an E term because in algebra there is a perfect relationship between X and Y. We often find such equations capture practical advice. For example, there are equations that calculate the minimum heart rate that people should achieve when exercising to get the full benefits of their workout. The minimum heart rate data is based on what is called Karbonen's formula. There is also an equation for the maximum heart rate you should strive for when exercising so as not to go too far. Of course, if you haven't been exercising, it's a good idea to check with a health care professional rather than using one of these formulas.

If we call P the minimum pulse rate you should get to when exercising and X your age, then the equation for the minimum rate for mild exercise is:

$$P = A + (B^*X) \qquad\qquad (2.4)$$

Exercise physiologists have suggested that for men, A = 167 and B = −0.8. So for a 20-year-old man, the target rate would be: $167 + (−0.8^*20) = 167 − 16 = 151$.

For women, A = 166 and B = −0.6, so for a 20-year-old woman:

$$P = 166 + (−0.6^*20) = 154 \qquad\qquad (2.5)$$

For men, each additional year of age decreases the target rate by 0.8 points. For women the decrease is 0.6 points per year.

Notice that we have different As and Bs for men and women. A and B are constant within a sex (the same for all men and for all women), but differ between sexes (different for men than for women). (We use the term sex here rather than gender on the presumption that the difference in target resting heart rate is based on biology rather than on culture.) So, A = 167 for all men and B = −0.8 for all men, but A and B are different for women. Looking at how constants vary across groups (that is, turning constants into variables) is an important way to study variation, as we will see in later chapters.

Because the equation is setting a target, there is no error term. But if we took actual heart rates for everyone in a group that was exercising, and also recorded everyone's ages, we could use the equation to predict the actual heart rate during exercise. Then we could compare the prediction for each person in the group, based on their age, to their actual resting heart rate. This would generate an E term for each person – the amount by which the prediction missed their actual heart rate. But for this example, there is no E.

E is added to the statistical equation to indicate that because of sampling error, randomization error or some other form of error we do not expect to be able to predict exactly the heart rate with age or the homicide rate with the poverty level. The prediction part of the function is A + B*X, and E is the error in prediction. We will return to the idea of separating a function for Y into a prediction and the error in the prediction in later chapters.

Units of Analysis

In this section, we discuss the observational elements or the units of analysis that are common in social science research. Most social scientists study people. But while some of us study individuals, others study groups of people, or the organizations people form, or different spatial aggregations of people, or the results of what people do. The unit of analysis in a study is the thing on which data were actually collected. In most polls and surveys, we collect data about individual people; thus the person is the unit of analysis. But it is also common to use a survey interview with a single person to find out information about households or families or couples. Then the household or the family or the couple might be the unit of analysis.

There is a large literature examining how people interact in small groups. In these studies, researchers convene groups, let them interact, and observe what happens. In such studies researchers will collect data on a number of groups, and the group becomes the unit of analysis. Other researchers are interested in how governments work, how nations differ from one another, why some firms have progressive child care policies and others do not. For such studies the unit of analysis might be governments, nations or firms. One of the first questions you should ask about a study is, "What is the unit of analysis?"

Answering this question tells you what has been studied. It can lead to a second, critical question – is the thing being studied the same thing described by the theory being used and the same thing that the authors are drawing conclusions about? A study can only legitimately draw conclusions about the unit of analysis used in that research. Conclusions about anything else require a leap of faith. The problem of studying one kind of thing and drawing conclusions about another has a special name – the **ecological fallacy**.

The *ecological fallacy* has nothing to do with ecology as the term is used at present, but the name persists. The idea first appeared in the social science literature in a study by sociologists William Ogburn and and Inez Goltra (1919). Women had just been given the right to vote in Oregon. Ogburn and Goltra wanted to know if women would vote differently than men.

Since Ogburn and Goltra didn't know how individual women voted, they used data on the percentage of women voting in each precinct in Portland, Oregon and the percentage of people voting "no" on various propositions on the ballot. They used this data to see if there were different voting patterns in precincts with a large proportion of women voting compared to those with a small percentage. But they noted that this kind of analysis doesn't necessarily prove that women tended to vote differently than men. One cannot draw that conclusion. One can conclude that precincts with many women voters have different voting patterns, but because the unit of analysis is precincts, not people, one cannot draw conclusions about people.[5]

We can imagine that there are precincts that are very liberal and those that are very conservative. Suppose that in the liberal precincts women turn out to vote and in the conservative ones they don't. Further, imagine that in the liberal precincts most people favor some bond issues on the ballot while in the conservative precincts most people oppose the bond issues. But it may be that within a precinct, liberal or conservative, women are no more likely to favor the bond issue than men. But when we look at the data, we will see that in the precincts with lots of women voting, the bond issues are favored. Bond issues are not favored in the precincts with few women voting.

Two special units of analysis are worth mentioning. One is when we do a study of events. We might collect data on the characteristics of wars, or the characteristics of judicial decisions, or of auto thefts. In each case, the event is the unit of analysis. We may gather data about the event by looking at official records or by interviewing people, but the thing being studied is still the event. Over the last few decades the study of events has become increasingly popular in the social sciences, in part because many kinds of events are important and theoretically interesting, but also because better and more powerful statistical tools have been developed to study events, called **event history analysis.**

The other unit of analysis that deserves special mention is time. For example, we can collect data on unemployment rates and homicide for various years for a county or a state or a nation. Then if we examine the relationship between unemployment and homicide and suicide, the unit of analysis is the year, so that we might have 25 years worth of unemployment, homicide, and suicide data for the state of Vermont.

If we are interested in studying social change, then data over time are very attractive. Such studies are called **time series analyses** (otherwise known as a **longitudinal** study). One reason they are popular is that they can help us solve the problem of **causal ordering**. Since time flows in only one direction, and no one has yet invented a practical device for time travel, we can assume that things that happened in the past are causes of, but are not caused by, things that happen later.[6] As we will see in chapter 6, being able to make assumptions about causality is very important in analyzing data.

There are some important problems with using time as a unit of analysis. Some of these are quite technical and beyond the scope of this book. But a simple and common problem is a tendency to think we see a link between two variables over time when in fact all we are seeing is a general trend that may be driven by a third variable we aren't thinking about. For example, if we plot the number of ministers in the US in the nineteenth century and the amount of rum consumed, using years as the unit of analysis, we'll find a strong positive relationship – years with lots of ministers will also be years in which lots of rum is consumed. Does this mean ministers are drinking a lot?

Not necessarily. A more reasonable explanation is that over time the population of the US grew, and as a result there were more ministers and more rum consumed, but there was no link between ministers and rum. This sort of **correlation** without a causal effect is called **spurious**. We will discuss spuriousness in more detail later. For now we simply want to note that many social variables have strong time trends in them, and those general trends can be mistaken for strong links between two variables in an analysis with time series data.

Data Structure

There are two basic data structures and one hybrid. The basic structures are *time series*, discussed above, and **cross-sectional**. Cross-sectional studies are those in which we collect data on many examples of the unit of observation at one point in time. In a cross-sectional study we might conduct a survey of people by interviewing all of them over a period of a few weeks (which, for the purposes of the study, can be considered interviewing them all at the same time).[7] Or we might collect data from official records on crime rates and unemployment rates for cities, using many cities and gathering data for the same year from each city. In contrast a time series study takes one object (say a country or a city) and collects data over time – every month or year.

One advantage of cross-sectional studies is that it is usually possible to collect more data than can be acquired for a time series study. And if you have a new idea for something to measure, you can collect those measurements on a cross-section in a short time and do your study. But if you use a time series, you of course have to collect your measurements over time. Thus if you can only make one measurement a year, you may have to wait a long time before you have enough data to analyze. This is why most time series studies use data from official records or other historical sources. But the disadvantages of a cross-sectional study is that the time ordering of **causation** is not available to provide leverage in determining what's causing what, so theory has to bear more of a burden in making assumptions about cause and effect.

The hybrid approach is called a **panel study** or pooled time series cross-section. Unlike the time series study, we collect data over time on many units. Unlike the cross-sectional study, you observe the units at several points in time. The panel design provides a large sample size and time ordering to help with assumptions about causality.[8] For example, a very famous study, the US Panel Study of Income Dynamics started by interviewing about 5,000 families in 1968. The families were re-interviewed every year until 1997 and are now being re-interviewed every two years. The data set has information on about 5,000 families for about 30 years, or 150,000 observations!

The US Panel Study of Income Dynamics has provided extremely valuable information about the economic life of American families. We have learned about how people move into and out of poverty, what role welfare has played in getting people out of poverty, how women's work outside the house has changed and how that change has influenced housework and childbearing, and many other important issues. Many of these analyses could not be done with a cross-sectional survey because while we would know how families differ at one point in time we wouldn't know how families change over time. Nor could following one or a few families over time give us an understanding of the diverse experiences that American families have faced over the last three decades.

We can also do panel studies when the unit of analysis is a state or nation or other unit for which official statistics have been collected for a long period of time. When someone else is collecting and storing the data, it can be rather easy to put the data together into a panel. But the researcher must be careful to check that the ways data have been collected have not changed over time.

In the example of state homicide rates, we could collect from the official data sources information on homicide rates and other variables for all the states for all the years for which the data exist. This would give us a larger sample size and also allow us to look at how things have changed over time, including looking at how changing economic conditions may have influenced homicide or how changes in laws like the increased use of the death penalty have influenced homicide rates. Again, these kinds of analysis either cannot be done with a single cross-section or are much less powerful.

The example of state homicide rates can make clear the differences between panels, cross-sections, and time series. The panel data set would use all states and all years for which data are available. The cross-sectional study (which is the design of our example) will use all states, but only one year (sometimes variables are not all from exactly the same year because of what's available when we do the study, but we treat the data as if it's all from the same point in time). A time series study would take one state and look at changes in homicide rates and other factors over time. Each design has its strengths and weaknesses, although panel data are generally the best if they can be collected with high quality. But that can be very expensive and time consuming.

Sample Size and Sample Selection

If we do not have data on every unit, we have a sample. The **sample** is a subset of all the units on which we would like to have data, and we refer to all those units as the **population**. How many observations are there and how were individual units selected to be in the sample? As we will see later, the more observations we have the better statistical tools work and the more likely we are to find interesting and subtle results. But even more important than the number of observations is the way observations were selected.

We need to be very careful about how we select individuals to be in our sample. In some studies, especially those relying on official statistics or historical records, researchers include all observations for which the data are available. By comparing these two kinds of studies, we can get a sense of why sample selection is important. In reading a study, one of the first questions to ask is: How was the sample selected?

In surveys, a great deal of care is taken to insure that the sample is representative and generated by a **random selection process** that is well understood by researchers. Recall from chapter 1 that a random selection process gives everyone in the population an equal chance of being included in the sample. It chooses people from the population rather like drawing balls from the hopper for a lottery or throwing a die. If the lottery is fair, every number (every numbered ball) has an equal chance of being selected. If the die isn't loaded, then each side has an equal chance of coming up.

We call randomly selected samples **probability samples** because we know the probability that each person in the population will be selected for the sample. We use random selection because it is representative and because we know how random samples behave and can use that understanding to make careful statements about the data and the population from which it was drawn. In other words, if we have data from a probability sample, we know how sampling error behaves and can account for it in our models.

When a historical record-keeping process generates the sample, the sample is drawn from a population of all units that might have such data. But it is not usually a random sample in which every unit has the same probability of being in the sample. This can introduce biases into our analysis. For example, if we use data on national emissions of pollutants, we find that mostly the richer countries gather and report such data. Thus our sample is mostly of rich countries and our conclusions must take that into account.

Some years ago there was an influential study of birth rates that used data from all countries from which data were available. It was a number of years before a critical reader checked the original data and found that the entire sample for Asia consisted of only two countries, Israel and Japan. Only Israel and Japan had reported data on the key variables. It is hard to draw valid conclusions about fertility patterns for the world if only two countries are used to represent the whole of Asia. No sample using historical records is perfect, just as no survey is perfect. We can learn a lot from imperfect samples if and only if we are careful to understand the flaws in the sampling process.[9]

Box 2.4　How Representative Are Surveys?

The US General Social Survey (GSS), one of the surveys participating in the ISSP (data used for Example 2 throughout the text), has a large proportion of women. The Census Bureau, which has the most accurate data on the US population, tells us that the proportion of women in the US population is 51.2 percent. The GSS is considered one of the higher quality data sets used on a routine basis in the social sciences. But in all surveys there are limitations. One of the limits of the methods used in the GSS is that there tend to be a higher proportion of female respondents and a lower proportion of male respondents than would be true if we interviewed everyone in the population. In the 2000 ISSP data set, 56.1 percent of American respondents are female.

Many government surveys that are used to make policy take special steps to try to insure that the sample is representative. There's an old saying "Close enough for government work," indicating that work for the government can be sloppy. But in the world of surveys the US government produces some of the highest quality data in the world. In this case "Close enough for government work" means the most precise work in the world. In contrast, many public opinion polls used by the news media and many marketing research surveys don't take much trouble to insure a representative sample. So it might be more accurate to say, "Close enough for the private sector."

Of course, there are exceptions to every rule. Some public opinion polls and marketing studies are done very carefully. When you are presented with results from a survey or other study, you should ask yourself: "Was this research done carefully? Are the results good enough for the purposes they are being used?" If the purpose is to get some general sense of public feeling to satisfy curiosity, not much precision is needed. If major decisions hinge on the data analysis, such as the allocation of federal funds to areas with high levels of poverty or unemployment, great care must be taken. That's why the federal government does such a good job at its statistical analyses – the results matter.

Box 2.5   Large and Small Samples

We have more confidence in results from larger samples than from smaller samples if both the large and small samples are random samples. But the quality of the process by which the data were collected is more important than the size of the sample. Here is one famous example. In 1936 the *Literary Digest*, a popular US magazine of the time, conducted a very large survey to predict the outcome of the US Presidential election in which Republican Alf Landon was running against incumbent Franklin Delano Roosevelt (Bryson, 1976). The *Literary Digest* sent questionnaires to 10 million people and got 2.3 million responses, a large sample by any standard. But they miscalled the election, predicting that Landon would win. At the same time, a young social statistician, George Gallup, did a survey of 10,000 people and not only called the election correctly but predicted the *Literary Digest* poll would be wrong. How did this happen?

The *Literary Digest* had plenty of data. But they drew their sample from lists of magazine subscribers, car owners and phone books, as well as a few lists of registered voters. In 1936, the world was in the midst of a depression, and so the *Literary Digest* lists were biased towards the wealthy. And by having such a low response rate (23%) they were probably getting people most concerned about the election and thus those most dissatisfied with Roosevelt. In contrast, Gallup used a sample similar to the random samples we use now, and got the right answers. How you draw the sample is more important than sample size.

What Have We Learned?

In this chapter we have explained the terms used in quantitative data analysis. In doing so, we have begun to introduce the concepts that are used in statistics. We have variables that take on different values for different observations. They may be at any of three levels of measurement (nominal, ordinal, or interval), and we have to be careful about what kinds of techniques we use, depending on what kinds of measurements we have.

We also have constants – numbers that are the same for everyone in the data set. Functions link independent and dependent variables using constants to show the relationship between the two. We can study a variety of different units of analysis, including people, groups, institutions, and events. And we can collect data from many units at one point in time, one unit at many points in time, or many units at many points in time.

But whichever data structure we have, we must pay attention to the processes by which our units of analysis ended up in the data set we are studying. Did we have a random sampling method? Did we use all of the data available? Given these various aspects of data, it is now time to begin to look at analysis tools that help us make sense of the data, starting with graphic displays.

Applications

In this Applications section we will look at the definitions, sources for, and properties of some of the key variables we use in each of our continuing examples. We will first define the key dependent variable in each of the examples, and we also describe one of the independent variables. We'll define other independent variables as we use them in later chapters. A description of all the variables we use and the data sets that contain them will be included in Appendix 1.

Example 1:   Why do some US states have higher than average rates of homicide?

The model we examined in chapter 1 was:

$$\text{homicide rate} = f(\text{poverty rate}) + E \tag{2.6}$$

Our idea was that the homicide rate might depend on the poverty rate. In the Applications for chapter 1 we also mentioned a number of other theories of what might cause homicide rates to vary from state to state.

Dependent variable: homicide rates

In most research that examines this topic, homicide rates are based on secondary data, data that were compiled for purposes other than our research. For example, in the Baron and Straus (1988) study, they explain that the homicide rates they used were collected by the Uniform Crime Reports and consist of the 1980 rate per 100,000 population of homicides known to police. In our example, we'll use the data from 2003 provided by the US Census Bureau. State data are measured in homicides per 100,000 population.

Independent variable: poverty

In chapter 1 and throughout the book we will look at the poverty rate as a possible cause of homicide rates. The US Census Bureau computes the poverty line using income "thresholds," which are cut-off income values for how much income is needed annually to afford the basics of day-to-day life. Rather than identifying one poverty line for everyone in the United States, there are several poverty lines, which vary by composition and size of families. In 2002, for instance, poverty was defined at an annual income of $14,393 or less for a three-person household with no children and at $22,509 or less for a five-person household with two children. Data on percent of the population in poverty for each state were taken from the US Census Bureau for 2002.

Labeling variables

Homicide rate is labeled homicide03, and poverty rate is labeled poverty02.

Units of analysis

The units of analysis are the 50 US states. The conclusions drawn from the study thus refer to US states and not individuals.

Levels of measurement

The dependent and independent variables, homicide and poverty rates, are measured at the interval level.

Data structure

The data structure is a cross-section. Most of the data are from the years 2000 to 2003 from the US Census Bureau.

Advanced Topic 2.1   Panel versus Cross Section for the State Data

Since most of the variables we might use are available for most years, it would be possible to construct a panel data set of all states and all years from, say 1980 to 2000. This would increase the sample size from 50 to 1,000 (20 years times 50 states) and would allow for some analyses that aren't possible with the cross-section for a single year. If we wanted to draw strong conclusions about homicide constructing the panel would be a good idea. But for purposes of understanding statistics, the cross-section is much simpler to use.

Sample selection
The entire population is in the study: all 50 US states. There is no need to estimate population parameters because we know the value of those parameters from the data collected.

Example 2: Why do people differ in their concern for animals?

Again, we propose a model:

$$\text{animal concern} = f(\text{gender}) + E \tag{2.7}$$

This model suggests that knowing an individual's gender will help us predict one's attitude towards animals. But the error term suggests that we don't expect the prediction will be perfect. To test this model, we need data. One of largest compilations of international data on the attitudes of residents in different countries in the world is the International Social Survey Programme (ISSP) (www.issp.org). Since 1985 many countries have conducted cross-sectional surveys with samples of their residents focusing on a mutually agreed upon research topic, using the same questions (after translation) in all participating countries. Some countries conduct interviews annually, others every other year, and some less frequently, so the countries participating in a given year vary. The focus of the survey changes from year to year, and has included environment, family and gender roles, religion, and social relationships. In 2000, the survey topic was attitudes on the environment and also included questions about basic demographics, including gender, ethnicity and income. One of the questions measured concern for animals. Twenty-six countries administered the environmental survey in 2000.

Dependent variable: animal concern
We have one question that taps respondents' views about animals:

> It is right to use animals for medical testing if it might save human lives.

Respondents were asked if they "strongly agreed, agreed, neither agreed nor disagreed, disagreed or strongly disagreed" with each statement. People who have strong concerns about animals will tend to disagree with this statement.

Advanced Topic 2.2   Multiple Measures

It is good practice in a survey to ask multiple questions on the same topic and to word the questions so that a person with strong views on the subject sometimes has to say "agree" and sometimes "disagree." The logic behind this is that if someone is not thinking much when answering the questions and is just saying "agree" to every question, she or he will seem to be pro-animal on one question and anti-animal on another other. When we combine the items in the scale, such a person will be in the middle rather than at either extreme. Unfortunately in this data set, we only have one survey question that measures animal concern. In constructing surveys there is always a tradeoff. Asking a lot of questions on the same topic allows for better measurement (and less measurement error) but increases respondent fatigue and the proportion of potential respondents who don't complete the survey (which increases sampling error).

If we were designing a study focused on concern with animals we would have asked at least a few questions, not just one. But in making their tradeoffs about the content of the survey the designers of the ISSP decided to go with just one question. This is often a problem with using "secondary data," – data that were collected by someone else. The researchers who designed the study may not have had the same interests that we do and may not include as many measures of what we are studying as we would like. But the advantage of such secondary data in this case is that we have data from a large international sample of high quality that would be very expensive for us to collect on our own.

We want our measure to have high scores for people who have pro-animal attitudes and low scores for those that have anti-animal attitudes. Individuals who disagreed with the question are more pro-animal than individuals who agreed. A person who "strongly agreed" has a value of 1 on this question, 2 for "agreed," 3 for "neither agreed nor disagreed," 4 for "disagreed," and 5 for "strongly disagreed," so we do not have to change the scoring.

Independent variable: gender

This is a variable that is almost always included in a survey. We call variables such as gender, age, and education "demographics." The interviewer recorded the "sex" of the respondent based on their judgment from the interview. So in a sense this is the interviewer's interpretation of the respondent's gender. The variable has two attributes: female and male.

Labeling variables

The dependent variable might be labeled "animalx" for animal concern index, and the independent variable could of course be labeled simply "sex."[10]

Units of analysis

The units of analysis in this problem are individuals, and the data are collected from individuals via surveys.

Levels of measurement

The animal concern question is measured at the ordinal level. Throughout the book, we often treat the item as if it is measured at the interval level, so that we can use this example with statistical procedures that require interval measurement.

Gender is a nominal level variable – the attributes of gender are labels or categories with no numerical value attached to them (female, male). Remember that we often assign a number to the categories of gender, i.e., male = 1, female = 2, but these numbers have no numerical meaning; they are only labels. (However, recall that when we define binary nominal variables with the values 0 and 1, we can do some kinds of arithmetic, liking taking the average, which is then just the proportion of people in the category labeled 1.)

Data structure
The data in the ISSP are cross-sectional.

Advanced Topic 2.3   Panel versus Repeated Surveys

The ISSP has compiled surveys every year since 1985. But the countries interview a different sample of people each time, so it is not a panel study. A panel study, like the US Panel Study of Income Dynamics, interviews the same people over and over again (see http://psidonline.isr. umich.edu/ for more details on this study). The panel study allows us to follow changes in an individual and within a family over time. But panel studies are much more complicated and expensive to conduct then a series of annual cross-sectional studies.

Sample selection
The 2000 ISSP survey consists of samples of residents in 26 countries: Austria, Bulgaria, Canada, Chile, Czech Republic, Denmark, Finland, Germany, Great Britain, Ireland, Israel, Japan, Latvia, Mexico, Netherlands, New Zealand, Northern Ireland, Norway, Philippines, Portugal, Russia, Slovenia, Spain, Sweden, Switzerland, and the United States. Each country translates the survey into its primary language and administers the survey. Sample sizes range from 745 residents in Northern Ireland to 1,705 residents in Russia. The combined sample size for all 26 countries is 31,042.[11]

Because we are using data from all the countries in the 2000 ISSP, we have to restrict our choice of variables to those that can be used meaningfully across such a broad range of countries. Gender is of course socially constructed so gender socialization, gender roles, the structural influences that impinge on men's and women's lives, and other gender-related factors will certainly differ across the countries in the data set. This will be true of other independent variables, such as marital status, education, and age that we will use to explain concern with animals in other chapters. After looking at the analyses that combine data from all countries, it would make sense in a research project to look at the data separately for each country and to explore how the effects of gender and other independent variables differ across countries and why those differences exist. Sophisticated methods for doing such analysis are the subject of a lot of research in the last few years (Byrk and Raudenbusch, 1992; Steenbergen and Jones, 2002; Western, 1998). These methods are beyond the scope of our presentation in this book. But in exercises for your class, it would be possible to pick one or a few countries and replicate the analyses we are doing for all countries to see how the models for animal concern might vary across countries.

In the rest of the book we will usually use the ISSP data on animal concern from all countries as a single data set rather than breaking it out by countries. This will give you a sense of how statistical tool work with a very large survey data set. But note that when we use the data for all countries at once we do not have a representative set of countries of the world, just the ones where social science researchers were interested in participating in the ISSP and were able to find the resources to do so. So we can't generalize to the world from the data set. And since the sample size for each country is not proportional to the populations of the country, the sample that uses data from all countries is not directly representative of the whole population of those 26 countries. So when we use the data for all countries with taking direct account of the individual countries in the analysis, we are doing something that is useful as an exercise in working with a large sample, but we would not do those kinds of "all countries" analyses for research. Rather, we would take explicit account of the countries to compare them. But you can replicate our "all countries" analyses with data for a single country and get results that are perfectly sound for that country. In the Applications at the end of Chapter 14 we do examine the effects on animal concerns of being in a particular country.

Advanced Topic 2.4   Sampling Approaches in the ISSP

The details of how each country generated their samples are quite complicated and vary by country. Some countries selected representative samples of the entire population, while others covered only major cities or are restricted in other ways. Some countries carried out face-to-face interviews, while others used mailed or self-administered surveys. The details on the sampling and data collection procedures are described in detail on the ISSP website at (http://www.gesis.org/en/data_service/issp/data/2000_Environment_II.htm).

Example 3:   Why are some countries more likely to ratify environmental treaties than others?

Here is our model:

$$\text{environmental treaty participation} = f(\text{voice and accountability}) + E \qquad (2.8)$$

One reason countries may differ in the extent they participate in environmental treaties is because they differ in how much a nation's citizens have "voice" and the extent to which a government is accountable to its citizens. But we don't expect voice and accountability to perfectly predict environmental treaty participation, so we also include an error term in the model.

Dependent variable: environmental treaty participation
Between 1946 and April 1999, there have been 22 multilateral international environmental treaties. Treaty topics include air pollution, oil pollution, and greenhouse

gases. The idea of looking at how many treaties a country has ratified was introduced by Dietz and Kalof (1992). In a more extensive analysis of this issue, the total number of treaties a country participated in was computed for 191 nations by Timmons Roberts, Parks and Velasquez (2004; see the article for a complete description of all the treaties and the sources of information the authors used to determine treaty participation).

Independent variable: voice and accountability

It may be that governments that are unaccountable and oppressive will be more likely to ignore the demands of environmentalists and the international community to act in environmentally responsible ways. To measure this, Kaufmann, Kraay, and Zoido-Lobaton (2002) created an index called "Voice and Accountability." This index includes numerous measures of citizens' political rights, civil liberties, aspects of the political process, and how independent the media are. Scores on the scale reflect the extent citizens of a country are able to participate in the selection of their government officials and have freedom of expression. The inclusion of how independent the media are serves to measure how closely authority figures are monitored and held publicly accountable for their behaviors. The index is created by several measures that come from various non-governmental organizations (NGOs), risk rating agencies, and think tanks. Scores can range from −2.5 to +2.5. While the computation of the index is quite complicated, what is important to note is that higher scores reflect greater levels of "voice and accountability," meaning more freedom of expression, a free press, and considerable citizen participation in government. Lower scores, on the other hand, reflect little voice and accountability.

Labeling variables

Typing (or saying) number of environmental treaties and voice and accountability rate is rather cumbersome. We could call the former X and the latter Y, but if we have lots of variables it can be hard to remember what is what. As we noted in chapter 1, sometimes it's convenient to use short strings of letters as names for variables. When we created this data set, we told the computer to call number of environmental treaties "envtreat" and the voice and accountability index "voice." This helps us remember what is what.

Units of analysis

Our units of analysis are nation states as recognized by the World Bank and other sources for our data. Our theory and any conclusions we draw should thus be about nations, not about individuals, communities or any other unit.

Levels of measurement

Both variables are interval level variables. Environmental treaty participation is measured from 0 to 22. The voice and accountability scale is scored from −1.93 to +1.73.

Data structure

Treaty participation is measured over six decades, so it might be thought of as a time series while voice and accountability is measured at one point in time. However, we are not looking at how participation in treaties changed over time but rather a summary measure of how many treaties a country had ratified by 1999.

Sample selection

We were able to get data on environmental treaty participation for 191 countries. Data were available on the voice and accountability scale for 169 of these countries. This includes many small countries; in fact 22 percent of the countries (N = 42) have populations of less than 1 million. Sometimes in analysis of cross-national data analysis researchers restrict their sample to large nations, for example only considering nations with populations of over 1 million. For some kinds of analyses it doesn't make sense to compare countries like China, the US, and the UK with very small countries like Trinidad and Tobago or Fiji. Since participating in environmental treaties is a political activity, and both large and small countries have political processes, we don't think the size restriction makes sense here. So we will use all countries for which we have data. Some researchers have theorized that the size of a political unit has a substantial effect on its political processes (Dahl and Tufte, 1973; Dietz, 1996/1997; Frey and Al-Mansour, 1995) so we could look at population size, population density or rate of population growth as predictors of treaty participation. We don't pursue those here but they might be interesting topics for further research.

Example 4: Why do people differ in their knowledge of how the AIDS virus is transmitted?

Our model is:

$$\text{AIDS knowledge} = f(\text{gender}) + E \tag{2.9}$$

This model suggests that knowing an individual's gender will help us predict his or her knowledge about AIDS transmission. But the error term suggests that we don't expect the prediction will be perfect. To test this model, we will use data from the 2000 Ugandan DHS dataset (see http://www.measuredhs.com/pubs/pdf/FR128/01Chapter1.pdf for a summary of the entire 2000 Ugandan-DHS survey, including background information on Uganda, how the sample was collected, and a summary of the survey data findings).

Dependent variable: AIDS knowledge

The interviewer asked each respondent to state spontaneously what they think can be done to prevent the transmission of the AIDS virus (common answers, for example, were abstinence from sex, having only one sexual partner, not engaging in prostitution). After respondents gave their complete list, they were also probed about various ways AIDS can or cannot be transmitted. We will focus on the transmission of AIDS and condom use and whether respondents knew that condom use can help prevent transmission of the disease.

The response categories were "yes," "no" and "don't know." Unless otherwise indicated, throughout this book, "no" and "don't know" responses are combined since both responses reflect a lack of knowledge that condoms can be used to prevent the transmission of AIDS.

Independent variable: gender

Again, this is one of those basic demographic variables that is almost always included in a data collection effort. The variable has two attributes: female and male.

Labeling variables

The dependent variable might be labeled "AIDScon" for knowledge that condoms can reduce the likelihood of transmitting AIDS, and the independent variable could of course be labeled simply "sex."[12]

Units of analysis

The units of analysis in this example are individuals, and the data are collected from individuals via face-to-face interviews.

Levels of measurement

If we combine "don't know" and "no" responses, we have a nominal level variable that could be scored 1 for those who said yes (i.e., knowledgeable about condom use and AIDS transmission) and 0 for those who said no/don't know (i.e., lacked knowledge about how to prevent AIDS transmission). We can calculate the mean (the proportion saying yes), which is 0.77 or 77 percent. That is, about three-fourths of respondents know that condom use can reduce the likelihood of transmitting AIDS.

Gender is also a nominal level variable – the attributes of gender are labels or categories with no numerical value attached to them (female, male).

Data structure

The data in the Ugandan-DHS survey are cross-sectional.

Sample selection

A stratified, clustered sampling design by region of the country was used to select the Ugandan-DHS sample. This is not a representative sample of Ugandan citizens though. Conducting a large-scale survey in a country like Uganda is quite challenging. First, some parts of the country were deemed too unsafe for interviewers to travel. Since most residents live in rural areas, urban area residents were oversampled. In addition, over three-fourths of the sample was female. Since other goals of the survey were to assess women's fertility behaviors and maternal and children's health, there was a particular interest in surveying women. The goal was to interview at least 6,500 women and 1,800 men.

Exercises

The purpose of these exercises is not only to practice the material in the chapter but also to stimulate thinking and discussion. We present the kind of information that often comes with codebooks describing data sets or brief summaries of research papers and reports. For each of the exercises, thinking through and discussing why you answered the way you did is as important as the answer itself, because this kind of thinking hones your skill at quantitative reasoning. In some cases, the information provided with data or research reports lends itself to more than one interpretation. Understanding the basis for multiple interpretations can provide the basis for deeper understanding of the concepts.

1. In the 2000 International General Social Survey that provides the data for our analysis of attitudes towards animals, there are a number of variables we might use as independent variables in a model predicting attitudes towards animals. For each of the following variables, indicate whether it is nominal, ordinal, or interval. Explain your reasoning. The word in parentheses is the name of the variable in the ISSP data set. The following information provided with the data set explains what each variable represents. In every case we have left out the codes for missing data (people who didn't answer the question).

A   Marital Status (marital)
    "Are you currently – married, widowed, divorced, separated, or have you never been married?"
    RANGE: 1 to 5:
    1   married;
    2   widowed;
    3   divorced;
    4   separate;
    5   never married

B   Age (age)
    Respondent's age
    RANGE: 18 to 99

C   Education (educ)
    What is the highest grade in elementary school or high school that you finished and got credit for?
    RANGE: 0 to 99

D   Highest degree earned (degree)
    Highest educational degree earned by respondent
    RANGE: 1–7:
    1   None; still in school;
    2   Incomplete primary school;
    3   Primary school completed;
    4   Incomplete secondary school;
    5   Secondary school completed;
    6   Semi-higher; incomplete university;
    7   University completed

E   Family income at age 16 (INCOM16)
    Thinking about the time when you were 16 years old, compared with American families in general then, would you say your family income was – far below average, below average, average, above average, or far above average?
    RANGE: 1 to 5:
    1   Far below average;
    2   Below average;
    3   Average;
    4   Above average;
    5   Far above average

F   Where you live (urbrural)
    Describe the place where you live
    Range: 1–5:
    1   Urban;
    2   Suburb;
    3   Small city or town;
    4   Country village;
    5   Farm, home in the country

2. Suppose we think the following model might explain homicide rates:

$$Y = A + BW + E \qquad (2.10)$$

Where Y is the state homicide rate, W is the percent urban and E is the error term. A good estimate of A and B for the US Census Bureau state data is A = 2.86 and B = 0.026. Table 2E.1 gives values for urbanization level and homicide rate for several states. Calculate the predicted value for homicide rate and the error value for each of these states.

Table 2E.1 Homicide rate, percent urbanization, and predicted homicide rate for five states

State	Homicide rate	Urbanization	Predicted homicide rate	Error
Georgia	7.6	71.6		
Maryland	9.5	86.1		
New Jersey	4.7	94.4		
South Carolina	7.2	60.5		
Wyoming	2.8	65.1		

3. For each of the following studies, identify the unit of analysis, the data structure, the independent variable(s), the dependent variable(s), and the levels of measurement of the variables.

A   McLaughlin, P. and Khawaja, M. 2000. The organization dynamics of the US environmental movement: Legitimation, resource mobilization, and political opportunity. *Rural Sociology* 65, 422–39.

This study models the number of environmental organizations founded in the United States each year from 1895 to 1994. The most important independent variable in the study is the number of books published that year on environmental topics, which the authors interpret as a measure of the ideological climate in a given year.

B   Gamson, W. 1975. *The Strategy of Social Protest*. New York: Dorsey.

This study draws a simple random sample of protest groups active in the US over a 75-year period. The dependent variable is whether or not the group was successful in achieving its goals. One of the key independent variables was whether or not the group used violent means.

C   Kalof, L. 2000. Vulnerability to sexual coercion among college women. *Gender Issues* 18(4), 47–58.

This study surveyed a random sample of undergraduate women at a university in up-state New York. In that survey the author measured women's gender attitudes and experiences with sexual coercion. Two years later she sent a follow-up survey to the women who were still enrolled at the university. She again examined gender attitudes and a new variable, experiences with sexual coercion since the first survey. She found that: 1) initial attitudes (as measured by the first survey) had no effect on experiences with sexual coercion over the 2-year period; 2) initial experiences with sexual coercion did not make women vulnerable to more sexual coercion over the 2-year period; and 3) attitudes were not changed by experiences with sexual coercion over the 2-year period.

D   St Lawrence, J. S. and Joyner, D. J. 1991. The effects of sexually violent rock music on males' acceptance of violence against women. *Psychology of Women Quarterly* 15, 49–63.

This study examined the effects of rock music on males' attitudes toward women. Undergraduate men were randomly assigned to listen to one of three types of music: sexually violent heavy-metal rock, Christian heavy-metal rock, or easy-listening classical. Participants were administered the Attitudes Toward Women Scale before and after exposure to the music. The authors found that exposure to either kind of heavy-metal rock music, regardless of lyrics about sexual violence, increased males' sex role stereotyping and negative attitudes toward women.

4. Here are some variables that have been used in the US General Social Survey and ISSP data sets. In exercises in later chapters, we will be analyzing many of these variables. The variable name is listed first, followed by a definition of the variable and how it is coded. Identify the level of measurement of each of these variables.

a) Childs = Number of children an individual has.
b) Worklife = "How successful are you in your work life?"
    1 not at all successful;
    2 not very successful;
    3 somewhat successful;
    4 very successful; and
    5 completely successful

c) Hrs2 = Number of hours usually worked in a week.
d) Wrkslf = Are you self-employed or do you work for somebody else?
    1 self-employed; and
    2 work for somebody else
e) Spwrksta = Spouse's work status:
    1 working full-time;
    2 working part-time;
    3 temporarily not working;
    4 unemployed/laid off;
    5 retired;
    6 in school;
    7 keeping house; and
    8 other.

---

5. In the previous chapter, we learned about independent and dependent variables. For each of the following hypothetical statements, identify the independent and dependent variables.

a) Canadians whose first language is English had higher average scores on a scale of patriotism compared to those Canadians whose first language is French.
b) The more hours spent each week in leisure activities, the higher a person's life satisfaction.

c) Higher infant mortality rates tend to be concentrated in countries that are less technologically advanced.
d) Children who played a video game that had violent content were more likely than children who were exposed to a non-violent video game to act aggressively in the two weeks following video game exposure.
e) The conservative country of Italy has had a much lower rate of cohabitation before marriage than the more liberal countries of Canada and the US.

References

Baron, L. and Straus, M. A. 1988. Cultural and economic sources of homicide in the United States. *The Sociological Quarterly* 29, 371–90.

Binder, A. 1984. Restrictions on statistics imposed by method of measurement – some reality, much mythology. *Journal of Criminal Justice* 12, 467–81.

Bollen, K. A. and Barb, K. H. 1981. Pearson's r and coarsely categorized measures. *American Sociological Review* 46, 232–9.

Bollen, K. A. and Barb, K. H. 1983. Collapsing variables and validity coefficients – reply. *American Sociological Review* 48, 286–7.

Bryson, M. C. 1976. The *Literary Digest* poll: Making of a statistical myth. *The American Statistician* 30, 184–5.

Byrk, A. S. and Raudenbusch, S. W. 1992. *Hiearchical Linear Models.* Newbury Park, CA: Sage.

Dahl, R. A. and Tufte, E. R. 1973. *Size and Democracy.* Stanford, CA: Stanford University Press.

Demographic and Health Surveys (DHS). 2000. *Ugandan-DHS 2000 Survey.* www.measuredhs.com. Calverton, MD.

Dietz, T. 1996/1997. The human ecology of population and environment: From Utopia to Topia. *Human Ecology Review* 3, 168–71.

Dietz, T. and Kalof, L. 1992. Environmentalism among nation states. *Social Indicators Research* 26, 353–66.

Ferrando, P. J. 1999. Likert scaling using continuous, censored, and graded response models: Effects on criterion-related validity. *Applied Psychological Measurement* 23, 161–75.

Frey, R. S. and Al-Mansour, I. 1995. The effects of development, dependence and population pressure on democracy: The cross-national evidence. *Sociological Spectrum* 15, 181–208.

Gaither, C. C. and Cavazos-Gaither, A. E. 1996. *Statistically Speaking: A Dictionary of Quotations*. Bristol, England: Institute of Physics Publishing.

Hattam, V. 2005. Ethnicity & the boundaries of race: Rereading Directive 15. *Daedalus* 134, 61–9.

Hollinger, D. 2005. The one drop rule & the one hate rule. *Daedalus* 134, 18–29.

International Social Survey Programme (ISSP). 2000. 2000 Environment II data set. www.issp.org. Catalog no. ZA 3440. Cologne, Germany: GESIS-ZA Central Archive for Empirical Research.

Kaufmann, D., Kraay, A., and Zoido-Lobaton, P. (Jan. 2002). *Governance Matters II: Updated Indicators for 2000/01*. Policy Research Working Paper no. 2772. The World Bank Research Development Group and World Bank Institute; Governance, Regulation and Finance Division (http://hdr.undp.org/reports/global/2002/en/).

King, G. 1997. *A Solution to the Ecological Inference Problem: Reconstructing Individual Behavior from Aggregate Data*. Princeton, NJ: Princeton University Press.

Krieg, E. F. 1999. Biases induced by coarse measurement scales. *Educational and Psychological Measurement* 59, 749–66.

O'Brien, R. M. 1983. Rank order versus rank category measures of continuous variables – comment. *American Sociological Review* 48, 284–6.

Ogburn, W. F. and Goltra, I. 1919. How women vote: A study of an election in Portland, Oregon. *Political Science Quarterly* 34, 413–33.

Prewitt, K. 2005. Racial classification in America: Where do we go from here? *Daedalus* 134, 5–17.

Roberts, J. T., Parks, B. C., and Vasquez, A. A. 2004. Who ratifies environmental treaties and why? Institutionalism, structuralism and participation of 192 nations in 22 treaties. Global Environmental Politics 4(3), 22–64.

Snipp, C. M. 2003. Racial measurement in the American census: Past practices and implications for the future. *Annual Review of Sociology* 29, 563–88.

Steenbergen, M. R. and Jones, B. S. 2002. Modeling multilevel data structures. *American Journal of Political Science* 46, 218–37.

Stevens, S. S. 1946. On the theory of scales of measurement. *Science* 103, 677–80.

Uganda Demographic and Health Surveys. 2001. Calverton, Maryland: UBOS and ORC Macro (http://www.measuredhs.com/pubs/pdf/FR128/00FrontMatter.pdf).

US Census Bureau. 2000. Table 33. Urban and rural population, and by state: 1990 and 2000 (http://www.census.gov/prod/cen2000/index.html).

US Census Bureau. 2002. Historical poverty tables: Table 21. Number of poor and poverty rate, by state: 1980 to 2006. Year 2002 (http://www.census.gov/hhes/www/poverty/histpov/hstpov21.html).

US Census Bureau. 2003. Table 295. Crime rates by state, 2002 and 2003, and by type, 2003 (http://www.census.gov/prod/2005pubs/06stata; www.census.gov/prod/2005pubs/06statab/law.pdf).

Velleman, P. F. and Wilkinson, L. 1993. Nominal, ordinal, interval and ratio typologies are misleading. *The American Statistician* 47(1), 65–72.

Western, B. 1998. Causal heterogeneity in comparative research: A Bayesian hierarchical modeling approach. *American Journal of Political Science* 42, 1233–59.

CHAPTER 3
DISPLAYING DATA ONE VARIABLE AT A TIME

Outline

The preliminary analysis of most data is facilitated by the use of diagrams. Diagrams prove nothing, but bring outstanding features readily to the eye; they are therefore no substitutes for such critical tests as may be applied to the data, but are valuable in suggesting such tests, and in explaining the conclusions founded upon them. R.A. Fisher, *Statistical Methods for Research Workers*, 1925 (Gaither and Cavazos-Gaither, 1996, p. 113).

As we noted in chapter 1, R.A. Fisher was one of the great geniuses of twentieth century science and had a hand in developing most of the statistical techniques we cover in this text. So we take his admonition to use graphics as compelling. Graphics are one of the most important aspects of data analysis because they allow us to visualize the data and the patterns in the data. Graphic display should always be the first step in statistical analysis. And sometimes we continue to use graphics as a critical tool in the application of some of the most advanced methods of data analysis.

In this chapter we will introduce some of the most common forms of data display for looking at one variable at a time. There are a few other methods for graphing one variable at a time we won't discuss until we have introduced some methods for summarizing a set of data with numbers in chapter 5. There also are very useful methods for examining the relationship between two variables that we'll discuss in Chapter 6. But first we need to look at some of the tools for data display for just one variable at a time. At the end of this chapter we will also consider some principles for displaying data that can guide you in making effective graphs.

Box 3.1 The History of Graphic Displays of Information

The earliest known map is a town map of Konya, Turkey, from about 6200 BC.[1] The first graph, as best we can determine, was a sort of bar graph from about 1350. Leonardo da Vinci and others in the 1500s began using graphs to investigate the physical world. By the 1600s graphs were being used to display social data, especially data on age. By the 1700s statistical graphing became quite popular. But change was slow, with few new graphing methods being developed, until the 1970s. Computers offered new possibilities for graphing. But the use of computers for statistical calculations also tended to keep data analysts removed from the data – something that was not the case when all statistical calculations were done by hand. (The term "calculator" originally meant not a machine but a clerical worker who did the arithmetic.) The combination of increased computer power and a concern that we always "look" at the data we are analyzing led to many new kinds of graphs, and a growing interest in graphs by professional statisticians. Now, "data visualization" as it is sometimes called, is a major research area in statistics.

Recall that in quantitative data analysis we are concerned with understanding variation. We want to know why people, states, countries, or other things we study vary from one to another. Graphics for a single variable at a time are a first step in understanding variation. They do not allow us to look at models that link dependent to independent variables. But they do allow us to get a sense of how much variation exists in the variable we are graphing, and what kind of shape that variation takes. As we will see in later chapters, one of the most important issues in statistical analyses is the pattern we find in the data – at what values there are many cases, at what values there are few cases. This is called the **distribution** of the data. The one variable plots we present here are guides to the distribution of the variables in our data. Sometimes these plots can reveal things that are substantively interesting, and they are a first step before building models to understand the variability in our data.

Graphic Display of Nominal and Ordinal Data

Nominal and ordinal data are often displayed in **pie charts** and bar graphs. Both are useful when the variable has only a few categories, such as gender, political affiliation, or race. Before drawing either a pie chart or bar graph, it is important to construct a **frequency table** of the data first, so you have a sense of how many cases there are and how they are distributed across categories of the variable. Carefully constructed tables can also be a very clear way of displaying data.

We created Table 3.1 using output from the statistical package SPSS. It shows how many people said "yes" to the question from the 2000 Ugandan DHS survey regarding whether using condoms during sex can prevent transmission of AIDS. The "Number" column shows the number of people who said "yes," "no," "don't know," and the total number of people who answered the question. Standard practice in presenting tables says we should always present the total number of people since, as we will see in later chapters, we have more confidence in larger samples than in smaller samples.

Table 3.1   Percentage responses to question about whether the chances of contracting AIDS are reduced by always using a condom during sex

Response	Number	Percent
Yes	6,420	77.3
No	819	9.9
Don't know	1,071	12.9
Total	8,310	100.0*

* Percentages do not always sum to 100 exactly, due to rounding
Data source: 2000 Uganda DHS data set, analyzed with SPSS.

> *Statistical packages for the social sciences* are discussed at the end of this chapter as an Advanced Topic.

The numbers in the table are straightforward. 6,420 people said they thought the transmission of AIDS can be reduced through condom use, out of a total of 8,310. We divide 6,420 by 8,310 to get 0.773 or 77.3 percent – the percent who answered "yes." We then divide 819 by 8,310 to get the percent that said "no" – 0.099 or 9.9 percent. Finally, we could divide the 1,071 who said "Don't know" by 8,310 to get 0.129. Alternatively we could add 0.099 to 0.773 to get 0.872, the proportion who answered either yes or no, and subtract that from 1.000 (100 percent) to get 0.128. While in principle either method should give us the same percentage saying "Don't know" in fact we have rounded each of our calculations to three decimal points from the actual calculation results. That builds in some rounding error. In statistical calculations we always have to be careful about rounding error as it can introduce small discrepancies into our calculations. If we used four decimal places instead of three, we would get the same result up to three decimal places. Either the division or the subtraction gives the same result up to the third decimal place, so some of the information in the table is redundant in that we can calculate the values in some of the cells in the table from the information in other cells.

Table 3.2 is the same information expressed in a slightly different form to minimize the amount of redundant information. Here we just give you the percentages saying "yes," "no," and "don't know" and the total sample size. You could calculate the number saying "Yes" by multiplying 0.773 by 8,310. (Remember that 77.3 *percent* is 77.3/100.)

Table 3.3 shows the variation in marital status in the 2000 Ugandan DHS data set. We introduce this variable because it may be that marital status is related to people's knowledge about how AIDS is transmitted. Tables 3.1 and 3.2 were based on SPSS output, but we "cleaned" them a bit. Table 3.3 is exactly what SPSS produces. Notice that the percent and valid percent columns are identical. This is because there are no "missing" data (data are called **missing** when people do not answer the question). If there were cases with missing data, the "percent" column would

Table 3.2 Percentage responses to question about whether the chances of contracting AIDS are reduced by always using a condom during sex

Response	*Percent*
Yes	77.3
No	9.9
Don't know	12.9
N	8,310

Data source: 2000 Uganda DHS data set, analyzed with SPSS.

Table 3.3   Marital status frequency table

			Frequency	Percent	Valid percent	Cumulative percent
			MARITAL: Marital status			
Valid	1	Never married	2,100	25.3	25.3	25.3
	2	Currently married	5,250	63.2	63.2	88.4
	3	Formerly married	960	11.6	11.6	100.0
		Total	8,310	100.0	100.0	

Data source: 2000 Ugandan DHS data set, analyzed with SPSS.

calculate the percentages for "yes," "no," "don't know" and "missing." The valid percent column would exclude the missing cases and give percentages for only those saying "yes," "no" and "don't know." The last column is the **cumulative frequency** – it adds the frequencies as we move down the column. In the data set 25.3 percent were never married, 88.4 percent were never married or were currently married, and so on. Why would we want this? In fact, with a nominal variable like marital status, we wouldn't. But that doesn't stop the computer from printing it out. As we'll see in the next example, if we are working with an ordinal variable, the **cumulative percentage** might be interesting. A common mistake by those new to data analysis is to try to interpret everything in the computer output. Often the standard output of the computer provides information that is not helpful, and sometimes not even meaningful, for the analysis we are doing.

Table 3.4 is from SPSS output on how people scored on the animal concern question. Here we can see one of the reasons the cumulative percentage might be

Table 3.4   Responses to the animal concern question (animalx)

		Score	Frequency	Percent	Valid percent	Cumulative percent
Valid	1	Strongly agree	5,520	17.8	18.7	18.7
	2	Agree	12,803	41.2	43.4	62.1
	3	Neither agree nor disagree	4,617	14.9	15.7	77.8
	4	Disagree	4,161	13.4	14.1	91.9
	5	Strongly disagree	2,385	7.7	8.1	100.0
		Total	29,486	95.0	100.0	
Missing	8	Can't choose, DK	1,218	3.9		
	9	NA, refused	338	1.1		
		Total	1,556	5.0		
Total			31,042	100.0		

Data source: 2000 Uganda DHS data set, analyzed with SPSS.

useful with an ordinal or interval variable. Cumulative frequencies or percentages tell us the number or percentage of people who fall below some value of the variable you are looking at. For the animal concern question, people who don't feel animals should be used in experiments have scores of 1 or 2. We know from the cumulative frequency column that about 62.1 percent of the respondents to the survey scored 2 or below and thus agreed with the item. We can also see that just over one-fifth of the respondents were at or above a score of 4 (13.4 percent of the sample had a score of 4 and 7.7 percent had a score of 5). We can also see that not everybody answered this question. 1,556 respondents did not have a valid answer to this question ("missing"), with 3.9 percent of the sample coded "8"or don't know and 1.1 percent coded "9" for not applicable or refused. The "percent" column includes these "missing" respondents in the percent calculations, while the "valid percent" column excludes them. Deciding which column of percents to report depends on whether you want to refer to the entire sample or just those respondents who answered that particular question.

Simple frequency tables give us some insight into the patterns in the data. They are also essential as a first step in working with a data set because they allow us to detect errors in the data. There are a number of steps in going from a survey interview or a collection of official statistics to a computer data set that can be analyzed by a statistical package and in many research projects several different people can be involved in the process. So we always look at frequency distributions of our variables to make sure there aren't inconsistencies. As we mentioned in chapter 2, it's important to make sure that the computer software knows that some people didn't answer a survey question and data may be missing for some countries or states. This is handled by giving the non-respondents a special "missing data" coding. For example, in the statistical package Stata, gender might be coded 0 for men, 1 for women and "." (a period) for those whose gender was not recorded. Stata knows in processing the data that the periods represent missing data and will leave those people out of any analysis using the gender variable. SPSS reports those with missing data as "System Missing" and also leaves them out of any analysis. In the case of the animal concern question, values of 8 and 9 were used to code cases that did not have valid responses. We want to check to make sure this is being handled properly – that the software is not making the "8" and "9" responses for people with especially strong feelings. Frequency tables are a good way to do this.

Pie chart

The pie chart was invented by William Playfair, a British scientist, in 1801. He also invented the bar chart in 1786 and a number of other forms of statistical graphics and maps for showing data. He was trying to understand how national economies worked. Recall that the word statistics has its origins in the Latin word for state. Many early statisticians were concerned with what we would now call public policy.[2]

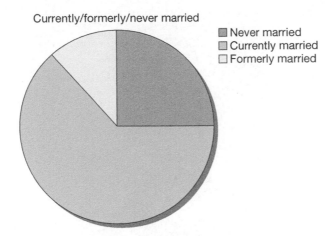

Figure 3.1   Pie chart of marital status, N = 8,310
Data source: 2000 Ugandan DHS data set, analyzed with SPSS.

Pie charts are often used to show frequency distributions. Figure 3.1 shows a pie chart of marital status from the 2000 Ugandan-DHS data set, the same data that were displayed in Table 3.3. In a pie chart, the size of a "slice" represents the number of cases in the category represented by that slice. The bigger the size of the slice, the more cases are in the slice. The proportion of cases thus determines the angle at the center of the pie that goes with the corresponding slice, and that angle determines the area of the slice.

Some pie charts are calculated using the percent of cases that fall into a particular category of the nominal variable. Others use the number of cases. But since the number of cases in a category divided by the number of cases for which we have data is the proportion in the category, and the proportion multiplied by 100 is the percent in the category, making a pie chart based on percentages in various categories and making one based on the number of cases in those categories produces the same pie chart. Only the labels will differ, and many pie charts don't include labels that tell you the exact percentage in each category. But pie charts and any other graphs should tell you how many cases were used to construct them. This is usually done in small print at the bottom of the chart. That's also a good place to indicate where the data used to make the graph came from.

Pie charts are very popular. But research on people's perceptions of graphs shows that pie charts are very hard to interpret accurately.[3] To understand a pie chart, you must comprehend angles and areas, because it is the angles and areas that show the proportion of cases in a category. This is not something most of us do well. But pie charts can be perceived accurately by most people when the important parts of the chart are around either one half or one quarter of the total chart. We tend to see the values less accurately when the critical "slices" of the pie are much larger or much smaller (Simkin and Hastie 1987).

Figure 3.2   Florence Nightingale
Source: http://commons.wikimedia.org/wiki/Image:Florence_Nightingale_1920_reproduction.jpg.

Pie charts can also work well when the data tell a dramatic story – when the differences are very clear. Florence Nightingale, who is usually considered the parent of modern nursing (see Figure 3.2), was also a pioneer in using statistical analysis to understand health problems.[4] Figure 3.3 is a redrawing of a famous pie chart

Figure 3.3   Florence Nightingale's pie chart of casualties in the Crimean War
Source: Redrawn from original and available online at: http://www.math.yorku.ca/SCS/Gallery/milestone/.

Box 3.2   How to Build a Pie Chart

Pie charts are a bit complicated to build, and since there are many circumstances in which they are not a good choice for displaying categorical data, we will only sketch their construction.

1   Find the proportion of cases in each category.
2   Multiply the proportion for each category by 360^0 to find out how big the angle associated with each category should be. (Pie charts represent data percentages by angles – this is one of the reasons they can be hard to read.)
3   Draw a circle of the size you want the pie chart to be.
4   Draw a straight line from the center of the circle to the edge.
5   With a protractor, measure an angle from the line drawn in step 4 of the size for the category you want to plot first, as calculated in step 2.
6   At that angle, draw another straight line from the center to the edge at the point indicated by the angle. This is the wedge of the pie for the first category.
7   From the line at the edge of that wedge, find the angle for the second category, then draw the line.
8   Continue until all the wedges are drawn.
9   Label the wedges.
10 Add labels indicating the title of the graph, and a note indicating the source of the data and the number of cases on which the graph is based.

---

she constructed to show British officials that the major cause of death in the Crimean war was not combat but death that came about because of disease. Here the pie chart is quite clear and dramatic. The pie chart by Nightingale actually contains information about three variables. Each "slice" of the pie represents one month of casualties, so the slices are representing time. The outer portion of each slice is for casualties not caused by battle, while the inner part of each slice shows the battle casualties. The area of each part of each slice indicates how many men died. So the pie chart shows the number of casualties, the month in which they occurred and whether they came from battle or from other causes. Nightingale called this type of chart a "coxcomb," and the technical name for them is "polar area chart." In mathematics, the term polar can refer to angles – that's why they are called "polar."

Bar chart

Figure 3.4 gives a bar chart of the data on marital status. The height of each bar indicates the proportion of people who reported belonging to that marital status. It seems to us that it is easier to see how large the smaller groups are than it was in the pie chart, but perceptions of graphs can vary from person to person.

Figure 3.4 is labeled with the percent of cases. Like pie charts, bar charts can be labeled in the percent of cases falling in each category, the proportion of cases in

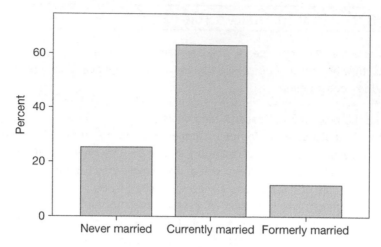

Figure 3.4   Vertical bar chart of marital status, N = 8,310
Data source: 2000 Ugandan DHS data set, analyzed with SPSS.

each category (which is just the percent divided by 100) or in the number of cases in each category. All three give the same graph, only the labels differ. Remember, for any graph, the number of cases used in constructing the graph and the source of the data should also be included. This is usually done in small print at the bottom.

Figure 3.5 shows the same information, but we've flipped the axes of the chart so that instead of looking up and down to understand the data, we now look left and right. We are used to scanning left to right, rather than bottom to top, so it is a bit easier to understand a graph in which the information can be understood by scanning horizontally. Remember, graphs are pictures, and there can be more than one way to paint a good picture of the data.

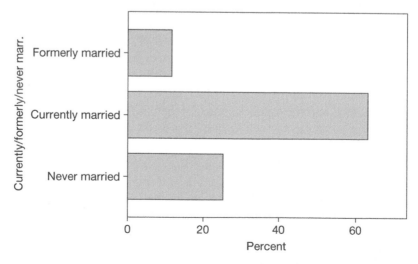

Figure 3.5   Horizontal bar chart of marital status, N = 8,310
Data source: 2000 Ugandan DHS data set, analyzed with SPSS.

Box 3.3   Drawing a Bar Chart

1   We begin the bar chart by drawing the vertical axis to represent the percentage of cases or number of cases that might fall in a category of the variable we are plotting. By tradition, the bottom of the vertical axis always starts at zero. This won't be true for all kinds of graphs, as we will see. We extend the vertical axis so that it is a bit higher than is needed to include the group that has the largest proportion or number of cases. We want the top tick mark to be a "nice" number, one that ends in "5" or "0" such as 20 percent or 25 percent.

2   Next we label the axis with "tick marks" (the little marks on the left just touching the vertical axis) at "nice" numbers – numbers that are multiples of ten, five, or two. These help us see the pattern more easily. Don't put the tick marks inside the axis, only things representing data are allowed there.

3   Now we draw the horizontal axis, making it big enough to have a bar for each category of the variable with a bit of space between each bar. In spacing the bars, remember that you will need room for labels for each category.

4   Now we mark the height for each category of the nominal variable we are working with based on the percent of observations in that category. For each category, find the appropriate height on the vertical axis and make a mark at that height above that category on the horizontal axis.

5   Then we simply draw a bar to that height, and do the same thing for every other category of the nominal variable. In a bar chart, the width of the bar and the distance between bars don't convey any information, they are chosen to make the graph look nice.

6   While not everyone does it, we follow Cleveland (1994) and always enclose graphs in rectangles (Cleveland calls this the data rectangle). This focuses the eye on the data, not on the labels.

7   As always, add labels and a note indicating the source of the data and the sample size.

Dotplot histogram

Figure 3.6 shows a **dotplot histogram** of marital status. There are actually several kinds of dotplots, as we will see. In this kind, each case gets a dot (really a small circle) on the graph. There is a "bin" for each category of the nominal variable, in this case for each marital status. Every never married person gets a dot in the "0" bin, every married person a dot in the "1" bin, and every former married person a dot in the "2" bin. Each dot goes a little to the right of the dots put in the graph before it. The result is much like a histogram (as we will see shortly), but it is constructed by simply adding dots to the graph. The number of cases is labeled in this dotplot. Remember that pie charts and bar charts can be labeled with the number

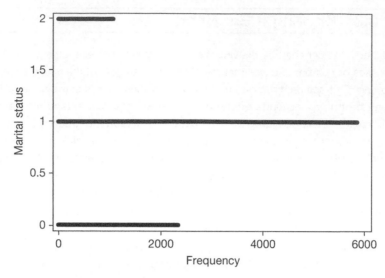

Figure 3.6   Dotplot histogram of marital status, N = 8,310
Data source: 2000 Ugandan DHS data set, analyzed with SPSS.

of cases, the percentage of cases or the proportion of cases. Dotplot histograms are always labeled with the number of cases. Because of the large sample size for the Ugandan DHS study, the individual dots are hard to see. In the Applications, we have included a dotplot histogram of whether or not a state was a member of the Confederacy to show how these graphs look with small samples.

Box 3.4   Making a Dotplot Histogram

1   First, on the vertical axis, mark a space, which will be a "bin" for each category of the variable being plotted. Each bin should occupy the same length on the axis, and as in making a histogram, it's important to think about how the labels for the categories will look on the outside of the graph. Remember, the labels for the bins go on the left side of the vertical line; the bins that will hold the data are on the right.

2   Draw a small circle (or an "x") in the bin that corresponds to the value of the first case on the variable being plotted. In our example, if the first case is a never married person, then draw a small circle in the never married bin.

3   Repeat this for every data point. In doing this by hand it is very important to make each dot (or "x" or other symbol) the same size and to "stack" them in the bin from left to right. This precision is necessary to make sure that the length of the stack of dots in the bin accurately shows how many cases are in the bin.

4   Close the data rectangle.

5   Add labels and the note with source of data and number of cases.

Graphic Display of Continuous Data

In this section we describe some basic graphic methods for continuous data: the one-way scatterplot, the Cleveland dotplot, the histogram, and the stem and leaf diagram. Here we restrict ourselves to the simplest version of these plots of numerical data, but later we will introduce some other graphics for these types of data.

One-way scatterplot

We used two-way scatterplots (i.e., plots of two variables), which we usually refer to just as scatterplots, in chapter 1. We will talk about them in detail in chapter 5. But a recently developed method called a one-way scatterplot can be useful in looking at the distribution of continuous variables. One-way scatterplots were developed by a number of statisticians during the 1970s and 1980s, though the first use of this approach seems to date to 1644 (StataCorp, 1999).

Figure 3.7 shows a one-way scatterplot of the state homicide rates. The one-way scatterplot has a continuous horizontal axis that covers the range of the variable being plotted – it starts with the lowest value and ends with the highest value. Unlike the bar chart and the dotplot for categorical variables, the scale on the horizontal axis for a one-way scatterplot is continuous – every possible value of the variable between the high and low values can, in theory, be plotted on the graph, although if there are many values, the marks on the graphs will overlap. Then a short vertical line is placed on the graph for each data point (in this case for each state).

One-way scatterplots work well when cases do not take on the same values. This is the case for the homicide data, where every state has its own value, though some states are rather close together. If a variable only has a limited set of values, such as the animal concern question (recall that this can only be a whole number between 1 and 5) then the lines on the one-way scatterplot fall on top of each other, and we don't see much.

The one-way scatterplot in Figure 3.7 suggests that there are a few states that are at the high end of homicide rates compared to most other states, and that there are two "bunches" of states, one bunch near the lowest end of values and one around the middle. Of course, different people will see somewhat different patterns in a graph.

As a preview of how even very simple plots can be used to show the relationship between variables, Figure 3.8 shows two one-way scatterplots of the homicide rate, one for states that were members of the Confederacy and one for states that were

1.2   homicide03   13

Figure 3.7 One-way scatterplot of state homicide data, N = 50
Data source: US Census Bureau 2003, analyzed with Stata.

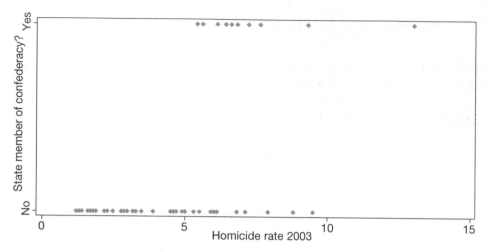

Figure 3.8 One-way scatterplots of homicide rate, by membership in Confederacy, N = 50
Data source: US Census Bureau 2003, analyzed with Stata.

Box 3.5 Making a One-Way Scatterplot

1 Construct a horizontal axis that starts at or a bit below the lowest value and extends to or a bit beyond the highest value of the variable being plotted.
2 Make a short vertical line above the horizontal axis for each case at the point on the axis that is the value of the variable for that case.
3 Add the graph title and note indicating data source and number of cases.

---

not. For ease of reading we have enclosed them in a box and used diamonds as a plotting symbol. Remember that one theory about variation in homicide rates focuses on why Southern states have higher homicide rates. We created a variable that was 1 for states in the Confederacy (as a way of defining "Southern" states) and 0 for all other states. We can see that the states that were in the Confederacy tended to have higher homicide rates in 2003, although some non-southern states have rates higher than some of the Confederate states. We will do much more work using graphs to examine the effects of independent variables on dependent variables in chapter 5.

Cleveland dotplot

This plot was proposed by William S. Cleveland (1993; 1994). In many texts it is called simply the dotplot, but since we are also using a dotplot histogram in this chapter, we will call it the Cleveland dotplot. Most of the other graphs in this chapter don't show us the value of the individual cases. The exceptions are the one-dimensional scatterplot, which shows us where each case is located but not which case it is, and the

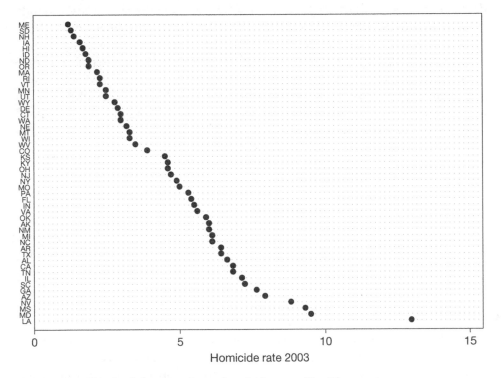

Figure 3.9 Cleveland dotplot of state homicide rates, N = 50
Data source: US Census Bureau 2003, analyzed with Stata.

Box 3.6   Building a Cleveland Dotplot

1 Construct a horizontal axis that runs from the lowest value of the variable being graphed or a bit lower to the highest value or a bit higher.

2 Sort the data in order from lowest to highest (or highest to lowest).

3 On the vertical axis, write a label for each case, from highest to lowest.

4 To the right of the label for each case, mark the point that corresponds to the value for that case on the variable being graphed by reading values from the scale on the horizontal axis.

5 Place a small circle or other symbol to mark the value for the case.

6 Place small faint dots that run from the left axis to the marker for the data point, and then the same small faint dots running from the data point to the right axis. The purpose of these dots is to create a horizontal line that the eye can follow. Some versions just run a line from the left axis to the point but most software does it using the dots that Cleveland preferred.

7 Close the data rectangle.

8 Add the title for the graph and the note indicating the data source and sample size.

dotplot histogram, in which we can plot symbols to represent the cases. The Cleveland dotplot shows us the value for each case while also showing the distribution of the data.

Figure 3.9 is a Cleveland dotplot. By convention, the data are arranged from lowest value to highest value (or vice versa). The name of the case is a label on the vertical axis. To make these readable, the Cleveland dotplot is best used with fewer than 30 data points. With 50 states, the labels are a bit closer to each other than is ideal. But the pattern is clear. Louisiana is at the very high end of homicide rates, with Nevada and Mississippi not far behind. Note that at the end of this chapter, we have included tables (Tables 3.10 and 3.11) listing the state and country abbreviations and names, which are used throughout the book.

Histogram

A histogram shows the number (or percentage) of cases that fall in a range of the variable being considered. A histogram is like a bar chart except that with a histogram we take a continuous variable and break it into categories, while with the bar chart the variable is already in categories. By tradition, there is a space between the bars in a bar chart, but there are no spaces between bars in a histogram to emphasize the continuous nature of the variable's measurement. Each bar in the histogram captures a range of values on the continuous value being plotted. As early as 1833, A. M. Guerry in France was using histograms to try to understand how crime rates varied by month of the year and by age group. The term "histogram" was introduced in 1895 by one of the major figures of modern statistics, Karl Pearson (Pearson, 1895).

Figure 3.10 is a histogram of the animal concern question. Since the scale runs from 1 to 5 and only whole number scores are possible, the bins created for the

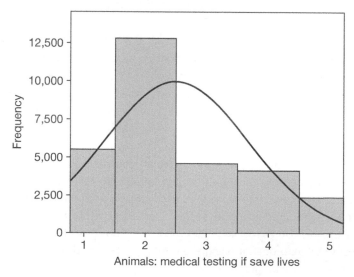

Figure 3.10   Histogram of animal concern question, N = 29,486
Data source: 2000 ISSP data set, analyzed with SPSS.

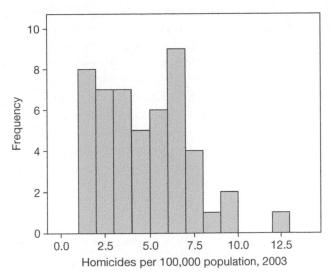

Figure 3.11 Histogram of homicide rates for states, N = 50
Data source: US Census Bureau 2000, analyzed with SPSS.

histogram correspond to the values 1, 2, 3, 4, and 5. Thus the histogram is the same as a bar chart in this particular example, except that in the histogram the bars touch and in a bar chart they don't. The middle of the scale would be at 3, since that's halfway between 1 and 5. We can see that responses seem to "lump up" at 2, and that taking an anti-animal stand (which is a score of 1 or 2) was a bit more common than taking a pro-animal stand on the question (a score of 4 or 5).

Figure 3.11 is a histogram of the homicide variable. Table 3.5 displays the data as a table. We told the computer to use 10 bins for the histogram. Remember that

Table 3.5 Proportion of cases in each range of values of state homicide rates

Range of homicide rate	Proportion of cases
1.00–1.99	0.16
2.00–2.99	0.14
3.00–3.99	0.14
4.00–4.99	0.10
5.00–5.99	0.12
6.00–6.99	0.18
7.00–7.99	0.08
8.00–8.99	0.02
9.00–9.99	0.04
10.00–10.99	0.00
11.00–11.99	0.00
12.00–12.99	0.00
13.00–13.99	0.02

a histogram is a picture, and the number of bins we use influences what we can see. If we use very few bins, each bin must cover a broader range of the variable being plotted and more cases will be lumped together. We lose detail. But if we use too many bins, we will have few or no cases in many bins and the histogram will look odd.

Here we see the same pattern we have noted before. A small number of states have homicide rates a good deal higher than most other states. For the bulk of the data there seem to be two places where we see a lot of states, one at homicide rates of around 2.5 and one at homicide rates between 5 and 7.5. These groupings of many cases around a few values are called "modes" – most often occurring scores. We discuss them further below.

Box 3.7 Building a Histogram

1 Decide on how many bins you will use. Table 3.6, which was developed for stem and leaf diagrams, gives general rules for how many bins to use. It also gives a starting point for histograms.
2 Divide the range of the variable being plotted by the number of bins. That gives the range that each bin covers. For example, homicide rates run from 1.2 to 13.0. We can call that range 1–13. If we want exactly ten bins, then each bin will cover 1.2 points. The first bin would run 1.00–2.20, the second from 2.21 to 4.40, the third from 4.41 to 5.60 and so on. On the computer this is easy; in drawing a histogram by hand we might want to use a range of just one point (1–2, 2–3, 3–4) and thus end up with twelve bins. Or we might want to use a two-point range for each bin (1–3, 3–5, 5–7, etc.) and have only six bins. (Note that if there are data points at numbers that match the bin boundaries, such as exactly 2.0, then we have to define the bin boundaries more precisely, for example, 1–2.9, 3.0–3.9, etc.)
3 Draw a horizontal axis and mark off the bins on the axis. Remember that even though we have to mark the horizontal axis in bins, the axis is continuous. If we had a very large amount of data we could have many, many bins and we can imagine having so many data points that we could have an infinite number of bins, each very tiny but still with a number of cases in it. Marking off a modest number of bins is a practical necessity when we have modest amounts of data.
4 Calculate the proportion (or percentage) of data points that fall into each bin.
5 Draw a horizontal line for each bin at the height that marks the proportion of cases that fall in that bin.
6 Draw the two vertical lines to the horizontal axis that define the box for each bin.
7 Add the labels and notes on data sources and sample size.

Stem and leaf diagram

The stem and leaf diagram is a simple and effective way of examining the distribution of variables, whether nominal, ordinal, or continuous. It was developed by John Tukey (1977). It shows the same information as a histogram, but is easier to construct with paper and pencil.

We will work through an example of making a stem and leaf in detail. Table 3.7 displays data for the homicide rate in 2000 for the 50 states.

Feature 3.1   John W. Tukey

Tukey was one of the great statisticians of the last half of the twentieth century. He was a pioneer in seeing how computers could change data analysis. He is credited with inventing the term "software" in an article in 1958. In 1970, he coined the term "bit" standing for "binary digit." He developed a series of methods we will discuss in later chapters that use computing power to guard against being misled by unusual data points – called outliers. But despite his fascination with computers, he was concerned that people might let computers do their thinking for them and stop looking carefully at their data. In the 1970s he invented a series of paper and pencil graphic displays for what he called "Exploratory Data Analysis." The idea of exploratory data analysis is to use graphs and simple calculations to identify patterns in the data. The stem and leaf diagram is one of these tools, and we will see several others in later chapters.

Tukey's expertise as a statistician led to invitations to serve on several important assessments of scientific research. He participated in an important study that criticized the Kinsey report, in which Alfred Kinsey attempted to draw conclusions about Amer-

ican sexual behavior based on a non-random sample. He chaired a US National Academy of Science committee that helped establish that chemicals released from spray cans were destroying the protective ozone layer of the earth and leading to dangerous levels of ultraviolet radiation in sunlight. More details on Tukey can be found at a website devoted to him: http:Stat.bell-labs.com/who/tukey.

Figure 3.12   John W. Tukey
Photograph by Paul R. Halmos, from *I Have a Photographic Memory*, © 1987 American Mathematical Society

Box 3.8   Building a Stem and Leaf Diagram

The first step in constructing the stem and leaf is to rearrange the data in order from highest to lowest. Then we have to decide how many "bins" will be used to display the data. For nominal data we would use the same number of bins as the number of categories in the data. For ordinal or interval data we have to decide how many bins to use based on what will make a graph that is easy to interpret. Two rules of thumb[5] have been suggested:

$$k = 10*(log10(N)) \text{ and } k = 2*\sqrt{N} \qquad\qquad (3.1)$$

where k is the number of bins and N the number of observations (the number of data points to be plotted) (Emerson and Hoaglin, 1983). The formula may look daunting if you do not remember logarithms and square roots, so a few examples are presented in Table 3.6. As you can see somewhere between 10 and 30 bins are appropriate. The log rule seems to overestimate the number of bins for small sample sizes while the square root rule probably suggests too many bins for large samples. Remember that this is just a rule of thumb. Constructing a good graphic is always an iterative process. We begin with a rough draft and then refine the graph.

Table 3.6   General rules for the number of bins in a stem and leaf plot

Number of cases	Number of bins (log rule)	Number of bins (square root rule)
25	14	10
50	17	14
75	19	17
100	20	20
150	22	24
200	23	28
250	24	32

The second step in creating a stem and leaf is to draw a vertical line. Then the left side of the line is divided into the appropriate number of bins, each of which occupies equal length on the line. Each bin is labeled with a number on the left-hand side of the line indicating the values that fall in that bin. This side of the diagram is the stem. The number for the stem side is the first digit of the cases that will go into that bin. Note that the same range of values of the variable being graphed (not the same number of cases) goes into each bin. Thus bins can hold 1 value, 2 values, 5 values, 10 values, or multiples of these.

Then each observation is assigned to the appropriate bin. This is done by printing a number on the right hand, or leaf side, opposite the appropriate bin label. The number printed should be the digit of the value being plotted that conveys the most information about that value, given that its placement in a particular bin already provides some information about its value. That is, if a value for homicide rate is falling into a bin that can only contain values from 8.00 and 8.99, then placing a 2 for a state with 8.2 years of education is appropriate, since there can be no ambiguity about the value. Once all observations have been placed in bins by printing an appropriate number on the right hand side of the line, numbers within bins are rearranged from highest to lowest, with the lowest numbers closest to the stem on the left.

Box 3.9   Common Errors in Building Stem and Leaf Diagrams

- Using more than one digit to represent a case in the leaf side. Each case gets exactly one digit on the leaf side.
- Recording the digit that represents the stem on the leaf side. That information is already conveyed in the bin label. The leaf side digit should be the next digit in the value for the case that conveys information that is not conveyed by the bin label.
- Leaving empty bins out. Even if a bin is empty (there are no cases that go there) the bin conveys part of the information about the distribution of the data. In the state income data, there are no states in the bin for 37 in thousands place (the bin for states, if there were any in 2002, with per capita incomes between $37,000–37,999). But if we leave that bin out of the stem and leaf, it distorts how far New Jersey and Connecticut (in the 39 and 42 in thousands place bin respectively) are from the other states.
- Making bins of unequal size. Each bin must cover the same range of values of the variable. For every change of one unit on the bin side, we must have 1, 2 or 5 bins, and it must be the same number of bins throughout the stem and leaf. If we don't do this, the distances between the values on the leaf side are distorted.
- Writing the ranges on the stem side. The stem and leaf is meant to be simple, with our focus on the data on the leaf side. If we label a bin with the full range, for example "$30,000–39,999" rather than just "30" the label becomes very "busy" and distracts our attention from the data.

Table 3.7 State homicide rates, N = 50

State	Homicide03: Homicides per 100,000 population, 2003
Alabama	6.6
Alaska	6.0
Arizona	7.9
Arkansas	6.4
California	6.8
Colorado	3.9
Connecticut	3.0
Delaware	2.9
Florida	5.4
Georgia	7.6
Hawaii	1.7
Idaho	1.8
Illinois	7.1
Indiana	5.5
Iowa	1.6
Kansas	4.5
Kentucky	4.6
Louisiana	13.0
Maine	1.2
Maryland	9.5
Massachusetts	2.2
Michigan	6.1
Minnesota	2.5
Mississippi	9.3
Missouri	5.0
Montana	3.3
Nebraska	3.2
Nevada	8.8
New Hampshire	1.4
New Jersey	4.7
New Mexico	6.0
New York	4.9
North Carolina	6.1
North Dakota	1.9
Ohio	4.6
Oklahoma	5.9
Oregon	1.9
Pennsylvania	5.3
Rhode Island	2.3
South Carolina	7.2
South Dakota	1.3
Tennessee	6.8
Texas	6.4
Utah	2.5
Vermont	2.3
Virginia	5.6
Washington	3.0
West Virginia	3.5
Wisconsin	3.3
Wyoming	2.8

Data source: US Census Bureau 2003, analyzed with Stata.

```
x 1 |  x 0.1
  1  |
  2  |
  3  |
  4  |
  5  |
  6  |
  7  |
  8  |
  9  |
 10  |
 11  |
 12  |
 13  |
```

Figure 3.13   Homicide rate stem and leaf diagram – first stage

Step 1. Building the bins

We have 50 data points. The range is from a little more than 1 to 13. The table of the general rules suggests around 14–17 bins. Let's try 13 though, one bin for each whole number. We need 13 because we'll start at one because the lowest homicide rate is 1.2. If the smallest homicide rate was between 0 and 1 then the first bin would be for 0 in units place. Remember that these are rates, so 1 means 1 homicide per 100,000 population, 10 means 10 per 100,000, and so on. It is good practice to put a number at the top of the stem and one at the top of the leaf indicating the scale of the numbers on the stem side and the leaf side. For this diagram the stem side will be times 1 and the leaf side will be times 0.1. It will look like Figure 3.13.

Step 2. Filling the bins

Now we need to insert states into bins. The first state is Alabama with a homicide rate of 6.6. This will go into the 6 bin. Since we know any state in that bin will have a rate between 6.0 and 6.9, the leaf side can convey the information after the decimal point. For Alabama, that is a 6. Looking at Figure 3.14, we know we have a state that has a homicide rate in the 6 range because it is in that bin, and that it is 6.6 because of the 6 on the leaf side. Note that we no longer know from the graph which state it is.

The next state is Alaska with a 6.0 (see Figure 3.15).

Filling in all the states gives the stem and leaf shown in Figure 3.16.

Step 3. Re-arranging within bins

Finally, for a proper stem and leaf, the cases must be rearranged within bins to go from the smallest number on the left to the highest on the right as in Figure 3.17.

x 1	x 0.1
1	
2	
3	
4	
5	
6	6
7	
8	
9	
10	
11	
12	
13	

Figure 3.14   Homicide rate stem and leaf diagram – first data entry

x 1	x 0.1
1	
2	
3	
4	
5	
6	60
7	
8	
9	
10	
11	
12	
13	

Figure 3.15   Homicide rate stem and leaf diagram – second data entry

x 1	x 0.1
1	78624993
2	9253538
3	9032053
4	56796
5	450936
6	604810184
7	9612
8	8
9	53
10	
11	
12	
13	0

Figure 3.16   Homicide rate stem and leaf diagram with all data entered

x 1	x 0.1
1	23467899
2	2335589
3	0023359
4	56679
5	034569
6	001144688
7	1269
8	8
9	35
10	
11	
12	
13	0

Figure 3.17   Homicide rate stem and leaf diagram – final version

Obviously, Louisiana is an outlier. The rest of the data seem to have one peak in the distribution around 2 (between 1 and 4.0) and another around 6.

Let's try a more complicated example. The data on per capita income in each state is in Table 3.8. Per capita income is a way of measuring how well off people in the state are. The percentage of the population below the poverty line focuses on the poor; this measure looks at the average. The incomes range from a low of $22,321 for Mississippi to a high of $42,505 in Connecticut. The general rule still suggests 14–17 bins. With 21 bins, as in Figure 3.18, we can create one bin for each $1,000, starting at 22 (for 22 in thousands place, or $22,000–$22,999) and ending with 42 (for 42 in thousands place, or $42,000–42,999).

Table 3.8 State per capita incomes, N = 50

State	Income02: Per capita income, 2002 (US$)
Alabama	25,409
Alaska	32,343
Arizona	26,507
Arkansas	23,363
California	32,803
Colorado	34,027
Connecticut	42,505
Delaware	32,925
Florida	29,709
Georgia	28,544
Hawaii	29,464
Idaho	25,185
Illinois	32,869
Indiana	28,023
Iowa	28,081
Kansas	28,980
Kentucky	25,404
Louisiana	25,194
Maine	27,756
Maryland	36,533
Massachusetts	38,985
Michigan	30,227
Minnesota	33,237
Mississippi	22,321
Missouri	28,358
Montana	25,065
Nebraska	29,182
Nevada	30,736
New Hampshire	34,043
New Jersey	39,296
New Mexico	24,246
New York	35,357
North Carolina	27,510
North Dakota	26,427
Ohio	29,212
Oklahoma	25,861
Oregon	28,924
Pennsylvania	31,016
Rhode Island	31,478
South Carolina	25,361
South Dakota	27,087
Tennessee	27,490
Texas	28,846
Utah	24,895
Vermont	29,291
Virginia	33,013
Washington	32,549
West Virginia	24,002
Wisconsin	30,025
Wyoming	30,986

Data source: US Census Bureau 2002.

Figure 3.18 shows the finished stem and leaf. There are a few points to note about how we constructed it. First, when making a stem and leaf, we don't round. John Tukey wanted something that could be constructed quickly, so he argued against rounding in building stem and leaf diagrams. Thus Illinois, which has a per capita income of $32,869, would normally be rounded to $32,900. But we are not rounding, so Illinois gets an 8 on the leaf side of the bin for 32 ($32,000–32,999). Second, note that we lose information. In the table we know that Illinois has a value of $32,869, but in the stem and leaf we only know that there is a state in the 32 bin with a value of 8 on the leaf side. We only know that the state falls between $32,800 and $32,899, not its exact value. Nor do we know from the stem and leaf which state it is.

Suppose we wanted to explore this data further, and we considered both New Jersey and Connecticut to be unusual data points. They might have very high incomes because they include many communities where residents work in New York City. They are perfectly valid data points, but the main thing we see in the stem and leaf is that they take on very high values. We could graph the state incomes with these two states removed to better see the patterns in the rest of the data.

Now we have 48 data points that range from Mississippi at $22,321 to Massachusetts at $38,985. To graph them we might use bins that cover a $500 range rather than the

x1000	x100
22	3
23	3
24	028
25	0113448
26	45
27	0457
28	0035899
29	12247
30	0279
31	04
32	35889
33	02
34	00
35	3
36	5
37	
38	9
39	2
40	
41	
42	5

Figure 3.18   Per capita income stem and leaf diagram

$1,000 range we have been using. Our first bin covers incomes from $22,000–22,499, the second $22,500–22,999, the third $23,000–23,499, the fourth $23,500–23,999, and so on. With only 48 data points, this will give us a lot of bins for the amount of data we have, but it should show more detail about how the 48 least affluent states vary (see Figure 3.19). We leave the second empty bin at 22 in thousands place because there are no states in the range $22,500–22,999. We also note at the bottom of the stem and leaf which states were left out. Which is the "correct" graph? Both – they are just different ways of looking at the same data.

We have now constructed stem and leaf diagrams where a change of one unit on the stem side is represented by one bin (for example, a bin for $23,000–23,999) and where such a change is represented by two bins (one for $23,000–23,499 and one for $23,500–23,999). We could also construct a stem and leaf where there are five bins for each change of one unit (one each for $23,000–23,199, $23,200–23,399, $23,400–23,599, $23,600–23,799 and $23,800–23,999). We must always use a number of bins that divides evenly into 10 in order to be sure that each bin covers the same range of the variable. That leaves three choices: 1, 2 or 5 bins for each increment of the units on the stem side.

Skew and mode

One reason to look at histograms is to see the shape of the distribution of the data. We often compare the shape to a special shape – the famous bell-shaped curve. There are actually many (technically an infinite number) curves that are roughly bell-shaped, but the one of importance in statistics is called the Normal distribution or the Gaussian distribution (because one of the greatest mathematicians, Karl Friederich Gauss, did important work on the curve).

Before we begin our discussion of comparing a histogram to a particular curve representing the Normal distribution, we need to explain the relationship between histograms. Figure 3.20 is another histogram of the homicide data (the first being in Figure 3.11). This time we used 16 bins to show the fine detail of the distribution of the data.

In later chapters we will discuss how the normal curve is used in statistics and in the next chapter will describe its history and how its interpretation is related to philosophical and political issues. It is important to understand what the **Normal curve** is, and what it is not as there is a lot of misunderstanding about this curve. Galileo, in his work on errors in measurement described earlier, noticed the bell-shaped pattern in his measurements. So at first the bell-shaped curve was considered a part of the "law of errors." If one is thinking of measurement errors, this might make sense. But often we don't want to think about the distribution of people on some variable as measurement error from some ideal.

The Normal curve matches a distribution we often see in the physical and biological sciences. But it less often matches the distribution we find in social science data, as we will see with several examples. Sometimes it is important to compare

x1000	x100
22	3
22	
23	3
23	
24	02
24	8
25	011344
25	8
26	4
26	5
27	04
27	57
28	003
28	5899
29	1224
29	7
30	02
30	79
31	04
31	
32	3
32	5889
33	02
33	
34	00
34	
35	3
35	
36	
36	5
37	
37	
38	
38	9

Extremes: Connecticut = $42,505; New Jersey = $39,296.

Figure 3.19   Per capita income stem and leaf diagram excluding two highest values

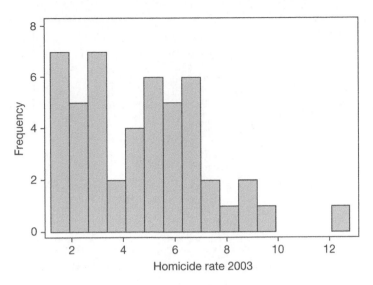

Figure 3.20 Histogram of state homicide rates, N = 50
Data source: US Census Bureau 2003, analyzed with Stata.

our data to the Normal distribution not because we think the Normal distribution is the way the world should be. Rather, if the population we are studying has a Normal distribution – if the distribution is bell-shaped – we can use certain statistical tools we cannot use if the data in the population is not Normally distributed. And, as we will see in later chapters, some statistical procedures produce results that can be interpreted using the Normal distribution. So it is often useful to compare our data to a Normal Distribution but not because we expect the data for some reason "should" have that distribution.

Figure 3.21 shows the histogram of the homicide rate for states with a Normal distribution plotted for comparison. As you can see, the data are not at all Normally distributed. If the data matched the Normal distribution, then the line for the Normal distribution would run through the center of the top of each bar in the histogram. There are too few states in the middle of the distribution (e.g., see around 5.0 in the graph) and too many at the high end of the homicide rate (e.g., see around 10.0) for the Normal curve to be a good picture of the pattern of the data.

When the distribution has more cases on the high end than we would expect, we say that the data have a **positive** or **right skew**. There are some cases that have values that are quite high compared to the rest of the data. Adding the Normal curve makes this clear, but our real basis for comparison is to the bulk of the data.

Figure 3.22 is a histogram of the percentage of the population living in urban areas for US states, again with a Normal curve added for comparison. Here we see that there are a few states that have very low levels of urbanization. (Vermont and Maine are both at or under 40 percent urban, making them the least urban states.) In this case the histogram seems as if it was pulled to the left compared to the Normal

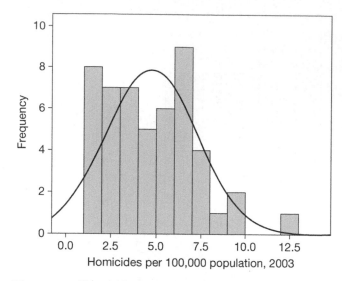

Figure 3.21   Histogram of homicide rate with Normal distribution, N = 50
Data source: US Census Bureau 2003, analyzed with SPSS.

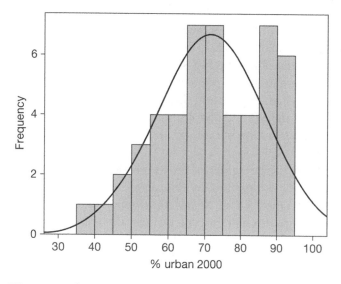

Figure 3.22   Histogram of percent urban with Normal distribution, N = 50
Data source: US Census Bureau 2000, analyzed with SPSS.

curve. We refer to this as **left** or **negative skew**. It means that there are some states that are rather low on this variable compared to the bulk of states.

There's nothing wrong with the data for percent urbanization in Vermont and Maine or with the data on homicide rates for Louisiana. It's just that these states are a bit different than most others. We sometimes refer to different/unique cases as **outliers**. When we see an outlier, we should check to make sure that there isn't an error in the data (e.g., that we didn't incorrectly type in a value or that the item was incorrectly coded), but once we've done that, the outliers can be the most interesting cases in the data set (Becker, 1998).[6]

The Normal curve has one peak in the center. The homicide data seem to have two peaks, as we noted above. With 10 and 16 bins we see two peaks. We refer to these peaks as **modes**. The Normal curve is **unimodal** because it has only one peak. The homicide data are **bimodal** because the distribution appears to have two peaks. And the urbanization data have two peaks, one around 70 percent urban and one around 90 percent.

Note that while the number of peaks we find may depend on the number of bins we use, the data and the basic shape of the distribution don't change. The more bins we use, the "lumpier" the data appear. This is just like looking at a painted wall from a distance, where it appears smooth, and then looking again very closely where we can see more unevenness.

Rules for Graphing

We close our discussion of graphics that display one variable at a time by quoting Edward Tufte (1982, p. 13), who has written three beautiful books on graphic display of information (Tufte, 1982; 1990; 1997): "Excellence in statistical graphics consists of complex ideas communicated with clarity, precision and efficiency. Graphical displays should:

- show the data;
- induce the viewer to think about the substance rather than about the methodology, graphic design, the technology of graphic production, or something else;
- avoid distorting what the data have to say;
- present many numbers in a small space;
- make a large data set coherent;
- encourage the eye to compare different pieces of data;
- reveal the data at several levels of detail, from a broad overview to the fine structure;
- serve a reasonable clear purpose: description; exploration; tabulation; or decoration; and
- be closely integrated with the statistical and verbal description of the data set."

Graphics are pictures. There are many good ways to display data, and like any good picture, a good statistical graphic can draw our attention to things we have not thought about before. And while drawing graphs by hand can be quite cumbersome, statistical software can produce very beautiful graphs that reveal much about the data quickly and easily. So some graphing is appropriate in every research project.

What Have We Learned?

We have described three ways of graphing nominal and ordinal data: pie charts, bar charts, and dotplot histograms. The pie chart is popular but can often be hard to interpret. The box chart is a very common alternative to the pie chart and is usually easier to interpret. The dotplot histogram is a simple graphic that provides the same kind of information as the bar chart but by using "stacks" of plotting symbols rather than bars.

We described four ways of plotting the distribution of continuous data: one-way scatterplots, Cleveland dotplots, histograms, and stem and leaf diagrams. One-way scatterplots use small vertical lines to show where in the range of the variable being graphed most cases fall. Cleveland dotplots show the values of individual cases, ranked from lowest to highest. The histogram divides a continuous variable into bins and plots a bar for each bin with the height of the bar indicating how many cases fell in that range of the variable being plotted. It is closely related to the bar chart but is used for continuous variables. Finally, the stem and leaf diagram uses a single digit to represent each case in a data set and yields a plot much like a histogram. While the histogram is probably the most common way to display continuous data, the choice of a graphical method depends on the character of the data, the purpose for which it is being graphed and the preferences and experience of the intended viewer.

We noted that it is sometimes useful to compare the distribution of our data to hypothetical distributions, such as the Normal distribution. We emphasized that we do not expect our data to be Normally distributed but simply use the Normal curve as a basis for making comparisons. The comparison can be made with a histogram that includes a line indicating what height we would expect bars of the histogram to take if the data were Normally distributed. If the data have more high values than is predicted by a Normal curve, we say the data have right or positive skew. If the data have more cases at low values than we would expect, we say the data have left or negative skew. It is also sometimes useful to count the number of "peaks" or "lumps" in the distribution of the data. If the data have only one peak when plotted as a histogram, then we say it is unimodal. If it has two peaks, we say it is bimodal, and with three peaks, trimodal.

Tukey's new way of tabulating is discussed below as an Advanced Topic.

Advanced Topic 3.1   Statistical Packages for the Social Sciences

We will use two software packages, SPSS and Stata, for the data analyses in the book. SPSS is one of the oldest statistical packages currently in use; it was first written in the 1960s when programs and data were stored on punch cards and often had to run overnight. It is also the most popular package in the business world. Stata (the manual tells us it rhymes with data, which, if you think about it, is a bit of a joke) is a research-oriented package developed in the 1980s. Each package has strengths and weaknesses. For some graphs and tables we will show you exactly what the programs produce; for others we will modify them as we might if we were putting them in a paper intended for publication or in a report. We want you to get a sense of both versions of graphs and tables, and we want you to realize that computer software often provides details that are important for you to look at as you conduct the research but that usually don't appear in publications from the research. Often, some of the detail that appears on you computer screen may need to be dropped from the final presentation of your results to prevent your report from being too cluttered.

Advanced Topic 3.2   Tukey's New Way of Tabulating

One of the most basic and ancient tasks in daily life is to write down marks to keep track of a count. Some of the earliest forms of writing seemed to be used for this purpose, and writing may have evolved from these marking strategies. Most of us use a traditional method of making four vertical strokes followed by a horizontal slash to mark the fifth item counted.

John Tukey was very creative and was always able to look at an old problem in new ways. In the case of counting, he was looking at one of the oldest mathematical problems. And yet he was able to propose a better way. Tukey proposed writing a dot for each of the first four objects counted, arranging them as the corners of a square. Then for the next four objects counted, the dots are connected to form the edges of the square. That gets us to eight objects. Finally, the ninth and tenth objects are represented by putting diagonal lines in the square. The result is a small box that represents 10 items counted. Tukey thought this would be less error prone than the 4 vertical and one diagonal line we traditionally use. It is easier to see his tool than to describe it. Table 3.9 shows his method compared to the traditional one.

Table 3.9   Tukey's method of counting objects

Number	Traditional method	Tukey's method
1	/	•
2	//	• •

Table 3.9 (cont'd)

Number	Traditional method	Tukey's method
3		
4		
5		
6		
7		
8		
9		
10		

Table 3.10 State names and abbreviations

State	Abbreviation
Alaska	AK
Alabama	AL
Arkansas	AR
Arizona	AZ
California	CA
Colorado	CO
Connecticut	CT
Delaware	DE
Florida	FL
Georgia	GA
Hawaii	HI
Iowa	IA
Idaho	ID
Illinois	IL
Indiana	IN
Kansas	KS
Kentucky	KY
Louisiana	LA
Massachusetts	MA
Maryland	MD
Maine	ME
Michigan	MI
Montana	MN
Minnesota	MN
Missouri	MO
Mississippi	MS
North Carolina	NC
North Dakota	ND
Nebraska	NE
New Hampshire	NH
New Jersey	NJ
New Mexico	NM
Nevada	NV
New York	NY
Ohio	OH
Oklahoma	OK
Oregon	OR
Pennsylvania	PA
Rhode Island	RI
South Carolina	SC
South Dakota	SD
Tennessee	TN
Texas	TX
Utah	UT
Virginia	VA
Vermont	VT
Washington	WA
Wisconsin	WI
West Virginia	WV
Wyoming	WY

Table 3.11 Country names and abbreviations

Country	Abbreviation
Afghanistan	AFG
Albania	ALB
Algeria	DZA
Angola	AGO
Antigua and Barbuda	ATG
Argentina	ARG
Armenia	ARM
Australia	AUS
Austria	AUT
Azerbaijan	AZE
Bahamas	BHS
Bahrain	BHR
Bangladesh	BGD
Barbados	BRB
Belarus	BLR
Belgium	BEL
Belize	BLZ
Benin	BEN
Bhutan	BTN
Bolivia	BOL
Bosnia and Herzegovina	BIH
Botswana	BWA
Brazil	BRA
Brunei Darussalam	BRN
Bulgaria	BGR
Burkina Faso	BFA
Burundi	BDI
Cambodia	KHM
Cameroon	CMR
Canada	CAN
Cape Verde	CPV
Central African Republic	CAF
Chad	TCD
Chile	CHL
China	CHN
Colombia	COL
Comoros	COM
Congo	COG
Cook Islands	COK
Costa Rica	CRI
Cote d'Ivoire	CIV
Croatia	HRV
Cuba	CUB
Cyprus	CYP
Czech Republic	CZE

Table 3.11   (*cont'd*)

Country	Abbreviation
Democratic People's Republic Korea	PRK
Democratic Republic of the Congo	ZAR
Denmark	DNK
Djibouti	DJI
Dominica	DMA
Dominican Republic	DOM
Ecuador	ECU
Egypt	EGY
El Salvador	SLV
Equatorial Guinea	GNQ
Eritrea	ERI
Estonia	EST
Ethiopia	ETH
Fiji	FJI
Finland	FIN
France	FRA
Gabon	GAB
Gambia	GMB
Georgia	GEO
Germany	DEU
Ghana	GHA
Greece	GRC
Grenada	GRD
Guatemala	GTM
Guinea	GIN
Guinea-Bissau	GNB
Guyana	GUY
Haiti	HTI
Honduras	HND
Hungary	HUN
Iceland	ISL
India	IND
Indonesia	IDN
Iran (Islamic Republic of)	IRN
Iraq	IRQ
Ireland	IRL
Israel	ISR
Italy	ITA
Jamaica	JAM
Japan	JPN
Jordan	JOR
Kazakhstan	KAZ
Kenya	KEN
Kiribati	KIR
Kuwait	KWT

Table 3.11   (*cont'd*)

Country	Abbreviation
Kyrgyzstan	KGZ
Lao People's Democratic Republic	LAO
Latvia	LVA
Lebanon	LBN
Lesotho	LSO
Liberia	LBR
Libyan Arab Jamahiriya	LBY
Liechtenstein	LIE
Lithuania	LTU
Luxembourg	LUX
Madagascar	MDG
Malawi	MWI
Malaysia	MYS
Maldives	MDV
Mali	MLI
Malta	MLT
Marshall Islands	MHL
Mauritania	MRT
Mauritius	MUS
Mexico	MEX
Micronesia, Federated States of	MIC
Monaco	MCO
Mongolia	MNG
Morocco	MAR
Mozambique	MOZ
Myanmar	MMR
Namibia	NAM
Nauru	NRU
Nepal	NPL
Netherlands	NLD
New Zealand	NZL
Nicaragua	NIC
Niger	NER
Nigeria	NGA
Niue	NIU
Norway	NOR
Oman	OMN
Pakistan	PAK
Palau	PLW
Panama	PAN
Papua New Guinea	PNG
Paraguay	PRY
Peru	PER
Philippines	PHL
Poland	POL

Table 3.11   (*cont'd*)

Country	Abbreviation
Portugal	PRT
Qatar	QAT
Republic of Macedonia	MKD
Republic of Moldova	MDA
Romania	ROM
Russian Federation	RUS
Rwanda	RWA
Saint Kitts and Nevis	KNA
Saint Lucia	LCA
Saint Vincent and the Grenadines	VCT
Samoa	WSM
San Marino	SMR
Sao Tome and Principe	STP
Saudi Arabia	SAU
Senegal	SEN
Seychelles	SYC
Sierra Leone	SLE
Singapore	SGP
Slovakia	SVK
Slovenia	SVN
Solomon Islands	SLB
Somalia	SOM
South Africa	ZAF
Spain	ESP
Sri Lanka	LKA
Sudan	SDN
Suriname	SUR
Swaziland	SWZ
Sweden	SWE
Switzerland	CHE
Syrian Arab Republic	SYR
Tajikistan	TJK
Thailand	THA
Togo	TGO
Tonga	TON
Trinidad and Tobago	TTO
Tunisia	TUN
Turkey	TUR
Turkmenistan	TKM
Tuvalu	TUV
Uganda	UGA
Ukraine	UKR
United Arab Emirates	ARE
United Kingdom	GBR
United Republic of Tanzania	TZA

Table 3.11 *(cont'd)*

Country	Abbreviation
United States	USA
Uruguay	URY
Uzbekistan	UZB
Vanuatu	VUT
Venezuela (Bolivarian Republic of)	VEN
Viet Nam	VNM
Yemen	YEM
Yugoslavia	YUG
Zaire	ZAR
Zambia	ZMB
Zimbabwe	ZWE

Applications

In this Applications section we will use the graphics tools developed in this chapter to examine the distribution of the dependent variables in each of our four applications.

Example 1:   Why do homicide rates vary?

We have used the homicide rate variable as an example throughout the chapter, so we won't repeat those graphs here.

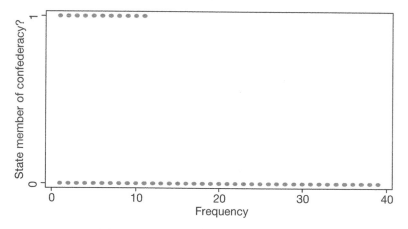

Figure 3A.1   Horizontal dotplot of membership in the Confederacy, N = 50
Data source: US Census Bureau, analyzed with Stata.

Figure 3A.1 is a dotplot histogram of whether or not a state was a member of the Confederacy. This shows that a relatively small number of states were scored "1" indicating that they were a member of the Confederacy while most states are scored "0" indicating that they were not.

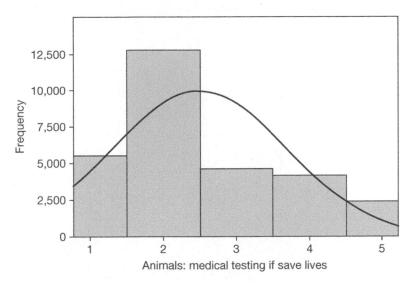

Figure 3A.2   Histogram of animal concern question, N = 29,486
Data source: 2000 ISSP data set; analyzed with SPSS.

Example 2:   Why do people vary in their concern with animals?

We are going to treat the animal concern question, which ranges from 1 to 5, as an ordinal variable. Because we have 29,486 people in the 2000 International General Social Survey who have scores on the question, graphic displays such as one-way scatterplots and Cleveland dotplots are not useful because the points overlap. Recall from Table 3.4 that the smallest category of the scale, a score of 5, has 2,385 people in it. In a one-way scatterplot or Cleveland dotplot we would have 2,385 lines or dots on top of one another. It would be worse for the categories with more respondents. The same problem arises in a stem and leaf diagram. The longest leaf, for the respondents who received a "2" would have 12,803 entries, which would not fit well on most computer screens or a printed page. So the best tool we have for studying the distribution of the animal concern variable is the histogram.

Since the animal attitudes score takes on only 5 values, we can use five bins. Figure 3A.2 is the histogram of the animal attitudes question. We had SPSS draw a Normal distribution on the histogram for comparison.

The distribution peaks at 2. It also has a **positive** or **right, skew** – there are more cases at high values than one would expect in a Normal distribution. Given the bulk of responses between 1 and 2 (from Table 3.4 that 62 percent of respondents had scores of 1 or 2), the folks who answered 5 (8 percent) seem unusual by comparison.

Not much more can be said about the animal attitudes question's distribution. But to preview how graphics can help us understand why animal attitudes may vary from person to person, we have created histograms separately for men and for women. It may be that gender is one reason for differences in animal attitudes. If that is the case then we would expect that the distribution displayed in the histogram of animal attitudes scores would differ across genders. Figure 3A.3 shows a separate histogram for men and for women.

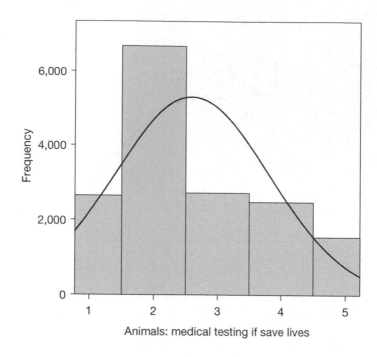

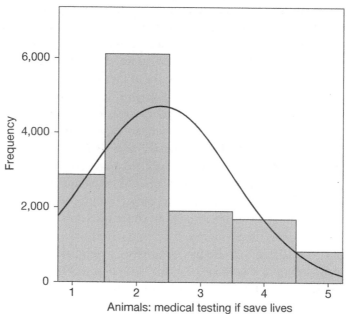

Figure 3A.3   Histograms of animal concern question by gender, for women, N = 16,062; for men, N = 13,412
Data source: 2000 ISSP data set, analyzed with SPSS.

0   envtreat   16

Figure 3A.4 One-way scatterplot of environmental treaty participation, N = 191
Data source: Roberts et al., 2004, analyzed with Stata.

We see that the two histograms peak at the same score, 2. And both are right skewed, with more cases than we would expect at 5. From the histograms, it appears that a similar proportion of women and men are at each score.

There are other ways of comparing the distribution of the animal attitudes score across genders that we will use in later chapters. But the stacked histograms give an initial impression that the data are not consistent with the model that uses gender to predict animal attitudes.

Example 3: Why do nations differ in environmental treaty participation?

The environmental treaty participation variable is a quantitative variable. Figure 3A.4 is a one-way scatterplot of environmental treaty participation. As can be seen, a one-way scatterplot is not useful when a variable has a limited range and many cases. This graph does not show the presence of multiple cases for each number of treaties, so the graph does not adequately show the variable's distribution.

In Figure 3A.5 we have a histogram. We selected 17 bins based on the general rules outlined in Table 3.6 and had SPSS add the Normal distribution on the plot. We could

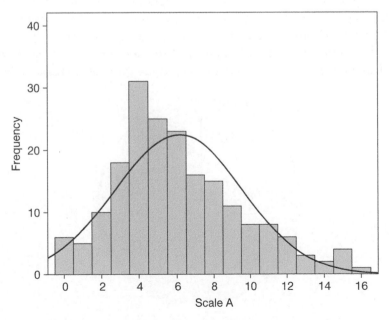

Figure 3A.5 Histogram of environmental treaty participation, N = 191
Data source: Roberts et al., 2004, analyzed with SPSS.

Scale A Stem-and-Leaf Plot

Frequency   Stem & Leaf

```
    6.00        0 . 000000
    5.00        1 . 00000
   10.00        2 . 0000000000
   18.00        3 . 000000000000000000
   31.00        4 . 0000000000000000000000000000000
   25.00        5 . 0000000000000000000000000
   23.00        6 . 00000000000000000000000
   16.00        7 . 0000000000000000
   15.00        8 . 000000000000000
   11.00        9 . 00000000000
    8.00       10 . 00000000
    8.00       11 . 00000000
    6.00       12 . 000000
    3.00       13 . 000
    2.00       14 . 00
    5.00  Extremes   (>=15.0)
```

Stem width: 1
Each leaf: 1 case (s)

Figure 3A.6   Stem and leaf diagram of environmental treaty participation, N = 191
Data source: Roberts et al., 2004, analyzed with SPSS.

interpret this histogram as having one mode. There is again the lump of countries at between four and six treaties. The countries' pattern is close to a Normal distribution.

Finally, Figure 3A.6 is the stem and leaf diagram of environmental treaty participation produced by SPSS. SPSS adds a frequency count to the left of the stem and leaf, and some other labels, all of which makes the diagram a bit cluttered. Because of the limited range, each number of treaties from 0 to 14 has a separate stem and all numbers on the leaf side are 0s (because the scale is measured in whole

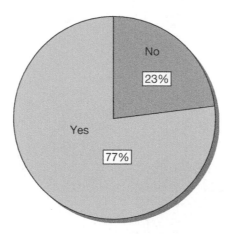

Figure 3A.7   Pie chart of AIDS knowledge, N = 8,310
Data source: 2000 Ugandan DHS data set, analyzed with SPSS.

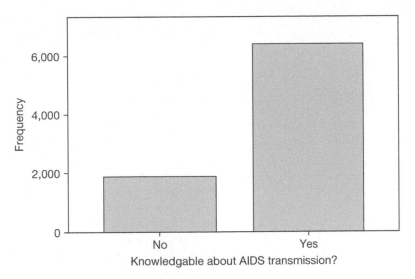

Figure 3A.8   Bar chart of AIDS knowledge, N = 8,310
Data source: 2000 Ugandan DHS data set, analyzed with SPSS.

numbers). The five countries that participated in 15 or 16 treaties are considered
extreme cases in this stem and leaf diagram.

Example 4:   Why do people differ in their knowledge of how the
AIDS virus is transmitted?

Our dependent variable here is a simple dichotomy, with people answering "Yes"
(who knew that condoms can reduce the likelihood of transmitting the AIDS virus)
or "No" (those who did not have knowledge about AIDS transmission). So we can
treat the variable as qualitative. Figure 3A.7 is a pie chart of the data, and Figure 3A.8
is a bar chart. For a simple variable like this one, we don't really need graphics to
see the pattern, a table such as Table 3.1 or 3.2 will suffice.

Exercises

1. In this chapter, you learned about many types
    of graphs for displaying data. For each variable
    listed, determine the type of graph that could
    be used to depict that variable. Explain why you
    selected that particular graph. Often there is
    more than one graph that could be selected, so
    be sure to note if there are multiple graphs that
    could be helpful in displaying the data.

    a) Type of government in a country: 1) demo-
        cratic; 2) socialist; 3) dictatorship; 4) other
        form of government; 99) missing; sample
        size = 100.

    b) Number of new friends made in the prior year
        – range is 0–30, 99 = missing; sample size = 450.

    c) Strength of belief that criminal activity
        decreases with age: 1) strongly disagree; 2)
        disagree; 3) slightly disagree; 4) slightly agree;
        5) agree; and 6) strongly agree; 0) missing; sam-
        ple size = 60.

d) Amount of depressive symptoms experienced in the past three months: scale range 0–40; 999 = missing; sample size = 90.

e) Volunteered in the prior year: 0) no; 1) yes; 8) missing; sample size = 175.

2. Table 3E.1 is a frequency table of the number of close friends who live near a respondent.

a) What is the most common number of friends living nearby? Note both the number of people who said this and the percent of the sample that did.

b) What percent of the sample said they have 1–2 friends?

c) What percent of the sample had missing data on this question?

d) How many people had ten or more friends living nearby?

Table 3E.1   Frequency table of number of close friends living nearby

			Frequency	Percent	Valid percent	Cumulative percent
Valid	0	None	11,588	31.2	32.4	32.4
	1	1 close friend	4,744	12.8	13.3	45.7
	2	2 friends	6,118	16.5	17.1	62.8
	3	3 friends	3,276	8.8	9.2	71.9
	4	4 friends	2,706	7.3	7.6	79.5
	5	5 friends	2,028	5.5	5.7	85.1
	6	6 friends	1,295	3.5	3.6	88.8
	7	7 friends	337	.9	.9	89.7
	8	8 friends	536	1.4	1.5	91.2
	9	9 friends	125	.3	.3	91.6
	10	10 friends+	3,022	8.1	8.4	100.0
	Total		35,775	96.2	100.0	
Missing	800	CDN: Few	1	.0		
	801	CDN: Many	6	.0		
	998	Don't know	179	.5		
	999	NA	1,175	3.2		
	System		52	.1		
	Total		1,413	3.8		
Total			37,188	100.0		

Data source: 2000 ISSP data set, analyzed with SPSS.

3. From the data presented in frequency Table 3E.2, would you say that people do or do not regularly discuss politics with their friends? Use specific information from the table when writing your response.

Table 3E.2 Frequency table of political discussions with friends

			Frequency	Percent	Valid percent	Cumulative percent
Valid	1	Almost all the time	1,366	3.7	5.4	5.4
	2	Most of the time	3,976	10.7	15.6	21.0
	3	Occasionally	11,675	31.4	45.9	66.9
	4	Almost never	8,427	22.7	33.1	100.0
		Total	25,444	68.4	100.0	
Missing	8	Can't choose	396	1.1		
	9	NA	441	1.2		
		System	10,907	29.3		
		Total	11,744	31.6		
Total			37,188	100.0		

Data source: 2000 ISSP data set, analyzed with SPSS.

4. Figure 3E.1 is a histogram of the number of hours worked in the week prior to the survey, with the normal curve drawn in.

a) Describe the distribution of data. Include in your response the number of hours that were most commonly worked, whether many people worked over the traditional 40-hour full-time work week, and whether many people worked part-time (often defined as fewer than 30 hours).

b) How close are the data to the normal curve? Do the data appear to be skewed?

c) Approximately what percentage of the sample worked 40 hours?

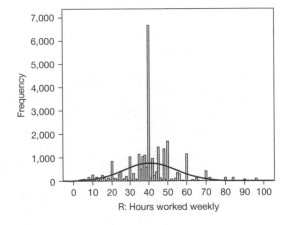

Figure 3E.1 Histogram of number of hours worked weekly, N = 26,544
Data source: 2000 ISSP data set, analyzed with SPSS.

5. Create a bar graph *or* pie graph of how happy people feel. Base your graph on the percentages shown in Table 3E.3.

Table 3E.3   Frequency table of happiness in general

		Frequency	Percent	Valid percent	Cumulative percent
Valid	1 Very happy	9,050	25.2	25.2	25.2
	2 Fairly happy	20,055	55.8	55.8	81.0
	3 Not very happy	5,197	14.5	14.5	95.4
	4 Not at all happy	1,648	4.6	4.6	100.0
	Total	35,950	100.0	100.0	

Data source: 2000 ISSP data set, analyzed with SPSS.

6. Figure 3E.2 is a stem and leaf plot of the weekly number of hours worked. To simplify things, a random sample of 117 respondents was selected.

a) How many people worked 45 hours a week? How many worked 60 hours?

b) What is the most number of hours worked, and how many people worked that number of hours?

c) Briefly summarize the distribution of hours worked based on the data presented. Include whether data tend to "clump" around certain scores in your response.

Frequency	Stem & Leaf
7.00	Extremes (=<13)
2.00	1 . 68
6.00	2 . 000012
8.00	2 . 55555888
5.00	3 . 00024
20.00	3 . 55556677777778888899
32.00	4 . 00000000000000000000000000222234
19.00	4 . 5555557888888888888
4.00	5 . 0000
1.00	5 . 5
4.00	6 . 0000
9.00	Extremes (>=68)

Stem width: 10
Each leaf: 1 case(s)

Figure 3E.2   Stem and leaf plot of hours worked weekly

Data source: 2000 ISSP data set, analyzed with SPSS.

7. Figure 3E.3 is a histogram of how often individuals attend religious services (to simplify, a random sample of 185 respondents was selected). Response options include: 1) at least several times a week; 2) once a week; 3) 2–3 times a week; 4) once a week; 5) several times a year; 6) once a year; 7) less frequently than once a year; and 8) never. Approximately what percent of respondents never attend religious services? Approximately how many people attend religious services once a week, and what percent of the sample is this?

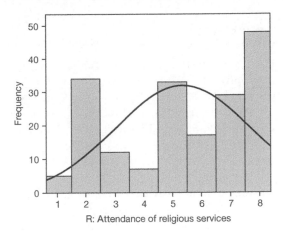

Figure 3E.3   Histogram of frequency of religious service attendance
Data source: 2000 ISSP data set, analyzed with SPSS.

8. Table 3E.4 is a frequency table of the number of children respondents have under the age of 18 (a random sample of 500 was selected to simplify). Create a histogram from this data.

Table 3E.4   Frequency table of number of children

			Frequency	Percent	Valid percent	Cumulative percent
Valid	0	None, no child under 18	210	42.0	54.7	54.7
	1	One child	81	16.2	21.1	75.8
	2	Two children	59	11.8	15.4	91.1
	3	Three children	21	4.2	5.5	96.6
	4	Four children	8	1.6	2.1	98.7
	5	Five children	5	1.0	1.3	100.0
		Total	384	76.8	100.0	
Missing	99	NA, refused	10	2.0		
		System	106	21.2		
		Total	116	23.2		
Total			500	100.0		

Data source: 2000 ISSP data set, analyzed with SPSS.

References

Becker, H. 1998. *Tricks of the Trade*. Chicago: University of Chicago Press.

Cleveland, W. S. 1985. *The Elements of Graphing Data*. Belmont CA: Wadsworth.

Cleveland, W. S. 1993. *Visualizing Data*. Summit, NJ: Hobart Press.

Cleveland, W. S. 1994. *The Elements of Graphing Data*. Summit, NJ: Hobart Press.

Emerson, J. D. and Hoaglin, D. C. 1983. Steam-and-leaf displays. In D. Hoaglin, F. Mosterller, and J. W. Tukey (eds), *Understanding Robust and Exploratory Data Analysis*, pp. 7–30. New York: John Wiley & Sons.

Fisher, R. A. 1925. *Statistical Methods for Research Workers*. Edinburgh: Oliver and Boyd.

Gaither, C. C. and Cavazos-Gaither, A. E. 1996. *Statistically Speaking: A Dictionary of Quotations*. Bristol, England: Institute of Physics Publishing.

International Social Survey Programme (ISSP). 2000. 2000 Environment II data set. www.issp.org. Catalog no. ZA 3440. Cologne, Germany: GESIS-ZA Central Archive for Empirical Research.

Pearson, K. 1895. Contributions to the mathematical theory of evolution – II. Skew variation in homogenous material. *Philosophical Transactions of the Royal Society of London, A* 186, 343–414.

Roberts, J. T., Parks, B. C., and Vasquez, A. A. 2004. Who ratifies environmental treaties and why? Institutionalism, structuralism and participation of 192 nations in 22 treaties. *Global Environmental Politics* 4(3), 22–64.

Simkin, D. and Hastie, R. 1987. An information processing analysis of graph perception. *Journal of the American Statistical Association* 82, 454–65.

StataCorp. 1999. *Stata Statistical Software: Release 6.0*. College Station, TX: Stata Corporation.

Tufte, E. R. 1982. *The Visual Display of Quantitative Information*. Chesire, CT: Graphics Press.

Tufte, E. R. 1990. *Envisioning Information*. Chesire, CT: Graphics Press.

Tufte, E. R. 1997. *Visual Explanations*. Chesire, CT: Graphics Press.

Tukey, J. W. 1977. *Exploratory Data Analysis*. Reading, MA: Addison-Wesley.

Uganda Demographic and Health Surveys. 2001. Calverton, Maryland: UBOS and ORC Macro. (http://www.measuredhs.com/pubs/pdf/FR128/00FrontMatter.pdf).

US Census Bureau 2000. Table 33. Urban and rural population, and by state: 1990 and 2000 (http://www.census.gov/prod/cen2000/index.html).

US Census Bureau 2002. Historical poverty tables: Table 21. Number of poor and poverty rate, by state: 1980 to 2006. Year 2002 (http://www.census.gov/hhes/www/poverty/histpov/hstpovv21.html).

US Census Bureau 2003. Table 295. Crime rates by state, 2002 and 2003, and by type, 2003 (http://www.census.gov/prod/2005pubs/06stata; www.census.gov/prod/2005pubs/06statab/law.pdf).

CHAPTER 4
DESCRIBING DATA

Outline

Descriptive Statistics and Exploratory Analysis

One of the purposes of statistics is to help us find patterns in data and to aid in the description of data. Indeed, many textbooks divide statistics into two subtopics – descriptive and inferential statistics. **Descriptive statistics**, as the name implies, describe the data. **Inferential statistics** are tools used to make statements in the face of error. For example, if we have a survey based on a sample of adults in a community, inferential statistics allow us to make valid statements about the whole population from data in the sample, at least if the sample was a probability sample. Descriptive statistics allow us to describe the sample itself.

In this chapter we discuss some of the most important descriptive statistics in current use, and raise some issues that are important to the proper interpretation of data. Until perhaps 25 years ago, professional statisticians did not pay much attention to descriptive statistics. But thanks to the pioneering efforts of John Tukey (Tukey, 1977) and others, great strides have been made in descriptive statistics.

It is still true that most professional statisticians direct their research towards inferential problems, primarily because inferential statistics are widely used in research that addresses causal relationships. But the old view of descriptive statistics as a set of simple tools that are mostly a stepping stone to inferential statistics has been replaced by the notion of exploratory statistics, where the tools employed are intended to help the researcher see patterns and deviations from patterns in an active way. Modern descriptive or exploratory statistics can be used to find patterns otherwise masked by perceptual error. With the advent of more powerful hardware and software, statistical graphics is certain to see rapid growth in the near future. One factor driving this will be the incredible increase in the amount of information available as a result of scientific advances and more extensive tracking of our lives through information technology. Indeed, the field of descriptive statistics has expanded so much that the term "data mining" is now used to describe methods that allow researchers to see patterns in large data sets.[1]

We begin the discussion of descriptive statistics in a traditional way, by reviewing measures of central tendency and variability that are used to summarize data. **Measures of central tendency** tell us what is typical in a set of data. **Measures of variability** tell us how different the observations are from one another. There are three reasons to start with measures of central tendency and variability. First, these measures are important tools of exploratory analysis, and underpin the more interesting recent developments in graphics that are the heart of data exploration. They can help us to see patterns in the data.

Second, summary measures are essential to statistical inference. It is not possible to learn every detail about a population from a sample (or make the equivalent inference in the face of measurement, randomization, or other sources of error). But methods *are* available to estimate the values of important summary descriptions of the population from sample information. If we have a simple random sample from a population, we can use the information in the sample to make very good guesses about the mean of the population, and even understand how far off our guess about the population mean is likely to be. Since summary statistics play such a central role in inference, it is important to appreciate their strengths and weaknesses.

Third, the basic summary measures are descriptions of a single variable. Thus they are one dimensional, or **univariate** (one variable). We can't answer complex questions about models by considering only a single variable. Our simplest model has two variables: a dependent variable and an independent variable. To see how well models explain our data, we'll have to do **bivariate analysis** that involves two variables and eventually **multivariate analysis** that involves more than two variables. But the simple univariate measures of central tendency and variation are the starting point for these more complex tools.

By thinking about how simple univariate statistics work, we lay the groundwork for understanding the more complicated and more powerful multivariate tools. Almost everyone will be able to visualize one dimension, most can visualize two dimensions, many three, but few can visualize higher dimensional spaces. We hope that the important properties of estimators in one dimension will be clear to our readers, who can then generalize to the higher dimensional problem even without directly visualizing it.

Measures of Central Tendency

The most basic (and in many ways the most important) descriptive statistics are measures of central tendency or location. These descriptors attempt to summarize the entire data set with one number that is "typical" of all the numbers in the data set. They indicate where the actual data are "located" among all possible values for the variable under study.

Is it reasonable to have a single number stand for a whole set of data? It depends on what kind of summary number we use, what the batch of data looks like, and how much we expect of the summary. As we note below, the Belgian researcher Adophe Quetelet felt the measure of central tendency – the typical value – was the most important aspect of a batch of data. As the following quote indicates, Charles Darwin's cousin, Francis Galton considered it only a starting point for looking at variability. The measure of central tendency is a useful starting point, but it's not the whole story. But we do need that starting point.

> It is difficult to understand why statisticians commonly limit their enquiries to Averages, and do not revel in more comprehensive views. Their souls seem as dull to the charm of variety as that of the native of one of our flat English counties, whose retrospect of Switzerland was that, if its mountains could be thrown into its lakes, two nuisances could be got rid of at once. (Francis Galton, *Natural Inheritance*, p. 62, quoted in Gaither and Cavazos-Gaither, 1996, p. 9)

We will discuss three common measures of central tendency: the mean, the median, and the mode. We will also introduce a less common but increasingly important measure of central tendency, the trimmed mean.

The mean

The arithmetic **mean**, or average, was discovered by classical civilizations.[2] You are already familiar with it as it appears in schools (e.g., grade point averages), business news (e.g., the Dow Jones Industrial Average) and sports (e.g., the batting average). These everyday examples capture the idea of the mean. Your grade point average is just the sum of the points you've earned with letter grades in courses divided by the number of courses. (Of course weighting 1, 2, 3, and 4 credit courses differently in calculating your GPA makes things a bit more complicated. In statistics this is called a weighted average.) One of the most commonly watched indicators of how the US stock market is doing, the Dow-Jones Industrial average, is just the sum of the prices of 30 stocks that are picked to include in the index divided by 30. (The stocks are picked by the Dow Jones company to represent the industrial part of the stock market.) In baseball, the batting average is just the number of hits a baseball player gets divided by the number of official times at bat.[3]

The mean is undoubtedly the most commonly used measure of central tendency. It has many nice statistical properties that allow it to be used in inference. And it is a good one-dimensional analog to many multivariate statistical methods, including regression analysis, perhaps the most commonly used tool in modern statistics. The mean of a set of data can be defined as[4]:

$$\bar{X} = \frac{\sum X}{N} \tag{4.1}$$

We would read this equation as: "X bar equals the sum over all N of X, divided by N." The interpretation of this formula is as follows: The Greek $\sum$ means "add up everything to the right." The N on top of it reminds us to add up all the data points we have, which is usually indicated by either a capital N or a lower case n. We will use capital N. So to calculate the mean following this equation, we first add up all the data points on the variable X and then divide by the number of data points. (Many textbooks use the subscript i in summations but we are avoiding it to keep things as simple as possible.)

Because the mean requires standard arithmetic to calculate it, it can be used only with interval level (continuous) data. But as we have noted before, there is a special exception. If you have a 0–1 variable (for example, gender recoded with women coded 1 and men coded 0), then the mean of the 0–1 variable is simply the proportion of women in the data set.

Deviations from the mean

For most inhabitants of industrialized societies, the use of the average is so ubiquitous that we don't think about why it is appropriate as a summary of a variable in a batch of data. But in fact the common usage of the mean has come about because of its interesting and valuable properties as a summary. Some of those properties can be examined by considering how observations in the data set differ from the mean:

$$X - \bar{X} \tag{4.2}$$

That is, we look at the difference between each observation in the data set and the mean of all those observations. These are the **deviations from the mean** of the individual cases. We are thinking about how people or countries differ from the average for all people or countries in our study. Again, you've encountered this in the educational system where you have been encouraged to compare your score on a test to the average. If you are above the average, that's good; if you are below the average, that's usually not good.

One interesting property of the mean is that

$$\sum (X - \bar{X}) = 0 \tag{4.3}$$

In English, the sum of all observations of the deviations from the mean equals zero. If on a test you calculated the deviation from the mean of each person, then added up all those deviations, the result will be zero.[5] The mean sits at the center of the data in the sense that it balances out positive deviations (data points above the mean) and negative deviations (data points below the mean).

To illustrate this, let's look at the homicide rates for five states we've picked at random (we are only using five to keep the example simple). Look at the first 3 columns in Table 4.1. The sum for the five states is 28.6. We divide this by 5 and we get a mean of 5.72. In the third column we subtract the mean from the score for each state. Then these sum to zero.

Table 4.1 Deviations from the mean homicide rate for 5 US states

State	Homicide rate	Deviations from mean	Squared deviations from the mean
Kansas	4.5	−1.22	1.49
North Dakota	1.9	−3.82	14.59
Alabama	6.6	0.88	0.77
Michigan	6.1	0.38	0.14
Maryland	9.5	3.78	14.29
Sum	28.6	0.0	31.28

Data source: US Census Bureau, 2003, analyzed with Stata.

Now we can define the **sum of the squared deviations from the mean**, or SSX:

$$SSX = \sum (X - \bar{X})^2 \qquad\qquad\qquad (4.4)$$

The last column in Table 4.1 shows the squared deviations from the mean. For example, for Kansas, $(-1.22)(-1.22) = 1.49$. Squaring each deviation eliminates minus signs – a positive number times a positive number is a positive number and a negative number times a negative number is also a positive number.

As we noted in chapter 1, science is the study of variability, and the sum of squares is an important measure of variability. Think about what we are doing in calculating it. We first find the average, a measure of what is typical. Then for each case, we take the value of the variable for that case and subtract the mean from it. For some cases, those with high values, the difference between the case and the mean will be a positive number. For other cases, those with lower values, the difference will be negative. Thus by subtracting the mean from each case we have created the deviation for the case, which is a positive or negative number indicating how far above or below the mean the case sits.

If we added up all the deviations, we would get a sum of zero, as noted above. Since the sum of the deviations from the mean is equal to zero, the mean of the deviations from the mean is also equal to zero. This is one of the reasons we square the deviations from the mean in calculating the sum of squares. For later work we will want a summary of the variability around the mean. It makes sense to think about variability around the mean in terms of the deviations from the mean. But the fact that they add up to zero (that positive and negative deviations cancel each other out when we add them up) suggests that we need to do something about the negative signs if we are to have a measure of variability based on deviations. Thus we square the deviations and add them up to calculate the sum of squares.

The sum of the squared deviations from the mean, which we often call just the **sum of squares** and abbreviate SSX, is important because it plays a role in many other calculations in statistics. It is a measure of variability in itself and more important, it is the starting point for calculating the most important measures of variability in statistics, the variance and the **standard deviation**. We will use the sum of squares of our variables throughout the rest of the book.

Squared deviations and their relationship to the mean are discussed at the end of this chapter as an Advanced Topic.

Effect of outliers

In the last two decades, statisticians have become concerned with a weakness of many common descriptive statistics – they can be strongly influenced by a few exceptional data points. The mean is subject to this problem. Suppose the data set consists of five numbers: 1, 2, 3, 4, and 5. The average of these numbers will be

15/5 = 3. This seems a reasonable summary. But suppose one unusual value is substituted for one of the previous values. For example, suppose 5 is replaced by 15. Now the average is 25/5 = 5. This seems a less reasonable summary, since all but one of the data points fall below this value.

The mean is strongly influenced by a single unusual data point. You are used to this notion in the form of a "curve breaker" – someone whose score on a test is much higher than that of everyone else. Such an outlying data point is called, not surprisingly, an **outlier**. One desirable property of a descriptive statistic is a high breakdown point. The **breakdown point** of a descriptive statistic is the proportion of the data set that can take on extreme values (in technical terms, outliers that are infinitely far from the rest of the data) without unduly affecting the summary. Because one bad data point pulls the mean away from the rest of the data, the mean has a breakdown point of zero – it cannot tolerate any extreme outliers. In general, descriptive statistics with high breakdown points are to be preferred to those with lower breakdown points, all other things being equal. Unfortunately, all other things are seldom equal. Some descriptive statistics that have a low breakdown point, like the mean, have a number of advantages that will be discussed in the context of inference.

In Table 4.1, we have a random sample of states. (We drew a random sample just so we could work with a very small data set where it is easy to see what we are doing. But it's the same process we use in drawing a sample when we cannot collect data on the whole population.) Suppose instead of Maryland with a homicide rate of 9.5 we had drawn Louisiana with a homicide rate of 13.0. Then the mean would be 32.1/5 = 6.42. The mean has been pulled up by 0.7 points. This is an illustration of how the mean is very sensitive to outlying data points.

We always want to be aware of outlying data points. But it is not always a good idea to simply eliminate them from the data set. We want to find out why the outliers are so far from the rest of the data. Think of the mean as a simple model for the data. We might say that the mean is what we expect, and the deviations show how far each data point is from that expectation.[6] Some data points may differ from the mean because a process different from the one that generated the numbers close to the mean generated them. Sometimes that process is simply a mistake – for example, someone made an error in coding the data or in entering the data into a computer file. But sometimes the process that generates outliers is important. If we could understand why Louisiana has such a high homicide rate, we could learn something about what causes homicide.

The median

The **median** is the second most commonly used measure of central tendency. Francis Galton developed it in the nineteenth century. The median is the fiftieth percentile, the value such that half the data points fall above the value and half below. If the data set contains an odd number of cases, the numeric value of the median will correspond to the value of the middle case. If there is an even number of cases the value of the median is the average of the values of the two cases at the mid-point.

To find the median, you must first place the cases in order from lowest to highest. Then the median is the value of the case in the position $(N + 1)/2$, where N is the number of cases. This is called the **median depth** because it is how far (how many cases) we have to count into the data to find the place where half the data points will be above and half below.

Because the median depends on the ordering of the data, it is sometimes called an order statistic, one of a family of statistics that depend on the order of the data. Because we do not perform any arithmetic on the data points in calculating the median, we can calculate the median for ordinal or interval level data. So the median is a bit more versatile than the mean. The median of the data for the five states in Table 4.1 would be found as follows:

1 List the values in order from lowest to highest: 1.9, 4.5, 6.1, 6.6, 9.5.
2 Find the position of the median with 5 data points (N = 5): $(5 + 1)/2 = 6/2 = 3$. So the median is at the third data point in order.
3 Find the value: Since the third data point in order has a value of 6.1, the value of the median is 6.1.

There are two common mistakes in thinking about the median. First, the term is *median* not *medium*. Think of the strip in the middle of the road that divides the road in half. Second, the value of the median is a number. It is not a data point. It is correct to say that for the five states the median is 6.1. It is not correct to say that the median is Michigan, but it is correct to say that Michigan has a value for homicide rate that is *at* the median of these five states. Michigan is a state and thus a data point, with a value of 6.1. The median is a number that summarizes the data.

Deviations from the median

As noted above, the sum of the squared deviations from the mean will be smaller than the sum of squared deviations from any other number. There is a parallel for the median. The sum of the absolute value of deviations from the median will be smaller than the sum of the absolute value of the deviations from any other number. That is, the median minimizes the sum of the absolute deviations.

$$\sum Absolute\ value\ of\ (X - Median) \tag{4.5}$$

The absolute value of a number is just the positive value of that number. If the number is a positive number, such as 3.0, then the absolute value is just the number. The absolute value of +3.0 is 3.0. If the number is negative, such as −9.0, the absolute value changes the sign to positive number. So the absolute value of −9.0 is 9.0. To calculate the sum of the absolute deviations, we would first subtract the median for the data from each data point. Then we would take the absolute value of each deviation. Finally, we would sum.

One way of conceptualizing the difference between the median and the mean as measures of central tendency is by the two different ways chosen to look at deviations from the center of the data. The deviations produced by subtracting data points from the measure of central tendency include both positive and negative numbers. These will tend to cancel each other out, so the negative signs must be dealt with before constructing a sum of deviations.

The mean is based on choosing the square of a deviation as an appropriate way to eliminate the negative signs. The median is based on choosing the absolute value of the deviation to eliminate negative signs. Thus, the mean can be thought of as the measure of central tendency based on minimizing the sum of the squared deviations, and the median can be thought of as minimizing the sum of the absolute values of the deviations. The mean and the median belong to two different "families" of statistics, one based on sums of squares, the other based on sums of absolute values.

Effect of outliers

Again, suppose the data points being analyzed are 1, 2, 3, 4, 5. The mean is 3 and the median is 3. Now suppose that 15 is again substituted for 5. Recall that the mean is now 25/5 = 5. But the median is still 3. In the case of substituting Louisiana (13.0) for Maryland (9.5), the median would still be 6.1.

While the ability of the median to resist outliers is an advantage, there are problems that keep the median from being the perfect measure of location. It is not as useful for inference as the mean because we don't have as many good tools for learning about the median of a population from sample data, as will be noted in later chapters. It can be argued that the median is *too* insensitive to outliers. As a result, a large number of alternative measures of central tendency have been proposed that attempt to incorporate some of the desirable features of the mean with the robustness of the median. We will discuss one of these alternative measures, the trimmed mean.

The trimmed mean

If outliers overly influence the mean, one solution is to calculate the mean for a data set from which outliers have been removed. While this seems attractive, it is rather arbitrary. How are outliers defined? If the researcher examines the data and discards observations that seem unusual by her subjective standards, the resulting analysis is heavily dependent on those standards. A theory cannot be tested with data from which observations that don't agree with the theory have been removed.

We need an alternative to discarding all observations that do not fit some *a priori* notion of what the data should look like. One approach is to set some objective standard for discounting observations before doing the analysis. The **trimmed mean** is based on such a standard – one that is applied before the data are analyzed and is not based on the expectations of a particular theory. A 10 percent trimmed mean is calculated by discarding the highest and lowest 10 percent of the observations in the data set, and calculating the mean of the 80 percent of the observations that remain. Exactly 10 percent of cases will be trimmed from the top and bottom of the data whether they seem like outliers in a subjective sense or not. Note that because we have to perform conventional arithmetic on the data (taking the mean of the middle data points) the trimmed mean cannot be used with nominal or ordinal data.

If we go back to the numbers 1, 2, 3, 4, 5, we calculate the trimmed mean as follows. We have five data points. Ten percent of five is 0.5. We will round this so we trim one data point from the high end of the data and one from the low end and average what is left. That is, we take the average of 2 + 3 + 4. The sum is 9, and 9 divided by 3 is 3. So the trimmed mean is the same as the mean. If we now substitute the outlier (15) for 5, we still will trim 10%, or one case, from top and bottom and get a trimmed mean of 3. So the trimmed mean protects against the outlier.

We can see the same thing with the state data. The mean with Maryland in the data was 5.7. When we substituted Louisiana, the mean became 6.4. Now if we trim one state from the top and one from the bottom of the original data set, we will drop from the calculation North Dakota at 1.9 and Maryland at 9.5. Then we add together 4.5 + 6.1 + 6.6 = 17.2. Dividing by 3 we get 5.7. If we substitute Louisiana for Maryland, we would still drop North Dakota as the lowest observation and Louisiana as the highest. The trimmed mean still adds together the three middle data points after we've dropped the highest and the lowest, so it is still 5.7. Notice that we might think of North Dakota as an outlier too, and the trimmed mean guards against its influence. By chance this random sample included the state with the lowest homicide rate.

The researcher chooses the trimming percentage. The most common percentages are 10 percent, 15 percent and 20 percent. Note that at 20 percent trimming, two-fifths, nearly half, of the data are ignored in calculating the measure of central tendency to guard against up to 20 percent of the data being outliers. The protection against outliers comes at the cost of ignoring a lot of the data. This is the major disadvantage of the trimmed mean. It ignores much of the data to guard against outliers. A second disadvantage is that the use of trimmed means is not as desirable for inference as conventional means.

Figure 4.1 shows a histogram of the number of non-governmental organizations (NGOs) for the nations we are using in the environmental treaty example. The mean number of NGOs is 848.68, the 10 percent trimmed mean is 686.07, and the median is 494.00.

Here we can see clearly that the data have a strong right or positive skew compared to the Normal distribution. There are a handful of nations with very large numbers of NGOs (>3000). When this is the case, the mean is very much "pulled" towards those data points. The data points that are outliers have a substantial influence on the mean.

In contrast, the median is much less influenced by the outliers. The trimmed mean falls in between the mean and median. Some of the countries with highest

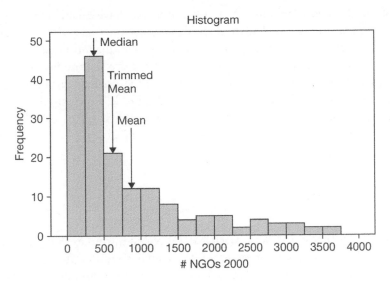

Figure 4.1   Histogram of non-governmental organizations, N = 170
Data source: United Nations Development Programme (UNDP), 2002, analyzed with SPSS.

values (10 percent of them) have been trimmed, as have some of the countries with lowest values (again, 10 percent), but there are still enough countries that have a large number of NGOs relative to the bulk of nations that the trimmed mean is larger than the median. If we used a higher trimming percentage, we would see the trimmed mean get closer to the median.

The relationship between the mean and the median gives us another way to think about skew. If data are right or positively skewed, as is the case in Figure 4.1, the mean usually will be larger than the median. If the data are left or negatively skewed, then the mean usually will be pulled towards those low data points and will be smaller than the median.[7]

Breakdown point is discussed at the end of this chapter as an Advanced Topic.

The mode

The **mode** is the score that occurs most frequently in the data. It is traditional to discuss the mode as a measure of central tendency. But the mode is more often used as a way of describing the shape of a distribution. When we use the mode in this more informal sense, we loosen the definition from the most common value in the data set to values around which cases cluster. In Figure 4.1, there appears to be a mode at between 250 and 500 NGOs.

If we have an interval measure that takes on many values, there may not be a mode until we group the data into categories. If there are many possible values for a variable, such as personal income, no two cases may have exactly the same score. So before finding a mode, we have to sort the data into bins, as we do when creating a stem and leaf diagram or a histogram. Then the bin with the most cases is the modal category, and the ranges of values that go in that bin are the modal values.

The mode does have the advantage that, unlike the median, mean or trimmed mean, it can be used with nominal data. But the mode is not used in statistical inference, or in any applications beyond describing the shape of the data set and the most common response on the variable being studied.

Measures of Variability

We argued above that much work in science is directed toward explaining variability – why individuals, nations, or other units of analysis differ from one another, and thus from a measure of central tendency. So we have numerous measures of variability. Indeed, in discussing the measures of central tendency and the measures of deviation they minimize, we have already introduced the fundamental logic of measures of variability.

Variance

The **variance** is the most commonly used measure of variability. It is based on the mean as a measure of central tendency, and like the mean, employs squares to eliminate negative signs in deviations from the mean. The variance is calculated as:

$$S^2 = \frac{\sum (X - \bar{X})^2}{N} \tag{4.6}$$

or in English, the average of the squared deviations from the mean. Earlier (see Table 4.1) we saw that the sum of squares (SSX) for the five state's homicide rate was 31.28. Then the variance is just $31.28/5 = 6.26$.

In many statistics books, the variance is described as the sum of squares divided, not by N, but by $N - 1$. Doing so obscures the fact that the variance is the average of the squared deviations from the mean. We divide the sum of squares by $N - 1$ when we use sample data to estimate population data, as will be noted later. But in simply describing the variability in a set of data, dividing by N and taking the average of the squared deviations from the mean makes intuitive sense, dividing by $N - 1$ does not. Thus our preference for defining the variance as the average squared deviation and dividing by N, leaving division by $N - 1$ for contexts in which we want to infer population properties from sample information. Note that most software assumes that data are a sample and that the user wants to draw inferences to the population, and so divides by $N - 1$.

Table 4.2 Variance calculations, with and without outlier, of homicide rates for 5 US states

State	Homicide rate	Deviations from mean	Squared deviations from the mean	Homicide rate	Deviations from the mean	Squared deviations from the mean
Kansas	4.5	−1.22	1.49	4.5	−1.9	3.61
North Dakota	1.9	−3.82	14.59	1.9	−4.5	20.25
Alabama	6.6	0.88	0.77	6.6	0.2	0.04
Michigan	6.1	0.38	0.14	6.1	−0.3	0.09
Maryland	9.5	3.78	14.29	–	–	–
Louisiana	–	–	–	13.0	6.6	43.56
Sum	28.6	0.00	31.28	32.1	0.0	67.55
Mean	5.7	–	6.26	6.42	–	13.51

The variance is based on the sum of the squared deviations around the mean. The variance thus follows in the mean family because it uses squares to eliminate the negative signs in deviations from the mean before adding them up. And like the mean, the variance is very heavily influenced by outliers.

It will be helpful to walk through the calculation of the variance using the data with Louisiana instead of Maryland, so we can see the steps in calculating the variance and so that we can compare the two variances. Table 4.2 and Box 4.1 show the steps.

Remember that the variance is the mean squared deviation from the mean. When we do these calculations we see that outlying data points (such as Louisiana in this data set) make a huge contribution to the variance. This is because the deviation is squared, which makes a large deviation into a huge squared deviation. Substituting Louisiana for Maryland increases the mean by about 0.7 points (from 5.7 to 6.4), but it caused the variance to more than double.

Standard deviation

One disadvantage of the variance is that, because it is the average of the *squared* deviations from the mean, it tends to be a very large number relative to the value of individual observations or of the mean. In addition, it is in squared units of measure, such as dollars squared or years of education squared or homicide rate squared. To correct for this, the standard deviation can also be used as a measure of variability. It is defined as:

$$S = \sqrt{S^2} \tag{4.7}$$

or the square root of the variance.[8] Like the variance on which it is based, outliers heavily influence the standard deviation.

Box 4.1   Calculating the Variance

1   Calculate the mean by summing all the cases and dividing by the number of cases.
2   Subtract the mean from each case to get the deviations from the mean.
3   Square the deviations from the mean.
4   Divide by the number of cases.

There's a shortcut for calculating the variance of a 0–1 variable. Since the variable has only two values, 0 and 1, the deviations from the mean only take on two values: $0 - \bar{X}$ and $1 - \bar{X}$. So some algebra that we won't bother to show you leads to a simpler formula for the variance in this case. If P is the proportion in one category (the proportion of people who said "Yes") then 1-P is the proportion in the other category (in this case the proportion of people who said "No"). It turns out that if we actually calculate the deviation from the mean for each case, square each of them, add them up and divide by the number of cases in the group, which is how we get the variance, we will get the same answer as the simple calculation $S^2 = P(1 - P)$. Note that this works if and only if you do the calculation in proportions. If you use percentages and forget to divide by 100, you will get a wrong answer.

The standard deviation plays some very important roles in inference. The standard deviation of the five states in our little sample is just the square root of 6.26, which equals 2.50. For the data where we substituted Louisiana for Maryland to see how an outlier affects things, the standard deviation is the square root of 13.51, which equals 3.67.

Median absolute deviation from the median (MAD)

Just as the variance is linked to the mean, the **median absolute deviation** is linked to the median. The MAD is defined as:

$$\text{Median of (the absolute value of } (X - \text{Median}))\tag{4.8}$$

Box 4.2 shows the steps in calculating the MAD. Table 4.3 shows the MAD for our five states. Remember that the median is 6.1. Also remember that the absolute values of the deviations in the last column have to be put in order from lowest to highest to find the median. Since we have five data points, the median depth is $(5 + 1)/2 = 6/2 = 3$, so the median will be equal to the value of the third case once they are ordered from low to high. The order would be: 0.0, 0.5, 1.6, 3.4, 4.2. So the median of the absolute values from the median is 1.6.

Box 4.2   Calculating the MAD

(Remember, do the work within parentheses first):

1 Calculate the median.
2 Subtract the median from each data point to get the deviations from the median.
3 Take the absolute value of each deviation from the median to get the absolute deviation from the median.
4 Sort the data in order by the size of the absolute deviations from the median. (This is the first step in finding the median of the absolute deviations from the median.) To find the median you may already have the data in order by the size of the variable itself. It has to be resorted to get the median of the absolute deviations.
5 Count to find the median of the absolute deviations from the median. This is the MAD.

Table 4.3   Calculation of the median of the absolute deviations from the median (MAD) homicide rate for 5 US states

State	Homicide rate	Deviations from median	Absolute value of deviations from the median
Kansas	4.5	−1.6	1.6
North Dakota	1.9	−4.2	4.2
Alabama	6.6	0.5	0.5
Michigan	6.1	0.0	0.0
Maryland	9.5	3.4	3.4
Median	6.1	–	1.6

Data source: US Census Bureau, 2003, analyzed with Stata.

If we used the data in which we substitute Louisiana for Maryland, the median depth doesn't change because we still have 5 data points. So the median is still 6.1, and the absolute deviations in order will be 0.0, 0.5, 1.6, 4.2, 6.9. So the median of the absolute deviations is 1.6. The MAD is unaffected by the single outlier.

Like the median, the MAD is highly robust, but is not as easy to calculate or to use for inference as is the variance.

Interquartile range

The MAD is part of the median family because it makes use of absolute values, which is how we deal with the negative signs in deviations from a measure of central tendency in the median family. But recall that the median is also an "order" statistic. Once we place the cases in order from lowest to highest, the median is in the middle (remember, the median is the fiftieth percentile).

We can also make other cuts in the data to aid in our descriptions. It is often useful to calculate the **quartiles** of a batch of data. The **upper quartile** is a value such that 25 percent of the cases fall above it and 75 percent fall below it. The **lower quartile** is a value such that 25 percent of the cases fall below it and 75 percent fall above it. Thus half the cases will fall between the upper quartile and the lower quartile, and 25 percent between each of the quartiles and the median. If we call the lower quartile the first quartile, then the upper quartile is the third quartile. This makes the median the second quartile.

A useful measure of variability is the **interquartile range** or IQR. The formula for the IQR is:

$$IQR = Upper\ quartile - Lower\ quartile \tag{4.9}$$

We can calculate the upper and lower quartiles by first placing the data in order from lowest to highest. We then calculate what is called the **quartile depth**. Tukey (1977) developed the notion of the depth of a data point in a data set. For the median, as we have seen, the depth is:

$$Median\ depth = (N + 1)/2 \tag{4.10}$$

Then the simplest way to find the quartile depth so we can find the upper and lower quartiles is:

$$Quartile\ depth = (Median\ depth + 1)/2 \tag{4.11}$$

For simplicity, use a whole number for the median depth by dropping a decimal point before calculating the median depth.[9] The quartiles can then be found by counting the quartile depth from the top and from the bottom.

Note that the upper and lower quartiles are numerical values in the same units as the variable we are measuring. The quartiles are not cases – we don't say the quartile is a particular state, although a state may fall on the quartile and thus that state and the quartile will have the same value. Nor is the quartile the number of cases you have to count in to find it, the quartile depth. The quartile is the value of the variable that we find at the quartile depth, not the quartile depth itself.

The interquartile range is the difference between the upper and lower quartiles of the data set. It is often used with the median. It is how far one must go to span the middle 50 percent of the data. Variances and standard deviation are based on squaring deviations. The MAD is based on taking absolute values. The interquartile range is based on the ordering of the data. Box 4.3 shows the steps involved in calculating the interquartile range.

For the five state data set, we would calculate the interquartile range as follows. The median depth is $(N + 1)/2 = 3$. Then the quartile depth is (Median Depth + 1)/2,

---

Box 4.3   Calculating the Interquartile Range

1   Calculate the quartile depth: Quartile depth = (Median depth + 1)/2
2   Arrange the data from highest to lowest.
3   Count in the quartile depth from the bottom. If the quartile depth is a whole number, then the value of the case that far from the lowest score is the lower quartile. If the quartile depth has a fraction, then the quartile lies between two cases. Average the values of the two cases adjacent to the location of the quartile. That number is the value of the lower quartile.
4   Count in the quartile depth from the top. If the quartile depth is a whole number, then the value of the case that far from the lowest score is the upper quartile. If the quartile depth has a fraction, then the quartile lies between two cases. Average the values of the two cases adjacent to the location of the quartile. That number is the value of the upper quartile.
5   Subtract the lower quartile from the upper quartile. The result is the inter-quartile range.

---

which is $(3 + 1)/2 = 2$. So the quartile of 5 cases is the value we find when we count in two cases from the top and two cases from the bottom. The state homicide rates are (in order): 1.9, 4.5, 6.1, 6.6, 9.5. So counting in two from the bottom, we find the value 4.5. That is the lower quartile. Counting in two from the top we find 6.6. That is the upper quartile value. Then the interquartile range is just the upper quartile value minus the lower quartile value, or $6.6 - 4.5 = 2.1$. So the interquartile range for the five states is 2.1. Again, this tells us that to cover the middle 50 percent of the data, we have to "travel" 2.1 units, where the units are homicides per 100,000.

If we substitute the data with Louisiana instead of Maryland, the data in order are 1.9, 4.5, 6.1, 6.6, 13.0. So the quartiles will be 4.5 and 6.6, and the interquartile range will still be 2.1. Both the IQR and the MAD are not very influenced by the single outlier. This is a general property of the median family – the measures in the median family are much less susceptible to outliers than those in the mean family.

Relationship among the measures

The relationship of one measure of variability to others will depend on the distribution of the data being analyzed. The key issue in comparing the mean and median is whether the data have a positive or negative skew. If the data have no skew, then we say the distribution is **symmetrical**. Technically, the term symmetrical means that if we placed a mirror in the center of the histogram for the data, the reflection of one side would be the same as the other side. Or if we folded a piece of paper with the histogram on it in half at the center of the histogram, the bars of the histogram on one side of the fold would fall exactly on top of bars on the other side. The theoretical Normal distribution is symmetrical, although when we place it on a histogram of real data, this can be hard to see because the values of the real variable may not run far enough to show all of one side of the Normal distribution plotted by the computer.

If the data are symmetrical, the mean and the median will be equal. Since the mean is more sensitive to outlying data points than the median, the mean will be pulled towards the long tails in a skewed distribution, as we saw in Figure 4.1. If the data are positively skewed, the mean will be greater than the median. If the data are negatively skewed, the mean will be less than the median.

There are fewer generalities that can be offered about relationships among measures of variability. The one exception is if the data actually have a Normal distribution. If the data follow the Normal distribution, then the standard deviation times 1.35 will equal the interquartile range. This is one way to assess whether the data are Normal. If the IQR divided by the standard deviation equals about 1.35, then the data are roughly Normal. But as we have already begun to see and will see in later chapters, most variables we use in research are not Normally distributed.

Table 4.4 shows all the measures of central tendency for both the five state sample and the sample with Louisiana substituted for Maryland.

What is a typical homicide rate? Looking at the original sample of five cases, the mean says about 6, the median about 6. Because we only have one rather low case in the original data, the 10 percent trimmed mean is very close to the median. With more outliers in a larger data set, the trimmed mean might be closer to the mean. The ability of the trimmed mean to resist influence by outliers depends on what fraction of the data is trimmed and how many data points are distant from the rest of the data. The data set with Louisiana as an outlier increases the mean by 0.7 points, but the median and trimmed mean don't change at all.

We are all familiar with the mean in its identity as the average and can use that as a starting point in thinking about the other two measures of central tendency. But we have less experience with measures of variability. This will come with practice. Here we can note that the variance is much higher in the data set with Louisiana as an outlier than in our original sample. The standard deviation is also substantially higher in the data set with the outlier. In contrast the MAD and IQR are not influenced by the outlier. But unlike the mean, median, and trimmed mean (which are all trying to tell us where the center of the distribution of the data is

Table 4.4 Comparison of measures of central tendency and variability of homicide rate for a sample of 5 US states

Sample	Mean	Median	10% trimmed mean	Variance	Standard deviation	MAD	IQR
Original	5.72	6.1	5.73	6.26	2.50	1.6	2.1
With outlier	6.42	6.1	5.73	13.51	3.67	1.6	2.1

Data source: US Census Bureau, 2003, analyzed with Stata.

located), the measures of variability tell us different things. The variance is the mean squared deviation from the mean, and the standard deviation is its square root. The MAD is the parallel measure for the median – we take absolute values of deviations from the median and then find the median to summarize them. And the IQR is something else again – it's the distance that we have to cover to "see" the middle 50 percent of the data.

Don't be concerned if for now you have trouble telling a story about the data from the measures of variability. It takes some experience with these measures to have a feel for them. Much of their use is in comparing data sets, a topic for later chapters.

Boxplot

While we sometimes see the **boxplot** attributed to Tukey (1977) who popularized it, the graphic manual for the statistical package Stata (StataCorp, 1999) suggests that it has been used at least since a paper by Crowe in 1933. It is also called a **box and whisker** diagram.

The boxplot displays five key numbers that summarize the data: the median, the upper quartile, the lower quartile, and two lines called **adjacent values**. As we will see below, the adjacent values are high and low data points selected so that anything beyond them can be considered an unusual data point, an outlier. We will explain how we pick the adjacent values below. Figure 4.2 shows a boxplot of the homicide data for all states. We have drawn in labels for each part of the boxplot. Normally we won't do this because it makes the graph very cluttered, but here the labels will help you understand the plot.

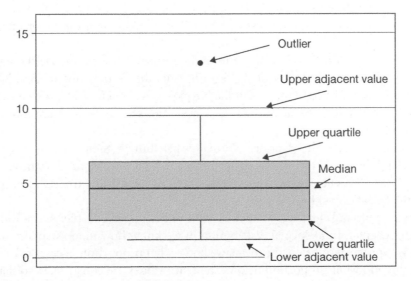

Figure 4.2   Boxplot of state homicide rates, N = 50
Data source: US Census Bureau, 2003, analyzed with Stata.

The vertical axis displays values of the variable we are graphing – the homicide rate. The horizontal axis does not have any meaning; it's just a horizontal line to hold the eye. Let's discuss the graph from the bottom to the top. Focus on the horizontal lines only – in a box plot the vertical lines are just there to help us see the picture. The bottom horizontal line in the graph is at the lower "adjacent value." The next line up, which is slightly longer than the fence line, is at the value of the lower quartile. The next horizontal line is the median. The next horizontal line is the upper quartile.

So the box in the middle of the boxplot shows the upper and lower quartiles at the bottom and top of the box and the median inside the box. Then at the top we have a shorter horizontal line for the upper adjacent value and finally a dot. The dot means that there was a data point above the adjacent value that should probably be considered an outlier.

The only complicated thing about a boxplot is the location of the top and bottom lines. We will discuss these and then show how to construct the boxplot. The idea of the adjacent values is to help identify unusual cases – outliers. We do this by constructing a fence. The rule for fences (which used to be called "whiskers" when the boxplot was called a "box and whisker" diagram) is that they are placed at:

Upper fence = Upper quartile + 1.5(IQR)

Lower fence = Lower quartile – 1.5(IQR)

We then plot the top line at the largest value inside the upper fence. We plot the bottom line and the lowest value inside the lower fence. So the top line for the upper adjacent value goes at the value of the largest case that is still smaller than the upper quartile plus 1.5*IQR. The bottom line for the lower adjacent value is placed at the value of the smallest case that is still larger than the lower quartile minus 1.5*IQR.

Why would we do this? It's an aid to finding outliers. Cases outside the fences are unusual. A simple version of the boxplot runs the vertical line to the highest and lowest cases. But common use of the boxplot uses the fences to define adjacent values. Then any data point outside the fence is an outlier.[10] These are indicated with small circles.

Remember that if the data are Normally distributed, then there is a special relationship between the interquartile range and the standard deviation. The IQR = 1.35 s. So when we are multiplying the IQR by 1.5 this translates into the standard deviation (s) by 1.5*1.35* s = 2.025 s.

So we are placing the fence at about 2 s above the upper quartile. As we will see in a later chapter, in a Normal distribution, if we know how many standard deviations we are above or below the mean, we know what proportion of cases fall above that point and what proportion below. The fence values are constructed so that, if the data were Normally distributed, only about 5 percent of the cases in a large data set would fall outside the fences.[11]

The boxplot shows the median, the middle 50 percent of the data (the box), the high and low adjacent values and any outliers.[12] The vertical lines convey no information directly, but are used to make it easier to perceive patterns. Boxplots convey less distributional information than the stem and leaf, but are easier to use and convey far more information than any single number. Box 4.4 lists the steps involved in drawing a box plot.

Box 4.4   Drawing a Boxplot

1 Calculate the median, upper and lower quartile, and interquartile range.
2 Calculate the value 1.5*IQR which is used to calculate the upper and lower fences (and identifies the upper and lower adjacent values).
3 Calculate the Upper Fence as: Upper Fence = Upper Quartile – (1.5*IQR).
4 Find the highest value in the data that is lower than the Upper Fence. This is the upper adjacent value.
5 Calculate the lower fence as: Lower Fence = Lower Quartile – (1.5*IQR)
6 Find the lowest value in the data that is higher than the Lower Fence. This is the lower adjacent value.
7 Identify any cases larger than the Upper Fence. These values are upper outliers.
8 Identify any cases smaller than the Lower Fence. These are lower outliers.
9 Draw a vertical scale. This is a continuous scale whose range allows inclusion of the highest and lowest values in the data set.
10 Draw a horizontal axis. This is not a scale but is simply a line to keep our eyes focused on the graph.
11 Draw a horizontal line at the value of the median on the vertical scale. It should be to the right of the vertical scale. Its length and distance from the vertical scale are determined by what will look pleasing. The distance from the vertical scale and the length of the horizontal lines do not convey any information.
12 Draw a horizontal line at the value of the upper quartile and one at the value of the lower quartile. These two horizontal lines should be the same length as the line that represents the median.
13 Use vertical lines to connect the upper quartile line, the median line and the lower quartile line. This completes the box.
14 Draw a horizontal line at the value of the high adjacent value. Draw a horizontal line at the value of the low adjacent value. These lines should be shorter than the horizontal lines used for the upper quartile, lower quartile and median.
15 Draw a vertical line from the center of upper quartile line to the upper adjacent value. Draw a vertical line from the center of the lower quartile line to the lower adjacent value line.
16 Add small circles or other plotting symbols for any data points above the Upper Fence or below the Lower Fence.
17 Add a title and information on the data source and sample size.

In Figure 4.2 we can see that the median is just below five. Since half of all cases fall between the lower and upper quartiles, we can see that most states have homicide rates in the single digits, and that none have a rate so low that it meets the criterion for a low outlier. But we also see that the homicide rates have a positive skew. The upper adjacent value stretches further from the upper quartile than the lower adjacent value does from the lower quartile. And we can see that one state is an upper outlier by the criteria we use. By convention this is plotted with a dot, but we could use a two letter state label and then we would see it is Louisiana, as we already know from examining the data.

Side-by-side boxplots

One of the most common uses of boxplots is to make comparisons between batches of data. Figure 4.3 shows the homicide rates plotted by whether each state was in the Confederacy. We can see that the medians and IQRs of Confederate and non-Confederate states differ. The lower quartile of the Confederate states is at about the same homicide rate as the upper quartile of the non-Confederate states, although it has a much smaller range.

But looking at the upper adjacent value for non-Confederate states, we also see that some of them have homicide rates that are as high as all but the highest Confederate state. The side-by-side boxplot is a relatively powerful tool for graphical comparison of groups. Its simplicity makes it easy to see patterns, but it still compares a great deal of information about the distribution of each group.

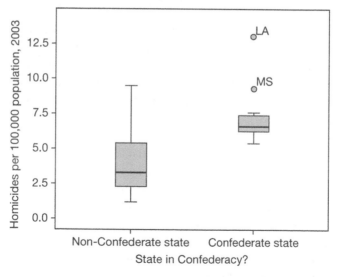

Figure 4.3　Boxplot of state homicide rates by membership in the Confederacy, N = 39 for non-Confederate states; N = 11 for Confederate states
Data source: US Census Bureau, 2003, analyzed with SPSS.

What Have We Learned?

In this chapter we've introduced the standard statistics that are used to summarize a batch of data. We have established two families. The mean family uses calculations based on the numerical values of the variable for all cases, and focuses attention on squared deviations from the mean. The mean of the squared deviations from the mean is the variance, and its square root is the standard deviation. These are very important statistics, but they have the limitation that they can only be used with interval data, and they are quite sensitive to outlying data points.

The alternative family starts with the median. It is based on counting the data once they are ordered, rather than doing standard arithmetic (for this reason statistics in the median family are sometimes called order statistics). The median family deals with variability in two ways. One is to take the absolute value, which leads to the median absolute deviation from the median as a measure of variability. The other is to look at percentiles. The median is the fiftieth percentile. The upper quartile is that value such that a quarter of the data are above it and three-quarters below, while the lower quartile is a value such that three-quarters of the data are above it and a quarter below. The distance between the two quartiles, the interquartile range, is also a measure of variability.

We have also learned a new graphic technique, the boxplot, which provides information about the median, the quartiles, and outliers in the data. This tool is very useful in comparing groups.

The mean family is the starting point for almost all statistical calculations. They introduce a key part of the logic of quantitative analysis. We begin with a summary, like the mean, which can be thought of as a simple model. Then we focus on deviations from the mean, the variability in the data. In the mean family we summarize this variability with the variance and standard deviation. Throughout the book, we will start with a model that summarizes or explains the data and then examine the variability around that summary.

Feature 4.1 Models, Error, and the Mean: A Brief History

In this chapter we have paid a lot of attention to deviations from the mean. The deviations play an important role in calculating statistics. But thinking about deviations can also help us understand why variability exists in our data. In some ways the history of statistics is about the development of different ways of thinking about deviations from the mean.

In chapter 1 we discussed Galileo's view of measurement error. Scientists of Galileo's time and after noticed that when they made repeated measurements of something, they found small differences from measurement to measurement. Further these measurements all clustered around a center, which was the average of all the measurements. And the shape of the distribution of measurements – the histogram – was bell-shaped. Subsequent work in statistics showed that many measurements errors followed the bell-shaped, or Normal distribution. Indeed, this is how science learned about the Normal distribution. As we will see later in the book, in the late 1700s, the great mathematician Gauss and others showed the equations that described the pattern of errors in data and established the Normal distribution, which they called "the law of errors."

The model that underpins this way of thinking about measurement error is:

Measured value of X = True value of X
+ Measurement error   (4.12)

The theoretical work showed that the average value of the measurements of X across many measurements would be a good guess of the true value of X. We will return to this idea in later chapters. For now, the point is that from the early 1600s on, variation in the values of a variable was seen as a deviation from the true value, with the mean representing the true value.

By the early 1800s, the Belgian scientist Lambert Adolphe Jacques Quetelet, after studying mathematics and astronomy, began to focus on "l'homme moyen" – the average man. To quote: "If an individual at any epoch in society possessed all the qualities of the average man, he would represent all that is great, good, or beautiful" (Stigler, 1986).

Quetelet saw as the goal of his science the analysis of distributions of data on people. This would allow him to uncover the "average man" or, in his later writings, the average man for different groups. If you have studied philosophy or otherwise read Plato, you will note that this notion of the average man is rather like the Platonic world of ideal things which underlies the real world of actual things. The actual is flawed but reflects the perfect ideal. Actual people are deviations from perfect average for their group.

He applied this approach to many human characteristics, most of them biological, such as the heights and chest circumferences of soldiers. As it turns out, many of these variables do in fact have the bell-shaped distribution that had emerged from the study of errors and that was to be called the Normal distribution. Quetelet was also interested in conviction rates for crimes and tried to use quantitative data to understand who was convicted and who was not. But the point for thinking about means and errors is that Quetelet was using a model something like this:

Value of X for a person = Average value
of X + Sources of deviation   (4.13)

Here the errors are the result of individual factors that cause people to differ from the ideal value, which is the average of X.[13] We have moved from a theory of measurement error to a theory of deviations from the mean as a result of idiosyncratic factors.

Francis Galton takes the next step in our thinking about variation. Galton was a prolific scientist who was fascinated with measuring human characteristics. Among many other things, he was one of the first to understand that fingerprints did not change over a person's life and thus could be used to identify people. His 1892 book "Fingerprints" was the key analysis that led, eventually, to the widespread use of fingerprints for identification. He also published books on Africa, where he led two expeditions, and on weather measurements, in addition to a vast amount of work on human beings, heredity, and statistics.

Francis Galton read Quetelet, but because he was so strongly influenced by his cousin Charles Darwin, his interest was not so much in the average as in the variation in data. Darwin's theory of evolution is a theory of variation. There is no ideal, and deviations from the average are not flaws compared to some ideal represented by the average. As the quote at the beginning of this chapter makes clear, variation was a delight to Galton in his research. For him, the average is removed from the pedestal of being either the true value we are trying to measure, as it was in applications in astronomy, or the ideal or essence, as it was for Quetelet. For Galton, the mean was simply the measure of the central tendency of the distribution. His model was something like:

Value of X for a person =
Average value of X + Effect of
many factors influencing X   (4.14)

This shifts us to a focus on what these other factors are and how they make X vary from person to person.[14] It allows us to think about the deviations from the mean of a variable as something we will try to understand, using models that explain why X may vary from person to person.

We cannot discuss Galton and his contributions to science without also noting that he was one of the founders of the eugenics movement. In fact, Galton coined the term "eugenics" in 1883.[15] The eugenics movement was concerned with "improving" humanity by encouraging the "best" people to have many children and encouraging the "less able" to have few or no children. They held the belief that much of the variability from person to person was the result of genetics, not life experiences and social conditions. So, in their view, a program of selective breeding would improve the human condition.

Not surprisingly, the definitions of ability and human quality tend to favor people like the proponents of eugenics – few of them ever argued that people like themselves should be the ones to have few children.

The eugenics movement had tremendous influence in the US and Europe until World War II. Then the horrors perpetrated by the Nazis made clear the ease with which eugenic ideas could be used to justify racism and atrocities on the part of the ultra-right wing. At the same time, better research on human genetics and on the other influences that shape individuals made clear that many eugenics arguments were not only culturally biased but incorrect in other aspects of their logic as well.[16] As a result, eugenics is viewed with appropriate skepticism, but arguments that human variation is mostly genetically determined reappear periodically, and even more common are policy arguments that are loosely grounded in some, often vague, notion that our genes determine who we are. Gould (1996) provides an excellent discussion of the problems with eugenics and other attempts to reduce what is human to the genetic.

Advanced Topic 4.1   Squared Deviations and Their Relationship to the Mean

The squared deviations have a special relationship to the mean. The sum of the squared deviations around the mean will be smaller than the sum of the squared deviations about any other number. We call this *minimizing the sum of squared deviations*. The fact that the mean minimizes the sum of squared deviations can be thought of as a search problem of a sort that is common in statistics. The object of the search is to find some number that, when substituted into an arithmetic expression as a "guess" makes the number that results from the arithmetic as small (or in some problems as large) as possible. Suppose you and your opponent have wagered on who can find the number G

which, when inserted into the calculation will make the sum of the squared deviations from that number G as small as possible. The equation would be:

$$\sum (X - G)^2 \tag{4.15}$$

If you are allowed to pick the mean as a guess for G, then you can never lose, for no number will yield a smaller value for the summation in Equation 4.15. If you get to choose the mean in this guessing game, the best an opponent can do is to tie you, and she can tie you only if she picks the mean as well. This is another sense in which the mean lies at the center of the data.

Advanced Topic 4.2   Breakdown Point

We've just seen that the mean is strongly influenced by a single data point. One desirable property of a descriptive statistic is that it not be unduly influenced by outliers. Statisticians describe this property of a summary measure as the *breakdown point*. The breakdown point of a descriptive statistic like the mean is the proportion of data that can take on extreme values (in technical terms, data points that are infinitely far from the rest of the data, but for our purposes we can think of data points that are just very distant from all the other data) without influencing the summary measure. For the mean, the breakdown point is zero – the mean is influenced by just one very unusual data point. If there are extreme outliers, the mean is pulled towards them. All other things being equal, we would like to have descriptive statistics with

high breakdown points. The median is a better measure of central tendency than the mean because it has a high breakdown point. But the mean still has many other advantages over the median.

While the mean is not *robust* with regard to outliers, the median is quite robust, with a breakdown point of 50 percent – fully half the data must take on extreme values before the median is heavily influenced. A higher breakdown point cannot be obtained. (If more than half the data is aberrant, who is to say which points are "good" and which "outliers?") The 20 percent trimmed mean has a breakdown point of 20 percent. If one more than 20 percent of the observations are outliers, then the trimmed mean will be pulled towards them, but 20 percent or fewer outliers will not influence the trimmed mean.

Applications

Example 1: Why do some US states have higher than average rates of homicide?

Table 4A.1 presents the mean, median, 10 percent trimmed mean, variance, standard deviation, MAD and IQR for the homicide rate and poverty level for all 50 US states.[17]

Note that it makes no sense to directly compare the homicide rate and the poverty level because they are measured on different scales. The poverty level is the percentage of the population in poverty. The homicide rate is the number of homicides per 100,000 population. We should not compare measures of central tendency and variability across variables unless the variables are measured in the same units.

The mean and median for the homicide rate are nearly the same so simply comparing these two doesn't show us the skew that we saw in the data from graphing it.

Table 4A.1 Measures of central tendency and variability for homicide rate and poverty level, N = 50

	Mean	Median	10% trimmed mean	Variance	Standard deviation	MAD	Interquartile range
Homicide rate	4.74	4.65	4.54	6.44	2.54	1.80	3.90
Poverty level	11.69	11.10	11.46	9.77	3.13	2.05	4.60

Data source: US Census Bureau, 2002, 2003, analyzed with Stata.

Box 4.5 Rounding

Small rounding errors can occur in statistical calculations, as we mentioned. If you are comparing results we have calculated to results you have calculated, small differences may be due to rounding error. If you get a result very different from ours, there are two possibilities. One is that you have done something wrong. The other possibility is that despite a lot of effort to make sure that there are no errors in the text, a typographical or other error might have been missed by the authors and editors. So, if you've tried to duplicate our results and get an answer that is quite a bit different, and you've double checked your work, it's a good idea to check with a colleague or your instructor before deciding that you don't understand statistics. (We will post any errors on the book website, so you should check there as well.)

If the mean and median are substantially different, it means the data have skew, but there can still be skew when the mean and median are about equal. The measures of variability are not directly comparable. Recall that earlier in this chapter we said that if the data have a Normal distribution, then 1.35 times the standard deviation will equal the IQR. Here 1.35*2.54 = 3.43, which is about 0.5 units from the value of the IQR.

The mean and the median for the poverty rate are different by about 0.6, which is a small difference but one that would make us suspect some skew. Since the mean is higher than the median, we suspect the skew is positive. But of course the most effective way to look for skew is to plot the data. Here 1.35 times the standard deviation is 1.35*3.13 = 4.23, which is fairly close to the IQR. But we'll have to look at plots to see how close to Normal the distribution of poverty rates really is. Comparing the standard deviation and the IQR can tell us when we *don't* have a Normal distribution, but it can't tell us when we do.

Figures 4A.1 and 4A.2 are boxplots of the homicide rate and poverty rate. In Figure 4A.1 we see that we have one outlier as noted by the circle, which is Louisiana, at 13 homicides per 100,000 per year. There are no lower outliers by the criteria used to make the fences and adjacent values. The median is a bit below 5 and the 50 percent of the states that fall between the upper and lower quartiles seem to have homicide rates between about 2.5 and about 7. (We draw

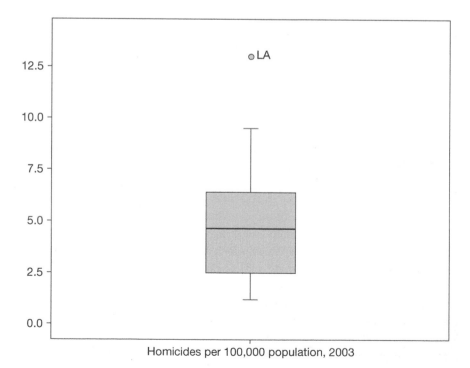

Figure 4A.1   Boxplot of homicide rate, N = 50
Data source: US Census Bureau, 2003, analyzed with Stata.

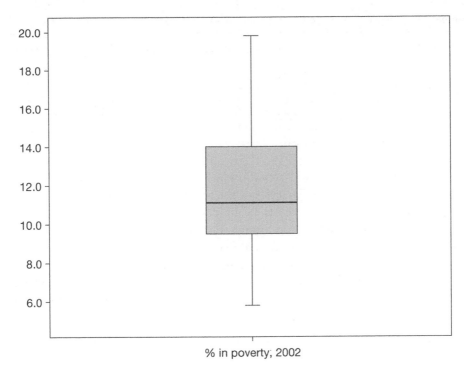

% in poverty, 2002

Figure 4A.2   Boxplot of poverty rate, N = 50
Data source: US Census Bureau, 2002, analyzed with Stata.

this conclusion by looking at where the top and bottom of the box is on the scale on the left.)

In Figure 4A.2, as we can see, there are no high or low outliers. The median seems to be just above 11 percent, and the middle 50 percent of the cases fall between slightly below 10 percent in poverty and slightly above 14 percent in poverty.

We have suggested that the poverty rate may be one of the causes of the homicide rate. To see if this **hypothesis** is consistent with the data, we have calculated the measures of central tendency and variability for two groups of states, those with higher poverty levels and those with lower poverty levels. If we split the states into groups at the median poverty level, then we will have 25 states in each group. This is convenient, although it is not necessary to have exactly equal groups when making comparisons of this sort. Table 4A.2 shows the summary statistics by groups of states.

We see in Table 4A.2 that each of the three measures of central tendency is lower for the states with lower poverty levels than for the states with higher poverty levels. In each case the lower states have a central tendency measure on homicide substantially less than the states with more poverty. So the theory is consistent with the data. The higher poverty level states seem to have a bit more variability in homicide than the lower poverty level states but the differences are not large. Note that our hypothesis linking poverty to homicide rates does not say anything about

Table 4A.2   Summary statistics for homicide rate for states with lower and higher poverty levels

	N	Mean	Median	10% trimmed mean	Variance	Standard deviation	MAD	Interquartile range
Higher poverty level	25	5.55	6.10	5.58	7.53	2.74	1.20	3.55
Lower poverty level	25	3.93	3.20	4.09	4.25	2.06	1.30	2.75

Data source: US Census Bureau, 2002, analyzed with Stata.

how variable homicide rates will be within a group of states, only that we expect, in general, for states with higher levels of poverty to have higher homicide rates.

One of the best uses of boxplots is for comparison. Figure 4A.3 shows the homicide rate for the resulting two groups of states.

The graph lends some support to the poverty/homicide theory. The lower quartile of homicide rates for the high poverty states falls above the median of the low poverty states. But some of the low poverty states (Nevada and Maryland) also have homicide rates greater than the upper quartile of the high poverty states, so the relationship is not perfect.

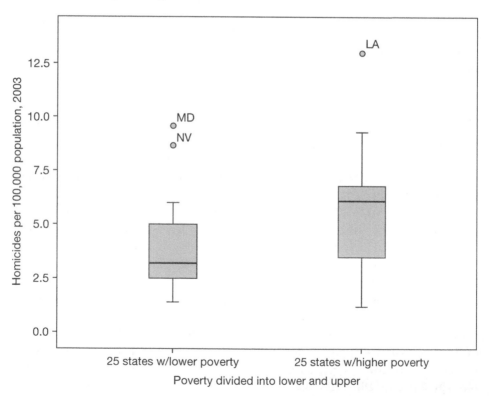

Figure 4A.3   Boxplots of homicide rate for states grouped by poverty rate, N = 25 per group
Data source: US Census Bureau, 2003, analyzed with SPSS.

Example 2:   Why do people differ in their concern for animals?

Table 4A.3 shows the measures of central tendency and location for the animal concerns question. Recall that this is a scale that runs from 1 to 5. The midpoint of the scale is at 3. The center of the distribution by any of our three measures is below this mid-point.

Figure 4A.4 makes the distribution clearer. There were 2,385 people who scored a "5" (strongly disagree) on the scale; we know this from looking at the frequency

Table 4A.3   Measures of central tendency and location for animal concern question, N = 29,486

	Mean	Median	10% trimmed mean	Variance	Standard deviation	MAD	Interquartile range
Animal concern scale	2.49	2.00	2.39	1.39	1.18	1.0	1.0

Data source: 2000 ISSP data set, analyzed with SPSS.

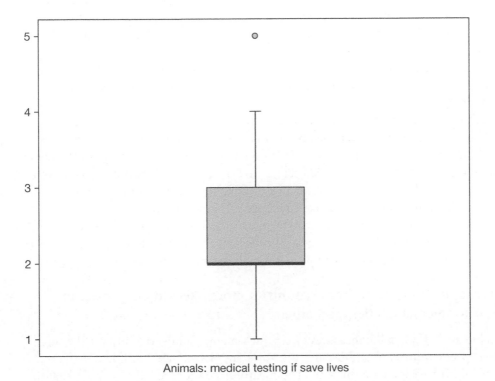

Animals: medical testing if save lives

Figure 4A.4   Boxplot of animal concern question, N = 29,486
Data source: 2000 ISSP data set, analyzed with SPSS.

Table 4A.4   Measures of central tendency and variability for animal concern by gender

	N	Mean	Median	10% trimmed mean	Variance	Standard deviation	MAD	Interquartile range
Men	13,412	2.37	2.00	2.31	1.29	1.13	1.00	1.00
Women	16,062	2.60	2.00	2.51	1.45	1.21	1.00	2.00

Data source: 2000 ISSP data set, analyzed with SPSS.

distribution table of the data. By the criteria we use in setting up fences and adjacent values, this score is seen as an outlier. The median (2.00) is the same as the lower quartile. When a variable has only a few values, the median sometimes has the same value as a quartile.

We have hypothesized that there may be gender differences in animal concern. Table 4A.4 shows the measures of central tendency and variation for men and for women.

Each of the measures of central tendency is higher for women than for men, although the medians are equal. Thus the hypothesis is consistent with the data. The measures of variability are about the same, except for the IQR, which is larger for women. This pattern may be clearer if we examine boxplots of the scale for men and women.

Figure 4A.5 shows the distribution of the scale scores by gender to see if the data are consistent with the hypothesis that women may be more concerned than men about animals.

The median for men and women is the same as the lower quartile and is identified with a thick black line. Both are equal to 2. By looking at frequency distributions of the scale for men and for women, we find that 46 percent of the men and 42 percent of the women had scores of 2 on the scale. When many cases of a variable bunch at a particular value, it's possible for the median and a quartile to fall at the same value, but it makes the boxplot harder to interpret.

The results do seem consistent with the theory, although the evidence is not strong. The men who scored a 5 (6 percent of men) are far enough from the other men to be treated as outliers, while this is not true for the women at 5.

Example 3:   Why are some countries more likely to participate in environmental treaties than others?

Table 4A.5 displays the measures of central tendency and variability for the number of environmental treaties and the voice and accountability index for nations for which we have data on both. The mean for environmental treaties is slightly larger than the median, indicating that some cases with high values may be pulling the mean towards higher values. In other words, the variable has some positive skew,

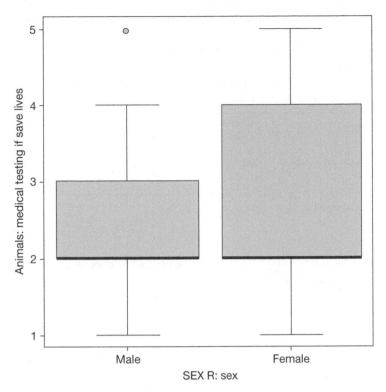

Figure 4A.5   Boxplot of animal concern question by gender, For men, N = 13,412; for women, N = 16,062

Data source: 2000 ISSP data set, analyzed with SPSS.

Table 4A.5   Measures of central tendency and variability for environmental treaties and voice and accountability, N = 169

	Mean	*Median*	*10% trimmed mean*	*Variance*	*Standard deviation*	*MAD*	*Interquartile range*
Environmental treaties	6.49	6.00	6.24	11.45	3.38	2.00	4.00
Voice and accountability	0.01	−0.07	0.01	0.91	0.95	0.80	1.71

Data source: Kaufmann et al., 2002; Roberts et al., 2004, analyzed with Stata.

although using the 10 percent trimmed mean only slightly reduces the mean. The mean and median for the voice and accountability scale, on the other hand, are relatively similar.

To look at the relationship between voice and accountability and environmental treaty participation, we have broken the countries into two groups, those with low voice and accountability and those with greater voice and accountability (see Table 4A.6). We make this distinction by splitting the countries into groups at the

Table 4A.6   Measures of central tendency and variability for environmental treaty participation by level of voice and accountability

	N	Mean	Median	10% trimmed mean	Variance	Standard deviation	MAD	Interquartile range
Greater voice and accountability	84	8.02	8.00	7.85	12.11	3.48	2.00	4.00
Lower voice and accountability	85	4.98	5.00	4.83	6.25	2.50	1.00	2.00

Data source: Kaufmann et al., 2002; Roberts et al., 2004, analyzed with Stata.

median score, which is -0.07. Since two countries fall exactly at the median, we have arbitrarily placed them in the lower voice and accountability group. This means the groups are unequal. It is not necessary to have exactly equal sized groups for these comparisons, but it is convenient to have groups of roughly the same size.

For all the measures of central tendency, the typical value of treaty participation is lower for the lower voice and accountability countries than for the countries with greater voice and accountability. This is consistent with the theory that countries with greater voice and accountability are more responsive to pressures from environmentalists and the international community. But we also note that all four measures of variability for the treaty participation measure are higher for the countries with greater voice, indicating there is considerable variation within that group.

The boxplot of treaty participation by low and high voice and accountability rate (Figure 4A.6) shows this variation. Countries with greater voice span the full range of treaty participation, from the lowest levels of treaty participation to the highest. The interquartile range of the lower voice countries falls just below the lower quartile of the greater voice countries. Countries with little voice that participated in more than nine treaties are considered outliers, such as Russia and Nigeria, while none of the countries with greater levels of voice and accountability are outliers. Countries with little voice that participated in no treaties were also considered outliers, so the range of treaty participation for countries with little voice has a smaller range and median than countries with greater voice.

Example 4:   Why do people differ in their knowledge of how the AIDS virus is transmitted?

The dependent variable for our AIDS knowledge example is a dichotomy, which we can score 0 if people said "No" (did not know the correct answer) and 1 if they answered "Yes" (respondents know that condoms can help prevent transmission of AIDS). It then makes sense to take a mean, which is just the proportion of people saying yes, and we can calculate the variance and standard deviation as well (see

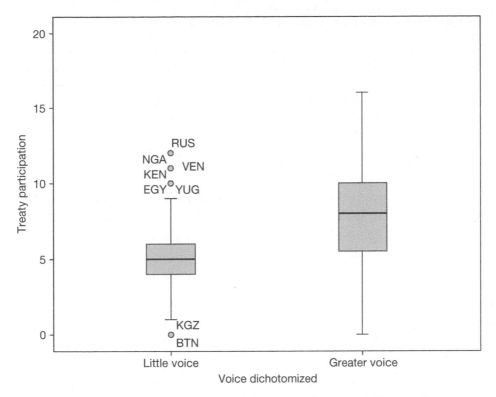

Figure 4A.6 Boxplot of treaty participation by voice and accountability rate. For countries with lower voice and accountability, N = 86; for countries with greater voice and accountability, N = 83

Data source: Kaufmann et al., 2002; Roberts et al., 2004, analyzed with SPSS.

Table 4A.7). However, since there are only two categories, the median will simply be the category with the most people in it – in this case the median and the mode are the same thing. It does not make sense to calculate any of the other measures of variability nor does the trimmed mean make any sense for a dichotomous variable.

One of our hypotheses is that there are gender differences in AIDS knowledge. Table 4A.8 shows the mean, variance, and standard deviation of the AIDS knowledge variable for men and for women. We should be clear about what we are doing when we calculate the mean and summary statistics for a 0–1 (zero–one) variable.

Table 4A.7 Measures of central tendency and variability for AIDS knowledge, N = 8,310

Mean	*Variance*	*Standard deviation*
0.77	0.176	0.419

Data source: 2000 Ugandan DHS data set, analyzed with SPSS.

Table 4A.8   Measures of central tendency and variability for AIDS knowledge by gender, N = 8,310

Gender	N	Mean	Variance	Standard deviation
Male	1,886	0.83	0.142	0.377
Female	6,424	0.76	0.184	0.429

Data source: 2000 Ugandan DHS data set, analyzed with SPSS.

We can see that a slightly larger proportion of men than women know that condoms can help prevent the transmission of AIDS. The variance and standard deviation for women are slightly larger.

Exercises

1. These exercises are intended to give you practice in calculating measures of central tendency and variability, graphing boxplots and using these descriptive statistics to evaluate models. Let's begin with the state homicide rate. One argument regarding homicide is that higher levels of urbanization lead to higher levels of homicide. Table 4E.1 provides the homicide and urbanization rates for a random sample of 20 states. We use the random sample rather than all 50 states simply to keep the calculations as simple as possible for you. Urbanization is defined as the percentage of urban residents in the state according to the 2000 US census.

a) Calculate the mean, median, and 10 percent trimmed mean for the homicide rate and for urbanization.

b) In a table, display the deviation from the mean, the squared deviation from the mean, the deviation from the median and the absolute value of the deviation from the median of the homicide rate for each state.

Table 4E.1   Homicide rate and percent urban for a random sample of 20 US states

State code	State name	Homicide rate	Percent urban 2000
1	VT	Vermont	2.3	38.2
2	MS	Mississippi	9.3	48.8
3	SD	South Dakota	1.3	51.9
4	NH	New Hampshire	1.4	59.3
5	MN	Montana	3.3	54.1
6	ND	North Dakota	1.9	55.9
7	AR	Arkansas	6.4	52.5
8	SC	South Carolina	7.2	60.5
9	AL	Alabama	6.6	55.4
10	IN	Indiana	5.5	70.8
11	WI	Wisconsin	3.3	68.3
12	NE	Nebraska	3.2	69.8
13	OK	Oklahoma	5.9	65.3
14	LA	Louisiana	13.0	72.6
15	KS	Kansas	4.5	71.4
16	MN	Minnesota	2.5	70.9
17	OR	Oregon	1.9	78.7
18	NM	New Mexico	6.0	75.0
19	OH	Ohio	4.6	77.4
20	FL	Florida	5.4	89.3

Data source: US Census Bureau, 2000, 2003, analyzed with SPSS.

c) Calculate the variance, standard deviation, MAD and IQR for the homicide rate.

d) Divide the data set into two groups of states – those with urbanization levels above and below the median.

e) For each group, display in a table the mean, median, 10 percent trimmed mean, variance, standard deviation, MAD and IQR of the homicide rate.

f) For each group plot a boxplot of the homicide rate. Place these boxplots on the same scale, side by side, to facilitate comparisons.

g) Using the table of descriptive statistics by level of urbanization and the boxplots, write a brief assessment of the plausibility of the theory that homicide rates vary with level of urbanization.

---

2. Table 4E.2 contains hypothetical data of confidence in the current leadership of a particular college for 115 students. Response options range from (1) "not at all confident" to (5) "very confident."

a) Calculate the mean, median, and mode.

b) Is there any reason to calculate a trimmed mean? Why or why not? In what instances would it be useful to calculate a trimmed mean?

Table 4E.2   Confidence in college leadership for 115 students

Response	Value	Frequency (N)
Not at all confident	1	25
A little confident	2	15
Somewhat confident	3	30
Confident	4	40
Very confident	5	5

---

3. A boxplot showing the distribution of expected retirement age among a sample of 279 individuals is presented in Figure 4E.1. A circle indicates an outlier, and an asterix indicates an extreme outlier.

a) What is the approximate median age at which people expect to retire?

b) The middle 50 percent of cases fall between what ages? How did you determine this? What is this range called?

c) Do the data appear to be highly skewed?

d) Are there any outlying cases? If there are outliers, are they young or old ages?

e) Based on the boxplot, would you conclude that there is considerable variation or little variation in when people expect to retire? Be sure to provide support for your answer.

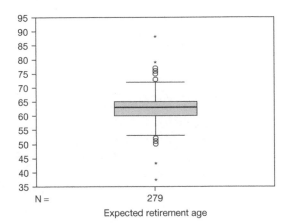

Figure 4E.1   Boxplot of the distribution of expected retirement age among a sample of 279 individuals

4. An employee of a local organization becomes concerned about the number of hours he and his colleagues are working. He asks 15 of his coworkers, who are paid full-time salaries, the average number of hours they have been working each week over the past three months. The hours are listed below. If you were this employee, which statistic (mean, median, etc.) would you report to your employer to emphasize your complaint about long work hours, and why did you select this statistic over others?

Hours: 22, 35, 42, 45, 55, 65, 65, 67, 68, 69, 70, 70, 70, 73, 75

5. Some demographic shifts have been occurring across the world. Table 4E.3 presents data on the percentage of the total population that is under age 15 for 24 countries on three continents: Africa, Europe and South America. To compare the three continents on their youthful populations, calculate: a) the mean; b) median; c) mode; d) variance; and e) standard deviation for each of the three continents, *as well as* for the three continents combined. After calculating this information, summarize the demographic differences you have found.

Table 4E.3 Percentage of the population under age 15 for 24 countries in Africa, Europe, and South America

	% *population* *under* *age 15*		% *population* *under* *age 15*
Africa		*Europe*	
Côte d'Ivoire	41	Austria	16
Ethiopia	46	France	19
Kenya	41	Germany	15
Liberia	47	Hungary	16
Rwanda	45	Italy	14
Somalia	48	Romania	17
Uganda	50	Spain	14
Zimbabwe	43	Switzerland	16
South America			
Argentina	27		
Brazil	28		
Chile	27		
Columbia	32		
Ecuador	33		
Peru	33		
Uruguay	24		
Venezuela	33		

Data source: The Statistics Division and Population Division of the United Nations Secretariat, 2003, website: http://unstats.un.org.

6. Tables 4E.4 and 4E.5 are two frequency tables of the perceived ability to balance work and family. The first table contains the frequency table as it originally appeared in the United States General Social Survey (GSS) dataset and corresponding statistics. The second table is a revised frequency table, with "missing" data recoded to no longer be "valid" values. Summarize how the improper coding of missing data as valid values in this example affects the *descriptive statistics* and consequent interpretation of findings.

Table 4E.4 Original frequency table for perceived ability to balance work and family (BALWKFAM How successful balancing work & family)

			Frequency	Percent	Valid percent	Cumulative percent
Valid	1	Not at all successful	22	0.8	2.2	2.2
	2	Not very successful	112	3.9	11.4	13.6
	3	Somewhat successful	479	16.5	48.7	62.4
	4	Very successful	310	10.7	31.5	93.9
	5	Completely successful	50	1.7	5.1	99.0
	9	NA	10	0.3	1.0	100.0
		Total	983	33.8	100.0	
Missing	8	Don't Know	11	0.4		
		System	1,910	65.8		
		Total	1,921	66.2		
Total			2,904	100.0		

Mean = 3.32; Median = 3.00; Mode = 3; Standard deviation = 0.99; Variance = 0.99
Data Source: US GSS, 1996.

Table 4E.5 Revised frequency table for perceived ability to balance work and family (BALWKFAM How successful balancing work & family)

			Frequency	Percent	Valid percent	Cumulative percent
Valid	1	Not at all successful	22	0.8	2.3	2.3
	2	Not very successful	112	3.9	11.5	13.8
	3	Somewhat successful	479	16.5	49.2	63.0
	4	Very successful	310	10.7	31.9	94.9
	5	Completely successful	50	1.7	5.1	100.0
		Total	973	33.5	100.0	
Missing	8	Don't Know	11	0.4		
	9	NA	10	0.3		
		System	1,910	65.8		
		Total	1,931	66.5		
Total			2,904	100.0		

Mean = 3.26; Median = 3.00; Mode = 3; Standard deviation = 0.81; Variance = 0.66
Data source: US GSS, 1996.

References

Bennett, D. J. 1998. *Randomness*. Cambridge, MA: Harvard University Press.

Demographic and Health Surveys (DHS). 2000. *Ugandan-DHS 2000 Survey*. Calverton, MD (www.measuredhs.com).

Frigge, M., Hoaglin, D. C., and Iglewicz, B. 1989. Some implementations of the boxplot. *The American Statistician* 53, 50–4.

Gaither, C. C. and Cavazos-Gaither, A. E. 1996. *Statistically Speaking: A Dictionary of Quotations*. Bristol, England: Institute of Physics Publishing.

Gould, S. J. 1996. *The Mismeasure of Man*. New York: W.W. Norton.

Hoaglin, D. C., Iglewicz, B., and Tukey, J. W. 1986. Performance of some resistant rules for outlier labeling. *Journal of the American Statistical Association* 81, 991–9.

International Social Survey Programme (ISSP). 2000. 2000 Environment II data set. www.issp.org. Catalog no. ZA 3440. Cologne, Germany: GESIS-ZA Central Archive for Empirical Research.

Kaufmann, D., Kraay, A., and Zoido-Lobaton, P. 2002, Jan. *Governance Matters II: Updated Indicators for 2000/01*. Policy Research Working Paper no 2772. The World Bank Research Development Group and World Bank Institute; Governance, Regulation and Finance Division (http://hdr.undp.org/reports/global/2002/en/).

Roberts, J. T., Parks, B. C., and Vásquez, A. A. 2004. Who ratifies environmental treaties and why? Institutionalism, structuralism and participation of 192 nations in 22 treaties. *Global Environmental Politics* 4(3), 22–64.

StataCorp. 2008. *Stata Statistical Software: Release 10.0*. College Station, TX: Stata Corporation.

Stigler, S. M. 1986. *The History of Statistics*. Cambridge, MA: Belknap.

Tukey, J. W. 1977. *Exploratory Data Analysis*. Reading, MA: Addison-Wesley.

United Nations Development Programme (UNDP). 2002. Human Development Report. Deepening Democracy in a Fragmented World. New York: Oxford University Press (http://hdr.undp.org/en/reports/global/hdr2002/).

US Census Bureau 2000. Table 33. Urban and rural population, and by state: 1990 and 2000. (www.census.gov/prod/cen2000/index.html).

US Census Bureau 2002. Historical poverty tables: Table 21. Number of poor and poverty rate, by state: 1980 to 2006. Year 2002 (http://www.census.gov/hhes/www/poverty/histpov/hstpov21.html).

US Census Bureau 2003. Table 295. Crime rates by state, 2002 and 2003, and by type, 2003 (http://www.census.gov/prod/2005pubs/06statab/law.pdf).

US General Social Survey (GSS). 1996. (http://www.norc.org/homepage.htm).

von Hippel, P. T. 2005. Mean, median, and skew: Correcting a textbook rule. *Journal of Statistics Education* 13 (www.amstat.org/publications/jse/v13n2/vonhippel.html).

CHAPTER 5
PLOTTING RELATIONSHIPS AND CONDITIONAL DISTRIBUTIONS

Outline

So far, we have looked at graphs that allow us to see the variability in a batch of data and learned how to calculate numbers, such as the variance or the MAD, that allow us to quantify the amount of variability. These are a prelude to our goal – explaining variability. We want to understand the variability in one variable (the dependent variable) as a function of the variability in other variables (the independent variables) that may be causes of the dependent variable. In the next chapter we will elaborate on how to think about causal models. But in this chapter we will introduce the basic tools for looking at how an independent variable influences a dependent variable.

Recall that our basic model of the relationship between an independent variable X and a dependent variable Y is:

$$Y = f(X) + E \qquad\qquad (5.1)$$

To see if Y really changes with differing values of X, we will want to develop plots and summaries that show the values of Y for different values of X. These are plots of what are called conditional values. The term **conditional value** means that the value of Y is conditioned by (i.e., influenced by) values of X. As with univariate graphs and summaries, there are several ways we can do this. We've already seen some of them by comparing univariate plots of a dependent variable, such as the boxplot, for groups of cases that have different values on the independent variable. We'll start with the most familiar graph of conditional values, the **scatterplot**, then move on to others.

Scatterplot

The scatterplot (or X–Y plot, or scattergram) is a plot that displays the joint distribution of two variables. While the stem-and-leaf and boxplot are univariate (one variable) graphs, the scatterplot is bivariate (two variables). As early as 1500, Leonardo da Vinci was using a plot with X and Y coordinates to understand better the velocity of moving objects. About 150 years later René Descartes began graphing mathematical functions of X–Y coordinates – the kind of plotting that is done in algebra. By the late 1600s, Edmund Halley had made bivariate scatterplots of the relationship between barometric pressure and altitude. So the scatterplot is one of the oldest graphical tools for understanding data. But it is still a very useful one, and new ideas about how to create scatterplots are still appearing in the literature (Cleveland and McGill, 1984).

The basic scatterplot consists of two axes, one for the dependent variable, usually indicated by Y. This is by tradition the vertical axis. The horizontal axis is for the independent variable, which is usually called X. The scales are continuous like the axis in a boxplot, not bins of the kind we use when we make stem and leaf diagrams. Following the suggestions of William Cleveland (1993; 1994), it is good practice to draw a top line parallel to the horizontal axis and a right line parallel to the vertical axis so that the whole scatterplot is enclosed in a box. This keeps the eye focused on the area in which the data are plotted.

Choosing scales for a scatterplot can be tricky. The basic principle is that the data points should fill the plot, leaving as little blank space in the plotting area as possible. To be sure that data points don't fall on the axes, which makes them hard to read, the scale on each axis should start a little below the lowest value in the data being plotted and end a little above the highest value. It is also conventional to make the labeled tick marks numbers that are familiar, such as numbers that end in 0, 1, 2 or 5. It is important, however, not to extend the scales too far beyond the range of the actual data. If the axes go beyond the data, there will be a tendency to think that the pattern in the data can be extended to areas where we have no data and thus no information. In his famous book, *How to Lie with Statistics*, Darrell Huff (1954) argues that starting the axes of scatterplots and other similar graphs at some value other than zero can be deceptive. This is true in the sense that any graph can be constructed to be deceptive. But, as William Cleveland (1994, pp. 92–3) notes, Huff assumes that readers won't look at tick marks and other information that labels the scale. If our data don't begin at zero, then there's usually no reason to make scales begin at zero. There can be exceptions to this when there's a special reason to compare the data to zero (or to any other specific number) but generally we want to set the scales so that the data fill the graph, and that means scales that go only a tiny amount above the range of the data.

The tick marks indicating points along the scales should be outside the plotting area, with nothing but data points inside. Labeling should be clear, letting the viewer know where each of the two scales starts and ends. Figure 5.1 shows a scatterplot

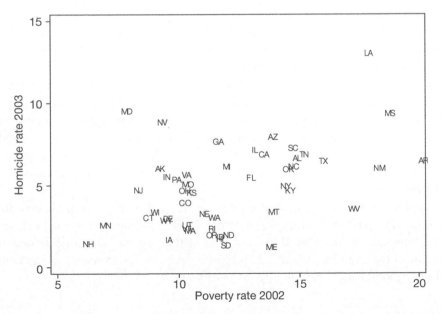

Figure 5.1 Scatterplot of the relationship between homicide rate and percent in poverty, N = 50
Data source: US Census Bureau, 2002, 2003, analyzed with Stata.

Box 5.1   Making a Scatterplot

1   Create a horizontal X axis that runs from a bit lower than the lowest X value to a bit higher than the highest X value. Place tick marks on the bottom of the axis at familiar values – those that end in 0, 1, 2, or 5.
2   Create a vertical Y axis that runs from a bit lower than the lowest Y value to a bit higher than the highest Y value. Place tick marks on the left side of the axis at familiar values – those that end in 0, 1, 2, or 5.
3   Draw the horizontal line at the top of the Y axis parallel to the X axis and the vertical line at the right side of the X axis parallel to the Y axis to close the box.
4   For each data point, move along the X axis until you find the place on the axis that matches its value. Then move up the Y axis to the places that matches its Y value. Place a plotting symbol at that point on the graph.
5   Add a title and source information to the graph.

---

we used in chapter 1. While most scatterplots use dots, small "Xs," little circles or other symbols for all the data points, it can sometimes be useful to use symbols that identify the data points. Here we use the two letter state codes developed by the US Postal Service.

The basic interpretation is straightforward. It appears that as poverty level increases, the homicide rate increases, but homicide rate is not perfectly predicted by poverty. This is the "scatter" in the scatterplot. If all the points fell exactly on a line – if the dependent variable were perfectly predictable by the dependent, then there would be no scatter. But here we see there is plenty of variability in homicide rates after we take account of the ability of poverty level to predict it.

There are ways to elaborate on the variability of the variables in a scatterplot. One common approach is to add a one-way scatterplot and boxplot to the top of the scatterplot to show the distribution of the independent variable and a one-way scatterplot and boxplot to the right hand side to show the distribution of the dependent variable. Figure 5.2 illustrates this. Here the statistical package Stata did not use whole numbers for the tick marks but instead shows the highest and lowest values in the data.

This plot emphasizes the limited number of states at high levels of poverty and homicide. If we look to the upper right corner, we see the dot representing Louisiana with a homicide rate of 13.0. If we then look at the boxplot for homicide rate on the right of the graph, we see that Louisiana is an outlier with regard to the fence of the boxplot for its homicide rate (but not its poverty level).

Scatterplot matrices are discussed at the end of this chapter as an Advanced Topic.

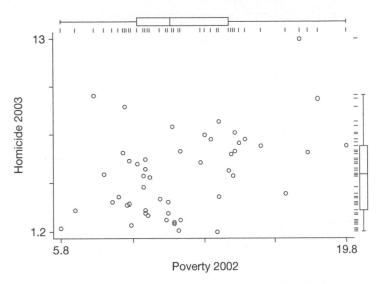

Figure 5.2   Scatterplot of homicide rate with poverty level with added one-way scatterplots and boxplots, N = 50
Data source: US Census Bureau, 2002, 2003, analyzed with Stata.

Another way of elaborating on the scatterplot information is to add some summary statistics for groups of variables. In the applications section of the last chapter we divided states into two groups based on the median of the independent variable, then examined measures of central tendency and variability for each of the two resulting groups. When we have enough data, we can divide the independent variable into more than two categories to get a better sense of patterns. We have sorted the states in order from those with the lowest poverty levels to those with the highest poverty levels. Then we broke the states into 10 groups of four to six states each (groups have different sizes since some states have the same poverty rates), from lowest poverty level to highest poverty level. Then for each group of states, we have calculated the mean for that group.

Table 5.1 displays the data. We can now add these mean homicide rates to the graph – we will be showing the mean homicide for the lowest five states in poverty level,

Table 5.1   Fifty states formed into ten groups from lowest to highest poverty level

Group (by poverty level)	1-Lowest poverty levels	2	3	4	5	6	7	8	9	10-Highest poverty levels
Mean homicide rate	4.22	4.88	3.85	3.52	2.98	3.40	5.13	5.32	6.62	7.64

N = 50, 4–6 per group
Data source: US Census Bureau, 2003, analyzed with SPSS.

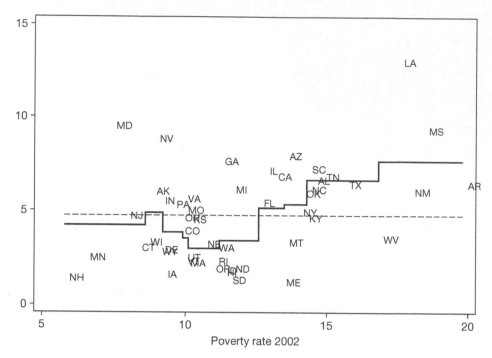

Figure 5.3   Scatterplot of homicide rate by poverty level, with group means and overall mean, N = 50; 4−6 per group; dashed line is overall mean
Data source: US Census Bureau, 2002, 2003, analyzed with Stata.

the mean homicide rate for the next lowest five states and so on. We are showing how the mean for homicide rate varies with the poverty level. The results are in Figure 5.3. We have also added a dashed line to represent the mean across all states, at 4.74.

There is always a tension between the number of cases we have to work with and how many categories we can use in looking at patterns in the data. Here we have divided 50 data points into ten groups of between four and six each. We would be reluctant to create more categories because we would have so few cases in each. We could create five groups of ten states each, but then we are lumping states into groups in ways that may hide patterns. But with computers it is possible to try alternative ways of looking at the data. Remember that while the data do not change, different ways of looking at it make clear different patterns in the data.

The graph is somewhat busy, but we see that the states below the mean homicide rate for all states tend to be clustered at the left side of the graph, indicating low poverty levels. We also see that the group means tend to get larger as we move from the left with the group with the lowest poverty levels, to the right, where the states have higher poverty levels. But the relationship is far from perfect.

Finally, we will remove the line for the mean of all states and add the straight line we used in chapter 1 that is a good fit to the data, based on methods we will learn later. This gives us Figure 5.4.

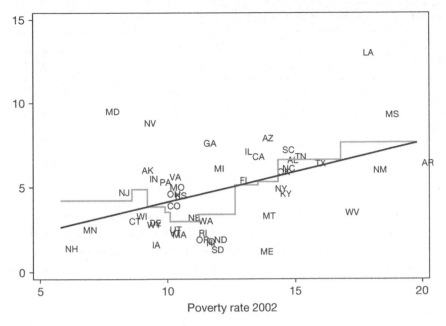

Figure 5.4 Scatterplot of homicide rate by poverty level, with group means and summary line, N = 50; 4–6 per group; straight line is the "best fitting" line
Data source: US Census Bureau, 2002, 2003, analyzed with Stata.

The summary line is:

$$f(X) = A + (B^*X) \tag{5.2}$$

f(X) is our prediction of the homicide rate, Y, using the independent variable poverty level. A good choice for A and B are A = 0.59 and B = 0.36. If we select values for X and then calculate f(X) using the equation and plot the f(X) points, we get the line we see on the graph. The line is the prediction of homicide rate based on the poverty level. We can see that it hits only a few states exactly; it is fairly far off for other states but seems to capture a general pattern of the data. If we compare it to the line of means for groups of states, it is sometimes above the mean and sometimes below it.

Table 5.2 should help you remember how we calculate the line. Here we have picked a few states at random just to illustrate the process. We take the poverty level, multiply it by 0.36 and add 0.59. This gives us the predicted homicide rate. If we plot a few of these predicted homicide rates, they will all lie on the straight line. The line is called a **"best-fit" line** because it comes as close to the data points as possible.

Ordinary least squares are discussed at the end of this chapter as an Advanced Topic.

Table 5.2   Calculating the summary line

State	Poverty level (X)	Homicide rate (Y)	Predicted homicide rate
Georgia	11.2	7.6	4.62
Idaho	11.3	1.8	4.66
Mississippi	18.4	9.3	7.21
North Carolina	14.3	6.1	5.74
Utah	9.9	2.5	4.15

One way to think about the prediction line is to imagine we had a huge number of cases, so that we could calculate means for many, many groups. Then we could plot the mean for each of those groups. As we do this for more and more groups, the summary line and the lines that show the means of each group will look more alike.

Smoothers are discussed at the end of this chapter as an Advanced Topic.

When we have large data sets with variables that have only a limited range of values, scatterplots are not as helpful in identifying the relationships between variables. Figure 5.5 shows an example where we plot the animal concern score against the age of respondents.

The problem here is that since there are many cases and only 5 values (1–5) for the animal concern scale, the dots overlap – each dot in the picture can represent

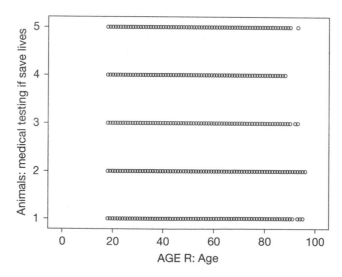

Figure 5.5   Scatterplot of animal concern score with age, N = 29,080
Data source: 2000 ISSP data set, analyzed with SPSS.

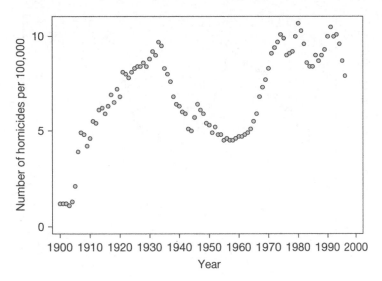

Figure 5.6   Homicide rate by year, N = 97
Data source: US Uniform Crime Reports, analyzed with Stata.

a different number of cases. In the next section we will show a better way to plot this data.

Time Series Graphs

Time series graphs are a special kind of scatterplot in which the independent variable plotted on the horizontal axis is time. Figure 5.6 is a time series plot of the homicide rate from 1900 to 2000.

We can see that the homicide rate has varied considerably over time, reaching highs in the Great Depression (1929 until the late 1930s), dropping during the mid 1950s through the mid-1960s, then rising again.

In an ordinary scatterplot, we never "connect the dots" because we can have multiple values of one variable for a single value of the other variable. This means there is no way to order the points. But in a time series plot, there is a natural ordering with one value of the dependent variable for each date. So it is often helpful to have lines linking the data points.

Figure 5.7 shows the same data with lines connecting the dots. This makes the swings in homicide rate easier for the eye to follow.

It is common to graph two time series variables on the same graph. Figure 5.8 does this for the homicide rate and the unemployment rate, from 1948 to 2000. We have let the Stata statistical package create the axis tick marks. Note that homicide rates are per 100,000, while unemployment rates are the percent of the active labor

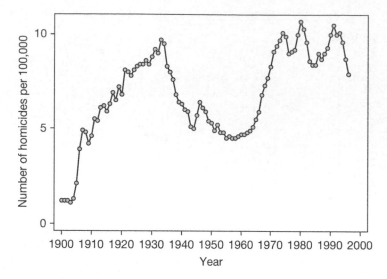

Figure 5.7   Homicide rate by year, points connected, N = 97
Data source: US Uniform Crime Reports, analyzed with Stata.

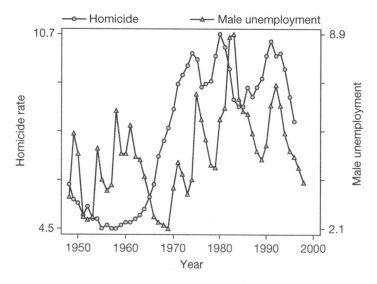

Figure 5.8   Time series graph of the homicide rate and the unemployment rate, N = 51
Data source: US Uniform Crime Reports, 2005, US Bureau of Labor Statistics, 2005, analyzed with Stata.

force not employed. The two variables are thus not measured on the same scales. Often the purpose of this kind of graph is to see if the two variables are moving in tandem, which might be evidence that unemployment leads to homicide. But patterns of this sort are often difficult to see in time series graphs of several variables.

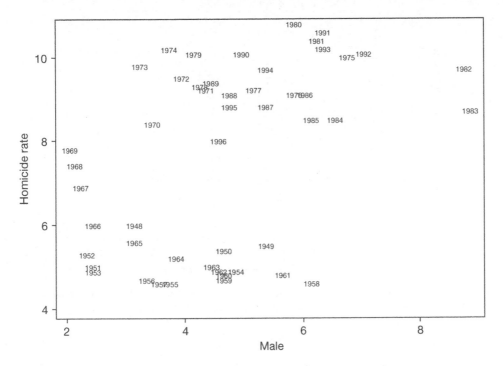

Figure 5.9   Scatterplot of homicide rate versus unemployment rate, N = 51
Data source: US Bureau of Labor Statistics, 2005, US Uniform Crime Reports, 2005, analyzed with Stata.

If we are interested in the relationship between homicide and unemployment, a simple scatterplot of the two variables might be preferable. Figure 5.9 provides such a plot.

It appears that low unemployment years do tend to have low homicide rates, while many of the years with the highest homicide rates also coincide with high unemployment rates. To our eyes, it is much easier to see this pattern in the scatterplot of the two variables than in the plot in which the time series of homicide rates and unemployment rates were both plotted against time. Time series plots are common and useful but may not be the best way to see the relationship between variables.

Bivariate Plot for Categorical or Grouped Independent Variables

When the independent variable is nominal and the dependent variable is continuous, we can plot histograms, stem and leaf diagrams or boxplots for each category of the independent variable. And we can always break a continuous variable into categories. This is what we did with the scatterplot that included the mean

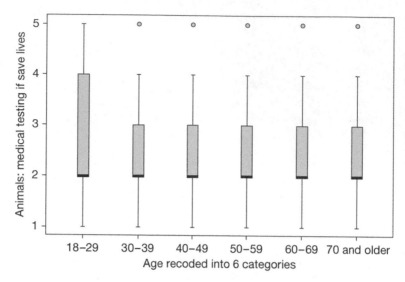

Figure 5.10   Boxplots of animal concern scores by age group, N = 29,193; N for age groups: 18–29 = 5,678; 30–39 = 5,984; 40–49 = 5,716; 50–59 = 4,822; 60–69 = 3,855; 70 and older = 3,138
Data source: 2000 ISSP data set, analyzed with SPSS.

homicide rate for groups of states, where we created the groups based on values of the poverty level.

Figure 5.10 shows boxplots of animal concern for age groups. Let's focus on the medians, the thick line in the boxes. The medians for the six age groups are all 2. This can be hard to see because the lower quartiles are also 2. The box and whiskers for five of the six age groups are identical, making it difficult to identify whether there is a relationship between age and animal concern. The box for the 18–29 year old age group is wider (ranging from scores of 2 to 4), suggesting more young people are pro-animal. But age is clearly not the only thing causing animal concern to vary.

We need to be careful showing descriptive statistics on a boxplot. Different groups may have different numbers of cases. This is true of Figure 5.10. We have 5,984 people in the 30–39 age group but only 3,138 people in the 70 and over age group. We have indicated this in the note at the bottom of the figure. All other things being equal, we would have more faith in results based on a larger rather than a smaller number of observations.

Remember that in making these plots, we have collapsed a continuous variable, age, into groups. In the process, we "lose" the information on exact age. This is often helpful in making plots, but it's possible that we are missing some subtle pattern. Graphing is always a compromise between a summary that may help us see things and the loss of information that comes with the simplification in the graph. But in the case of a large data set, like the ISSP, a scatterplot or other graphic that displays more of the detail of the data can be very unclear because so many data points can overlap.

A Historical Example

Good graphic display can be tremendously important. Edward Tufte (1997, pp. 38–52), in his book, *Visual Explanations*, describes how poor use of graphics to display data contributed to the Challenger tragedy. On January 28, 1986, the US space shuttle Challenger exploded soon after launch and seven astronauts were killed. The problem was that an "O-ring," a kind of gasket, on the solid fuel boosters became brittle because they were too cold. This caused them to leak and led to the explosion that destroyed the Challenger. It took many months and a great deal of work on the part of the famous physicist Richard Feynman to convince everyone that the O-rings were the cause of the disaster. But good graphics make it evident that the O-rings might have been a problem.

The engineers at Morton Thiokol, the company that built the boosters, were concerned that the shuttle was to be launched on a very cold day for Florida (in the low 30°s F). The engineers prepared a series of graphs of the data they had on how the O-rings behaved at various temperatures and tried to use that to convince NASA to cancel the launch. But NASA was reluctant to postpone the launch because the White House wanted Ronald Reagan, who was then President, to be able to talk with Christa McAuliffe (the teacher who was on the Challenger) while she was in space, during his "State of the Union" address the night after the launch.

On the night before the launch, the engineers faxed a series of tables and charts to NASA about the potential problem with the O-rings, but NASA was not convinced. Tufte shows a number of these faxes in his book, and it is clear that it is very hard to understand the relationship between temperature and damage to the O-rings from them, although the engineers had understood the problem. Politics seemed to override the concerns of the engineers.

Figure 5.11 is a chart prepared by Morton Thiokol for the Presidential Commission that investigated the Challenger disaster. This is much clearer than the graphs the engineers were struggling with the night before the launch, but it is still hard to see the relationship between temperature and damage to the O-rings.

Table 5.3 displays the same data. The damage index was based on examining O-rings after launch, the "?" represents missing data from a launch where the boosters were lost at sea and couldn't be examined for damage. The problem becomes a bit clearer here.

Figure 5.12 is a simple scatterplot of the data. Remember that the usual rule of scaling the axes says to only go a little below the lowest value and a little above the highest. But here it is necessary to extrapolate from the lowest temperature for which data were available, 53 degrees, to the temperature expected at the Challenger launch, somewhere between 26 and 29 degrees (marked with the vertical line in the scatterplot).

It is fairly easy to see that there is a negative relationship between damage and temperature. It is also easy to see that deciding to launch at a temperature in the high 20s was extrapolating far beyond the available data. And it is easy to see what a great risk was being taken – one that led to the tragedy.

History of O-Ring Damage in Field Joints (Cont)

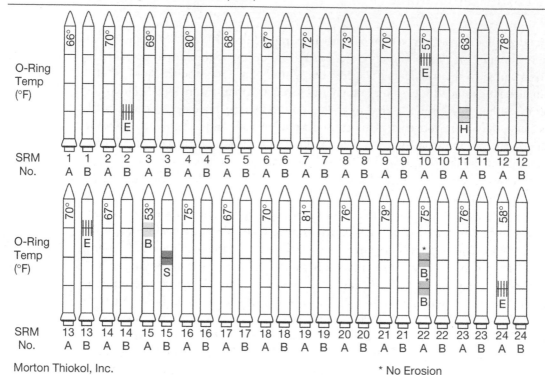

Morton Thiokol, Inc.   * No Erosion

Figure 5.11   History of O-ring damage
Source: Tufte (1997, p. 47).

Table 5.3   Temperature and damage data for 24 shuttle launches

Temperature	Damage	Temperature	Damage
53	11	70	4
57	4	70	0
58	4	72	0
63	2	73	0
66	0	75	0
67	0	75	4
67	0	76	0
67	0	76	0
68	0	78	0
69	0	79	0
70	4	80	?
70	0	81	0

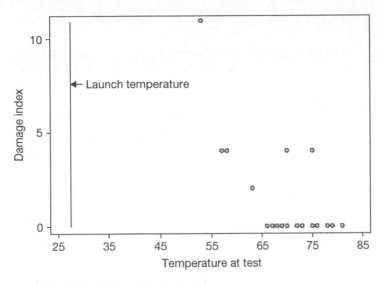

Figure 5.12   Scatterplot of damage index and temperature

What Have We Learned?

The goal in most applications of statistics in the social sciences is to understand how one variable affects another. By looking at the values of the dependent variable that correspond to various values of the independent variable, we can assess whether the independent actually does predict the dependent. We can thus develop graphs or tables that correspond to our theoretical models, and use the graphs and tables to assess how well the models fit.

We have seen several ways to do this. One of the most common is the scatterplot. We can enhance the scatterplot by adding boxplots to show the distribution of each variable. We can use plotting symbols that identify individual cases or groups of cases. We can add lines indicating summary statistics, such as the overall mean or the mean of subgroups. And we can add a line summarizing the data, though the methods for finding that line won't be discussed until later in the book.

Time series plots are a special case of a scatterplot in which time is the independent variable. These plots are very common, but they can be confusing if several variables are plotted at once. Time series plots are usually not as helpful as scatterplots in looking at the relationship between a dependent and independent variable.

When we have categorical independent variables and continuous dependent variables, we can also plot summary graphs, such as a boxplot, for each category of the independent variable to see how values of the dependent variable change across categories of the independent variable. And of course we can also create a categorical variable from a continuous variable, so we can use the same approach to plotting the distribution of a continuous dependent variable against an independent variable.

Advanced Topic 5.1   Scatterplot Matrices

Scatterplots are sometimes stacked in what is called a **scatterplot matrix** to show the inter-relations among several variables. Figure 5.13 is a matrix showing homicide rates, poverty levels, and percent of the population over 15 who are single. It might be argued that single people are more likely to engage in violent crime. Or it may be that domestic violence leads to homicide. In either case we might find a relationship between homicide and the percent single population. We have created this variable by adding together the percent divorced and the percent over 15 never married.

The scatterplot matrix simply displays all the possible scatterplots among all the variables being used. Our theory assumes that homicide is the dependent variable while percent single and percent in poverty are independent variables. This means we would focus on graphs where the homicide rate is on the vertical axis, since by convention this is where we put the dependent variable. (To make the graphs less complicated, we have plotted dots rather than the state abbreviation, but we could use the abbreviations if we liked.)

The scatterplot matrix shows the two graphs of homicide versus percent poverty (top center graph – the same graph we have already seen several times) and homicide rate versus percent single (top right graph). It also gives us the graph of percent in poverty versus percent single (bottom center). This could be used to see how closely related the two independent variables are to each other. On the middle left, bottom left and middle right, the matrix gives us the scatterplots with the dependent and independent variables reversed. Usually we are not interested in these, but the scatterplot matrix command gives them to us just in case we might be interested.

Looking at the top row, which are the graphs with homicide rate on the vertical axis, we

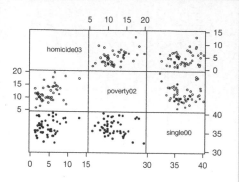

Figure 5.13   Scatterplot matrix of homicide rate, percent single population, and percent in poverty, N = 50
Data source: US Census Bureau, analyzed with Stata.

get a sense that homicide rate is more closely related to poverty level than to percent single. At least to our eyes, it appears that there is less "scatter" in the graph of homicide rate versus percent in poverty than in the graph of homicide rate versus percent single. So the poverty rate seems a better predictor of homicide rate. Looking at the scatterplot on the right center (or the scatterplot on the bottom center) we don't see a strong relationship between poverty and percent single. As we will see later, if two independent variables are highly correlated, it can be hard to figure out the effects of each on the dependent variable. But that is not the case here.

It is harder to follow the scales on the scatterplot matrix than on a single scatterplot. The scatterplot matrix is intended to give a quick look at several variables at a time, while single scatterplots, augmented by one-way scatterplots and boxplots, are better for focusing on the relationship between two variables.

Advanced Topic 5.2   Ordinary Least Squares

Just as there is more than one way to measure the center of the data – we can use means or median or trimmed means – there is more than one way to define a "best fit." Here we use the most common criteria, which we will discuss in detail in the chapter on bivariate regression. It is called **ordinary least squares**. We find the line to make as small as possible the sum of the squared errors. That is, we can pick A and B and that defines a line. Then we can calculate the predicted values of Y. Then we can subtract to get the error – the amount by which the predicted value of Y misses the actual value of Y.

Some of these errors will be positive, some negative. To get rid of the negatives, we can square the errors – the same logic used in the mean. Then we add together all the squared errors to get the sum of the squared errors. We try to find A and B to make the sum of the squared errors as small as possible. The mathematics works out so we don't actually have to guess; rather we have equations to tell us what A and B should be to give us the "best fit." If this sounds a bit like our earlier discussion of deviations from the mean, it should. The line we draw is in the mean family.

Advanced Topic 5.3   Smoothers

The values for A and B in the best fit line we displayed in Figure 5.4 were calculated by taking all the data we are using into account. This is because it tries to find the predictions for Y that make the sum of the squared errors as small as possible, and in doing so takes into account the error in predicting every value of Y in the data set. So the line is a summary of the data based on the pattern of all the data points.

The traditional scatterplot without any lines provides a summary of the data where only the individual values are taken into account in presenting the data. We plot each point based on its value of X and Y for one observation, paying no attention to the other observations when we place points on the graph.

Statisticians have developed a number of methods for summarizing data that are compromises between plotting one point at a time and plotting lines based on all the data points in a data set. The plot of the mean of five points at a time is one way to do this, but it has the disadvantage that the line that results jumps abruptly when we move from one set of five points to the next but doesn't change at all within a set of five values on X.

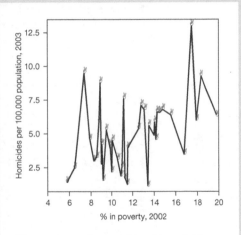

Figure 5.14   Scatterplot of homicide rate and poverty level with points connected

In Figure 5.14, for purposes of illustration, we simply connect the data points in order from the lowest value on poverty level to the highest value. This is a bad idea, as the line is very jumpy and distracts us from any pattern in the data more than it reveals the pattern. (We'd also have trouble if two data points had exactly the same value on X, as this would mean we wouldn't know in what order we should connect the points.)

We'd like a summary line that is "smoother" than the line that just connects the points but more "local" – in the sense that it doesn't try to summarize all the data at once – than the best fit line. One way to approach this is to calculate, for each value of X, a "local" summary statistic for Y. That is, for each value of X we create a small "window" and summarize the values of Y in that window (you might think of the window as a neighborhood around a particular X value). We will create such a "window" around each state. Table 5.4 shows the results of such a cal-

culation. (Note that we only show two decimal places for the poverty and homicide rates, so it appears that some states are "tied" in the poverty rate when with more decimal places they are ordered.)

First we place the states in order from lowest to highest based on the value of their poverty level. Then we take the homicide rate for the state we focus on itself, the state just below it, and the state just above it (that is 3 data points, called a window of size 3). For the lowest state on the poverty level, New Hampshire, we can't do the calculation because there is no state below it (the same thing will happen with Arkansas, the highest state because there's nothing above it). But for Minnesota we will add

Table 5.4 Calculations of local summary statistic, or a smoothed value

State	Poverty level	Homicide rate	Mean homicide rate of window of size 3
New Hampshire	5.8	1.4	–
Minnesota	6.5	2.5	4.47
Maryland	7.4	9.5	5.57
New Jersey	7.9	4.7	5.73
Connecticut	8.3	3.0	3.67
Wisconsin	8.6	3.3	4.10
Alaska	8.8	6.0	6.03
Nevada	8.9	8.8	5.87
Wyoming	9.0	2.8	5.70
Indiana	9.1	5.5	3.73
Delaware	9.1	2.9	3.33
Iowa	9.2	1.6	3.27
Pennsylvania	9.5	5.3	3.60
Colorado	9.8	3.9	4.60
Ohio	9.8	4.6	4.50
Missouri	9.9	5.0	4.03
Utah	9.9	2.5	4.37
Virginia	9.9	5.6	3.47
Vermont	9.9	2.3	3.37
Massachusetts	10.0	2.2	3.00
Kansas	10.1	4.5	3.30
Nebraska	10.6	3.2	3.20
Oregon	10.9	1.9	2.70
Washington	11.0	3.0	2.40
Rhode Island	11.0	2.3	4.30
Georgia	11.2	7.6	3.87
Hawaii	11.3	1.7	3.70
Idaho	11.3	1.8	1.60
South Dakota	11.5	1.3	1.67
North Dakota	11.6	1.9	3.1
Michigan	11.6	6.1	4.47
Florida	12.6	5.4	6.20
Illinois	12.8	7.1	6.43
California	13.1	6.8	5.03
Maine	13.4	1.2	5.30
Arizona	13.5	7.9	4.13
Montana	13.5	3.3	5.37
New York	14.0	4.9	4.70
Oklahoma	14.1	5.9	5.13
Kentucky	14.2	4.6	5.90
South Carolina	14.3	7.2	5.97
North Carolina	14.3	6.1	6.63
Alabama	14.5	6.6	6.50
Tennessee	14.8	6.8	6.60
Texas	15.6	6.4	5.57
West Virginia	16.8	3.5	7.63
Louisiana	17.5	13.0	7.50
New Mexico	17.9	6.0	9.43
Mississippi	18.4	9.3	7.23
Arkansas	19.8	6.4	–

together 1.4 for New Hampshire, 2.5 for Minnesota, and 9.5 for Maryland to get a sum of 13.4. Then dividing by 3 we get 4.47.

This is a "smoothed" value that takes into account the values of the two adjacent states, once the states are ordered on the independent variable. We have used the mean as the summary statistic and a "window" of size 3 to get the smoothed value. Figure 5.15 plots this smoothed value along with the best fit line.

There are many other choices. We could have plotted the median for each window, for example. Or we could weight the neighboring cases' values on the Y variable less than the Y value for the case on which we are centered. Or we can use a different size window. In Figure 5.16, we have used a window of size 9 for smoothing but again using the mean as our measure of location. So we take the mean of the four cases below the one we are looking at, the case itself, and the four above.

Here we don't have values of the smoother for the four highest and lowest cases on poverty level. This line is less "local" than the line with a window of 3, and as a result is a bit smoother.

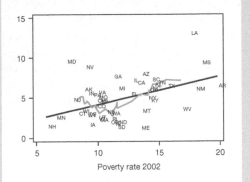

Figure 5.16 Scatterplot of homicide rate versus poverty level with best fit line and line of smoothed mean, window = 9, N = 50
Data source: US Census Bureau, 2002, 2003, analyzed with Stata, smoothed mean with window of 9.

Both the line based on a window of 3 and that based on a window of 9 show the general relationship between poverty level and homicide rate – states with high poverty levels tend to have high homicide rates, but there is substantial variation around that pattern. This is because the smoother lines give influence to data points in their "neighborhood" (ones that have nearby values for the poverty rate) but no influence to more distant data points (those with values for the poverty rate further away). In contrast the overall summary line tries to show the pattern in all the data, not just what is in the neighborhood.

The idea of a smoothed value is straightforward. We choose a window size, a summary measure, and the weight to assign to the data points in the window we have chosen. In the statistical literature, discussions of smoothers can get quite complicated because of elaborate weighting schemes, methods for handling points at the top and bottom of the data set for which the calculation can't be applied directly, and complex summary statistics. And we can smooth the smoothed values by applying the smoother again. For example, we could apply the window

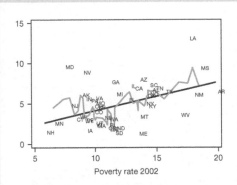

Figure 5.15 Scatterplot of homicide rate versus poverty level with best fit line and line of smoothed mean, window = 3, N = 50
Data source: US Census Bureau, 2002, 2003, analyzed with Stata, smoothed mean with window of 3.

of 3 mean smoother to the last column in Table 5.4 to get double smoothing.

In dealing with these complications, the statistical literature has produced a number of smoothers that work better than the two simple mean-based smoothers we have used. But the principles are the same. We want to have a value that is not as "local" as a single data point's value, but not so "global" as the best fit line based on all the data points. There is no single correct way to build such a set of values for graphing, but many choices that allow us to graph a summary line that helps identify the patterns in the data.

Applications

Example 1:   Why do homicide rates vary?

Here we examine the relationship between homicide and some state level variables that we haven't used before. Figure 5A.1 shows a scatterplot of the relationship between homicide and divorce per 100,000 population. The relationship is positive – as the divorce rate increases, the homicide rate increases. But of course, there is still much

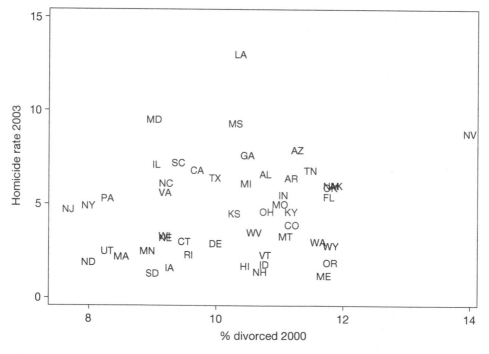

Figure 5A.1   Scatterplot of the relationship between homicide rate and divorce rate, N = 50
Data source: US Census Bureau, analyzed with Stata.

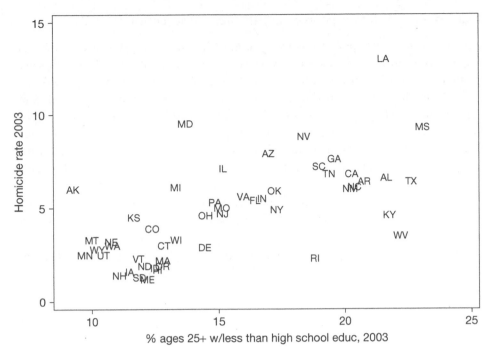

Figure 5A.2   Scatterplot of the relationship between homicide rate and percent of persons who left high school without graduating, N = 50
Data source: US Census Bureau, analyzed with Stata.

variability in homicide that is not accounted for by divorce. The wide scatter of the data points is an indication that there is only a moderate relationship between the two variables (remember that a perfect relationship between homicide and divorce would give us a straight line plot of data points). And if Nevada with its very high divorce rate and relatively high homicide rate weren't in the data set, we might not see a pattern at all.

Figure 5A.2 is a scatterplot of the relationship between state homicide rate and percentage of people who did not graduate from high school. Here the relationship appears to be a bit stronger, with the data points clustered closer together than in Figure 5A.1. Once again we see a positive relationship, with increases in the percent leaving high school without graduating associated with increases in the homicide rate.

It appears that there is a reasonably strong relationship between the two variables, such that states with a high proportion of adults without a high school education also have relatively high homicide rates, while the lower homicide rates are mostly in states where relatively few people have less than a high school education. Does this mean that the less educated are those committing homicide? Remember that we should only draw conclusions at the same level as our unit of analysis. This data allows us to draw conclusions about states, not about people.

Example 2:   Why do people vary in their concern for animals?

Remember that scatterplots are not very useful when the variables take on a limited number of values and the dataset is very large. This is because the data points are plotted on top of each other. So boxplots are probably a better way to look for patterns in the animal concern data. We examined boxplots of age and animal concern in this chapter. Here we examine a different idea about why people might vary in their concern for animals. It may be that people who live in rural settings have different views about animals because they will have more contact with animals, more contact with farming and more contact with people who hunt and/or fish.

Figure 5A.3 is a boxplot of the animal concern question with the place where people live. Place of residence was divided into three categories: urban; rural; and suburb/city/town.

Here we find little support for the idea that size of place of residence influences concern for animals. The median and IQR for the three groups are identical at 2 and 2–3, respectively.

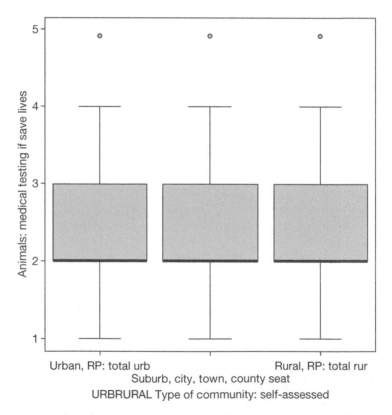

Figure 5A.3   Boxplot of animal concern question by size of place of residence, Urban N = 11,363; Suburb/town = 7,780; Rural = 8,116
Data source: 2000 ISSP data set, analyzed with SPSS.

Example 3:   Why do nations differ in environmental treaty participation?

It has been hypothesized that participation in environmental treaties may be influenced by the extent citizens have voice and governments are accountable to its citizens. Figure 5A.4 shows a scatterplot of environmental treaty participation versus the voice and accountability rate. (Here we have used the country abbreviations rather than the full country names.)

The scatterplot is a bit difficult to read because some of the country abbreviations overlap at various points. There appears to be a positive relationship between voice and accountability rate and environmental treaty participation. Countries with lower voice and accountability seem to participate in fewer environmental treaties, while countries with greater voice and accountability tend to have participated in treaties. This relationship is far from perfect though, with countries like Pakistan ("PAK") having little voice (about −1.3 on the graph) but relatively high (7) rates of environmental treaty participation and countries like Sao Tome and Principe ("STP") having considerable voice and accountability (1.1 on the graph) but low (no) treaty participation (0).

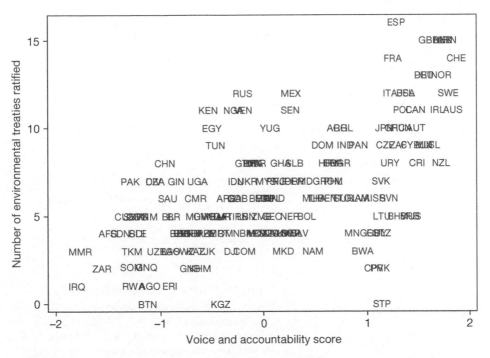

Figure 5A.4   Scatterplot of environmental treaty participation and voice and accountability

Data source: Kaufmann et al., 2002, Roberts et al., 2004, analyzed with Stata.

Example 4:   Why do people differ in their knowledge about how AIDS is transmitted?

The AIDS knowledge variable is nominal, with only two categories, "Yes" and "No." As a result, the graphic techniques of this chapter aren't really useful in analyzing it.

Exercises

1. Figure 5E.1 shows boxplots of the male and female illiteracy rates for 60 countries. Illiteracy rates are taken from the United Nations Educational, Scientific, and Cultural Organization (http://unstats.un.org/) and are typically from 2000 (some countries last reported rates a few years earlier). Illiteracy rates are based on the percent of males or females ages 15+ who cannot read or who cannot write.

a) What is the approximate median illiteracy rate for men and women?

b) The middle 50 percent of cases lie between what percentages for women? For men?

c) Based on the data for the 60 countries presented here, what conclusions can you make about illiteracy rates? And how do rates differ for men and women?

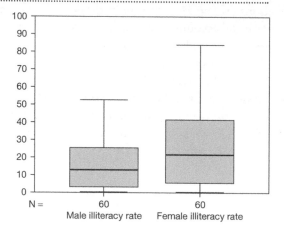

Figure 5E.1   Boxplot of illiteracy rate for 60 countries by gender

2. Figure 5E.2 is a time-series graph charting the savings behavior of Americans from 1929 to 2002. Savings behavior is measured as the average percent of disposable income that Americans save in a given year.

Provide a summary of the savings behaviors of Americans over this 74-year period. Be sure to reference specific years or periods of years and statistics in your answer.

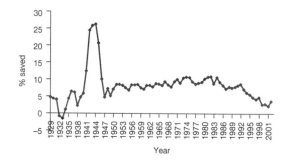

Figure 5E.2   Line graph of US saving as a percentage of disposable income (1929–2001)
Data source: National Income and Product Account information, a popular measure of savings, taken from the US Commerce Department Bureau of Economic Analysis; Table 2.1 – Personal income and its disposition (www.bea.doc.gov/bea/dn/nipaweb/Index.asp).

3. Figure 5E.3 is a scatterplot of the relationship between completed years of education and age at which one's first child was born. Data were taken from the 1996 US General Social Survey data file (number of cases = 2,039).

Does there appear to be a relationship between education and age of one's first child? Why or why not? If there is a relationship, does it appear to be strong? Does the distribution of data points appear at all unexpected or unusual to you? Explain.

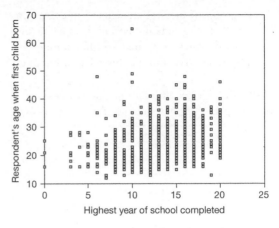

Figure 5E.3   Scatterplot of the relationship between completed years of education and age at which one's first child was born

4. Figure 5E.4 is a scatterplot of years of education by self-assessed social class (1 = lower class to 6 = upper class). The data were taken from the 2000 ISSP dataset and represent 131 individuals.

First, summarize the range of years of education. Does there appear to be a relationship between these two variables? How did you determine this?

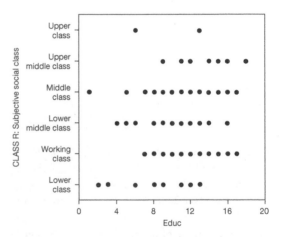

Figure 5E.4   Scatterplot of years of education and subjective social class

5. A group of medical students is interested in whether the infant mortality rate in the United States is related to the number of physicians. To examine this relationship, create a scatterplot using the data for the 25 states presented in Table 5E.1. Does there appear to be a relationship between infant mortality and doctors?

Table 5E.1 Table of infant mortality rate and doctors per 100,000 population for 25 US states

State	Doctors per 100,000	Infant mortality rate
Alaska	179	6.8
Arkansas	188	8.4
Colorado	234	6.2
Delaware	235	9.2
Florida	234	7.0
Hawaii	265	8.1
Illinois	263	8.5
Iowa	174	6.5
Kentucky	210	7.2
Louisiana	250	9.0
Maryland	373	7.6
Michigan	229	8.2
Mississippi	164	10.7
Montana	201	6.1
Nevada	172	6.5
New Jersey	298	6.3
New York	380	6.4
North Dakota	220	8.1
Oregon	227	5.6
Rhode Island	328	6.3
South Dakota	191	5.5
Texas	201	5.7
Vermont	327	6.0
Washington	240	5.2
Wisconsin	232	6.6

The infant mortality rate represents deaths of infants under 1 year old per 1,000 live births (excludes fetal deaths); the doctor data represent the number of doctors per 100,000 population. Data exclude doctors of osteopathy, federally employed persons, and physicians with unknown addresses
Data source: US National Center for Health Statistics; 2000 data (http://www.cdc.gov/nchs/default.htm).

6. Draw a scatterplot for the data in Table 5E.2. You should begin the tick marks on the X axis at 22,000 and increase to 36,000, using increments of 2,000. For the Y axis, start at zero and increase to 14, using increments of 2. Remember to label the plot. Interpret the relationship between homicide and per capita money income for these 20 states.

Table 5E.2 Table of homicide rate and income per capita for a random sample of 20 states

	State	Homicides per 100,000 (2003)	Per capita income (2002)
1	Vermont	2.3	29,291
2	Mississippi	9.3	22,321
3	South Dakota	1.3	27,087
4	New Hampshire	1.4	34,043
5	Montana	3.3	25,065
6	North Dakota	1.9	26,427
7	Arkansas	6.4	23,363
8	South Carolina	7.2	25,361
9	Alabama	6.6	25,409
10	Indiana	5.5	28,023
11	Wisconsin	3.3	30,025
12	Nebraska	3.2	29,182
13	Oklahoma	5.9	25,861
14	Louisiana	13.0	25,194
15	Kansas	4.5	28,980
16	Minnesota	2.5	33,237
17	Oregon	1.9	28,924
18	New Mexico	6.0	24,246
19	Ohio	4.6	29,212
20	Florida	5.4	29,709

7. Calculate the overall mean for the homicide rate
    for the 20 states in Table 5E.2, and draw a hor-
    izontal line on the graph at the mean. Identify

five states that are above the mean and five states
below the mean.

References

Cleveland, W. S. 1993. *Visualizing Data*. Summit,
   NJ: Hobart Press.

Cleveland, W. S. 1994. *The Elements of Graphing
   Data*. Summit, NJ: Hobart Press.

Cleveland, W. S and McGill, R. 1984. The many faces
   of a scatterplot. *Journal of the American
   Statistical Association* 79, 807–22.

Feynman, R. P. 1988. *What do you Care What Other
   People Think? Further Adventures of a Curious
   Character*. New York: W.W. Norton & Co.

Huff, D. 1954. *How to Lie with Statistics*. New
   York: W.W. Norton.

International Social Survey Programme (ISSP).
   2000. 2000 Environment II data set.
   www.issp.org. Catalog no. ZA 3440. Cologne,
   Germany: GESIS-ZA Central Archive for
   Empirical Research.

Kaufmann, D., Kraay, A., and Zoido-Lobaton, P.
   (Jan. 2002). Governance Matters II: Updated
   Indicators for 2000/01. Policy Research
   Working Paper no. 2772. The World Bank
   Research Development Group and World
   Bank Institute; Governance, Regulation
   and Finance Division (http://hdr.undp.org/
   reports/global/2002/en/).

Roberts, J. T., Parks, B. C., and Vasquez, A. A. 2004.
   Who ratifies environmental treaties and why?
   Institutionalism, structuralism and participa-
   tion of 192 nations in 22 treaties. *Global
   Environmental Politics* 4(3), 22–64.

Tufte, E. R. 1997. *Visual Explanations*. Chesire,
   CT: Graphics Press.

US Bureau of Labor Statistics. 2005a. *Employment
   Status of the Civilian Noninstitutionalized
   Population, 1940 to Date* (http://www.bls.gov/
   cps/cpsaat1.pdf).

US Bureau of Labor Statistics. 2005b. *Employment
   Status of the Civilian Noninstitutionalized
   Population, by Age, Sex and Race* (http://
   www.bls.gov).

US Census Bureau 2000. Table 33. Urban and
   rural population, and by state: 1990 and
   2000. ("http://www.census.gov/prod/cen2000/
   index.html"\t "_blank"www.census.gov/prod/
   cen2000/index.html).

US Census Bureau 2002. Historical poverty tables:
   Table 21. Number of poor and poverty rate,
   by state: 1980 to 2006. Year 2002 ("http://
   www.census.gov/hhes/www/poverty/histpov/"
   \t "_blank"www.census.gov/hhes/www/poverty/
   histpov/hstpov21.html).

US Census Bureau 2003. Table 295. Crime rates by
   state, 2002 and 2003, and by type, 2003 (http://
   www.census.gov/prod/2005pubs/06statab/law.
   pdf).

US Uniform Crime Reports. 2005. *Homicide
   Trends in the U.S.* Bureau of Justice Statistics
   (http://www.ojp.usdoj.gov/bjs/).

CHAPTER 6
CAUSATION AND MODELS OF CAUSAL EFFECTS

Outline

The social sciences, along with evolutionary biology, geology and astronomy, are, for the most part, historical sciences. They take as their subject the broad sweep of history. They study changes in societies, changes in the forms and patterns of life on earth, changes in the structure of the earth itself and the structure and dynamics of the universe as it has evolved over time. In all these sciences, our ability to conduct experiments is sharply limited. We cannot rerun human history changing one variable to see what happens.[1] We cannot rerun the history of life on earth without the asteroid strike that helped push the dinosaurs to extinction, nor see how earth's geology would evolve if the mass of the earth were just a bit larger or smaller.

The challenge for the historical sciences comes in trying to understand causation without being able to do experiments. Experiments with control groups and randomization are a very powerful tool for understanding causation, but experiments are impractical or unethical for many of the most important questions in science – those posed by the historical sciences. So we are faced with trying to understand what causes what without experiments. Fortunately, some very powerful methods have been developed to understand causation even without the luxury of experiments. Some of these methods are at the cutting edge of modern statistics. But the logic that underpins them is clear and can be practiced even with the most straightforward statistical tools – simple graphs and tables of the sort we've seen in the last three chapters. In this chapter we will examine the logic of studying causation using that logic and the tools we've already developed.

Feature 6.1   What is an Experiment?

In statistics, an experiment means a research design in which the researcher divides the objects to be studied, called "subjects," into groups that are then treated differently – "treatment" and "control" groups. The researcher then tries to give the **treatment group** (sometimes called the experimental group) and the control group exactly the same experience except for the one thing being studied. In the example we used in chapter 1, the experimental group watched one type of video while the treatment group watched another. In educational experiments, the experimental group might be placed in a new curriculum while the control group stays in the traditional curriculum. The idea is that by comparing the two groups afterwards, we can learn what effect the one thing that is different has had. We have learned that the best way to design a simple experiment is to divide subjects into groups using a random process, like the flip of a coin or the toss of dice.

Why is random assignment a good idea? Suppose we use a coin toss to divide subjects into two groups. Then, repeating the example discussed earlier, one group watches a music video with a highly gendered content while the other group watches a video with no such content. After watching the video, each group is given a survey questionnaire that measures gender attitudes. We find there are differences between the two groups. To what can we attribute the differences? There are two reasonable explanations. First, it may be that at the start, the two groups differed in their attitudes. Or it may be that the video had an effect.

Since we know how coin flips behave, we can calculate, using methods to be discussed in chapter 7, how large a difference in attitudes between the two groups is likely due to chance (by luck in flipping the coin). If the difference between the two groups is substantially larger than what a coin toss would usually generate, then we conclude the differences didn't occur by luck but rather because the video made a difference.

If we sort by some other method that is not random – by gender, by age, by time of day the students are free to watch the video (which may be related to their major and what classes they are taking), then we can't separate the effect of the video from these other things. But the coin flip (a random selection process) is powerful because it should make the groups roughly the same in all characteristics. What's more since we

know a lot about how coin flips behave, we can predict how big a difference should occur between the two groups if the experiment had no effect. This provides a basis of comparison for the differences between the two groups that will occur by luck and what might be an effect of the experiment.

The value of the experiment with random assignment to treatment and control groups is that we can establish that the experimental treatment caused the change in the subjects. It does this because the coin flip provides a basis of comparison between what we see and what would happen by chance. And because we formed the groups by chance, all differences between the groups must be the result of either chance or the experimental treatment of the two groups. If the differences between the two groups are too large to take as chance, then they must be due to the experimental treatment. To paraphrase Sherlock Holmes: "When you have eliminated the improbable, whatever remains must be the truth." Or at least what is likely to be the truth.

The causes of events are ever more interesting than the events themselves. (Cicero, *Epistolae ad Atticum*, Book IX, Section 5, quoted in Gaither and Cavazos-Gaither, 1996, p. 20)

There is no result in nature without a cause; understand the cause and you will have no need of the experiment. (Leonardo da Vinci, quoted in Gaither and Cavazos-Gaither, 1996, p. 21)

Causation and Correlation

Remember that we usually refer to a cause as X and to an effect (the thing caused) as Y.[2] When do we think of X as a cause of Y? A variable X is considered a cause of variable Y when changes in X (all other things remaining constant) lead to changes in Y. In an experiment, we would manipulate X in a design that ensures nothing else changes and then watch what happens to Y. If the difference between the values of Y across various levels of X is greater than can plausibly be attributed to chance, we can assume that the variation in Y must be due to the manipulation of X.

What do we do when we can't experiment? How do we determine if poverty causes homicide, or if gender is a cause of AIDS transmission knowledge? We have to turn to evidence that comes from observations without experiments. We can't change the poverty rates in a random sample of states and see what happens to crime rates nor change the gender of a random sample of people and see what their knowledge is. Instead, we have to work with the correlations between variables in our data sets.

There's an old saying "correlation is not causation." In a strict sense that's true. But like many aphorisms, the saying is too general. Correlation can be strong evidence about causation. Suppose we assert that X causes Y, that, in the language of our models, $Y = f(X) + E$. This assertion has implications for what we expect to see when we measure X and Y and look at how the relate to each other. If X causes Y we would expect X to be correlated (co-related) to Y. We all understand this, and we have used it in many of the examples in the previous chapters to examine ideas about the causes of homicide, animal concerns, AIDS knowledge, and environmental treaty participation. In this chapter we expand on the basic intuition that if X causes Y, X should allow us to predict Y, and we should see evidence of that linkage in our data. We will look at other explanations for an observed correlation between X and Y and see how we can use data to examine the plausibility for those other explanations.

We will never be able to "prove" that one variable causes another. In the language of causal analysis there are many "threats to validity" in establishing causation. But the difficulties have been carefully studied, and we have many tools that let us counter these threats.[3] In this chapter we stay with simple tools, yet we can still make rather strong arguments for causation. But by the end of the book, we will have discussed much more powerful tools that allow us to develop considerable confidence in our conclusions about causality.

Three points should be kept in mind when discussing the use of correlation to provide evidence for causation. First, when we consider X a cause of Y, it is usually not the only cause of Y, so Y will not be perfectly predicted by X. Thus all our models will have error (E) terms. As discussed in chapter 1, the E term in a model can represent everything other than X that can affect Y. Later we will discuss and quantify the amount of observed variation in Y that can be attributed to X. But for now it will suffice to say that we are discussing "averages" or "tendencies." For example, homicide rates tend to be higher in states with high levels of poverty. In chapter 5 we saw a number of ways of looking at these relations. For example, we found that the mean homicide rate for the five states with the lowest poverty rates was 4.22, while the mean homicide rate for the states with the highest poverty rates was 7.64.

Feature 6.2   What is Correlation?

In chapter 13, we will show you how the correlation between two variables can be measured. But the common sense idea of correlation, or co-relation, is sufficient for the work we will do in this chapter. The idea of correlation hinges on one thing being associated with another, and the term "association" can be used as a synonym for correlation.

If high values of Y tend to occur mostly when the values of X are high, and low values of Y occur mostly when the values of X are low, then we can say that there is an association or correlation between X and Y. In fact, we call this a **positive correlation**, not because it's a good thing, but because if we drew a summary line through the data, the slope (the B coefficient) would have a positive sign to indicate that large values of X are associated with large values of Y and small values of X are associated with small values of Y. We saw this pattern in the

relationship between poverty rates (X) and homicide rates (Y).

Negative correlations occur when large values of Y are associated with small values of X and small values of Y are associated with large values of X. We call this negative because the slope of the line summarizing the data would be negative.

The term correlation appears in the scientific literature at least as early as 1846 when the physicist W. R. Grove published a book entitled *The Correlation of Physical Forces*. But the concept was imprecise. The term "co-relations" was used by Francis Galton in 1885. Galton and the distinguished statistician, Francis Ysidro Edgeworth, struggled to find a number that could be calculated from data to represent the relationship between two variables. Pearson solved this problem in 1895–7 when he developed what we now call Pearson's correlation coefficient.

Sometimes we will discuss the other causes of Y explicitly, and label them with letters such as W, Z, and so on. Or X1, X2, X3 and so on could be used to represent the causes of Y. When we do this we have several independent variables. So we might think about both poverty and Southern culture causing homicide rates or about both gender and place of residence as causing AIDS knowledge. And E, the error term, representing all the causes of Y we haven't otherwise included in a model, will always be present.

The second point to remember elaborates on our earlier discussion. Correlation does not prove causation, but under proper circumstances it can be very persuasive evidence. In an experiment, since the only sources of difference across groups are random assignment and the value of X, then a correlation between X and Y greater than that which would occur by chance through random assignment provides strong evidence of causation. The circumstances under which causation can be inferred from correlation with non-experimental data are more complex, though they still rely on a thorough and rigorous application of logic and theory (or common sense in simple cases).[4]

Finally, evidence of causation developed with non-experimental data is never as strong or as clear as that derived from experimentation. But systematic non-experimental evidence is much more persuasive than casual observation, anecdotes, or the strongly held opinions of politicians or professors (Stern and Kalof, 1996). And the replication of studies by different researchers and using different methods builds our confidence about conclusions. Every study has flaws, but when many careful studies reach the same conclusion, we can have considerable faith in the overall conclusions.

Causation in Non-Experimental Data

Generally, three criteria are considered essential to make a plausible case for causation from non-experimental data. First, the variables involved must be correlated. In other words, changes in one variable must be associated with changes in another variable. In previous chapters we used graphs and summary statistics to examine the relationship between two variables. Graphs and tables can provide empirical evidence in support of causation based on this first criterion.

Second, we must be able to establish causal order on theoretical grounds. That is, we must be able to determine which variable is independent and which is dependent. This is sometimes a difficult task with many social science variables. The strongest case can be made for what are often called **subject variables**, such as gender, race, and age. While these characteristics are socially constructed in significant ways, it is reasonable to assume that gender may be a cause of an individual's behavior or attitudes, and that behavior or attitudes are not the cause of gender. Another way to establish causal order is to determine the time order between variables. The independent variable comes first in time (e.g., exposure to a stimulus, such as a stereotyped video) and causes changes in a dependent variable (e.g., gender attitudes). For example, it's just not reasonable to think of the current homicide rate in a state as a cause of whether the state is in the South, but we can offer the hypothesis, that may or may not be correct, that being in the South leads the state to have a higher homicide rate.[5]

Finally, we must establish that the correlation between the variables is not spurious. A **spurious correlation** occurs when two variables are correlated, but that correlation comes about because both are caused by a third variable. Neither of the two variables involved in the spurious association is actually causally related to each other. Not including the third variable in the analysis can give the mistaken impression that the two variables are causally related. For example, we find a correlation between number of NGOs and environmental treaty participation. But it may be that both the presence of NGOs in a country and the level of environmental treaty participation are driven by the extent of freedom citizens are given in a country. The correlation between NGOs and treaty participation is real but it could be spurious if it is the result not of a causal relationship between these two variables but a link of both of them to voice and accountability. In the last chapter we looked at only two variables at a time so we couldn't address spuriousness, but we will develop methods in this chapter and later chapters that allow us to look at several independent variables at once.

Statistical tools like those in the last chapter and others covered later help establish correlation. They can also help to establish whether an association is spurious. This ability usually depends on knowledge about the causal ordering, and that ultimately rests on theory, other empirical evidence or logic. Techniques called structural equation modeling can help disentangle what variables are causes and what variables are effects, but they do so only because the causal ordering of other variables are known from theory or previous research. Thus the ability of statistical analysis to establish causal ordering with non-experimental data always rests to a substantial degree on the knowledge (or assumptions) brought to the analysis.

Explanatory and Extraneous Variables

Several terms are used to describe variables when trying to understand causation. Independent and dependent variables have already been discussed. We use the term **explanatory variables** for the independent and dependent variables Xs (independent variables) and the Ys (dependent variables) on which we have data and on which we are focused.[6] **Extraneous variables** are all other possible variables that could potentially affect the dependent variable in our research. For example, in a study of high school grade point average and college grade point average, the explanatory variables are high school grade point average (the independent variable, X) and college grade point average (the dependent variable, Y). There are many other variables that could have an effect on variation in college grade point average (other than high school grade point average): gender, college extracurricular activities, number of science courses taken in high school, and so on. These are all extraneous to the focus of the research.

There are two types of extraneous variables, controlled and uncontrolled. **Controlled variables** are held constant – their effects are dealt with in statistical analysis. We have three ways of controlling for the effects of an extraneous variable. First, when we have randomization in an experiment, we know that all subject variables (except those manipulated by the experiment, such as video images) are going to be equivalent across the experimental and control groups. Any differences between the groups will be the result of a random process such as the flip of a coin. The random process will limit the size of the effects of the extraneous variables. For example, in the experiment examining the effects of music videos on gender attitudes, we know that the two groups are unlikely to differ much in initial attitudes because of the random process used to create the groups. Initial attitudes (and gender, college major, family size, religion and so on) are extraneous variables, while the viewing or not viewing of the gendered video is an explanatory variable. The experimental design controls for the extraneous variables by randomization. This is one of the reasons why randomized experiments are so powerful – they control for all extraneous subject variables.

However, often we can't assign the things we are studying to experimental and control groups. For example, we can't randomly assign states to poverty levels or to regions of the country with different political, economic, and cultural histories. So we must use one of the two other ways of controlling for extraneous variables. One is to limit the scope of our analysis by only looking at one group. Suppose we are interested in the relationship between poverty rate and homicide rate, but we acknowledge the plausibility of the theories arguing that there are cultural reasons why Southern states have higher homicide rates. Then we could proceed by studying the relationship between homicide rate and poverty rate only among Southern states. We have "held constant" being in the South. Of course we could also do an analysis looking only at Northern states.

Finally, as we will see later in the book, we can calculate measures of the effect of one independent variable on the dependent variable while taking into account the effects of another, or several other, independent variables. We can also do this with tables when we use two independent variables (e.g., homicide rates for subgroups of states selected based on both poverty and region). But there are other statistical tools for making such comparisons, and multiple regression is the most powerful of these.

Controlling for variables gives us a better sense of what is causal. Let's consider an example in which we think of factors that might influence AIDS knowledge. Gender might have an effect. Place of residence may have an effect. Table 6.1 provides the proportion of men and women living in rural and urban areas, who know that condom use can reduce the likelihood of AIDS transmission.

We will discuss how to interpret these data later in the chapter. For now, note that our dependent variable is the response on the AIDS transmission question. The explanatory independent variable on which we are focused is gender. But we control for another independent variable (here treated as extraneous), whether the

Table 6.1 Proportion of respondents with AIDS knowledge, by gender and place of residence

	Men in urban areas	Women in urban areas	Men in rural areas	Women in rural areas
Proportion who are informed	0.87	0.85	0.81	0.70
Total N	588	2,269	1,298	4,155

Data source: 2000 Ugandan DHS data set, analyzed with Stata.

person lives in an urban or rural area, by looking at the relationship between place of residence and AIDS knowledge within gender.

What makes one independent variable an extraneous variable for which we control and another an explanatory variable? The focus of our research determines what is explanatory and what is extraneous. The concepts of explanatory and extraneous variables emerge from the literature on experimentation. Experiments are designed to provide as much information as possible about the effects of the explanatory variables that are the focus of the study, but care must be taken to control for the extraneous variables. Thus in experiments, explanatory and extraneous variables are determined by the interests of the researcher and may be handled in different ways in the experimental design. In non-experimental data, an extraneous variable can be handled in the same way as the explanatory variable, but differs from it because of the researcher's focus. We use the term to help make the link between experimental and non-experimental research.

We could try to control for all other important variables in our research, but that is impossible in the social sciences. Some variables are always left uncontrolled and thus are potential alternative explanations for findings, or the sources of variation in the dependent variable. In Table 6.1 we have left uncontrolled many variables that could affect AIDS knowledge, such as level of education and age. Since we have data on these variables, we can control for them, especially once we learn techniques for handling multiple independent variables. The literature on AIDS knowledge indicates that being exposed to AIDS educational programs highly influences AIDS knowledge, but that information is not available in our data set (although it should be noted that place of residence may be highly related to exposure to AIDS educational programs, since efforts have been concentrated in urban areas). So it will have to remain uncontrolled.

Causal Notation

Using path diagrams and path analysis to evaluate causal relations among variables was invented by the geneticist Sewall Wright (Wright, 1921), refined by Herbert Simon (Simon, 1952; 1953; 1954) and developed in detail in sociology by Hubert Blalock (Blalock, 1960) and Otis Dudley Duncan (Duncan, 1966).[7] **Path diagrams** are a very useful way of expressing theories. Path diagrams can make models clear to the reader, show flaws in logic, and make it obvious how the theory translates into empirical implications that can be tested with data. The major weakness of these diagrams is that some theoretical models are difficult to express using path analysis, though most fit without too much trouble. Indeed often theories that cannot be placed in the path diagram format are often ill-conceived, confused, or imprecise.

Generally, arrows are used to indicate causal relationships, so Figure 6.1 shows X is a cause of Y. For example, X might be gender and Y AIDS knowledge. The diagram implies a gender difference in AIDS knowledge and that the effect of gender is causal.

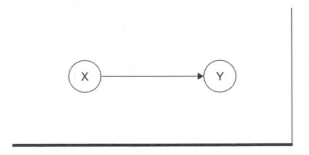

Figure 6.1   Path diagram for a single cause of Y

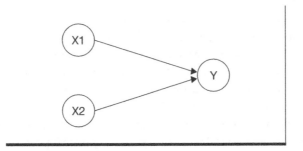

Figure 6.2   Path diagram for two causes of Y

Figure 6.2 shows two causes of Y (AIDS knowledge), X1 and X2. X1 could be gender and X2 place of residence (living in rural versus urban areas).

This implies that both gender and place of residence influence an individual's level of AIDS knowledge. A double-headed arrow or a line without arrow-heads indicates a correlation between two variables, a situation in which the variables are assumed to be associated but the causal structure of their relationship is not considered in the model. But for simplicity, we usually will not include the double-headed arrows or lines without heads in our examples.

When we draw an arrow from gender to AIDS knowledge and from place of residence to AIDS knowledge we are assuming that gender may be a cause of AIDS knowledge and that place of residence may be a cause of AIDS knowledge. The lack of an arrow from AIDS knowledge to gender or to place of residence indicates that we don't think what one knows about AIDS changes one's gender or one's place of residence. In this example, those assumptions seem plausible. But we can imagine other analyses where the assumption that causality flows in only one direction is not plausible. For example, we might study knowledge about AIDS, attitudes towards people with AIDS and trust in Western medicine. With these three variables it is harder to make plausible assumptions that any one of them has no causal effect on one of the others. If we collected data on the same people over time (recall this is a panel design) we could see how knowledge, attitudes, and trust changed and that could help us untangle what is causing what. But unlike the gender, place of

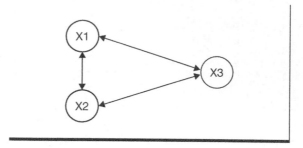

Figure 6.3   Path diagram for three variables that are correlated

residence, and knowledge example, there is no logic that seems to say that one variable could not cause another.

In such situations, the path diagram, such as the one depicted in Figure 6.3, shows that, theoretically, everything is causing everything else. We do this either with one single-headed arrow for each link or to make the diagram a little simpler, with a double-headed arrow. The double-headed arrow means that we think the variables are correlated but we are not saying anything about what causes what. Figure 6.3 suggests that knowledge about AIDS (X1), attitudes towards people with AIDS (X2), and trust in Western medicine (X3) are all correlated, but there is no assumption that any of the variables are influencing (causing) another variable. The techniques we use can show us whether the variables are correlated but we can't get at causation because everything is plausibly causing everything else. Sometimes just learning that the data indicate that there are correlations is helpful. But usually we are trying to get at causation. And the data can't tell us about causation unless we have some theory that guides us with regard to which causal paths are plausible and which ones are not. So we re-emphasize a point we mentioned earlier – that statistics and theory always work together in analyzing data.

In our notation, a circle around a variable indicates that it is directly observed, as an item on a questionnaire (gender, place of residence, AIDS knowledge) or a statistical record (number of treaties ratified, homicide rate). These are called **observed variables** because they are measured. Since we do not believe Y is perfectly determined by X1 and X2, it can be useful to include E as a cause of Y, the error variable to remind us of this. E is not directly observed, as a cause of Y. In our path notation, we will place variables not directly observed in a box to distinguish them from the directly observed variables that are enclosed in circles. This is shown in Figure 6.4.

More elaborate causal notation is discussed at the end of this chapter as an Advanced Topic.

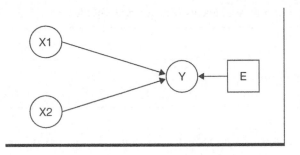

Figure 6.4 Path diagram for two causes of Y that includes error term

Social science journals sometimes publish path diagrams as part of research papers. One advantage of this notation is that you can work through rather complex statistical models visually. Authors may differ slightly in the details of their notation, but almost everyone follows the general pattern described here – straight one-directional arrows indicate causation. Sometimes curved arrows and/or two-headed arrows are used to indicate correlations that are not examined as causal in the model being diagrammed. And most path diagrams use different symbols for observed and unobserved variables just as we have used circles for observed variables and boxes for unobserved variables. Finally, in some path models, letters are placed next to the arrows to indicate that there is a parameter that corresponds to the path indicated by the error.

Assessing Causality: Elaboration and Controls

To begin thinking about causal analysis, a simple proportion table can be used to test some of the theories diagrammed above. Consider the relationship between gender, place of residence, and AIDS knowledge. Table 6.2 shows AIDS knowledge broken down by gender and then broken down by place of residence. That is, we are looking at AIDS knowledge, first conditional on gender (on the left side), then conditional on place of residence (on the right side of the table). Here and in all

Table 6.2 Proportion of respondents who have AIDS knowledge, by gender and by place of residence

	Men	Women	Urban dwellers	Rural dwellers
Proportion correctly knowing about condoms as preventive	0.83	0.76	0.86	0.73
N	1,886	6,424	2,857	5,453

Table 6.3 Proportion of respondents who are rural dwellers

	Men	Women
Proportion living in rural areas	0.69	0.67
N	1,962	7,246

the tables in this chapter we follow the convention that dependent variables are in the rows and independent variables define the columns.

The obvious conclusion to be drawn from the table is that more men are informed about ways of preventing AIDS transmission and people who live in urban areas are more informed. (In this chapter we won't address the issue of how large a difference is enough to be considered more than just sampling error. That will be handled in later chapters.) We would not expect gender differences in the propensity to live in rural or urban areas, but we will examine the relationship to be sure. Table 6.3 shows that relationship.[8] There does not appear to be an association between the variables. The path diagram depicted in Figure 6.2, which does not show any relationship between X1 (gender) and X2 (place of residence), is correct.

Let's suppose (for illustrative purposes) though that gender and place of residence are correlated. Figure 6.5 presents the path diagram for this example.

If we assume a relationship between gender and place of residence, what's causing what? In developing countries like Uganda, it is not uncommon for rural dwelling men to leave their families to find work in a larger city and send money home to their spouses and children. Using this logic, we could make an argument that gender influences place of residence. Realistically, we do not have a strong causal claim, but for purposes of illustration we will hypothesize this causal ordering. Now let's look at the data to see if the data are correlated in the pattern the hypothesized causal links imply. The rest of this section will examine how we can do that.

Several models might explain the relationships between gender, place of residence, and AIDS knowledge and still be consistent with the basic logic of causal order we just discussed. The first is a situation in which both independent variables have a causal influence on AIDS knowledge but where we don't specify any relationship

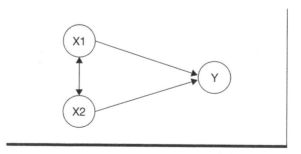

Figure 6.5 Path diagram for two causes of Y that are correlated

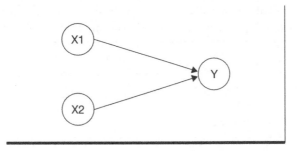

Figure 6.6   Path diagram for two causes of Y and no correlation between independent variables

of causation between gender and place of residence (this is most realistically the case with gender and place of residence and AIDS knowledge). That is, we treat place of residence and gender as causes of AIDS knowledge, but take no account of the argument that gender might cause place of residence. This could be diagrammed as in Figure 6.6.

As an equation, this would look like:

$$Y = f(X1, X2) + E \tag{6.1}$$

A second possibility is a spurious correlation in which gender (X1) is a cause of both place of residence (X2) and AIDS knowledge (Y). Since it is a cause of both, it will cause X2 and Y to vary together (covary) and thus to be correlated even when there is really no causal relationship between X2 and Y. This is a case of spurious correlation. It is diagrammed as Figure 6.7.

Because this model has two dependent variables, we will need two equations to describe the model:

$$Y = f(X1) + E1 \tag{6.2}$$

$$X2 = g(X1) + E2 \tag{6.3}$$

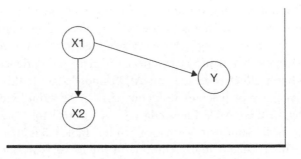

Figure 6.7   Path diagram for causal effect of X1 and spurious effect of X2

Table 6.4 Hypothetical data on AIDS knowledge, place of residence, and gender, showing causal relationship of gender and spurious relationship of place of residence

	Men		Women	
	Urban	Rural	Urban	Rural
Proportion with AIDS knowledge	0.65	0.65	0.55	0.55

Since we have two equations, we have to use f and g to indicate that there are two functions, one linking X1 to Y and the other linking X1 to X2. E1 indicates the error term in the equation predicting Y, and E2 indicates the error term in the equation predicting X2.

Thus when we control for gender (X1), there should be no association between place of residence (X2) and AIDS knowledge (Y). We would say that the partial relationship between place of residence and AIDS knowledge, controlling for gender, should be zero. By **partial relationship** we mean the relationship between an independent variable and a dependent variable when the effects of one or more other independent variables are held constant. We can check this with a table where we see if AIDS knowledge changes across levels of place of residence *within* a level of gender. Such a table might look like Table 6.4. In the classic language of "elaboration" in tables, this is called an explanation, because we have explained away the relationship between place of residence and AIDS knowledge.[9]

Note that the data in the table are just hypothetical and one example of the pattern we would expect to see if gender is a cause of both place of residence and AIDS knowledge, and there is no effect of place of residence on AIDS knowledge when gender is controlled. The key point is that the proportion with AIDS knowledge varies across genders within a category of place of residence, while place of residence has no effect within gender. We will use these hypothetical tables in this chapter to help you think about what to expect with differing patterns of data.

Another possibility is that place of residence (X2) is an **intervening variable** (sometimes called a **mediating variable**) that is caused by gender (X1) and in turn causes AIDS knowledge (Y). That is, gender affects place of residence, and place of residence affects AIDS knowledge. But in this model, gender has no **direct effect** on AIDS knowledge other than via its effect on place of residence. This pattern is called an **indirect effect** of gender on AIDS knowledge. If this model is correct, within a category of place of residence (urban or rural) men and women would not differ in their AIDS knowledge. If we control for place of residence, AIDS knowledge is the same across levels of gender. Table 6.5 is an example of such a relationship. This is called "interpretation" in the classic vocabulary and would be diagrammed as in Figure 6.8.

Table 6.5   Hypothetical data on AIDS knowledge, with direct effect of place of residence and indirect effect of gender

	Men		*Women*	
	Urban	*Rural*	*Urban*	*Rural*
Proportion with AIDS knowledge	0.50	0.60	0.50	0.60

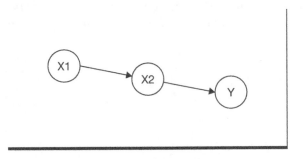

Figure 6.8   Path diagram for direct effect of X2 and indirect effect of X1

For this diagram, the equations would be:

$$Y = f(X2) + E1 \tag{6.4}$$

$$X2 = g(X1) + E2 \tag{6.5}$$

A closely related possibility is shown in Figure 6.9. X1 has a direct effect on X2 and an indirect effect on Y through X2 (as in Figure 6.8). But now X1 also has a direct effect on Y, controlling for X2. That is, in the table, we would find that Y changes with X1, controlling for X2, and Y changes with X2 controlling for X1. In the classic

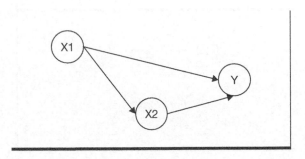

Figure 6.9   Path diagram for direct effect of X2 and direct and indirect effect of X1

Table 6.6 Hypothetical data on AIDS knowledge, showing both direct and indirect effects of gender and direct effect of place of residence

	Men		Women	
	Urban	*Rural*	*Urban*	*Rural*
Proportion with AIDS knowledge	0.55	0.65	0.50	0.60

language this would be called a "replication of the relationship" between X1 and Y within levels of X2 and a "replication of the effect" of X2 on Y within levels of X1. The equations would be:

$$Y = f(X1) + g(X2) + E \tag{6.6}$$

$$X2 = h(X1) + R \tag{6.7}$$

This relationship is shown in Table 6.6.

The most complicated situation is the **interaction effect**. In an interaction, the effect of X2 on Y depends on the value of X1. For example, place of residence may have an effect on AIDS knowledge for men but not for women, or it may have a stronger effect for one group (i.e. one gender), or a different direction of effect. Interactions are not easily diagrammed, but there are two possibilities shown in Figure 6.10.

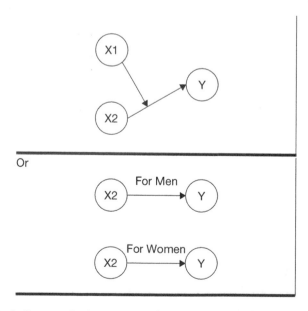

Figure 6.10 Path diagram for interaction effect of X1 and X2

Table 6.7 Hypothetical data on AIDS knowledge showing interaction between gender and place of residence

	Men		Women	
	Urban	*Rural*	*Urban*	*Rural*
Proportion with AIDS knowledge	0.60	0.50	0.50	0.60

Table 6.8 Observed (actual) data on AIDS knowledge by place of residence and gender

	Men in urban areas	*Men in rural areas*	*Women in urban areas*	*Women in rural areas*
Proportion with AIDS knowledge	0.871	0.810	0.852	0.704
N	588	1,298	2,269	4,155

Table 6.7 shows one kind of interaction effect. Here, for men, living in urban areas increases AIDS knowledge while for women living in urban areas decreases AIDS knowledge.

The actual data from the survey are displayed in Table 6.8. Here we have the mean of the AIDS knowledge question (proportion yes) displayed as a function of both place of residence and gender. For women, living in a rural area reduces AIDS knowledge by about 15 percent. For men, the difference is about 6 percent. So we find an effect of place of residence even when gender is controlled, and the effect is stronger for women than for men. When we compare genders within a category of place of residence, we find that the difference between men who live in urban areas and women who live in urban areas is nearly 2 percent (positive because in this group men have slightly more knowledge than women). When we compare men and women who live in rural areas, we find a difference of about 11 percent. So we might conclude that there is a relationship between gender and AIDS knowledge but only among those living in rural areas.

Therefore, the model that fits these data best seems to be one that assumes that gender has an interaction effect on the relationship between place of residence, where gender differences exist only among those living in rural areas (Figure 6.10).

Another Example

An example that illustrates causality, elaboration, and control variables comes from a study of status and social distance in the faculty office (Seyfrit and Martin,

Table 6.9   Desk placement by academic rank

Desk placement	Full Professor	Associate Professor	Assistant Professor	Instructor
Desk between (%)	48.3	24.0	12.7	20.0
Desk not-between (%)	51.7	76.0	87.3	80.0
(N)	(35)	(55)	(50)	(35)

Table 6.10   Desk placement by gender

Desk placement	Male faculty	Female faculty
Desk between (%)	34	8
Desk not-between (%)	66	92
(N)	(149)	(49)

1985). Data were collected on the placement of faculty desks – whether the desk was between the faculty member's chair and the student's chair. The desk-between arrangement is interpreted as a way of creating social distance between the faculty member and the students. The study attempted to determine if faculty of higher academic status (rank and tenure) were more likely than other faculty to use a desk-between arrangement of office furniture to create social distance between themselves and their students. Since the authors' believed that issues of academic power and status also involve gender, they also collected data on the faculty members' gender. Tables 6.9 and 6.10 show the results of the first stage of the analysis.

Table 6.9 shows that faculty with higher academic rank were more likely to place their desks between themselves and their students. Almost half of the full professors used a desk-between design in their offices compared to less than one-fourth for any other academic rank. In addition, Table 6.10 shows that male faculty were much more likely than female faculty to create social distance in desk placement.

Here we have a correlation between academic rank (X1) and desk placement (Y) and between gender (X2) and desk placement (Y). Also, there is a correlation between gender and academic rank (data not presented here): 32 percent of the males were full professors, compared to 18 percent of the females, and 14 percent of the males were instructors, compared to 29 percent of the females.

Now, what about causality? It is not plausible that gender is caused by academic rank or desk placement. But since rank is correlated with gender, for the purpose of this example, we can say gender causes rank. Further, desk placement can be assumed to have no effect on academic rank. So, it may be that academic rank intervenes between gender and desk placement, thus the causal order might be as in Figure 6.11.

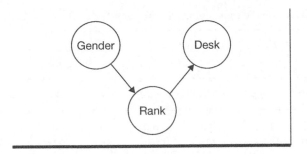

Figure 6.11   Path diagram of academic rank intervening between gender and desk placement

Thus, gender has an influence on desk placement only through its influence on academic rank (i.e., an indirect effect only). In that case, we would expect to find no gender difference within a level of rank, but rank differences within gender.

Or, perhaps, gender determines both academic rank and desk placement, but rank has no direct effect on desk placement. Because of the influence of gender on both, rank and desk placement covary, and there is a spurious correlation between them (i.e. no true causal relationship). If this model is true, then we would find gender differences within rank, but no rank differences within gender (see Figure 6.12). Finally, it may be that gender and academic rank interact so that the effect of rank on desk placement depends on gender. This interaction is shown in Figure 6.13.

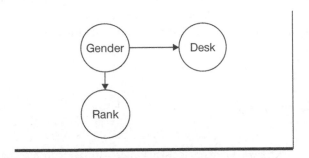

Figure 6.12   Path diagram showing that gender influences both academic rank and desk placement, but rank has no direct effect on desk placement

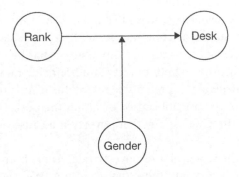

Figure 6.13   Interaction of gender on relationship between rank and desk placement

Table 6.11   Desk placement by academic rank, controlling for gender

Desk placement	Full Professor		Associate Professor		Assistant Professor		Instructor	
	Male	Female	Male	Female	Male	Female	Male	Female
Desk between (%)	55.1	11.1	30.0	0.0	15.4	6.2	23.8	14.3
Desk not-between (%)	44.9	89.9	70.0	100.0	84.6	93.8	76.2	85.7
(N)	(49)	(9)	(40)	(10)	(39)	(16)	(21)	(14)

If there is an interaction, the effect of rank on desk placement will have one pattern for men and a different pattern for women.

Table 6.11 is the three variable table showing desk placement broken down by rank and gender. For men, it appears that there are substantial differences across rank – higher ranking faculty are more likely to use the "desk between" placement, though instructors are an exception to this pattern. Yet there seems to be no pattern for women faculty, indicating an interaction between gender and rank. (In a later chapter we will discuss how to test hypotheses about the patterns in contingency tables.) You see that when controlling for gender, the simple relationship between academic rank and desk placement disappears. Higher academic status is related to social distance between faculty and students for male professors and not for female professors.

An Example with Continuous Variables

So far we have worked with categorical variables. But we can also examine causal relationships with continuous variables. Later chapters will discuss several methods of doing this. Here we will work with scatterplots to get a sense of how to examine causation with continuous variables.

Let's consider two independent variables that might drive the homicide rate: poverty and being a Southern state. For convenience, we'll define a state as "Southern" if it was a member of the Confederacy. We can assume that being a member of the Confederacy is causally prior to both poverty and homicide rate in the 2000s – that is, being in the Confederacy represents a set of historical and cultural factors that might be causes of but are not caused by either the poverty rates or homicide rates in the 2000s. In turn, we assume that poverty is a cause of homicide and not vice versa.[10]

Figure 6.14 shows the scatterplot of poverty rate versus homicide rate for states that were in the Confederacy and those that were not. We have set the same axes

a) For non-Confederate states (N = 39)

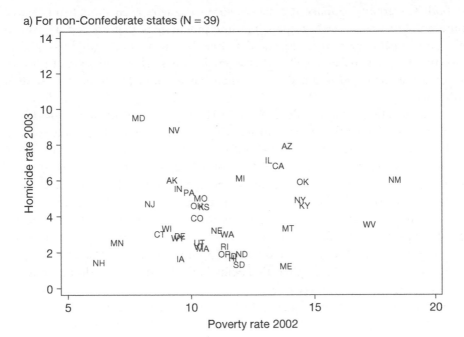

b) For Confederate states (N = 11)

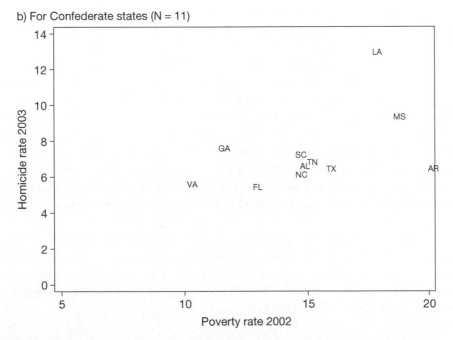

Figure 6.14 Scatterplots of homicide rate versus poverty level for Confederate states and non-Confederate states
Data source: US Census Bureau, 2002; 2003, analyzed with Stata.

on both scatterplots to make it easier to compare them, even though this leaves some empty space in each graph. Our interpretation is that we see a positive relationship between poverty rate and homicide rate in both graphs – higher poverty rates are associated with higher homicide rates. This suggests that poverty does have an effect on homicide even when having been a Confederate state is controlled.

We also find that the homicide rates tend to be higher in the former Confederate states (though this would be easier to see with a stem and leaf plot, of the sort we used earlier, for Confederate and non-Confederate states). To get a sense of the effect of being in the Confederacy, look at the states in each group with poverty rates around 15. In the non-Confederate states these are Kentucky (homicide rate about 5), New York (5), and Oklahoma (6). (Remember these are only roughly correct from reading the graph – to get the exact homicide rates we'd have to look at the actual data.) In the former Confederate states at a poverty rate of around 15 we have South Carolina (homicide rate about 7.5), Tennessee (7), Alabama (7), and North Carolina (6). So this suggests that being in the South has an effect even when we take account of the poverty rate.

So both poverty rate and being in the South have an effect when the other is controlled. Neither effect seems to be spurious. And our first impression is that the effect of poverty rate may be about the same in the Southern and non-Southern states – it appears that the data move from lower left to upper right in very roughly the same way, although there are no Confederate states with very low levels of poverty. So we will conclude that there is no interaction effect, although we might want to investigate this more carefully in a larger research project. All of this leads us to believe we have a model like that in Figure 6.6 or 6.9. Both membership in the Confederacy and poverty rate have an influence on homicide rate. Recall that in Figure 6.6 we simply say that the two independent variables are correlated while in Figure 6.9 shows the effect of X1 (being in the South) on X2 (poverty rate). To look quickly at the effect of being in the South on poverty rate, we note that for Southern states the mean poverty rate is 14.8 and the median is 14.5. For non-Southern states the mean is 10.8 and the median is 10.1. This suggests that being in the South influences poverty, so the model specified in Figure 6.9 seems a plausible description of the data.

Note that the scales for the independent and dependent variables go well beyond the range of the data in the above scatterplots. Generally, this is not a good idea but in this case we want to compare the scatterplot for non-Southern states with the scatterplot for Southern states. To make comparisons like this it is very important to have the same scales on the two plots.

Further ideas in assessing causality are discussed below as an Advanced Topic.

What Have We Learned?

We have learned that fairly simple diagrams using variables and arrows linking them can display theories about which variables are causes of other variables. Diagrams can tell us whether the relationships between variables are proposed to be causally related or correlated and how variables may be related. These diagrams help us understand better the relationships between variables and to convey proposed relationships to others. When we know which variables might be causes and which might be effects – when we can make assumptions about causal order, then we can translate those theories into what we would expect to find in tables and other summaries relating variables to one another. This allows us to use data to assess the plausibility of causal theories.

Advanced Topic 6.1   More Elaborate Causal Notation

Causal notation can also be used to think about problems of measurement. Remember that the 2000 ISSP data set only had one question measuring concern for animals. Say there were two questions though. Suppose we label as W1 and W2 the two questions that address concern for animals. Then we might consider W1 and W2 as indicators of an underlying concern for animal welfare which we could label F. We can conceptualize underlying concern for animal welfare as a general set of beliefs that elicit certain responses to specific questions, including W1 and W2 in the survey. W1 and W2 each have their own idiosyncratic causes, labeled E1 and E2. Neither the underlying concern for the animal welfare variable (F) nor the errors are directly observed, so the diagram would look like Figure 6.15.

Suppose we think that the underlying concern for animal welfare is determined in part by gender and education (X1 and X2). Then the diagram would look like Figure 6.16 (leaving out the Ws and Es for a moment to keep things as simple as possible).

B1 represents the causal effect of gender on concern for animal welfare, controlling for education, while B2 represents the causal effect

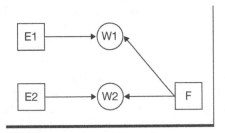

Figure 6.15   Path diagram for unobserved variable

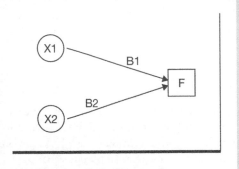

Figure 6.16   Path diagram for two causes of unobserved variable

of education on concern for animal welfare, controlling for gender.

If we add the measurement model, and let C1 and C2 be the links between the two animal concern questions and the underlying concern for animal welfare (F), the result would be Figure 6.17.

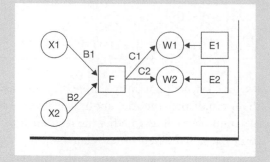

Figure 6.17 Path diagram for two causes of unobserved variable with measurement model

Advanced Topic 6.2   Further Ideas in Assessing Causality

In several of the path diagrams, some variables have been both independent and dependent. For example, in looking at the effects of gender and place of residence on AIDS knowledge, we considered place of residence as a cause of knowledge but a consequence, in part, of gender. In causal modeling, we use the terms **endogenous** (for "in the system") and **exogenous** (for "outside the system") rather than independent and dependent to account for this. In the AIDS knowledge example, gender is exogenous because it is in our model but nothing in the model explains it. Both place of residence and knowledge are endogenous because the model includes variables we think are causes of them. In the homicide, poverty, and Confederacy example, membership in the Confederacy is exogenous because nothing causes it, while both the poverty level and homicides were endogenous because they have possible causes in the models we were considering. In a path diagram of the sort we have used, an exogenous variable has no arrows point to it, while an endogenous variable does have arrows pointing to it, but this simple rule doesn't follow when we include error terms or in more complicated path models.

Experiments can offer strong evidence of causality but the experimental method has two limitations. First, we can't do experiments on many subjects of great importance because it would be impractical or unethical. Second, experiments are always somewhat artificial in how they are constructed. It can be hard to generalize from an experimental study in which great attention has been paid to providing the treatment to the subjects to the rest of the world – in the language of methodology experiments often lack **external validity**. Consider the Kalof experiment with MTV videos. This was a well-designed study and makes a strong case for causal effects of the videos on gender-stereotyped attitudes. But we have to note that the circumstances in which the students watched the videos – together in a classroom setting – may have made the effects stronger or weaker than if such videos are watched by small single gender groups or by individuals alone in their dorm rooms. So efforts to disentangle causal relationships from non-experimental data are very important. No single non-experimental study is ever adequate for firmly establishing causal relationships among variables – every study is flawed. But when many studies using different data sets and different approaches lead to the same general conclusion, we can be quite confident we have developed a reasonably good model of the world.

The panel data discussed in chapter 2 are very helpful in uncovering causal relationships. While the methods used for analyzing panel data can become quite complex, the basic idea is that by watching things unfold over time among multiple people, organizations, or countries, we can get a better sense of what is changing and why and in the process learn much more about causality. As we mentioned, panel studies can be hard to conduct but they often provide very large payoffs in a better understanding of causality.

Applications

Since we had a thorough discussion of AIDS knowledge and of homicide rates in the chapter, we won't develop examples for those topics in the applications. We focus on applications using the animal concern example and the environmental treaty participation example.

Example 2:   Why do people differ in their concern for animals?

Tables 6A.1 to 6A.5 provide the mean animal concern score by two plausible independent variables: 1) belief that humans developed or evolved from earlier species of animals; and 2) age. It is reasonable to assume that those who feel some evolutionary kinship with other animals might have higher scores on the animal concern question. The causal order among these two variables, however, is unclear. Concern for animal welfare and belief in evolution could develop simultaneously. For example, a respondent might have taken a course in animal ethics that addressed our evolutionary history, our kinship with animals, and the need to be concerned about the welfare of other species. Thus attitudes and beliefs could develop together, perhaps as the result of some event, as in our example. And it is at least plausible that concern for animals develops early in life before people learn about Darwinian evolution. For the purposes of examining causal relations among the variables, we will proceed as if belief in evolution is causally prior to animal concern but remember that this is only a working assumption, and not something we strongly assert. However, age almost certainly is causally prior to animal concern score and belief in evolution – neither animal concern nor belief in evolution can determine a person's age. This is a common situation in analysis of non-experimental data. We have clear causal priority for some variables but only tentative working assumptions about causal order for other variables.

Table 6A.1 shows the association between belief in human evolution and age. To simplify this example, we will divide age into two categories at the median age: people ages 45 and under and people ages 46 and older. This is a more complicated table than we have shown before and it may be confusing. It gives the percent saying "yes," the number saying "yes," the percent saying "no," the number saying "no," the total number of people who answered the question and the sum of the percent saying "yes" and percent saying "no" (which must always be 100% within rounding error). You will sometimes see this kind of table in computer output and in technical reports. It can be helpful when we are first looking at data, but we have to be careful not to get lost in the details. Notice that if we know the percent saying "yes" and the overall number of people who answered the question for each group, we could calculate the percent saying "no," the number saying "yes," the number saying "no" and the sum of the two percents. We can analyze the data simply by looking at the percent saying "yes." About 73% of people ages 45 and under and 68% of people ages 46 and older believe in human evolution. Since these percentages

Table 6A.1   Percentage of respondents who believe that humans developed from earlier species of animals, by age group

Humans developed from earlier species of animals	Under age 46	Ages 46+	Total
Yes (%)	73.2	68.0	70.8
Total	10,095	7,993	18,088
No (%)	26.8	32.0	29.2
Total	3,693	3,762	7,455
Total	13,788	11,755	25,543
	100%	100%	100%

Data source: 2000 ISSP data set, analyzed with SPSS.

are relatively similar, it does not seem like there is a relationship between age and belief in evolution.

Tables 6A.2 and 6A.3 show the mean animal concern score by belief in evolution and by age group, respectively. Contrary to our hypothesis, those who believe in evolution do not have higher animal concern scores than those who do not. However, younger people do have higher mean animal concern scores than older people. So it appears that there is an association between age and animal concern but not between belief in evolution and animal concern.

Table 6A.2   Mean animal concern score by belief in evolution

Humans developed from earlier species of animals	Mean	N
Yes	2.49	17,737
No	2.50	7,260
Total	2.49	24,997

Data source: 2000 ISSP data set, analyzed with SPSS.

Table 6A.3   Mean animal concern score by age group

Age group	Mean	N
Ages 45 and under	2.54	15,239
Ages 46 and older	2.43	13,841
Total	2.49	29,080

Data source: 2000 ISSP data set, analyzed with SPSS.

Table 6A.4 Mean animal concern score by age group and belief in evolution

Ages 45 and younger		Ages 46 and older	
Believe in evolution	Do not believe in evolution	Believe in evolution	Do not believe in evolution
2.54 (N = 10,095)	2.51 (N = 3,693)	2.42 (N = 7,993)	2.48 (N = 3,762)

Data source: 2000 ISSP data set, analyzed with SPSS.

Table 6A.4 shows the mean animal concern score for the two age groups by those who do and do not believe in human evolution. Table 6A.5 redisplays the information to show that there are always multiple ways to display the same information.

You can see from these tables of means that younger people who do and do not believe in evolution have higher mean levels of animal concern than both older people who do and do not believe in evolution. Also, of those who believe in evolution, younger people have higher mean levels of animal concern. Among those who do not believe in evolution though, there are no apparent age differences. We have previously concluded that age does not appear to influence beliefs about evolution (Table 6A.1) and that evolutionary beliefs do not themselves affect animal concern (Table 6A.2). So we could conclude that only age directly influences animal concern. However, we see in Tables 6A.4 and 6A.5 that age and evolutionary beliefs have an interaction effect because the effects of evolutionary beliefs on animal concern

Table 6A.5 Mean animal concern score by belief that humans developed from earlier species of animals and age group

Humans developed from earlier species of animals	Age group	Mean animal concern score	N
Yes	45 and under	2.54	10,095
	46 and older	2.42	7,993
	Total	2.49	18,088
No	45 and under	2.51	3,693
	46 and older	2.48	3,762
	Total	2.50	7,455
Total	45 and under	2.54	13,788
	46 and older	2.43	15,755
	Total	2.49	29,543

Data source: 2000 ISSP data set, analyzed with SPSS.

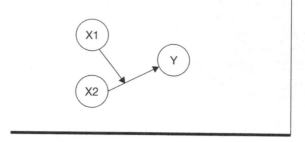

Figure 6A.1   Interaction effect of X1 and X2 on Y

differ by age group. In fact, there are only age differences in animal concern among those who believe in evolution. We can alternatively say that the effects of age on animal concern differ by beliefs in evolution; when describing an interaction effect age and evolutionary beliefs can be either X1 or X2. So the model that best fits the data is the interaction effect depicted in Figure 6.10. To show the causal order, we reproduce the relevant path diagram from Figure 6.10 (as Figure 6A.1), which shows an interaction effect of X1 and X2 on Y.

Here we have proceeded on the assumption that small differences in means and percentages are just "noise." A major theme of the rest of the book is developing methods that tell us when a difference between two groups is probably just a result of random error and when it should be taken seriously. Or put in the language of models, we will explore when the effect of an independent variable is large enough that the data support a model of causation and when the effects are so small that the data do not lend support to the model.

Example 3:   Why are some countries more likely to participate in environmental treaties than others?

In this application, we consider two independent variables that could influence treaty participation: the number of NGOs and level of voice and accountability in nations. Here again we have to be cautious about what we assume about causal ordering. We will assume that environmental treaty participation does not affect the number of NGOs or level of voice and accountability in a country. We consider these two variables causally prior to environmental treaty participation. It is expected that NGOs may exert pressure on countries to participate in treaties, so a larger number of NGOs may increase the number of treaties signed by a nation's government. Additionally, governments that allow greater voice among their citizens and are more accountable to their citizens may be more likely to participate in environmental treaties due to pressure from citizens. The causal order between the two independent variables is harder to establish on theoretical grounds. It is plausible that having an open government would permit more NGOs to have a presence, but it is also plausible that NGOs could exert pressure on governments to give citizens rights

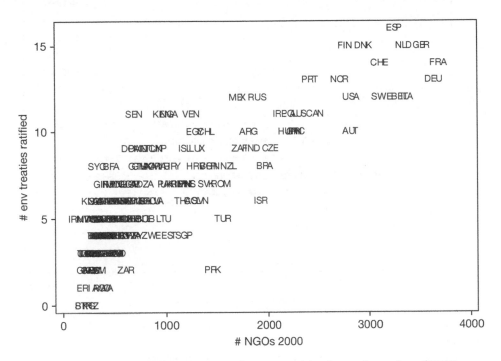

Figure 6A.2 Scatterplot of environmental treaty participation and number of NGOs
Data source: Roberts et al., 2004, UNDP, 2002, analyzed with Stata.

and participate in government. So we will make no assumptions about the causal order between number of NGOs and level of voice and accountability.

Figures 6A.2, 6A.3, and 6A.4 show scatterplots for the three variables. These locate each data point with the country abbreviation. While we recognize it is difficult to read all the country abbreviations on the scatterplots, it is useful for illustrative purposes to mark the countries. We could have decided to use a small circle to mark the data points instead. The patterns of relationships between number of NGOs and treaty participation and between voice and accountability and treaty participation are relatively strong. The pattern on the scatterplot between NGOs and voice and accountability is not as clear, with nations with a large number of NGOs having high voice and accountability rates but nations with smaller numbers of NGOs having the full range of scores on the voice and accountability index. Now we can break the countries into two groups, those with voice and accountability rates above the median and those below the median and look at scatterplots of environmental treaty participation versus number of NGOs.

Figure 6A.5 shows the scatterplot of treaty participation and number of NGOs for countries with voice and accountability rates less than the median. There is a positive relationship between environmental treaty participation and number of NGOs, with more NGOs associated with participating in more environmental treaties. Figure 6A.6 is a scatterplot of treaty participation and number of NGOS for those countries above the median voice and accountability rate of −0.07.

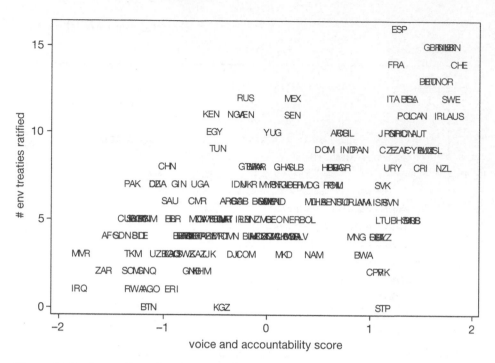

Figure 6A.3 Scatterplot of environmental treaty participation and voice and accountability rate

Data source: Kaufmann et al., 2002, Roberts et al., 2004, analyzed with Stata.

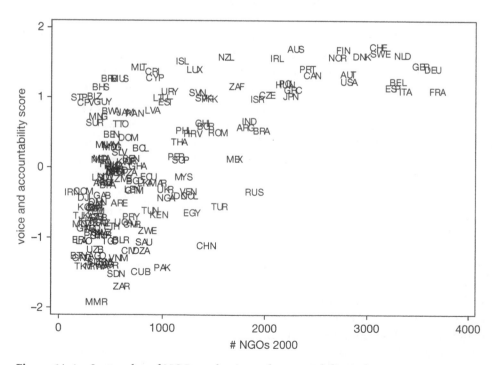

Figure 6A.4 Scatterplot of NGOs and voice and accountability index

Data source: Kaufmann et al., 2002, UNDP, 2002, analyzed with Stata.

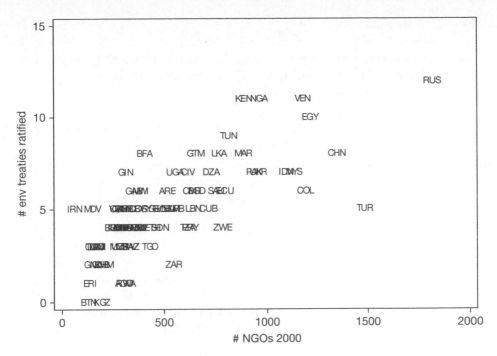

Figure 6A.5 Scatterplot of environmental treaty participation and number of NGOs for countries with voice and accountability rates less than the median of −0.07
Data source: Kaufmann et al., 2002, Roberts et al., 2004, UNDP, 2002, analyzed with Stata.

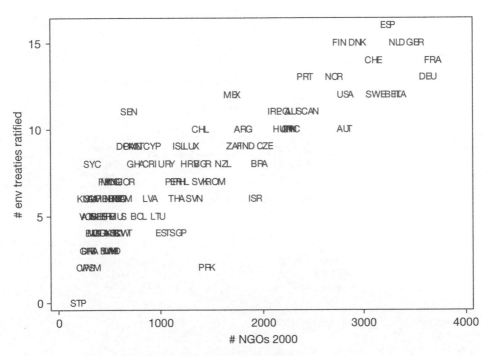

Figure 6A.6 Scatterplot of environmental treaty participation and number of NGOs for countries with voice and accountability rates greater than the median of −0.07
Data source: Kaufmann et al., 2002, Roberts et al., 2004, UNDP, 2002, analyzed with Stata.

Similar to the graph for countries under the median voice and accountability rate, the relationship between NGOs and treaty participation is positive for countries with greater voice and accountability.

Exercises

1. Draw a path diagram to reflect each of these statements.

a) Time spent using the Internet is negatively correlated with time spent interacting with people face-to-face.

b) Childhood poverty is predictive of lower educational achievement and poverty in adulthood. Lower educational achievement is predictive of a greater likelihood of experiencing poverty as an adult.

c) Durkheim's structural functionalist theory states that religion is a key social ingredient in decreasing the rate of suicide caused by the disruption of the nuclear family.

d) Having an unstable family situation is predictive of a greater likelihood of drinking excessively, and in turn excessive drinking is predictive of more sick days taken from work.

e) The economic prosperity of a country predicts individuals' life satisfaction.

f) Poor people living in poor countries are less satisfied with their lives than are poor people living in affluent countries.

2. Write out the relationships that are depicted in the path diagrams in Figures 6E.1, 6E.2, and 6E.3.

a)

Figure 6E.1   Age and internet shopping

b)

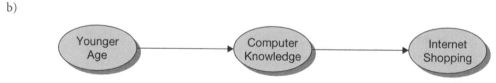

Figure 6E.2   Age, computer knowledge, and internet shopping, Model 1

c)

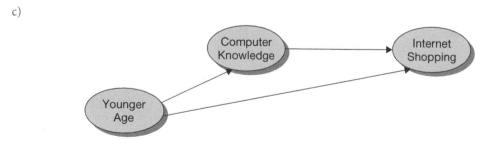

Figure 6E.3   Age, computer knowledge, and internet shopping, Model 2

3. A researcher is interested in studying the relationships between the following four variables: weekly average number of hours spent working, childhood exposure to volunteer activities, gender, and weekly number of hours spent volunteering. Draw a path diagram of the most logical causal ordering of these variables. Briefly write a justification for your causal diagram.

4. Let's consider whether marital status and gender relate to the belief that "divorce is the best solution to marital problems." Responses to this question were: 1 "strongly disagree;" 2 "disagree;" 3 "neither agree nor disagree;" 4 "agree;" and 5 "strongly agree." Data on the relationships between these variables are taken from the 2000 ISSP data set. Summaries are shown in Tables 6E.1, 6E.2, 6E.3, and 6E.4. Be sure to include in your interpretation the following: Which variables are associated? How did you determine that these variables are or are not associated? Is there an interaction? On what basis did you determine this?

Table 6E.1 Marital status by gender (%)

	Men	Women	Total
Married	60.4	56.2	58.0
Widowed	3.7	11.8	8.2
Divorced	5.0	7.4	6.4
Separated (but married)	2.0	2.4	2.2
Never married	28.9	22.2	25.2
Total	100.0	100.0	100.0

Data source: 2000 ISSP data set, analyzed with SPSS.

Table 6E.2 Mean scores on belief about divorce by gender

Gender	Mean	N
Men	3.60	19,549
Women	3.64	24,693
Total	3.62	44,284

Data source: 2000 ISSP data set, analyzed with SPSS.

Table 6E.3 Mean scores on belief about divorce by marital status

Marital status	Mean	N
Married	3.56	25,528
Widowed	3.67	3,529
Divorced	3.94	2,877
Separated	3.90	1,004
Never married	3.65	10,992
Total	3.62	43,930

Data source: 2000 ISSP data set, analyzed with SPSS.

Table 6E.4 Mean scores on belief about divorce by gender and marital status

Gender	Marital status	Mean divorce belief score	N
Men	Married	3.55	11,756
	Widowed	3.70	700
	Divorced	3.87	1,001
	Separated	3.84	406
	Never married	3.64	5,572
	Total	3.60	19,435
Women	Married	3.56	13,757
	Widowed	3.66	2,828
	Divorced	3.98	1,871
	Separated	3.93	597
	Never married	3.66	5,413
	Total	3.64	24,466

Data source: 2000 ISSP data set, analyzed with SPSS.

5. A researcher is interested in whether there are gender differences in the propensity to spend evenings in a bar. Data from the 1996 General Social Survey on this topic are presented in the Table 6E.5. The frequency of bar attendance is coded as (1) "never" to (7) "almost daily". The researcher then wonders if the gender differences will still remain when age is considered. To examine this question, ages were broadly collapsed into two categories: (1) ages 40 and younger; and (2) ages 41 and older. Based on the data presented in Tables 6E.5, 6E.6, and 6E.7, summarize the relationships between gender, age, and frequency of bar attendance.

Table 6E.5   Mean frequency of bar attendance by gender

Gender	Mean	N
Men	2.72	857
Women	2.17	1,094

Table 6E.6   Mean frequency of bar attendance by age group

Age category	Mean	N
Aged 40 or under	2.86	905
Aged 41 or older	2.03	1,042

Table 6E.7   Mean frequency of bar attendance by age group and gender

		Mean	N
Men	Age 40 or younger	3.15	405
	Age 41 or older	2.34	452
Women	Age 40 or younger	2.62	500
	Age 41 or older	1.79	590

References

Achen, C. H. 1986. *The Statistical Analysis of Quasi-Experiments*. Berkeley, CA: University of California.

Blalock, H. M., Jr. 1960. *Social Statistics*. New York: McGraw Hill.

Blalock, H. M., Jr. 1985. *Causal Models in the Social Sciences*. Hawthorne, NY: Aldine.

Campbell, D. T. and Stanley, J. C. 1963. *Experimental and Quasi-Experimental Designs for Research*. Chicago: Rand McNally.

Cook, T. D. and Campbell, D. T. 1979. *Quasi-Experimentation*. Chicago: Rand McNally.

Duncan, O. D. 1966. Path analysis: Sociological examples. *American Journal of Sociology* 72, 1–16.

Gaither, C. C. and Cavazos-Gaither, A. E. 1996. *Statistically Speaking: A Dictionary of Quotations*. Bristol, England: Institute of Physics Publishing.

International Social Survey Programme (ISSP). 2000. 2000 Environment II data set. www.issp.org. Catalog no. ZA 3440. Cologne, Germany: GESIS-ZA Central Archive for Empirical Research.

Kaufmann, D., Kraay, A., and Zoido-Lobaton, P. 2002, Jan. *Governance Matters II: Updated Indicators for 2000/01*. Policy Research Working Paper no 2772. The World Bank Research Development Group and World Bank Institute; Governance, Regulation and Finance Division (http://hdr.undp.org/reports/global/2002/en/).

Roberts, J. T., Parks, B. C., and Vásquez, A. A. 2004. Who ratifies environmental treaties and why? Institutionalism, structuralism and participation of 192 nations in 22 treaties. *Global Environmental Politics* 4(3), 22–64.

Seyfrit, C. L. and Martin, J. P. 1985. Status and social distance in the faculty office: A replication. *West Georgia College Review* 17, 21–7.

Simon, H. A. 1953. Causal ordering and identifiability. In W. Hood and T. Koopmans (eds), *Studies in Econometric Methods.* New York: John Wiley, pp. 49–74.

Simon, H. A. 1952. On the definition of the causal relation. *Journal of Philosophy* 49, 517–28.

Simon, H. A. 1954. Spurious correlation: A causal interpretation. *Journal of the American Statistical Association* 49, 467–92.

Sobel, M. E. 1995. Causal inference in the social and behavioral sciences. In G. Arminger, C. C. Clogg, and M. Sobel (eds), *Handbook of Statistical Modeling for the Social and Behavioral Sciences.* New York: Plenum.

Sobel, M. E. 1996. An introduction to causal inference. *Sociological Methods and Research* 24, 353–79.

Stern, P. C. and Kalof, L. 1996. *Evaluating Social Science Research.* 2nd edn. New York: Oxford University Press.

Stouffer, S. 1962. *Social Research to Test Ideas.* New York: Free Press.

Uganda Demographic and Health Surveys (DHS). 2000. *Ugandan-DHS 2000 Survey.* Calverton, MD (www.measuredhs.com).

United Nations Development Programme (UNDP). 2002. Human Development Report. Deepening Democracy in a Fragmented World. New York: Oxford University Press (http://hdr.undp.org/en/reports/global/hdr2002/).

US Census Bureau (2002). Historical poverty tables: Table 21. Number of poor and poverty rate, by state: 1980 to 2006. Year 2002 (http://www.census.gov/hhes/www/poverty/histpov/hstpov21.html).

US Census Bureau (2003). Table 295. Crime rates by state, 2002 and 2003, and by type, 2003 (http://www.census.gov/prod/2005pubs/06statab/law.pdf).

Wright, S. 1921. Correlation and causation. *Journal of Agricultural Research* 20, 557–85.

CHAPTER 7
PROBABILITY

Outline

To this point, we have discussed descriptive statistics. If statisticians had developed only descriptive statistics, you would probably not be required to take a statistics course. The material covered so far could be included in a methods course or in substantive courses in sociology and other disciplines. But over the last three centuries, statisticians have developed a tremendously useful conceptual framework for dealing with luck and error, and many tools for working within that framework. The major challenge in learning statistics is to understand that framework. When you understand how statisticians think about chance – the random part of life and science – you understand the essence of statistics. Unfortunately, many people learn the tools without understanding the concepts behind them. As a result, the tools are applied blindly.

In this chapter, we will help you understand how statisticians think about random events, and how randomness is linked to our efforts to build models that describe the world. We begin by talking about luck and random errors in models. In the next chapter we describe the fundamental concept of statistics: the sampling distribution. Then we show how the idea of the sampling distribution leads to tools for estimating what is true in the face of random error, and for testing hypotheses that are assertions about what is true.

Life is filled with uncertainty. Nearly every action we take in day-to-day life is based on assessing uncertainty. Will my professor be on time for class? Will it rain? Much of what happens in the world cannot be predicted. Quantum mechanics posits that random events are at the very foundation of the natural world. Chaos and complexity theory suggest that even when mathematical laws drive a set of events, if those equations are even slightly complex, the events cannot be predicted in any reasonable way.[1]

Despite the fact that uncertainty permeates human life, we are uncomfortable with it. Einstein won a Nobel Prize for work that underpins quantum theory, but he never reconciled the randomness at the heart of quantum mechanics with his own philosophy of science. He is reputed to have said: "God does not play dice with the universe."[2] As noted in chapter 1, learning to deal with uncertainty comes later in human cognitive development than other quantitative reasoning skills.

Many studies show that people have difficulty thinking through problems that involve random events. Perhaps the most telling evidence of this is that lotteries and casinos continue to flourish even though their profits rely on people having expectations about games of chance that deviate from how random events really work. For example, some gamblers claim to have "a feeling" or "intuition" about whether the next card will be an ace or a king, and so on. An even more common error that we will examine below is the belief that some result is "due." If a coin has come up heads three times in a row, it seems reasonable to think that a tail must be next, if red hasn't come up on a roulette wheel in four spins, surely it must come up next. These beliefs are wrong, and can be very costly if you wager large sums based on them.

Given how hard it is to think clearly about random events, the development of statistics over the last three centuries should be viewed as one of the great triumphs of science. It seems especially profound because chance impinges so heavily on all aspects of human life. An ability to deal with uncertainty in a systematic way may be essential to material wealth so the evolution of our understanding of probability is an important element in the growth of affluence in recent human history (Bernstein, 1998). But even scientists often see statistics as a dull set of tools, necessary to master for writing publishable papers, but not particularly profound. We disagree. A solid understanding of statistics leads us to the heart of how the world works. The problem is that people too often learn the tools without understanding the view of the world that underpins those tools.

What is Random?

Random events are those we cannot predict with certainty. This intuitive approach suggests that there are degrees of randomness. Things that can be predicted perfectly without much trouble are clearly not random, while things that cannot be predicted at all are totally random. In between are things that can be predicted, but not with certainty.[3]

The most common example of a random event is the result of flipping a coin. While there are certainly physical forces that determine the outcome of the coin flip – the characteristics of the coin itself, details of the forces applied when the coin leaves the hand, air currents, characteristics of the surface on which the coin lands, etc. – in practice the outcome cannot be predicted. Thus the coin flip can be seen as a purely random event, with no deterministic factor that aids our ability to predict how it will come out.

Probability

All possible definitions of probability fall short of the actual practice. (William Feller, *An Introduction to Probability Theory and Its Applications*, p. 19, quoted in Gaither and Cavazos-Gaither, 1996, p. 167)

It is remarkable that a science that began by considering games of chance should itself be raised to the rank of the most important subject of human knowledge. (Pierre-Simon Laplace, *A Philosophical Essay on Probabilities*, p. 123, Quoted in Gaither and Cavazos-Gaither, 1996, p. 172)

Bayesian statistics are discussed at the end of this chapter as an Advanced Topic.

Feature 7.1   A Brief Note on Thomas Bayes

Thomas Bayes was a member of the English clergy whose argument about uncertainty is central to the modern approach to probability. We know very little about Bayes, even though he was a Fellow of the Royal Society, a considerable honor. We know of only two papers written by him, both published after his death in 1761. It took a while for them to be noticed, but when they were, they helped clarify thinking about probability. In fact, one approach to probability is called "Bayesian" in his honor. We will discuss the Bayesian approach in Advanced Topic 7.1.

Figure 7.1   Thomas Bayes
Source: http://en.wikipedia.org/wiki/ThomasBayes (includes a discussion of this portrait)

Feature 7.2   Making Decisions about Uncertainty

One of the places where Bayes' work is most often used is in the analysis of uncertainty when decisions have to be made. Work by cognitive psychologists has shown that when people are given probability problems, they don't handle them very well. Consider the following problem, one that is like many that come up in medicine.

A pregnant woman has taken a test to see if it is likely that the child she carries has Down's syndrome. Unfortunately, the test came back positive indicating Down's. The health care worker who is going to counsel her knows that 0.15 percent (that is 0.0015 as a proportion) of babies born have Down's syndrome. And she knows the tests are not

perfect. For pregnancies that will lead to a child with Down's syndrome, the test says they have Down's syndrome 80 percent of the time, and thus misses detecting Down's 20 percent of the time. The test also can give a false positive – 8 percent of the women who the test says may give birth to a child with Down's in fact won't. So, what should the health care worker tell the pregnant woman about the chances that the child she would bear would have Down's?

If you are not sure how to figure this out, you are not alone. When this kind of problem is given to medical students and even to physicians, three-quarters of them get the answer wrong, and in fact, very wrong. They can be off by a factor of ten (Hoffrange, Lindsey, and Hertwig, 2000). The correct answer is that the health care worker should tell the woman that the chances a baby would have Down's syndrome is 1.5 percent. Below we will show you how to solve this problem.

This problem and others like it have led to a substantial literature showing that people have a very difficult time working with information on probability.[4] This has led to arguments that people's judgments about things that involve risk, such as environmental problems and technology, shouldn't be trusted and that perhaps public participations in such decisions should be limited. We disagree. Jaeger, Renn, Rosa, and Webler (2001) review the arguments about risk and make compelling arguments for keeping the public involved even in decisions that involve complex probabilities.

The problem with the genetic testing case is that the information is presented in terms of probabilities. A number of studies have shown that while people have a lot of trouble working with probabilities, we do reasonably well with frequencies! This seems odd, given that frequencies and probabilities can be converted from one to the other. But it may make evolutionary sense. We have evolved dealing with how many fruit trees might be found in a grove, how many lions may be stalking us and other problems that emerge as counts rather than percentages or proportions that represent probabilities.[5]

What happens when we couch the problem in frequencies instead of probabilities? Figure 7.2 will show us. Suppose we start with 10,000 pregnant women (and to keep things simple we will ignore that some of these women would have twins). The medical literature says that 0.15 percent will have Down's Syndrome, so that's 10,000 times 0.0015, or 15. That means that 9,985 won't have Down's. This is the second row in the figure. For the 15 who will have Down's, on the left side of the figure, the test picks it up and makes the right assessment 80 percent of the time. That means 0.80 times 15 or 12 women whose babies will have Down's will get a positive test result. For the 9,985 women, the test gives a "false positive" and says the baby will have Down's when it won't 8 percent of the time. That means the 9,985 times 0.08, or about 799 women will get false positives. The remaining 9,186 women will get a test result that says the baby won't have Down's when that is correct.

Now look at the last row. Two groups of women have a positive test result. Twelve women at the far left of the row have a

Equal Probability and Independence

The examples of random processes that come to mind for most people are processes in which all outcomes have the same probability. In tossing a fair coin, heads is as likely as tails (this is what we mean by a fair coin). In tossing fair dice, each of the six sides has the same chance of coming up. But random processes do not have to have all outcomes with equal probability. A coin might be loaded to come up heads 2/3 of the time, or dice loaded so that 6 comes up more often than any other side. While the loaded coin or dice would still qualify as random processes,

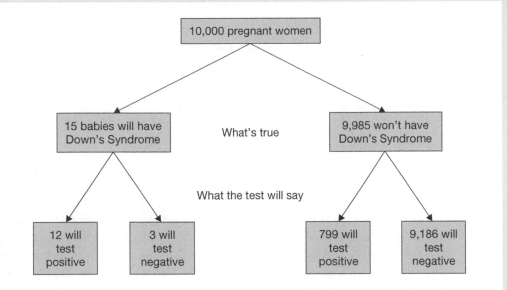

Figure 7.2   Frequency of Down's test results for 10,000 women

positive test, and it is unfortunately accurate in that their babies will have Down's. But 799 women have a positive test result that is a mistake. So if a woman has a positive test, what are the chances that her baby will have Downs? It's 12 (the number of correct positives), divided the number of positives altogether, which is 799 + 12 = 811. Those chances are 12/811 or about 1.5 percent. So the health professional can tell the woman that while in the general population of pregnant women the chances of a baby with Down's is about 0.15, her chances are ten times that great of having a baby with Down's. But the chances are still small.

Notice that there are two ways a test can be wrong when things are uncertain. It can tell us there is a problem when the baby would be healthy (a false positive). Or it can say everything is okay when it is not (a false negative). We will see these two kinds of errors again later in the book.

they are not equal probability processes. And if we think back to the definition of random as "unpredictable," the loaded coin or dice are less random than the fair coin or dice in that the outcomes, while not perfectly predictable, are more predictable than the equal probability process.

Repeating equal probability processes can lead to outcomes that are not equally probable. If we toss a coin once, we have a 50 percent probability of getting a head and a 50 percent probability of getting a tail. If we toss the coin twice and keep track of the order in which we get heads or tails, there are four outcomes: HH, HT, TH, and TT. Each of these is equally probable. But if we ignore the order of the toss and just count heads and tails, then there is one way in four of getting two

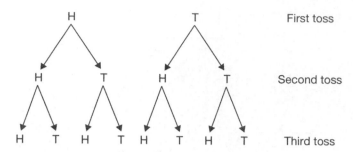

First toss

Second toss

Third toss

Figure 7.3   Probability of outcomes for three tosses of a fair coin

heads (1/4 or 25 percent), one way in four of getting two tails (1/4 or 25 percent), but two ways out of four of getting a head and a tail (2/4 or 50 percent). So getting two heads or two tails is less likely than getting one of each. The "tree" in Figure 7.3 shows the probability of getting various outcomes. The second row describes two tosses, and the third row describes three tosses. Each branch is equally probable, but if you ignore the order in which heads and tails come up, some mixes of heads and tails are more probable than others.

For three tosses, there are $2 \times 2 \times 2 = 8$ possible outcomes. There is only one way to get all heads, the furthest left path. Thus the chances of getting all heads is one out of eight, or 0.125. The only way to get all tails is the further right path, so the probability of all tails is also 0.125. But there are three ways to get two heads and one tail: HHT, HTH, and THH, so the chances of two heads in three tosses is 3 out of 8 or 0.375. The chances of two tails and a head are also 0.375. Note that this covers all the possibilities, and thus sums to 1 ($0.125 + 0.125 + 0.375 + 0.375 = 1$).

Independent events

In most statistical analyses, we will assume that the errors generated by the random part of the model are independent. **Independence** between events is an idea with a special statistical meaning. It is *not* the same idea as that of an independent variable. If two events are independent, it means that knowing one does not allow you to predict the other. Statisticians sometimes refer to these as processes that have no memory. In other words, if the events occur one after another, the second event takes no account of what happened on the first.

This is a hard concept for most people to grasp, at least if human behavior is any evidence. As noted above, people in casinos have the feeling that a slot machine is "ready" to pay off, or that a certain number is "due" on the roulette table. But for that to be true the slot machine or the roulette ball would have to have a memory of what has come before and know that a particular set of symbols or numbers has not come up for some time, and then somehow make those symbols or that number appear. Think about a coin toss. If we have four heads in a row, we may think that tails are "due" since with a fair coin, tails should come up about

Table 7.1   Results of 5,000 tosses of a coin

Number of heads in a row	Prediction from theory as to how often this will happen	Actual proportion	Proportion of times tails came on the toss after this many heads
2	0.250	0.259	0.512
3	0.125	0.124	0.480
4	0.0625	0.060	0.486
5	0.03125	0.030	0.490
6	0.015625	0.015	0.507
7	0.0078125	0.008	0.533
8	0.00390625	0.003	0.425
9	0.001953125	0.002	0.529

the same amount as heads. But the coin doesn't keep a record of what has happened, so each time the coin is flipped the chances of a head or a tail is still 50:50.

To demonstrate this, we had the computer simulate flipping a coin 5,000 times, then counted how many times we got a string of heads. If statistical theory is correct, no matter how many heads have come in a row, heads and tails will each come up on the next flip around 50 percent of the time. If gamblers are correct that the event that has not occurred recently is "due" to happen, then we should see tails more than 50 percent of the time after a long string of heads.

Table 7.1 shows the results, including the predictions from theory as to how often a certain run of heads will occur, the actual occurrence of runs of heads in the 5,000 tosses, and how often a tail came up after a run of heads. In the first row, we are looking at how many times the theory says we should get two heads in a row – one-half times one-half or 0.25. That's the second column. The third column tells us how often we actually got two heads in a row, 0.259 – pretty close to the theory. The third row tells us how often, after getting two heads in a row, the coin came up tails. Theory says it should come up half the time after two heads. In fact it came up 51.2 percent of the time – again, pretty close.

Look at the last row. Theory says we should get nine heads in a row a bit less often than 0.2 percent (0.002) of the time. In fact, we got 0.2 percent, just a bit more than theory predicts. Imagine you were playing a game based on flipping the coins. You'd probably be convinced that after nine heads in a row, surely a tail is due. In fact, in this experiment, a tail comes up after 9 heads about 53 percent of the time, very close to the 50 percent predicted by theory and not at all as if a tail is "due." If we did the experiment over again, we might find that "heads" would come up a bit more than 50 percent of the time or a bit less but it would almost certainly be around 50 percent. It's pretty clear that our simulated "coin" is not remembering what has happened in past flips.

As you can see, statistical theory does a pretty good job of predicting how often a certain run of heads will occur. Furthermore, there seems to be no increase in the chances a tail will come up after a long sequence of heads. The proportion is always around 50 percent and does not get systematically higher as the number of heads in a row increases. The coin does not know that the last eight flips in a row were heads, so it is as likely to come up heads as tails.[6]

As we mentioned above, the fact that a coin (or the universe) has no memory of what has just happened has a special name in statistics. Statisticians call events that cannot predict one another independent. The outcomes of coin flips, rolls of dice, spins of the roulette wheel, and the random numbers generated by computers are independent events.[7]

There are also many everyday examples of events that are not independent, although they are still random. In most state lotteries, numbers are picked by placing about 40 balls, each with a number from 01 to 40, in a hopper. Then six balls are drawn at random after the balls are bounced around a lot. Since there are 40 balls, and each number from 1 to 40 appears on one ball, the chances that any particular number appears on the first ball drawn is 1 out of 40 or .025 (2.5 percent). But once that ball is drawn, the odds for any particular number that was not drawn as the first number is now 1 out of 39, or about 2.6 percent, since there are only 39 balls left. By the sixth draw, there are five numbers already drawn and 35 balls left, so the chances for a number not yet drawn are 1/35 or 2.8 percent. Here the process has a "memory" because the balls are not tossed back in the hopper and therefore cannot be picked again. This is different from a roulette wheel where there is no "memory" of what number has come up in the past.

One of the human limits in thinking about random events is that we seem to assume that random processes have memory, while many processes in the modern world do not. Most gambling processes are designed not to have memory.[8] But it makes evolutionary sense that we would process information about uncertain occurrences as if the random process had memory. For most of human history, we were food foragers dealing with plants, prey, predators, and weather. While the distribution of plants, prey, predators, and weather cannot be predicted with certainty, and thus are random processes, they are not processes without memory. If you have evidence that a lion was hiding in a patch of brush yesterday, it may be a good idea to avoid that area today. The lion is not a random number generator but a creature with a memory too. So our history has biased us towards thinking that random processes have memory.

Statistical tools exist for dealing with both independent and dependent events, but the mathematics for independent events is much simpler. As a result, we have a broader array of tools for independent events. Our general strategy in this book will be to examine tools that apply to independent events so as to build an understanding around the simpler problems in statistics.

In applying statistics that assume we are studying random events without memory, it is important to understand the process that generated the random element in our data to insure that it does come from a process without memory. Indeed, as we will see below, we often design studies to ensure that the random components in our data are independent – that they come from a process without memory.

The toss of a fair coin is the simplest example of an independent random process. There are only two outcomes, heads and tails, and each has equal probability – 50 percent. Statisticians call a random process with only two outcomes a **binomial process**. The plot of the outcomes of that process is called a **binomial distribution**. The flip of a fair coin has the added feature of equal probability. But a binomial process does not have to have equal probability. It can deal equally well with an unfair coin as long as we know the probability of heads and the probability of tails. For example, we can have a binomial process where the chance of one outcome is 0.75 and the other 0.25. But just as in a random process where independent events are easier to analyze than non-independent processes, equal probability processes are easier to manage than those with unequal probabilities.

The toss of a single dice has six possible outcomes. When a process has a countable number of possible outcomes, but more than two, statisticians call it a **multinomial process**. The multinomial process, like the binomial, can have equal probabilities for each outcome, or the outcomes can have different probabilities. With fair dice, every side has the same chances of coming up, so every number is equally likely. We call this a **uniform distribution**.

But while all numbers have an equal chance of coming up with one dice, things change with two dice. We still have a multinomial process, but the sum of the figures of the two sides that come up are not uniform. For example, suppose we add up the faces of two dice tossed at once. There are 11 possible values for the sum, so this is a multinomial process with 11 outcomes. Table 7.2 shows the outcomes.

Table 7.3 summarizes these results, showing how often each sum will occur in the long run, that is, how probable it is. Remember that there are 36 different outcomes for two different dice but only 11 different sums, so the probability is the number of times a sum occurs divided by 36. Getting a sum of seven is the most probable occurrence, popping up one-sixth of the time. Following the classic "7 come 11" of craps, 11 will come up only about 6 percent of the time. Since two tosses of the two dice are independent, 11 has a 6 percent chance of coming up whether or not it was preceded by a 7, assuming of course the dice are not loaded. A histogram of the results is displayed in Figure 7.4.

The results of tossing a single dice are a flat, uniform distribution. However, when we toss two dice and add together the numbers coming up, the resulting distribution of results is not flat, but takes on what will become a familiar bell-shaped form. We return to this pattern below. It represents one of the most important results in statistics. For the moment, there is one point to note. Adding together the faces that come up on two tossed dice produces a multinomial distribution with 11 values (2, 3, 4, 5, 6, 7, 8, 9, 10, 11, and 12) but not all those values have equal probability.[9] So distributions with unequal probability can come from combinations

Table 7.2 Results of the toss of two dice

First dice	Second dice	Sum
1	1	2
1	2	3
1	3	4
1	4	5
1	5	6
1	6	7
2	1	3
2	2	4
2	3	5
2	4	6
2	5	7
2	6	8
3	1	4
3	2	5
3	3	6
3	4	7
3	5	8
3	6	9
4	1	5
4	2	6
4	3	7
4	4	8
4	5	9
4	6	10
5	1	6
5	2	7
5	3	8
5	4	9
5	5	10
5	6	11
6	1	7
6	2	8
6	3	9
6	4	10
6	5	11
6	6	12

of distributions in which every event is equally probable. Note also that the sum divided by the number of dice we added (two) would be the mean score on the dice. So the distribution of the mean score would also not be uniform. Again, we will make use of this fact in later chapters.

Table 7.3   Results of summing two dice

Sum of two dice	Number of ways this sum can occur	Probability of this sum
2	1	0.028
3	2	0.056
4	3	0.083
5	4	0.111
6	5	0.139
7	6	0.167
8	5	0.139
9	4	0.111
10	3	0.083
11	2	0.056
12	1	0.028

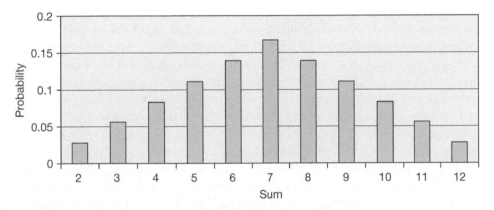

Figure 7.4   Probability of various numbers as the sum of two dice

Random Variables and Models of Error

The kinds of models described previously are helpful here. We can think of an event as having two components, a certain part and a random part.

$$\text{Observed event} = \text{deterministic factors} + \text{random factors} \qquad (7.1)$$

If the event is not at all random, then the random factors play no role in predicting the event. Many of the laws of physics have this character, at least at the level of human experience. If you drop a ball from a height and there is no wind, then the time the ball takes to reach the ground and the speed it will be going when it lands are determined with no random factor. Of course, if you can't measure the time and speed you will have measurement error that will lead to a random component.

In the realms studied by the social and biological sciences, few things occur that do not have a random component. If the event is totally random, then deterministic factors play no role. Most events in the world fall in between, with some deterministic factors that let us predict but also some random factors that keep those predictions from being perfect.

Statisticians sometimes speak of random variables, by which they mean a variable in which the random factors are important or even in which there are no deterministic factors. The outcome of flipping a coin or tossing one dice or several dice can be thought of as purely random variables. Because most things we study have some random component, an understanding of that random component can help us understand the non-random part as well. In statistics, we use knowledge of how random variables behave to understand our data.

Probability distributions

The frequency distribution of a random variable is called a **probability distribution** because the histogram shows us what kinds of outcomes are common and what kinds are rare – it shows the probability of different outcomes. Above we considered three probability distributions. The first was the binomial distribution, which is just the probability of one event or the other when there are two outcomes. We then looked at a multinomial distribution with equal probability – the results of tossing a single dice. The third was a multinomial distribution with unequal probabilities – the sums of the sides that come up when we toss two dice.

These are called *discrete distributions* because we know all the possible results (heads or tails, one of six sides of a single dice, one of 11 possible sums from tossing two dice and adding the sums of what comes up), and there are a countable number of them – two for the toss of one coin, six for the toss of one dice, 11 for the tossing two dice and adding the faces that come up. Thus these distributions correspond to qualitative or discrete variables. We can also have probability distributions that are continuous, like the continuous variables discussed earlier. We can think of these continuous distributions as a situation where the discrete distribution takes on so many values that the frequency plotted in a histogram looks like a smooth curve rather than the set of bars.

Statisticians think of distributions as the product of equations that predict the probability of getting a number for any value of that number. However, an easier strategy is to think of a continuous probability distribution as just a frequency distribution in which there are many, many possible values, each with a probability of occurring.

Probability distributions are used in describing random processes. Because some things that happen in the world are truly random, and many things have some random component, probability distributions are sometimes useful in describing data. But more important, probability distributions help us to understand how large the random component is in things we observe, thus learning more about the non-random component.

Returning to the logic of models, we can think of the errors that make our observations differ from the truth as numbers drawn from some random distribution. It is as if nature added (or subtracted) a random number that comes from the flip of a coin or the toss of a dice, or the sum of tossing many dice, to the truth. So what we observe is the truth plus some random error. By knowing how the random part of our observation behaves, we get a better understanding of the truth. We will return to this point at the end of the chapter.

The Normal Distribution and the Law of Large Numbers

As we have seen earlier, the most famous continuous probability distribution is the *Normal distribution*. It first was discovered as astronomers were learning to deal with measurement error. When an astronomer measures some variable describing a star or planet, she will get slightly different results each time she does the measurement. Galileo noted that: "1) these errors seemed unavoidable; 2) small errors are more likely than large ones; 3) measurement errors are symmetrical (equally inclined to overestimate as to underestimate); and 4) the true value of the observed constant is in the vicinity of the greatest concentration of measurements" (Bennett, 1998, p. 90).

In 1778, the mathematician Marquis Pierre-Simon Laplace discovered something we noted above: the sums of uniform random numbers take on a bell-shape that looks like the kind of distribution Galileo was describing. In 1808 an American, Robert Adrian, described this bell-shape mathematically. In 1809 the German scientist Carl Friedrich Gauss independently made the same discovery. Even though Adrian made the discovery first, it has come to be associated with Gauss, thus the distribution is often called the Gaussian distribution, though we will refer to it as the Normal distribution.

Finally, by 1812 LaPlace had put together his work on the sums of random variables with Gauss's work on the mathematical description of observational errors. He produced what is called the **Central Limit Theorem** or the **Law of Large Numbers**. It says that if you add together a large number of independent random numbers (like the results of tossing a dice), the result is a very specific bell-shape. This theorem also applies to the mean, since the mean is just the sum of a set of numbers divided by how many numbers there are.

This meant the measurement error in astronomical observations which took on that bell-shaped curve could be described as a mix of the actual value of the variable being measured and an error that can be thought of as drawn from a

Figure 7.5 Pierre-Simon Laplace
Source: http://commons.wikimedia.org/wiki/File:P_S_Laplace.jpg

normal probability distribution. It made sense to think of the measurement error as a number drawn from a Normal distribution because it could be considered the result of many small things. Two measurements of the position of a star might differ because of many factors – slight changes in the breeze, the temperature, the fatigue of the observer, and so on.

Because many kinds of measurements being made in the sciences in the nineteenth century seemed to have this pattern of error, the distribution was usually called the *Normal distribution*. Figure 7.7 shows a Normal distribution of 5,000 observations. The statisticians' Normal distribution is a theoretical curve that has an infinite number of observations. This curve is superimposed on the histogram for the 5,000 data points.

The Law of Large Numbers is quite remarkable. It says that when you add together many random numbers (or take their average) the result is not chaos, as you might expect, but something quite orderly, in fact more orderly than your original random numbers. It says that these sums or averages of random numbers take on a very specific pattern, one that can be described quite precisely with mathematics.

Let's see how well it works. We drew independent pairs of values from a uniform distribution and added them together. Remember, the **uniform distribution** is one where every number in some range is equally likely to show up. It's rather like an **equiprobability** binomial or multinomial distribution, except there are an unlimited number of numbers rather than some countable number (like two for coins, six for dice or 52 for cards). This uniform distribution is limited to the range between

Figure 7.6 Carl Friedrich Gauss
Source: http://commons.wikimedia.org/wiki/File: Carl_Friedrich_Gauss.jpg

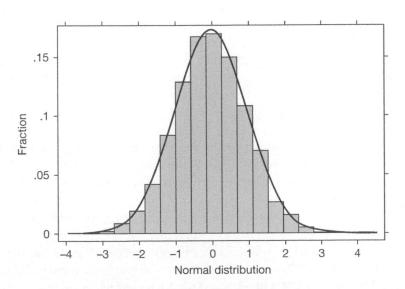

Figure 7.7 A Normal distribution of 5,000 observations

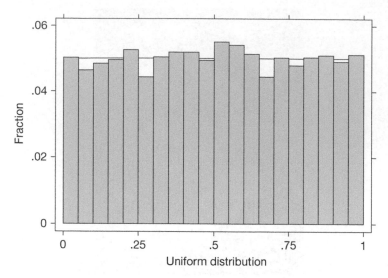

Figure 7.8   A uniform distribution of observations

0 and 1. So any number between 0 and 1 has equal chances of popping up. We did this 5,000 times. Figure 7.8 shows the uniform distribution we used.

In this little experiment, we asked the computer to draw two numbers and add them together. Figure 7.9 shows a histogram of the result, the sum of two continuous, uniform random variables. The solid line is the theoretical Normal distribution.

Remember that the Law of Large Numbers is supposed to work only when we add together lots of numbers (not just two). But as you can see the histogram based on just adding together two numbers is a pretty good match to the Normal. It is light at the tails, low in the center and high in the shoulders. But even with only two random numbers added together, the Normal distribution is a reasonable approximation of what we get.

Figure 7.10 shows the results of adding together five numbers drawn from the same uniform random distribution. Before, the sum ranged from 0 to 2 because we were adding two numbers that range from 0 to 1. Now we are adding five numbers, so the range is from 0 to 5.

Again, the Normal is a pretty close fit to the histogram, even though five hardly qualifies as "large." Recall what we are doing. We are choosing five random numbers that range from 0 to 1, where every number between 0 and 1 has the same chances of being chosen. Then we add those five numbers together to get the sum. We do this little process 5,000 times to see what happens when we try the experiment of drawing and adding many times. We plot these 5,000 sums in the histogram. The Law of Large Numbers says that when we add together many numbers (not five) drawn independently, and plot the histogram, it will have the special bell-shape called a Normal distribution. Here we find that even though we don't have many numbers, the Normal comes pretty close to matching the histogram.

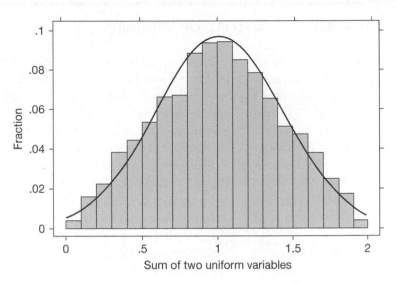

Figure 7.9   Distribution of the sum of two uniform variables

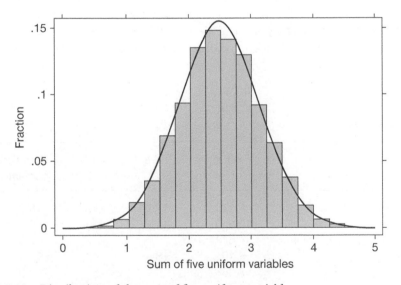

Figure 7.10   Distribution of the sum of five uniform variables

Feature 7.3   Cumulative Probability Distributions

Sometimes it's useful to ask how many observations (such as people, states, or countries of the world) fall above or below a particular value in a distribution. If we are working with a probability distribution then the number is the probability of being above or below that value. Recall that earlier in the book we looked at frequency tables that had a cumulative percent column in them to give this information. The table works well with a small number of values. For example, Table 7.4 gives the number and percent of people at particular values on the animal concern question, and the percent of people at or below each score.

Note that the table includes in the "Frequency" and "Percent" columns the count and percentage of people who have missing data because they did not answer the question. We put them in the table to remind you that you must always pay attention to the amount of missing data in surveys.

But let's focus on the last two columns that list the percent of those who answered the two questions at each value on the scale and the cumulative percent. Recall that the midpoint on the scale is 3. We find that 62.1 percent of the respondents fell below 3, 77.8 percent fell at 3 or below. We can subtract and find that 22.2 percent of respondents had a score above 3. Or we can see how far into the data we have to go to find a particular percentage. For example, nearly 92 percent fell at 4 or below. Figure 7.11 shows a bar chart with the same data.

Figure 7.12 presents the same information but instead of having a bar for each value, we have a line that runs as a summary of the percentage of cases that fall below each value on the X-axis. This is called a **cumulative distribution graph**. If you look at a score of 3 and then look across to the percentage you can see that just under 80 percent of respondents fell at 3 or below. Of course we

Table 7.4   Frequency distribution for animal concern question

			Frequency	Percent	Valid percent	Cumulative percent
Valid	1	Strongly agree	5,520	17.8	18.7	18.7
	2	Agree	12,803	41.2	43.4	62.1
	3	Neither agree nor disagree	4,617	14.9	15.7	77.8
	4	Disagree	4,161	13.4	14.1	91.9
	5	Strongly disagree	2,385	7.7	8.1	100.0
		Total	29,486	95.0	100.0	
Missing	8	Can't choose, DK	1,218	3.9		
	9	NA, refused	338	1.1		
		Total	1,556	5.0		
Total			31,042	100.0		

Source: 2000 ISSP data set, analyzed by SPSS

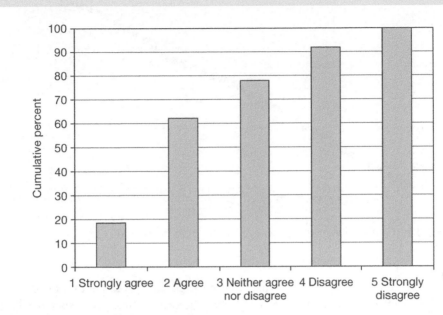

Figure 7.11   Bar chart of animal concern question

Figure 7.12   Cumulative distribution graph of the animal concern question

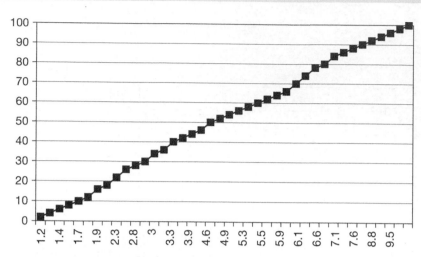

Figure 7.13 Cumulative distribution graph for state homicide rate

already know that from the table, but the graph and cumulative distribution bar chart are other ways of looking at the same thing. And if we remember that "percent" means proportion, we can say that the probability that any one person, drawn at random from our data set, falls at 3 or below is nearly 80 percent.

With an ordinal variable or a continuous variable that only has a few values, the table and bar chart give us precise answers about how many people fell where on the distribution. But with continuous variables with many values, the variable has to be lumped into groups before making a table or else the table will be very large with one row for each case. Figure 7.13 is the cumulative distribution graph for the state homicide rate.

If we go up the vertical axis to 50 percent then across to the line representing the cumulative percentage and down to the value of the homicide rate, we find that the corresponding homicide rate is at about 4.7. This means that about half the states should have a homicide rate of 4.7 or below. Of course, we already know that the median is 4.65, which tells us the same thing with more precision. But with the cumulative distribution plot we can work with any percentile within the limits of our ability to read the graph. Again if we think in terms of probability, if you draw a state at random from our data set, the chances are about 50 percent that the state will have a homicide rate below 4.7 and thus 50 percent of the states will be above that value.

As we will see later in the book, this regularity allows us to say some very precise things about random error in our observations, at least when we have many observations. Of course, there are many research problems where we don't have a large number of observations. Statisticians have also been able to develop tools that allow us to work with small samples, as we will see in later chapters.

Random Variables and Data Analysis

How does this excursion into ideas of probability help us understand data? It will be helpful to now review some of the sources of error described in chapter 1 and describe how random factors affect error.

Sampling error

Suppose we want to know what proportion of people in Uganda knows that condoms can help prevent the transmission of AIDS. The DHS asked 8,310 people this question and found that 77.3 percent of them (0.773 as a proportion) knew this. How accurate is this result? We certainly don't expect it to be exactly the exact result we would get if we asked everyone. But how far off is it likely to be?

Statisticians suggest that we can answer that question by thinking about our survey average in the following way (remember that the proportion saying yes is just the mean of a zero-one variable with 0 for no and 1 for yes):

Sample average = population mean + sampling error   (7.2)

Once again, we can think of the sampling error as some number drawn from a random distribution that is added to or subtracted from the population mean to create the sample average. (Or, to be more precise, it is a number drawn from a random distribution that includes both positive and negative numbers.) If we are careful about how we draw the sample, we can describe the sampling error with great accuracy, and thus develop a sense of how far off our sample average may be from the population mean.

Randomization error

Recall that in controlled experiments, we use a random distribution to assign subjects to experimental and control conditions. If there are only two conditions, one experimental and one control, then we can literally use the flip of a coin to make assignments. With more experimental conditions, we might use dice, or a random number generator on a computer, or a table of random numbers.[10] In Kalof's (1999) study of the effect of music video imagery on gender attitudes, she assigned students to experimental and control groups using a coin toss. The experimental group watched a video with stereotypical gender images, and the control group watched a video without gender stereotypes.

After the study, she found that the mean attitude score differed between the experimental and control groups. Why? There are two possible reasons. One is that the experimental condition influenced the people in that group. The other

Feature 7.4 Kinds of Samples and Probability

Statisticians have developed powerful methods for drawing samples from populations and using the sample data to provide accurate information about the population. The simplest and in many ways the most useful kind of sample is one in which every person in the population has an equal probability of being included in the sample (the equiprobability property mentioned above). We also design samples so that draws into the sample are **independent** – knowing who is already in the sample doesn't change the probability of who else will be in the sample. Samples based on drawing observations from the population with equal probability and independence are called **simple random samples**. They allow the simplest mathematical treatment of sampling error and thus are what we usually teach in a statistics course.

We can draw probability samples in other ways so that each person in the population has a defined probability of being in the sample but those probabilities are not all equal. Sometimes we want to oversample groups who are a small portion of the population to compare them to other groups. This is called **stratified sampling**. We might, for example, design our study to oversample African Americans, Asian Americans, Hispanic Americans and Native Americans because in a simple random sample too few members of these groups will appear to allow us to make comparisons. As long as we pick people from groups with known (but not equal) probability, we can still analyze the sampling error in the data. For example, we might design the sample so that each of these minority groups has twice the chance of appearing in the sample as European Americans, to boost the number of minorities. If we want to talk about the whole US population (the "average American") we would weight the data to take account of the oversampling.

In face-to-face interviews where travel costs matter, rather than drawing a random sample of households, as we do in telephone or mail surveys, we draw a random sample of counties in such a way that heavily populated counties have more chances of being in the sample. Then we interview people living in the counties in our sample, and we don't interview anyone in the counties that were not included in the sample. This **cluster sampling** "clusters" interviews in a smaller number of counties that would happen in the simple random sample. Again, as long as we know the probabilities we used to pick counties and people within counties, there are statistical tools to handle the sampling error.

In our applications, we are using the data as if they were a simple random sample. There are slightly more complex formulas for the statistical tools we will present in later chapters that take account of the violation of the equiprobability assumption. But we will only present the simpler formulas that apply to independent, equiprobability samples because they are easier to learn and give results that are very close to those from the more complex formulas.

Suppose instead we draw a convenience sample by interviewing the first 300 people we can find in a big city. This can be useful, but because there we have no idea what kind of random process drew people into the sample, we cannot say anything about the sampling error involved. Convenience samples can be useful (particularly in the early stages of research) because they are relatively inexpensive. But we have to be very careful in using statistical tools based on random sampling when analyzing convenience samples.

is luck of the draw – by chance, more people with stereotyped attitudes ended up in the experimental group than in the control group. We can model these explanations as:

Experimental result = experimental effect + randomization error   (7.3)

If we use a random process to assign people to groups, we can use statistical tools to determine how large the randomization error might be. We can compare the differences between the two groups to what would probably occur if only randomization error is creating the differences. If the difference between the two groups is substantially larger than we would expect from random differences, we assume that the experimental treatment had an effect. If the differences between the experimental and control group is the sort of thing that would occur because of randomization, we conclude that the experimental effect is zero, that is, the experimental treatment had no effect.

Measurement error

Scientists trying to understand measurement error developed important statistical theory. And measurement error remains an important motivation for much contemporary statistical research. We can see even in simple examples how statistical methods help us deal with measurement error.

We can think of measurement error in measuring animal concern. As we have noted in previous chapters, while the ISSP only asked one question to tap animal concern, the US survey asked a second question: "Animals have the same moral rights as humans." This was in addition to the question we have been using: "It is right to use animals for medical testing if it might save human lives." Respondents could "Strongly agree," "Agree," "Neither agree nor disagree," "Disagree," or "Strongly disagree" with each of these statements. If we were only using the US data, we could combine both questions into a scale of animal concern. The first question would be scored as 5, 4, 3, 2, 1 and the second as 1, 2, 3, 4, 5 so that high scores on each represent concern with the rights and welfare of animals.

A respondent might misinterpret the question or have an unexpected reaction to the terms used in the question.[11] Some of those factors may be the same on each question, and some may be different across the two questions. Let's call the first question X1 and the second X2. We are saying that both X1 and X2 may measure attitudes about animals but each may do so with error. This gives us the model:

Question response = actual attitude + random measurement error   (7.4)

or

X1 = actual attitude + random measurement error for X1   (7.5)

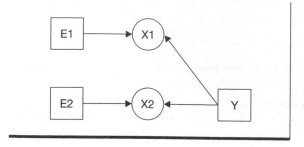

Figure 7.14   Path diagram for animal concern (Y), two questions about animal welfare (X1 and X2), and two error terms (E1 and E2)

and

$$X2 = \text{actual attitude} + \text{random measurement error for X2} \qquad (7.6)$$

In this example, measurement error can be thought of as a multinomial process. We think of the question response as starting with the person's real attitude, and then having a number added to or subtracted from it (with the constraint that the result has to be between 1 and 5). Statistical theory tells us that if this is a reasonable model for measurement error, then a good way to estimate someone's attitude is to add together the scores on the individual items.[12] The path diagram for this kind of model is in Figure 7.14, with E1 and E2 used as symbols for the two errors. The idea is that if the Es are really random, then we get a better estimate of the attitude by adding X1 and X2 together. This idea goes back to Galileo, who found that the measured values centered around the true value.

Students are very familiar with this approach to measurement error. It is the logic that underpins scoring on all tests, and especially on standardized tests. The professor (or the testing service) is trying to estimate how thoroughly students have mastered some material. Any single question is fallible, it may get at one of the few things a particular student didn't understand, or one of the few things some other student did understand. So each single question has measurement error. But if it's possible to ask a large number of questions, and if the measurement error is really behaving like drawing a random number, averaging the responses to each question can make the errors average out and provide a better estimate of what the student knows.[13]

Excluded variables

In several chapters, we presented a scatterplot of the relationship between homicide rate and poverty rate for US states. States with higher levels of poverty tend to have higher levels of homicide. But the relationship is hardly lockstep. If poverty perfectly predicted homicide we would expect to see all the states tightly clustered along a summary line. Instead they are scattered around. This scatter might be related to measurement error in homicide – certainly the data aren't perfect. But we would be naïve to think that poverty is the only factor driving homicide. Many other factors are at work. It can be useful to think of these factors using our general model:

$$Y = f(X) + E \tag{7.7}$$

In this case, E represents the effects of every variable that influences homicide other than the poverty rate. We can sometimes assume that all the other variables in the world act rather like numbers drawn from a random distribution, one E for each state. For some states, the E will be a positive number and push the homicide rate above what we would expect from poverty alone. For others, the E will be negative and push the homicide rate below what the level of poverty for that state would suggest.

Again, we can use statistical tools to see how much effect E (the random component) has and compare that to the independent variable. If the independent variable (X) does not behave much differently than a random number, it is hard to sustain the idea that poverty causes homicide. But if X does not behave like a purely random variable, then the theory is consistent with the data.

Recall the example of admission to graduate school by gender where this kind of logic was used explicitly. Permutation errors are much like the left-out variables described above. Suppose we want to know if admissions to graduate school are driven by gender (a binary variable). Then our model might look just like the model for left-out variables:

$$Y = f(X) + E \tag{7.8}$$

For each applicant in the data set we know whether she or he was admitted (Y), his or her gender (X) and once we predict admission with f(gender) we can have an error – did we predict their admission status correctly or incorrectly?

$$Admission = f(gender) + Error \tag{7.9}$$

Then we can ask how much gender and random error contribute to admissions. If gender contributes nothing special (that is, if there is no gender bias), then gender will behave just like a random variable. Only if gender predicts admissions better than a random number generated by a flip of a coin will we suspect bias.

This takes us to the heart of why statistics is so useful. We know our models have error. In many cases, that error can be considered random. Probability theory tells a great deal about random error. Statistics show us how to apply that knowledge of randomness to our models. We cannot make error go away, but if we correctly understand the process that generated the error, we can learn how large it is likely to be and draw better conclusions about the relationship between our model and the world we are trying to understand.

Box 7.1 Working with the Normal Distribution

The Normal distribution seems to match the distribution of many variables we observe in the world. It also emerges from theoretical statistics when we try to understand how error may influence our models and how to take account of that error. We would not be exaggerating by much to say that the Normal distribution is the most important distribution in statistics. As a result, it's worth looking at it in a little more detail.

When statisticians talk about the Normal distribution, they are thinking about an equation that tells them how likely it is to get a particular value of the variable when the distribution of the variable is described by the Normal distribution. The equation is a bit complex.[14] However, if we look at a bell-shaped curve as a frequency distribution, we see that most of the observations are clustered near the center of the distribution, that is, near the mean. (The normal distribution is symmetric, not skewed, so the mean and the median will be equal.) The further we get from the mean, moving either towards high scores or low scores, the less likely we are to have observations.

One of the features of the Normal distribution that makes it so interesting to statisticians is that the equation says that the probability of a particular value of X, based on knowing just X, is the mean of all the Xs and the standard deviation of all the Xs. That is, the probability of a particular value of X depends on how far X is from the mean of the all the Xs. The distance is measured relative to the standard deviation of the Xs. The **Z score**, or **standardized score**, for X is:

$$Z = (X - \bar{X})/s.$$

where $\bar{X}$ is the mean of the Normal distribution and s is its standard deviation

This is exactly what determines how likely it is to get a particular value of X in the Normal distribution. In fact, if we convert all the values of the variable we are interested in to Z scores, then we can use what is called the *standard Normal distribution* to assess the chances of getting a Z (and thus an X) with a particular value. As we will see in later chapters, this becomes a very useful tool for assessing the uncertainty that error brings into our models. We will develop those tools in the next two chapters.

Figure 7.15 shows the standard Normal distribution. The vertical axis is labeled "fraction" and is just the proportion of the data. Sometimes this is called "density." We see that the most frequently occurring values of X, which are those closest to the mean, appear around 40 percent of the time (a "fraction" or "density" of 0.4). Remember that in a standard Normal distribution the X values have been converted to Z scores, so a Z score of 0 means that the difference between the X value and the mean is 0, so the value on the horizontal axis labeled 0 is the same as the mean.

We have placed tick marks to indicate how many standard deviations above or below the mean a particular value of X might fall. For example, an X value

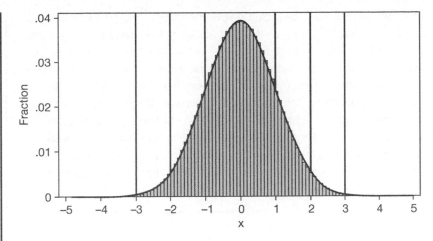

Figure 7.15   The standard Normal distribution

that is two standard deviations above the mean would get a score of 2, that is, $(X - \overline{X})/s = 2$. An X value three standard deviations below the mean would get a Z value of –3, that is $(X - \overline{X})/s = -3$.

In the figure we have drawn vertical lines to mark the places where an X value is three, two, and one standard deviations below the mean (Z scores of –3, –2, –1) and one, two and three standard deviations above the mean (Z scores of +1, +2, +3). Because the equation for the Normal distribution tells us how often we will find values that far from the mean, we can calculate what proportion of values are further away from the mean than a particular Z score, or closer than that Z score. Table 7.5 gives those calculations for Z scores of –3, –2, –1, +1, +2, and +3. In later chapters we will often think about what proportion of values are further away from the mean that a particular value.

We can see that about 16 percent of values in a standard normal distribution are more than one standard deviation above the mean, and about 16 percent

Table 7.5   Calculations for Z scores of –3, –2, –1, +1, +2, and +3

Z-score (distance from the mean in standard deviations	What proportion of values are further away than this	What proportion of values lie between this value and the mean
–3	0.0013	0.4987
–2	0.0228	0.4772
–1	0.1587	0.3414
+1	0.1587	0.3413
+2	0.0228	0.4772
+3	0.0013	0.4987

are more than one standard deviation below the mean. When we move out to two standard deviations, the percentage of values further away from the mean drops to a bit over 2 percent, and when we get to three standard deviations, only 1 value in a thousand is further away. Sometimes we talk informally about something being "three standard deviations from the mean." This implies that it's a pretty unusual thing, far from what is typical. Sometimes in business mention is made of a "six sigma" approach. Since "sigma" is a symbol for the standard deviation, this means an approach that reduces errors, such as faculty parts, so that they are as rare as being six standard deviations from the mean. The chance of a value being that far out on a Normal distribution is 0.00003, or 3 chances in 100,000 – very small indeed.

The arithmetic of probability is discussed below as an Advanced Topic.

What Have We Learned?

Random events are unpredictable. But combinations of random events exhibit strong patterns. For example we can't predict the results of a single toss of two dice, but we know that 7 is the most likely sum, and that sums of 2 or 12 are pretty rare. Of particular interest is the fact that, if we have many random numbers added together, the pattern of results is very predictable – a Normal distribution.

Results about random processes are useful because often the error in our models of the world can be thought of as behaving like random numbers. Sampling error, experimental error, measurement error, and the effects of variables not in the model can often be thought of as taking the real value (which we don't get to see in our research) and adding a random number to it to produce what we observe. Then knowledge of random numbers helps us to understand real data. Statistics applies what has been learned about random processes to models to understand better the relationship between the model and the world it is trying to describe.

Advanced Topic 7.1 Bayesian Statistics

There is a long and deep debate in statistics as to how best to think about probability. It is mostly a debate between two views. One, called the **frequentist approach**, closely matches our common sense understanding of probability. The probability of an outcome is thought of as the proportion of times that the outcome occurs when the event is repeated many times. We think of the probability of getting a head in tossing a fair coin as 0.5. The frequentist approach to probability says that a probability of 0.5 for coming up heads means that if we tossed the coin many, many times, it would come up heads half the time.

This approach is similar to the conventional understanding that we use throughout the rest of the book. But we want to note that there is an alternative view, called the **Bayesian** or **subjectivist approach to probability**. A subjectivist says probabilities are something inside our heads. When we say a coin has a probability of coming up heads 50 percent of the time we are really saying that our best guess is that we don't know whether it will come up heads or tails, and there is no reason to predict either heads rather than tails or tails rather than heads.

Many statisticians find the idea of thinking in terms of subjective probability appealing. What we mean by probability seems related to personal knowledge of the world. For example, Edward Beardsworth (1999, p. 2449) recalls a demonstration by Ron Howard of Stanford University that illustrates the subjective character of probability:

At one point in the class he flipped a coin and held it covered in his hand. He asked the class to estimate the probability of "heads," and of course we confidently replied 50 percent. Then he peeked, paused a moment, and asked us again. For him, the probability had changed, to either 0 or 100 percent, but for the rest of us, it was still 50 percent – as the best information we had available to us.

The probability of heads was different for different people in the room. But in the **frequentist view** the probability is always 50 percent and can't change just because someone looks at a coin. As Beardsworth notes, "Frequentism is a special case of Bayesian statistics where there is no prior knowledge" (1999, p. 2449).

There have always been applications of the subjectivist or Bayesian approach to situations where we have information about the outcomes of processes that have a random component beyond what is in the data. Until recently there have been few applications of the Bayesian approach in social science data analysis. This is changing as many statisticians have come to realize that Bayesian approaches may allow us to develop analysis tools that would be hard to construct using frequentist logic.[15] Most introductory texts ignore Bayesian statistics. We believe most students learn key concepts most easily using frequentist ideas. For many simple problems the Bayesian approach gives answers that are quite similar to the frequentist approach. So we will emphasize the frequentist approach not because we believe it to be more correct but because it is easier to learn as a beginning statistics student.

There are some simple rules that can help us see how to do arithmetic on probabilities when we have various kinds of linked events.

The probability of both of two independent events occurring

By definition, independent events are things that don't influence one another. Suppose two of us flip fair coins at the same time. How one coin comes up has no influence on what happens with the other coin. If we look at the decision trees we've done, we find that there are four possibilities. We can both get heads, you can get a head and I can get a tail, you can get a tail and I can get a head or we can both get tails. So there is one way to get two heads out of four things that might happen, so the chances of two heads is just $1/4$ or 0.25.

Since with fair coins the chances of a head are 0.50, the chances of two heads are $0.5 \times 0.5 = 0.25$. In general, when we two have independent events, the chances of a particular outcome is just the probability of the outcomes. If we call you getting a head A and my getting a head B, then the probability of A and B (two heads), can be written as:

$$P(A \& B) = P(A) * P(B).$$

If three of us flipped coins, and we called the third person getting a head C, then the probability of three heads is:

$$P(A \& B \& C) = P(A) * P(B) * P(C).$$

With fair coins, that's $0.5 \times 0.5 \times 0.5$ or 0.125.

The probability of either of two independent events occurring

Suppose we are interested in the chances of either of us getting a head. If we go back to the possible outcomes (two heads, you get a head and I get a tail, you get a tail and I get a head, two tails) then we see that out of four possible outcomes, three of them give us at least one head.

The easiest way to do the arithmetic is to consider how we could fail to get the outcome we are interested in – that is, what is the probability of neither of us getting a head. That can occur only if we both get tails, and with fair coins those chances are the same as getting two heads, or, as we have just seen, $1/4$. So the odds of getting at least one head is everything else, or $1 - 0.25 = 0.75$. Writing this as an equation, when we have independent events:

$$P(A \text{ OR } B) = 1 - P(A\&B)$$
$$= 1 - (P(A)*P(B))$$

If we have more than two independent events, the same logic follows. For example, to find out the chances of getting at least one head from three people tossing fair coins:

$$P(A \text{ OR } B \text{ OR } C) = 1 - (P(A \& B \& C)$$
$$= 1 - (P(A) * P(B) * P(C))$$
$$= 1 - 0.125 = 0.875$$

Note that these calculations all assume that the events don't influence one another – that is what we mean by independence. As we will see in later chapters, by comparing these kinds of calculations that assume independence with the actual frequencies that we find in our data, we can find out if it is likely that what we are seeing is the action of two independent variables. For example, if gender and animal concern are unrelated to each other, then we would expect that the probability of a person being a woman and having an above average score on the animal concern scale could be calculated by just multiplying the probability of being a woman by the probability of being above average on the scale. If the two things are unrelated, then the proportion in the sample that have both properties (being a woman and being above the average on the scale) should be just the product of the two proportions. If we find something different in the data, we might suspect that being a woman influences animal concern – that the two things are not independent. We will develop this logic in much more detail in later chapters.

Applications

Animal concern and AIDS knowledge were used as examples in the chapter, so the other two topics will be explored in this section.

Example 1:   Why do homicide rates vary?

At the end of the chapter we discussed the use of standardized or Z scores. We can use the homicide data to understand Z scores better. Remember that we have data on all states, so this is population rather than sample data. Figure 7A.1 shows a histogram of the homicide rate, with a Normal curve added. Figure 7A.2 is also a histogram of homicide rate, but turning them into Z scores has standardized the data. Remember that we calculate a Z score by subtracting the value of the mean from each case and dividing the result by the standard deviation. As you can see, when the data are standardized, the shape of the distribution does not change so we can look at the shape of the distribution using either the original data or the data transformed into Z scores. The Z score is useful for comparing across data sets or for seeing just how closely the data set matches a Normal distribution. The homicide data have a distribution that is not too different from a Normal curve.

Table 7A.1 is a frequency table of the standardized values, showing how many standard deviations states fall from the mean. Remember that in a Normal distribution, we know what proportion of cases will fall a certain distance in Z scores above and below the mean. Twelve percent of states (N = 6 states) have homicide rates at least one standard deviation above the mean and 18 percent of states (N = 9 states)

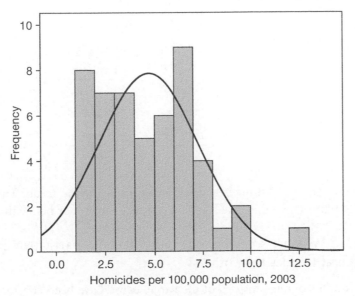

Figure 7A.1   Histogram of homicide rate with Normal curve

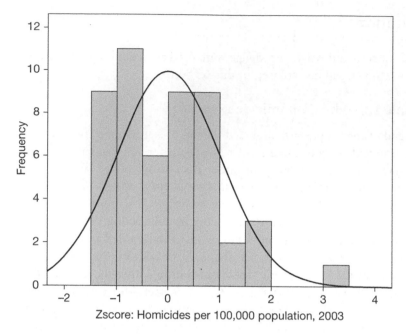

Figure 7A.2   Histogram of standardized homicide rate with Normal curve

Table 7A.1   Frequency table of standardized homicide data

Standard deviations from the mean	Percentage of states
−3.01+	0
−3.0 to −2.01	0
−2.0 to −1.01	18.0
−1.0 to 0	34.0
0 to .099	36.0
1.0 to 1.99	10.0
2.00 to 2.99	0
3.00+	2.0

have homicide rates at least one standard deviation below the mean. The bulk of the states (N = 35) fall within one standard deviation above or below the mean.

Example 3:   Why are some nations more likely to participate in environmental treaties than others?

We can look at histograms for two of the variables we have been considering that may impact countries' environmental treaty participation – voice and accountability

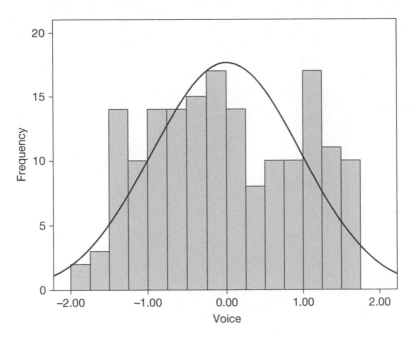

Figure 7A.3   Histogram of voice and accountability with Normal curve

and number of NGOs in a county. We have imposed a Normal curve on these histograms to see how closely the distributions are to being Normally distributed. Again, in this example, we are working with data for the population, rather than sample data. As can be seen in Figure 7A.3, the voice and accountability data are not too far off from being Normally distributed. There are more countries with scores between 1 and about 1.75 than in a Normal distribution and fewer countries with scores between 0 and 1. When this histogram is compared to the histogram in Figure 7A.4 for number of NGOs, we see the voice data are closer to Normally distributed than the NGO data are. Countries tend to be more heavily clustered at the low number of NGOs (less than 750) than in a Normal distribution, with fewer countries with NGOs between 750 and 2000 than in a Normal distribution. So we see that the distribution of some variables look rather like the Normal distribution for other variables the Normal distribution is not a good "sketch" of the data. We will see in later chapters that we will sometimes want to assess whether or not the data look roughly like a Normal distribution.

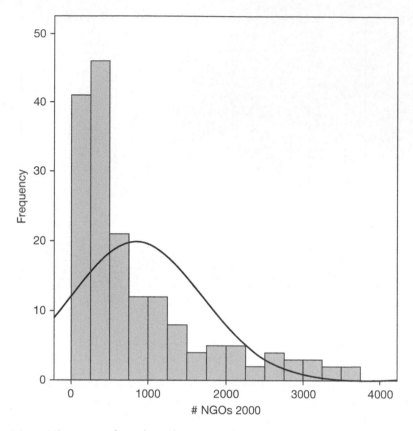

Figure 7A.4   Histogram of number of NGOs with Normal curve

Exercises

1. We talked about probabilities in this chapter and how to calculate the probability of a particular outcome occurring. Let's calculate some probabilities using a standard deck of cards with 52 cards. Be sure to show your calculations.

    a) What is the probability of drawing a black card when drawing one random card from the deck?

    b) What is the probability of drawing the six of hearts when drawing one random card from the deck?

    c) Now let's say we draw one card randomly from a deck of cards and then draw one card randomly from a second deck of cards. What is the probability that both cards will be black?

    d) If we again select one card randomly from a deck of cards and then draw one card randomly from a second deck of cards, what is the probability that the two cards will be a different color?

2. Figure 7E1 is a cumulative distribution graph for the question from the ISSP data set measuring beliefs about whether humans developed from animals. There were four responses: "definitely true" (1), "probably true" (2), "probably untrue" (3), and "definitely untrue" (4).

a) What is the probability that if a case were drawn randomly from the sample that the case would have a score of "definitely true" or "probably true"?

b) What is the probability that if a case were drawn randomly from the sample that the case would have a score of "definitely true," "probably true," or "probably untrue"?

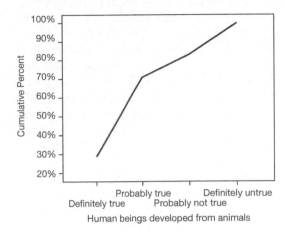

Figure 7E.1 Cumulative distribution graph for the question from the ISSP data set measuring beliefs about whether humans developed from animals

References

Beardsworth, E. 1999. The Bayesian way. *Science* 286(5449), 2449.

Bennett, D. J. 1998. *Randomness*. Cambridge, MA: Harvard University Press.

Berk, R. A., Western, B., and Weiss, R. E. 1995. Statistical inference for apparent populations. In P. V. Marsden (ed.), *Sociological Methodology 1995*, pp. 421–58. Washington, DC: American Sociological Association.

Bernstein, P. L. 1998. *Against the Gods: The Remarkable Story of Risk*. New York: John Wiley & Sons.

Cosmides, L. and Tooby, J. 1996. Are humans good intuitive statisticians after all? Rethinking some conclusions from the literature on judgment under uncertainty. *Cognition* 58, 1–73.

Gaither, C. C. and Cavazos-Gaither, A. E. 1996. *Statistically Speaking: A Dictionary of Quotations*. Bristol, England: Institute of Physics Publishing.

Gigerenzer, G. 1998. Ecological rationality: An adaptation for frequencies. In D. D. Cummins

and C. Allen (eds), *The Evolution of Mind*, pp. 9–29. New York: Oxford University Press.

Gigerenzer, G. and Hoffrage, U. 1995. How to improve Bayesian reasoning without instruction: Frequency formats. *Psychological Review* 102, 684–704.

Hoffrange, U., Lindsey, S., and Hertwig, R. 2000. Medicine: Communicating statistical information. *Science* 290, 2261–2.

International Social Survey Programme (ISSP). 2000. 2000 Environment II data set. www.issp.org. Catalog no. ZA 3440. Cologne, Germany: GESIS-ZA Central Archive for Empirical Research.

Isaacson, W. 2007. *Einstein: His Life and Universe*. New York: Simon & Schuster.

Jaeger, C., Renn, O., Rosa, E. A., and Webler, T. 2001. *Risk, Uncertainly and Rational Action*. London: Earthscan.

Kahneman, D., Slovic, P., and Tversky, A. 1982. *Judgement Under Uncertainty: Heuristics and Biases*. Cambridge, England: Cambridge University Press.

Kalof, L. 1999. The effects of gender and music video imagery on sexual attitudes. *The Journal of Social Psychology* 139(3), 378–85.

Kaufmann, D., Kraay, A., and Zoido-Lobaton, P. (Jan. 2002). Governance Matters II: Updated Indicators for 2000/01. Policy Research Working Paper no. 2772. The World Bank Research Development Group and World Bank Institute; Governance, Regulation and Finance Division (http://hdr.undp.org/reports/global/2002/en/).

Kurzenhauser, S. and Hoffrage, U. nd. Teaching Bayesian reasoning: An evaluation of a classroom tutorial. Berlin: Max Planck Institute for Human Development.

Langton, C. G. 1992. Life at the edge of chaos. In C. G. Langton, C. Taylor, J. D. Farmer, and S. Rasmussen (eds), *Artificial Life II*, pp. 41–91. Redwood City, CA: Addison Wesley.

Langton, C. G. 1995. *Artificial Life: An Overview.* Cambridge, MA: The MIT Press.

Roberts, J. T., Parks, B. C., and Vasquez, A. A. 2004. Who ratifies environmental treaties and why? Institutionalism, structuralism and participation of 192 nations in 22 treaties. *Global Environmental Politics* 4(3), 22–64.

United Nations Development Programme (UNDP). 2002. Human Development Report. Deepening Democracy in a Fragmented World. New York: Oxford University Press (http://hdr.undp.org/en/reports/global/hdr2002/).

Western, B. 1996. Vauge theory and model uncertainty in macrosociology. In A. E. Raftery (ed.), *Sociological Methodology 1996*, pp. 165–92. Washington, DC: American Sociological Association.

CHAPTER 8
SAMPLING DISTRIBUTIONS AND INFERENCE

Outline

In the last chapter we examined what happens when random events occur. In particular, we saw that repeated random events can lead to probability distributions. You may have found that interesting but wondered how it could be useful for anything other than gambling. This chapter introduces the **sampling experiment** where we draw samples from the population over and over again to understand how samples behave. Sampling experiments are just like the probability distributions described in the last chapter but now the random event isn't the toss of a coin or the throw of a dice but the random process of drawing a sample. The sampling experiment is the conceptual tool that statisticians use to draw conclusions about populations based on sample data. Furthermore, the sampling experiment helps us draw conclusions in the face of all sorts of error – not just sampling error, but randomization error, measurement error, error from the effects of excluded variables, and permutation error which can be thought of as sampling from a hypothetical population. So it is a very powerful tool.

When we do a sampling experiment we get results for each of the many samples we create and then plot the results for all the samples. The plot of the results from many samples is called a sampling distribution. The sampling distribution is a probability distribution that shows what happens when we repeat the sampling process many times. The sampling distribution shows what happens when a random element is involved in generating data just as the probability distributions in the last chapter showed what happened when we repeatedly flipped coins, tossed dice, or generated sums from a uniform distribution.

The Logic of Inference

One of the countries for which animal concern data are available in the 2000 ISSP data set is the United States (again, the US data are known as the General Social Survey). Among other things, people were asked how many years of formal education they have. The mean number of years of education for those who answered the question was 13.63 years. For most research projects, we are interested not simply in the sample, but in the characteristics of the population from which the sample was drawn – in this case the population of US adults.[1] In fact, our only interest in a random sample is that it provides information that can be generalized to the larger population.

The mean of the sample will not be exactly equal to the mean of the population except by the most extraordinary of chances. By luck of the draw, the sample may include a few too many well-educated people compared with the overall population. Or we might have too few. Thus we expect that the sample mean provides some information about the population mean, but it is not free from error. This can be represented as an equation

$$\bar{X} = \mu + e \qquad\qquad\qquad (8.1)$$

where $\bar{X}$ is the sample mean, μ (the Greek letter "mu" – rhymes with "new") is the population mean and e is sampling error.

Remember that e can take on positive or negative values and thus make the sample mean larger or smaller than the population mean. We have shifted to a lower case e from the capital E that used in earlier chapters because we are now making distinctions between populations and samples. Typically values for the population are indicated with capital letters or Greek letters (or Greek capital letters) and values for the sample with lower case English letters. The equation can be written in general as:

Sample Statistic = Population Parameter + Sampling Error   (8.2)

What is sampling error and how can we think about it? In drawing a particular sample we may get a slightly larger proportion of younger people than we might find in the population as a whole. As a result, the mean for that sample will be lower than the mean of the population. In another sample, by chance, we might get a larger proportion of older people than in the population as a whole. The mean of that sample would be a bit larger than the mean of the population. If we draw many samples, each will have a mean but we expect those means to differ from one another and from the population mean by random chance.

It's helpful to think of a hypothetical process – a sort of thought experiment that helps us understand sampling error and its link to random processes. In the thought experiment, for each sample we draw a random number (which can be positive or negative) to represent sampling error. We then add that number to the true population mean. (We don't actually know the population mean, but this is a thought experiment to help us think through sampling error.) The result is the mean of a sample. For each sample in the sampling experiment, we draw a different random number representing sampling error and add that number to the true population mean (remember that some of the numbers representing sampling error will be negative and some positive, so some sample means will be larger and some smaller than the population mean.)

That is, we can think of each sample mean as a mixture of the true population mean that we would like to know plus some random error that comes from the sampling process. When we have an experiment, the error comes from the randomization process. When we are dealing with measurement error, the random error is the measurement error. So while this discussion focuses on sampling error, the same logic applies in any application of statistics to error.

The statistical problem is to understand the error. If we knew the exact amount of error for our actual sample, we could subtract it from the sample mean and know the value of the population mean. The tools we use for statistical inference do not allow exact calculation of the size of sampling error for the particular research sample we have. But they do allow us to estimate the likely size of the error, and thus estimate a likely value for the population mean. The process by which this is done involves several steps. Understanding these steps is critical to an understanding of inferential statistics and to successfully understanding and using statistical tools.

It's very important to keep in mind the distinction between the actual data being analyzed and the data generated by the sampling experiment, which is not real but just a guide to understanding sampling. The ISSP data on animal concern, the state data on homicide, the Ugandan-DHS data on AIDS knowledge, and the data on nations' participation in environmental treaties are collected by observing and describing the world. But to understand error we use the computer or just our minds to conduct sampling experiments. These are not experiments in the sense we have discussed in chapter 1 in which there is an experimental and control group. Rather the sampling experiment is a process we think about to understand samples. In the sampling experiment, we pretend we know the population mean and can draw many samples from the population and compare their means with the population mean. In our actual research we don't know the population mean (it's what we are trying to find out), and we usually can afford to draw only one sample. But in our imagination or on the computer we can create an imaginary population and observe what happens when we take samples from it. The sampling experiment may be artificial, but it's a powerful tool for understanding the samples we really draw from the world.

The sampling experiment

If we knew the exact amount by which the mean of the sample we are using in our research differed from the population mean – that is, we knew the exact value of the error – then we could calculate the population mean. But we can't determine that error for our research sample. The sampling experiment provides a way to understand what usually happens when we draw a sample from a population. We do this by studying how samples behave – how various samples differ from the mean of the population we are studying.

A sampling experiment consists of repeating the process that generated the data over and over to see what happens. Suppose we have drawn a simple random sample of size 1,000 from a real population to get our data, and we are interested in using the sample to find out the mean of the population. In the sampling experiment, we would have the computer generate a population that we think looks like the population from which our real sample was drawn. We would have the computer draw a sample of size 1,000 from the computer population, and calculate its mean. (Below we will talk about the kinds of populations the computer might generate. For the moment just imagine that the computer is generating many numbers to represent people's ages and we are then drawing samples of those ages.) Then we would have the computer draw another sample of size 1,000 and calculate the mean of that sample. We would continue this process many times, so that we have many samples drawn from the computer-based artificial population.

By examining how the means of the samples behave, we learn something about the likely relationship between any particular sample and the population from which it was drawn. We can find out how often the process gives us sample means that are close to the population mean and how often we get sample means that are quite different from the population mean. If most of the sample means are pretty close to the population mean, that would suggest that for most samples we might draw, the sample mean is a reasonable indication of the population mean.

If we don't have a sample but an experiment, then the random assignment of subjects to experimental and control groups will produce differences between the two groups from randomization error. We can imagine a sampling experiment in which the real experiment was run over and over, and the randomization error will generate differences between the experimental and control groups each time. Much of the thinking about these issues first developed in the context of measurement error, where the sampling experiment was a matter of taking repeated measurements of the same object. Remember that when we talk about a sampling experiment we don't mean the kind of experiment that has experimental and control groups. The term sampling experiment just means a process of drawing many samples from an imaginary population to see what happens when samples are drawn.

Many pioneering researchers in statistics in the late nineteenth and early twentieth century did sampling experiments by hand – they wrote down numbers on sheets of cardboard, put them in a box, shuffled them extensively and drew a random sample of the slips, calculated the mean of the numbers on the slips, put the slips back in the box, and repeated the process hundreds of times. Now we do this on the computer, and it is called a *Monte Carlo simulation* after the casinos of Monte Carlo where gambling losses inspired much important statistical work.

Of course, it usually isn't practical to conduct the sampling experiment by hand with slips of paper. It is sometimes possible to use mathematics and probability theory to determine the sampling distribution by solving equations. We call this approach an analytic solution, in contrast to doing the experiment by hand or on the computer. For most of the techniques covered in this book, sampling distributions were determined analytically. But we will often use the computer to do sampling experiments to see how the process works.

Sampling with replacement and sampling from large populations are discussed at the end of this chapter as an Advanced Topic.

Bootstrapping is discussed at the end of this chapter as an Advanced Topic.

In principle, a sampling distribution can be generated for any type of sampling procedure, but in practice, we can generate sampling distributions only for probability samples. If we use a convenience sample, there is no way to know the relationship between the sample and population and thus no way to speak about the population. But if we have a simple random sample, we know a great deal about how the error behaves. A sampling distribution can be developed, at least in theory, for any number (statistic) that can be calculated from sample data. For most of the simple statistics in the mean family, the sampling distribution can be developed analytically. But for the median family, we often don't have analytic results and have to rely on sampling experiments done on the computer. So in most of the rest of the book, we will focus on the mean family because that's where we have the best understanding of sampling distributions.

Even within the domain of probability samples, sampling distributions are difficult to generate for complex sample designs such as cluster and stratified sampling. There are equations that describe the sampling distributions for many statistics such as the mean when we use a complicated probability sampling method. But they are more complex than those for a simple random sample. Here we will focus on simple random samples for simplicity. If you understand the concepts of sampling distributions and sampling experiments for simple random samples, then you have the basic tools for understanding more complex samples.

The sampling distribution

Suppose we have calculated (or more realistically, had the computer calculate) the means of many samples (we might use 1,000 as a large but practical number). Each of these samples in the sampling experiment is a simple random sample of size equal to the size of the actual sample we are studying. We could then ask the computer to plot the stem-and-leaf diagram or a histogram of sample means. This diagram will show the sampling distribution of the sample means for all the simple random samples in our sampling experiment. This display shows the results of the sampling experiment. The display shows us: 1) how often we would get samples whose mean is greater than a particular number; or 2) smaller than a particular number; or 3) that lie between two numbers; or 4) that are equal to a particular number.

We will never know the exact size of the error that causes the mean of a particular research sample to differ from the real population from which it was drawn. That is, we never know which of the samples in the sampling distribution matches the actual sample we are using for our sampling experiment. But with the sampling distribution we can tell how often we get samples with particular values, or, by looking at these occurrences as probabilities, how likely we are to get such values. It can tell us what results are unlikely as a result of sampling error, and what kinds of results sampling error could easily generate.

Let's try some computer-generated sampling experiments to look at the problem of determining from sample data the mean number of years of education for the US population. We can easily get the computer to draw simple random samples from a population of numbers on the computer and to calculate the means of those samples. But since we don't know what the distribution of the actual population looks like, how can we construct a computer population from which to draw samples?

First, we might assume that education in the US has a uniform distribution, with roughly equal numbers of people having each possible year of education, 0 to 20 (the range in our survey). This may not be very realistic, as it would seem that some years of education are more likely than others – people stop going to school after 12 years or 14 years or 16 years of education more often than they stop without getting a high school diploma or with just one year of college. We'll try other distributions next, but the uniform distribution, even if it isn't realistic, is an easy place to start. Note that if we use a uniform distribution, the mean of the population is going to be 10, because every number of years of education between 0 and 20 is equally likely.

The first step in generating the sampling distribution is to create a population on the computer that ranges from 0–20 with every number of years equally represented. Then we have the computer draw a sample of 901 from the population, calculate the mean, and store that sample mean. It then repeats the process of drawing samples of 901, calculating the mean and storing it many times. Remember we are sampling with replacement, so all the hypothetical people drawn in one sample go back into the population and might get drawn in another sample in the experiment. We do this many times to identify the patterns. The size of each sample is a key feature of the sampling experiment. In this case it's 901 to match our actual sample of the US population. (Note that the sample size for this data set – and for many others – varies due to missing data. If we were to examine those who answered the animal concern question and a question about age instead of education, our sample size would be different due to missing data.) The number of times we repeat the experiment just has to be large enough to do the plots and see the patterns. Here we have done 1,000 replications because it's large enough to show the patterns but not so large as to burden our computer. The exact number of samples we draw in the sampling experiment doesn't matter as long as it's fairly large. But as we will see, the size of each sample is very important and must always match the size of the real sample we are trying to understand. The result is in Figure 8.1, where a Normal curve is drawn over the histogram for comparison.

We get a bell-shaped frequency distribution with a center around 10, the population mean. The solid line is the Normal distribution. It matches the histogram rather well, for reasons we'll discuss below. The mean of all these sample means was 10.01, very close to the population mean. Many of the samples had means quite close to the true population mean. When we ask the computer to describe the sampling distribution, we find that the lower quartile is 9.86 and the upper quartile is 10.14, so half of the samples (those with means between the two quartiles of the sampling distribution) were within 0.15 years of education of the population mean. That is, of the 1,000 samples that we took (each of size 901), 500 were within about 0.15 years of education of the actual population mean of 10 years of education. If we ask the computer to count how many samples had means more than half a year of education below the population mean of 10, we find that there were less than ten that far away. And if we ask how many were more than half a year of education above the population mean, it turns out that it's again less than ten. That is,

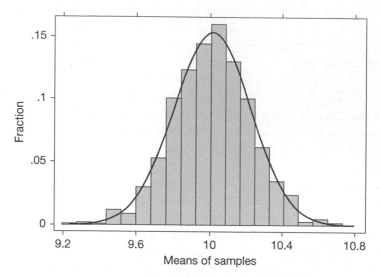

Figure 8.1 Sampling distribution of means from a uniform population, N = 901
Data source: Simulation of sampling from a uniform distribution from 0 to 20

only 20 samples of the 1,000, or 2 percent, we drew were more than half a year of education away for the true population mean. So it would seem that using the mean of a sample as a guess of the population mean is not a bad strategy – most sample means are close to the population mean, very few are far away from the population mean, and on average, we get the right answer – the mean of the sample means is equal to the population mean.

A skeptic might argue that this is a result of drawing samples from a uniform distribution. What if the population has several categories that occur much more commonly than others? This can happen if the distribution of the population is not uniform but has several modal categories with many cases compared to other categories. This is likely true for education since people tend to stop their schooling after finishing requirements for diplomas. One simple distribution that is plausible is shown in Table 8.1. Twelve years of high school is the most common level of education, followed by four years of college (16 years of education). People are equally likely to complete 6, 8, 14, and 20 years of education, but no one falls into any other level of education.

Table 8.1 Hypothetical distribution of years of education

Years of education	1	2	3	4	5	6	7	8	9	10	11	12	13	14	15	16	17	18	19	20
Percent of population	0	0	0	0	0	10	0	10	0	0	0	30	0	10	0	20	0	10	0	10

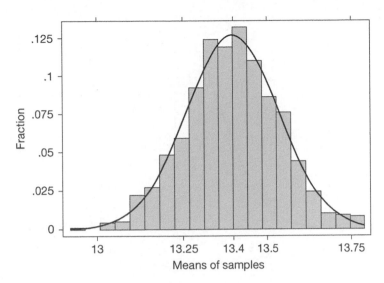

Figure 8.2   Sampling distribution of means from a "lumpy" population, N = 901
Data source: Simulation

We had the computer generate this population as well. It has a mean (the population mean for our sampling experiment) of 13.4 years. Remember that this is not really the distribution of education in the US population. It is an artificial population we have created on the computer. We are using these hypothetical populations to see what happens when we draw samples and calculate their means. We again generated a sampling distribution by drawing repeated random samples of 901, calculating the mean of each one and plotting the results. This sampling distribution is presented in Figure 8.2.

Despite the fact that the samples were drawn from a distribution with a rather messy shape, we once again get a bell-shaped curve that is pretty well matched by the Normal curve drawn as a line on the histogram. The mean of all the sample means is 13.397, quite close to the population mean of 13.4. The lower quartile is 13.30 and the upper quartile is 13.49, so half of the samples (those with means between the two quartiles of the sampling distribution) were within 0.1 years of education of the population mean. That is, of the 1,000 samples (each of size 901), 500 had means within 0.1 years of education of the actual population mean of 13.4 years of education. Less than 10 of the 1,000 samples had means one-third of a year of education below the population mean of ten, and just about 10 samples had means one-third of a point above. So once again, we find that the sample mean provides a good guess of the population mean – usually close, very seldom far off and correct on average.

We did these experiments with a fairly large sample, 901. Many social science studies work with samples of 1,000 or larger. But many other studies use much smaller samples. What would happen if we had used a smaller sample, say 100? Figures 8.3 and 8.4 give the sampling distributions for sample means of samples of size 100 drawn from the Uniform distribution and our "lumpy" distribution.

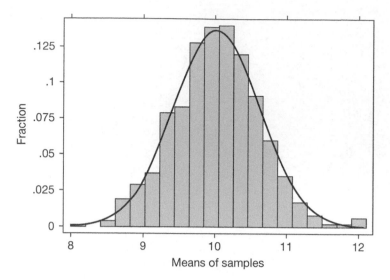

Figure 8.3   Sampling distribution of means from a uniform population, N = 100
Data source: Simulation

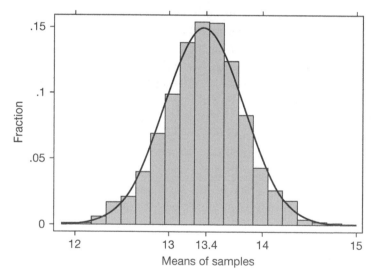

Figure 8.4   Sampling distribution of means from a "lumpy" population, N = 100
Data source: Simulation

The sample means in Figure 8.3 still cluster around the population mean of ten, but they are much more spread out (have more variance) than when the sample size was 901. The histogram for samples of 901 only needed to range from about 9.4 to about 10.6 to cover all the 1,000 sample means. Now we have a range from about 8 to about 12 to cover all the sample means we got in the experiment with samples of 100 cases. So with a small sample size we have more samples with means distant from the population mean. The mean of all the sample means is about 10.02, so it is still very close to the population mean of 10. But now the upper quartile is 10.41 and the lower quartile is 9.64, so half of the samples are within about 0.40 years of education. Remember that with the sample size of 901 half of the samples were within 0.15 years of education. That is, we have to go nearly three times as far to cover the middle half of the samples with the smaller sample size. And now if we go to the 10 highest and lowest sample means, they are about 1.5 points away from the population mean. So there is still clustering around the true value, but with many more chances of getting a sample with a mean far from the population mean. Again, the histogram is rather like a Normal distribution.

In Figure 8.4 we have samples of 100 from the "lumpy" distribution. Here again the sample means mostly are close to the population mean of 13.4. The mean of all the means is 13.372. With the sample size of 901 we had to go from a bit under 13 to about 13.75 to cover the all 1000 sample means. Now we have to go from 12 to 15, so again with the smaller sample there is a greater chance that a sample will have a mean further away from the population mean. The upper quartile of the means in this sampling experiment is at 13.66 and the lower quartile is at 13.1, so half of the cases are within about 0.3 years of education of the population mean of 13.4. The ten samples of the 1,000 with the lowest means are more than 2 years of education away from the population mean while the 10 with the highest means are nearly 2 years above it. So again, most sample means are fairly close to the population mean, and again we get a bell-shaped curve for the sampling distribution of sample means. But with the smaller sample size, there are more sample means further away from the population mean. We will return to the importance of sample size later in the chapter.

The expected value and the standard error

We can calculate the mean of the sampling distribution, just as we can calculate the mean of any of data. The difference is that the sampling distribution is not composed of data about the social world but data from an experiment we did on the computer or with slips of paper. If we have generated a sampling distribution of sample means, we are calculating the mean of the means. If we instead had calculated the variance or the median of each sample, then calculating the mean of the sampling distribution would be the mean of the variances or the mean of the medians. The mean of the sampling distribution is particularly important and is called the **expected value**.[2] The expected value is a value typical of samples, just as the mean of a particular sample is typical of the numbers in that sample. To keep the example simple for now, we will focus on sampling distributions of sample means. Later on we will talk about sampling distributions of other statistics.

Since not all samples will have means identical to the mean of the sampling distribution, it is also useful to calculate the variance and standard deviation of the sampling distribution. The standard deviation of the sampling distribution is

Table 8.2   Means, variances, and standard deviations from four sampling experiments

Population and sample size	Mean of sample means	Variance of sample means	Standard deviation of sample means (standard error)
Uniform distribution, N = 901, Mean = 10	10.010	0.0428	0.2069
Uniform distribution, N = 100, Mean = 10	10.021	0.3581	0.5984
"Lumpy" distribution, N = 901, Mean = 13.4	13.397	0.0186	0.1364
"Lumpy" distribution, N = 100, Mean = 13.4	13.372	0.1757	0.4192

called the **standard error**. Thus, for the example involving means, we can use the data from the sampling experiment to calculate the expected value of the sample mean and the standard error of the sample mean.

Table 8.2 shows the means, variances, and standard deviations of the sampling distributions from the four sampling experiments we have conducted. Notice that the expected value (the mean of the means) is close to the population mean in each of our four experiments. But also notice that the standard deviation of the means (the standard error of the means) is around three times as large in the small sample than it was in the larger sample. This is what we saw in the histograms of the sampling experiment results. The smaller the sample, the more likely it is to get samples far from the population mean and thus the larger the variance and standard deviation of the sampling experiment.

The Law of Large Numbers

In chapter 7, we described the Law of Large Numbers, which is also known as the Central Limit Theorem. If we take a large number of random numbers and calculate their sum, then generate more random numbers and take their sum, and keep doing this sampling experiment over and over, we find that the frequency distribution of the sums has a Normal distribution.

Since the mean is just the sum of the cases divided by the number of cases, the Law of Large Numbers also applies to the mean. Here we can restate the law more formally to describe a sampling distribution for the means of simple random samples. It can be stated as a formal theorem – a statement in an "if, then" form that has been shown to be true by mathematical proof. The theorem is:

If large simple random samples are drawn from a population, then the sampling distribution of sample means will be Normal in shape and have a mean (the mean of all the means) equal to the population mean (μ) and variance (the variance of all the sample means) equal to the population variance divided by the sample size (σ^2)/N.

This is quite a bit to understand, so let's look at it one piece at a time.

1 The situation is one in which we are drawing simple random samples (samples in which each person in the population is equally likely to appear and in which the probability that one person is drawn does not change the probability another person will be drawn).
2 These samples are large.
3 When we do the sampling experiment of drawing many such large samples and we calculate the mean of each sample, the theorem tells us what will happen (the part after "then").
4 The sampling distribution will have a special shape, the Normal distribution.
5 The mean of the sampling distribution (the mean of all the means) will be equal to the population mean.
6 The variance of the sampling distribution will be equal to the variance of the population divided by the size of the samples we have been drawing.

One of the remarkable features of the Law of Large Numbers is that it applies whenever we take a large simple random sample – no matter what the distribution of the population may be. In our four sampling experiments, we started with two populations, one that had a uniform distribution and one that had a non-uniform "lumpy" distribution. The Normal curve on the histogram did a pretty good job of matching the results of our sampling experiments from each of these populations, especially when the sample size was 901. But we might wonder what would happen if we did the sampling experiment drawing samples from a population with a different shape.

The Law of Large Numbers gives us the answer – no matter what the shape of the population, as long as we take large samples, we always get a Normal distribution. This is why statisticians like mathematical theorems for sampling distributions – it tells us what will happen in general. A sampling experiment can only tell what happened with a certain population and a certain set of samples. It's a useful guide, but the theorems are more powerful. Of course, if we don't have any theorems, then the sampling experiment done on the computer may be the only guidance we have.

The Law of Large Numbers tells us more than the shape of the sampling distribution of sample means. It also says that the mean of the sample means will be equal to the population mean and that the variance of the sample means will equal the population variance divided by the sample size. Table 8.4 compares these predictions with what we actually got in our four sampling experiments.

Box 8.1 Keeping Things Straight

We now have three distributions or sets of information we are dealing with, and it can get confusing. One is the population, the second is the sample and the third is the sampling distribution. The population is what we are ultimately interested in – when we conduct research and especially when we draw a sample, we do so in order to talk about the population. The sample is the data we have in hand, what we are calling the research sample. The sampling distribution is an odd thing until you get used to thinking statistically. It is a frequency distribution for the results of the sampling experiment. In the sampling experiment, we draw many samples of the same size as our research sample. We calculate the statistic of interest (in this case the mean) for each sample. Then we plot all those means in the sampling distribution. Since we think about doing the sampling experiment many, many times to generate the sampling distribution, we can think of the sampling distribution as a probability distribution since it tells us how often, in the long run, we would get a sample with a certain set of values.

Table 8.3, which shows the symbols we use for the key characteristics of each of these three distributions, may help.

Table 8.3 Characteristics of population, sample, and sampling distribution for the Law of Large Numbers

Characteristic	Population	Sample	Sampling distribution
Size	Very large so sample is a small fraction of it	N (large)	Very large so the graph of the distribution is clear
Shape	Anything	Whatever it comes up	Normal
Mean	μ	$\bar{X}$	μ
Variance	σ^2	s^2	σ^2/N

Table 8.4 Results of sampling experiments compared to predictions from the Law of Large Numbers

Population	Sample size	Predicted mean (equal to the population mean (μ))	Actual mean of sampling distribution	Predicted variance (σ^2/N)	Actual variance of sampling distribution
Uniform	901	10	10.010	0.0406	0.0428
Uniform	100	10	10.021	0.3667	0.3581
Lumpy	901	13.4	13.397	0.0187	0.0186
Lumpy	100	13.4	13.372	0.1684	0.1757

The means of the two populations were 10 and 13.4 and the variances were 36.6684 and 16.8408. The Law of Large Numbers predicts that the mean of the sampling distribution is just the mean of the population and that the variance of the sampling distribution is the variance of the population divided by the sample size. As we can see in the table, the Law of Large Numbers does an accurate job of describing the results of the sampling experiment.

Feature 8.1   Who invented the Law of Large Numbers?

The history of statistics indicates that the Law of Large Numbers was an idea that evolved over time. As it evolved, a variety of names were applied to different versions of it. While we will call it the Law of Large Numbers, statisticians also refer to it as the Central Limit Theorem, Bernoulli's Theorem, Chebyshev's Inequality and a variety of other names as numerous researchers worked on various aspects of it. Galileo had the basic idea in his discussions of errors in observation. The mathematician Abraham DeMoivre developed a formula for it, and it seems Jacob Bernoulli did as well. Pierre Simon Laplace did fundamental work on it, and of course, Carl Friederich Gauss whose name is associated with the curve, did as well.

Understanding the importance of the Normal curve beyond the realm of pure probability or dealing with measurement error involved contributions by Quetelet, Galton, Pearson, Gosset, and Fisher. In the twentieth century literally dozens of statisticians have worked on the underlying mathematical theory. Jarl Lindeberg from Finland and Paul Levy from France developed what is considered the modern, theoretically sound, foundation for the Law of Large Numbers. Many brilliant researchers were involved. Statistics, like most of science, is not a matter of a single breakthrough. Rather a community of scholars each made contributions and understanding evolved as a result of their discourse.

Small samples and estimating the population variance

After the Law of Large Numbers was discovered, statisticians knew that with a large sample the sample variance was a good estimate of the population variance and thus could be used with the Law of Large Numbers to describe the sampling distribution. As our two sampling experiments have shown, it works pretty well. But what if you don't have a large sample?

The problem of dealing with random error in small samples came up in a critical area of human welfare – the need to make better beer. William Gosset worked for Guinness and was concerned about their quality control methods. A critical problem was estimating how much yeast to add to the mash. Too little yeast and the fermentation would not complete, too much and the beer became bitter. The idea was to draw an aliquot (a small volume) of water from a vat and count the number of yeast cells under a microscope. But one or a few draws of water from the vat might not represent the yeast concentration accurately, nor was it practical to draw several hundred samples and use the Law of Large Numbers.

William Gosset realized he had two problems to solve, or rather two steps to take in solving the problem of working with small samples. Finding either step alone would have been a major contribution to statistics so it is remarkable he managed both. But, as we note in a later feature in this chapter, Gosset was by all accounts a pleasant and unpretentious person. Indeed, most of his work today is known under the name "Student." A few years before Gosset began to work at Guinness, one of

Feature 8.2   Pearson to Gosset to Fisher: The Coevolution of Statistical and Scientific Thinking

Karl Pearson, William Gosset and R.A. Fisher were the three great statisticians who developed most of what we cover in this book, and thereby laid the foundations for the modern use of statistics. Each was an interesting person and strikingly different from the others. Yet their work in response to one another led to modern statistics. Pearson and Fisher could not tolerate each other and were not on speaking terms for most of their lives, while Gosset mediated between the two intellectually. We will sketch the biography of each below. But first it is useful to emphasize the evolution of statistical and scientific thinking and the interaction between them.

Probability theory and the ideas that underpinned the Law of Large Numbers first emerged in efforts to understand games of chance. But scientists from Galileo on used these same ideas to describe the relationship between the measurements they were making and the "true" value of the thing they were measuring. The core idea underpinning their science was that there was a true value for the brightness of a star or the speed with which a ball ran down a slope, or the temperature at which some metal melted. If multiple measurements were made, they would each give

slightly different results because the measurements were flawed. Then the "law of errors" showed that the Law of Large Numbers often applied to these measurements, so the mean of the measurements was a good guess at the true value.

This was a reasonable way to think about physics. It was consistent with Christian theological views that had borrowed from the pagan Plato and other Mediterranean philosophers the idea that there are true essences that we observe with error. In the Christian view these true things were the way God constructed the universe. Science was attempting to read the mind of God. This was something like the view that Quetelet brought to the study of the human world – the important thing was to understand the "average man" because the deviations from that average represented some sort of error, not the essence that should be the focus of science.

Darwin changed the focus for the historical sciences, including the social sciences, evolutionary biology, and geology. It is not coincidence that Galton, who led the shift in thinking, was Darwin's cousin, and that Fisher founded the mathematical population genetics that placed Darwin's arguments into equations. The statistical revolution was directly connected to the Darwinian revolution. For Darwinian evolutionary theory, variability is not a matter of deviations from some ideal, not a matter of error. Rather, variation is the essence of life itself. Galton emphasized variation as "normal" and focused much of his attention on it. With Karl Pearson and Raphael Weldon, he founded the journal *Biometrika* to document variation in the world. It was only later that *Biometrika* began to emphasize theoretical statistics rather than papers that reported the variation in the beak lengths of birds or the heights of soldiers. Today this fascination with variation seems odd, but at the time it was part of the shift in ways of thinking about science.

Pearson made many contributions to statistics, as we will see. But for the development of inferential statistics, his key insight was to show that one could use a few numbers (called parameters, which is Greek for "almost measurements") to summarize a distribution. He showed that by changing the value of four numbers – the mean, the variance, the skew (a number that measures the extent to which the distribution is pulled left or right), and the kurtosis (a number that shows the degree to which the distribution is peaked or flat) – one can calculate many different shaped distributions. His goal was to use these numbers to characterize the distributions of variables. In a sense, he viewed science as learning as much as possible about the distributions that describe the world. The Normal distribution was one of these distributions.

In this work, Galton and Pearson, like most scientists of the time, were satisfied if they had lots of data with which to characterize distributions. Large samples were the rule, but not much concern was given to how a sample was taken. They used what we would call convenience samples or whole populations.

Gosset was a chemist and mathematician by training. He knew that scientists often had to work with very small samples and began to develop procedures that could be used with small samples. Fisher, who was the best mathematician of this group, put Gosset's work on a stronger mathematical footing and also showed the importance of careful design in data collection. He worked mostly with experiments where he showed the importance of random assignment to experimental and control groups. From that came our current understanding of drawing samples. Gosset and Fisher were characterizing distributions, as had Galton and especially Pearson, but they focused on sampling distributions that showed how we can make valid statements about populations based on samples.

So there was an intellectual evolution. First statistical methods were used to sort error from truth. Then variation was seen as a thing to be studied itself. Today, we retain an interest in variation but also understand how data collection methods – not just measurement error but also sampling and experimental design – can generate variation and that statistical tools can help us draw conclusions in the face of that variation. Of course, the evolution of the ideas that underpin modern statistics was not as smooth as this summary makes it seem. In addition, Pearson, Gosset, and Fisher were each very productive scholars so in their writings one can find complexity far beyond what this sketch suggests. But thinking a bit about the history of the ideas behind statistics helps in understanding how we use them today.

the master brewers published a scientific paper that revealed some secrets about the ingredients in one of the Guinness beers, so Guinness, although a pioneer in hiring scientists to help improve their product, had established a policy that prohibited their employees from publishing. So several of the most important papers in the history of statistics were published not under Gosset's name but as "Student" to pre-serve his anonymity.

Let's look at Gosset's contributions one at a time. Remember that the Law of Large Numbers tells us what the variance of the sampling distribution will be by using the variance of the population as a starting point. What it tells us is that the variance of the sampling distribution will be the variance of the population divided by the sample size. Yet in nearly every scientific problem, we don't know the variance of the population – if we did we'd know so much about the popula-tion we wouldn't worry about sampling. So for most problems, the Law of Large Numbers doesn't provide practical help.

The practice of the time was to use the variance of the sample as the estimate of the variance of the population. Here is Gosset's first major contribution. He showed that we would be better off estimating the variance of the population with something other than the variance of the sample. The Law of Large Numbers tells us that the sample mean is a good guess of the population mean in that the average of all guesses will be the correct number (the mean of the sampling distribution of sample means is the population mean). As we have seen in our four sampling experi-ments, most sample means will be close to the population mean so the chances of getting a sample mean far from the population mean are small.

But the variance of the sample is not a good guess of the variance of the popu-lation. The mean of the sampling distribution of sample variances is not equal to the population variance but is smaller, so using the sample variance as an estimate of the population variance tends to underestimate. Figures 8.5 to 8.8 show the sampling distributions of sample variances for our uniform and lumpy distribu-tions in samples of 901 and 100. Table 8.5 displays the mean of the variances as well as the actual population variance.

We can see some tendency for the sample variances to be smaller than the population variance. But even 100 is a fairly large sample compared to the small samples that many researchers use. Let's see what happens when we draw samples of only 10 (see Figures 8.9 and 8.10).

While the problem in using the sample variance to estimate the population vari-ance was subtle in the larger sample, the histograms of the sampling distributions for samples of 10 make the problem quite clear. The sample variance tends to be a very inaccurate estimate, and quite often below the population variance.

Gosset found that a better way to estimate the population variance is to divide the sum of squares by $N - 1$, rather than by N. Remember that we think of the variance as the average squared deviation from the mean and in taking the aver-age we divide by N. But to estimate the population variance from sample data, Gosset argued that it is better to divide by $N - 1$ and use the result as the estimated population variance. That is, his estimate of the population variance is:

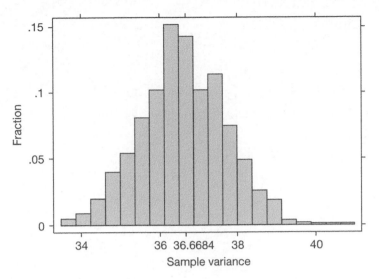

Figure 8.5 Sampling distribution of sample variances from a uniform population, N = 901
Data source: Simulation

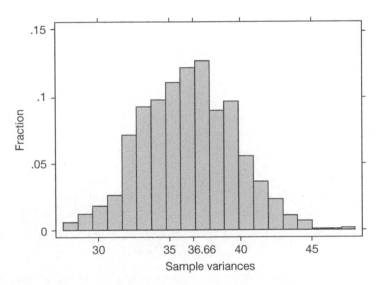

Figure 8.6 Sampling distribution of sample variances from a uniform population, N = 100
Data source: Simulation

$$\hat{\sigma}^2 = \frac{\sum (x_i - \bar{x})^2}{(N-1)} \tag{8.3}$$

where the hat ($\wedge$) over σ^2 means an estimate of σ^2.

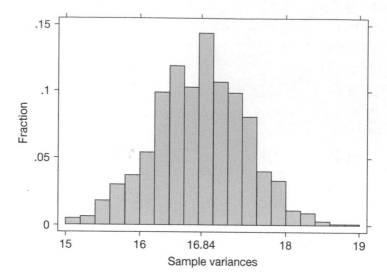

Figure 8.7 Sampling distribution of sample variances from a "lumpy" population, N = 901
Data source: Simulation.

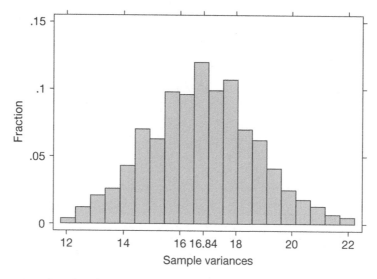

Figure 8.8 Sampling distribution of sample variances from a "lumpy" population, N = 100
Data source: Simulation.

Or, if we've already calculated the variance of the sample, the formula can be thought of as a correction to the sample variance as in the following:

$$\hat{\sigma}^2 = \left(\frac{N}{N-1}\right)s^2 \tag{8.4}$$

where s^2 is the sample variance. That is, multiply the sample variance by $(N/N - 1)$.

Table 8.5 Mean of the variances from sampling experiments and the actual population variance

Population and sample	Population variance	Mean of sample variances
Uniform, N = 901	36.6684	36.6157
Uniform, N = 100	36.6684	36.3582
Uniform, N = 10	36.6684	33.2262
"Lumpy", N = 901	16.8408	16.8221
"Lumpy", N = 100	16.8408	16.7868
"Lumpy", N = 10	16.8408	15.1986

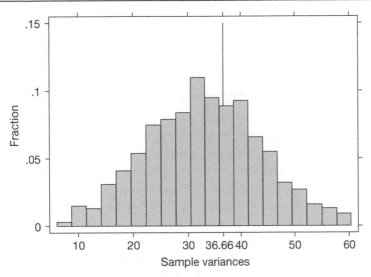

Figure 8.9 Sampling distribution of sample variances from a uniform population, N = 10
Data source: Simulation

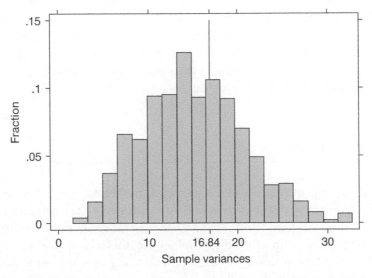

Figure 8.10 Sampling distribution of sample variances from a "lumpy" population, N = 10
Data source: Simulation

> ### Box 8.2   "Hat" Notation for Estimates
>
> In statistics we are often estimating some number for the population. For example, the Law of Large Numbers tells us that if we have large simple random samples, the sample mean is a good guess of the population mean. In symbols we are saying that $\bar{X}$ is a good guess of μ. These two symbols $\bar{X}$ for the sample mean and μ for the population mean, are used in almost all statistics texts. Statisticians frequently indicate an estimate of something in the population, rather than the actual population value by putting a little "hat" on top of the symbol for the population value. So if the population variance is symbolized by σ^2 (pronounced "sigma squared") then an estimate of the population variance based on the sample would be indicated by $\hat{\sigma}^2$ (pronounced "sigma squared hat"). (Outside of statistics, the "hat" is called a caret.)

Dividing by N − 1 instead of N works nicely – it takes the systematic error out of the estimate of the population variance. The mathematical theory shows that the sum of squared deviations from the mean divided by N − 1 gives a good estimate of the population variance. We can offer some rough insight into why this is so. In calculating the sums of squares of the sample (and thus in calculating the sample variance) we start with the mean of the sample. But the mean of the sample is also calculated from the same data set we are using in the variance calculations. So one of the terms in calculating the sum of squares, the sample mean, is based on the rest of the data. It is an estimate of the population mean rather than the actual population mean. Data in the tails of the sample will pull the sample mean toward them and as a result they will contribute less to the sum of squares than they would if we calculated the sum of squares using the population mean as the starting point. But of course we don't know the population mean, so we calculate the sum of squares and thus the sample variance using a starting point that is based on the same data as the sum of squares. Dividing by N − 1 is just the right correction for this problem.

The t Distribution

Gosset made more contributions to statistics after this breakthrough. He went on to show what the sampling distribution of sample means looks like when we have small samples. But there is a limit to our ability to do this. He could show how to do this *only* if the population from which the samples were drawn is Normally distributed. We gain the advantage of being able to work with small samples, but at the cost of only being able to work with small samples that have a Normal distribution – the ones where if we plotted the histogram for the whole population it would look like the Normal curve. In contrast, the Law of Large Numbers required large samples but didn't care what the population looked like – it applied to any population. Gosset's theorem is:

If simple random samples of size N are drawn from a Normal population with mean, μ, and variance, σ^2, the sampling distribution of sample means will be a **t distribution** with N − 1 degrees of freedom, a mean of μ and a variance of σ^2/N.

Again, this is quite a bit to understand, so let's look at it one piece at a time.

1   The situation is one in which we are drawing simple random samples (samples in which each person in the population is equally likely to appear and in which the probability that one person is drawn does not change the probability that another person will be drawn).
2   These samples don't have to be large but the population has to be Normally distributed.
3   When we do the sampling experiment of drawing many such large samples and we calculate the mean of each sample, the theorem tells us what will happen (the part after "then").
4   The sampling distribution will have a special shape, the t distribution with N − 1 degrees of freedom.
5   The mean of the sampling distribution (the mean of all the means) will be equal to the population mean.
6   The variance of the sampling distribution will be equal to the variance of the population divided by the size of the samples we have been drawing.

This looks like the Law of Large Numbers except for two differences. First, it only applies when the samples are drawn from a Normal population. Second, instead of having a sampling distribution that is Normal in shape, the sampling distribution has a new shape, something called a t distribution with N − 1 degrees of freedom. The exact shape of the t distribution changes with the number of degrees of freedom and since the number of degrees of freedom for this theorem is the sample size minus one, the shape changes with the sample size. With the Law of Large Numbers, we get one distribution, one shape, for any sized sample. But with Gosset's theorem, the shape of the sampling distribution changes a bit from one sample size to another.

The **t distribution** is bell-shaped, but it has longer tails than a Normal distribution. The smaller the number of degrees of freedom, the "heavier the tails" of the distribution, which is a statistician's way of saying that more sample means will fall farther away from the mean. As the number of degrees of freedom gets large, t looks more and more like the Normal distribution. A t distribution with infinite degrees of freedom is a Normal distribution. Figure 8.11 shows histograms of t distributions with 4, 20, 100, and infinite degrees of freedom (that is, a Normal distribution). It is obvious that with small degrees of freedom, the t has far more outliers than a Normal distribution or a t distribution with a large number of degrees of freedom.

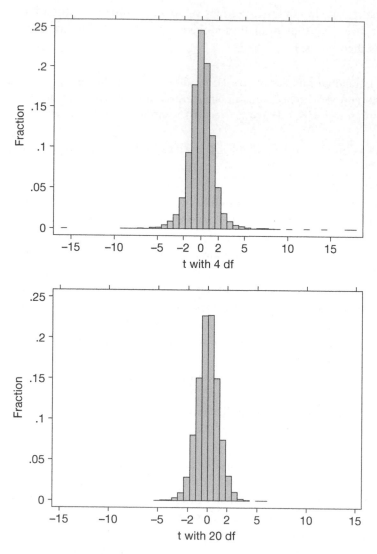

Figure 8.11   Comparison of t distributions with 4, 20, and 100 degrees of freedom, and Normal distribution

What does this mean when we are working with a research sample or when we are designing a study? Remember that under Gosset's theorem, the sampling distribution of sample means of random samples drawn from a Normal population will be a t distribution with N − 1 degrees of freedom. So the four distributions plotted correspond to the sampling distribution of sample means drawn from a Normal population with sample sizes of 5, 21, 101 and a large sample. With a sample size of only 5, it is not uncommon to get sample means rather far from the population mean. For example, out of 10,000 samples in a sampling distribution, we would get means greater than 2 about 500 times and means less than −2 about

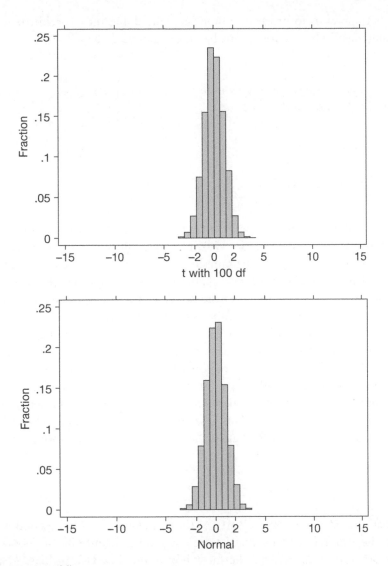

Figure 8.11   (*cont'd*)

500 times. That is, we would be off by 2 in one direction or another 1,000 times out of 10,000, or 10 percent of the time. In contrast, if we had a sample size of 101, we would get sample means greater than +2 only about 200 times and less than −2 only about 200 times. So with a sample size of 101 our chances of having a sample mean off by 2 from the population mean are only about 400 times out of 10,000, or about 4 percent. So we are less likely to have an unusual sample if we have a large sample size, and with small samples we must be cautious when we use the sample mean as our estimate of the population mean. The next chapter will discuss how we can use the information in sampling experiments and the sampling

theory that allows us to proceed without actually doing the experiment to make careful statements about populations based on sample data.

> *Small sample versus large sample estimators* are discussed at the end of this chapter as an Advanced Topic.

Properties of Estimators

The purpose of the sampling experiment is to learn how to use the information in a sample to guess at the population. We use sample information to estimate population parameters. The logic of the sampling experiment can be used to define the properties of estimators. We have more than one strategy of guessing (estimating) the population parameter, so we need a way to decide which strategy is best. For example, if we want to know the average number of years of education in the population, we can guess using the sample mean. That is why we did the sampling experiments above – to see how well the sample mean does as an estimate of the population mean. But we might also use the median, or something less traditional, like the average of the highest and lowest numbers, as our guess of the population mean. To decide which sample statistic provides the best guess of the population parameter, we must decide what we mean by "best." This is what properties of estimators are: criteria for what constitutes a good or bad guess that allows us to choose among estimators.

Bias

The statistics used to estimate features of the population should be good guesses. Most of the properties of estimators are essentially definitions of "good." The simplest property (and perhaps the most important) is a lack of bias. A sample estimate of a population parameter is **unbiased** if the mean of the sampling distribution for the estimator (called the **expected value** of the estimator) is equal to the population value being estimated. For example, for large simple random samples, the mean of the sampling distributions of sample means is equal to the population mean, so for such samples the sample mean is an unbiased estimator of the population mean. The estimator is correct "on average" even though any particular sample may have a mean quite different than the population mean.

Not all estimators are unbiased. In small samples, the sample variance is a biased estimate of the population variance because the mean of sample variances for all samples in a sampling distribution will be smaller than the population variance. The difference between the population value and the expected value of the estimator is the bias:

Bias = Population Value − Expected Value of Estimator   (8.5)

The expected value of the sampling distribution of means of simple random samples will be the population mean. So we can calculate its bias as:

Bias = Expected value of $(\bar{X}) - \mu = \mu - \mu = 0$ (8.6)

where μ is the population mean, and thus also the expected value of sample means. Thus the sample mean has a bias of zero when used to estimate the population mean, it is an **unbiased estimator** of the population mean.

However, theory tells us that the expected value of the sample variances will be

$[(N - 1)/(N)] * \sigma^2 = (N\sigma^2 - \sigma^2)/N = \sigma^2 - \sigma^2/N$ (8.7)

where σ^2 is the population variance. So the difference between the expected value (the mean of the sample variances across all the samples in the sampling experiment) and the value we are trying to estimate is $-\sigma^2/N$. That is the bias if we use the sample variance to estimate the population variance.

In large samples (large N), the term $-\sigma^2/N$ will be close to zero and thus can be ignored as was done in practice until Gosset's work. But it is better practice to always use the unbiased estimator developed by Gosset and Fisher. We use the unbiased estimator:

$$\hat{\sigma} = \frac{1}{N-1}\sum(X - \bar{X})^2$$ (8.8)

In fact, most computer software and statistics books only use the estimate – they don't include the definitional formula for the sample variance – only the estimate using $N - 1$.

Efficiency

There are often several unbiased estimators of a population parameter. The second criterion for choosing an estimator is **efficiency**. In comparing two unbiased estimators of a population parameter, the one with the smallest variance for its sampling distribution is said to be more efficient. All other things being equal, we prefer estimates that have small sampling variance because a smaller variance in the sampling distribution indicates that more sample estimates will be close to the population mean. The sample median, like the sample mean, is an unbiased estimate of the population mean when the population is normally distributed. But the median has a larger sampling variance than the mean, so the mean is more efficient and is preferred for normal populations and/or large samples.

There are more properties of estimators that we sometimes want to think about in doing analysis. We discuss these in Advanced Topic 8.4.

Feature 8.3 William Sealy Gosset

In the history of statistics, Gosset was both a major innovator and a bridge between the two other key figures of his time, Karl Pearson and R. A. Fisher. This latter role was important. Both Pearson and Fisher were apparently difficult people and from nearly their first intellectual encounter they disliked each other. This dislike grew and became extreme over the years. In contrast, Gosset was apparently well liked by all and helped to keep ideas circulating between Fisher and Pearson.

Gosset completed a degree in chemistry and mathematics at Oxford (statistics was not yet a field of specialization in universities) and in 1899 at age 23 was hired by the Guinness Brewing Company of Dublin. Guinness was a pioneer in trying to use science in business. Gosset worked at Guinness for the rest of his life, contributing not only as a scientist but as an administrator, eventually running the Guinness operation in the London area.

As we have noted in the text of the chapter, he quickly began to apply statistical ideas to brewing, and in the process developed some procedures and concepts that underpin modern statistics. His work draws on the ideas that were then being developed by Karl Pearson, especially Pearson's focus on the importance of distributions. Gosset's work was put on a firmer mathematical footing by Fisher. Fisher also developed mathematical models based on ideas first raised by Pearson, but the tone of that interchange was of a critic and an author under attack. In contrast, Gosset was always humble about his contributions and impressed with Fisher's abilities to develop rigorous mathematical approaches to statistical problems. Gosset is quoted as saying "my own investigations

[provide] only a rough idea of the thing . . ." and "Fisher really worked out the complete mathematics . . ." (Salsburg, 2001, p. 28). Gosset published his first paper in *Biometrika* in 1904 and became friends with Pearson. He then convinced Guinness to let him spend a year working with Pearson in the Galton Biometrical Laboratory. In 1908 he published in *Biometrika* "On the Probable Error of the Mean" the paper that is the foundation of much of modern statistics. Gosset met Fisher in 1912 while Fisher was an undergraduate at Cambridge. He introduced Fisher's work to Pearson, but the two quickly perceived each other as rivals and began the feud that is so famous in the history of statistics.

Gosset did most of his statistical work in his leisure time, and published as "A Student" because Guinness forbad its employees from publishing. It is said that Guinness did not know that the famous statistician "Student" was in their employ until after his death when the company was asked to contribute funds to support publication of a book containing Gosset's collected papers. He was not only modest but amazing in his diligence, often conducting sampling experiments by hand – drawing samples of slips of cardboard with numbers written on them, calculating the mean, drawing another sample, and so on. For his classic paper on small sample means, he developed 750 samples of size 4. Since he couldn't use data from Guinness, he used data that Galton had published on the height and finger length of criminals.[3] He chose these data because Galton had noted that they had roughly a Normal distribution. Remember there were no electronic calculators to do the arithmetic. This may have been one of the first efforts to use what we today call a Monte Carlo simulation to understand sampling distributions.

The history of statistics seems filled with both kindly and irascible characters. It is especially pleasant to be aware of Gosset, whose contributions were immense, who managed to make them while much of his professional time was spent on other matters and who was a friend to so many major figures.

In addition to maintaining a good relationship with both Fisher and Pearson, he was a strong supporter of F. N. Nightingale in her early days in statistics, and suggested in correspondence some key ideas on hypothesis testing to Egon Pearson, Karl's son. As we will see later in the book, these ideas are another major contribution to modern statistical thinking, though they would be developed by Egon Pearson and Jerzy Neyman.

What Have We Learned?

When we calculate the mean of a sample, we can think of that sample mean as the mean of the population plus some sampling error. If we draw many samples from the population, each will have a slightly different value for the error and thus a slightly different value for the sample mean. When we plot all these sample means, we have a sampling distribution of sample means. The Law of Large Numbers shows that if we have large samples, the sampling distribution of sample means will be Normal with the mean of all the samples equal to the population mean and the variance of the sample means equal to the variance of the population divided by the size of the samples. When we have small samples drawn from a population that has a Normal distribution, the sampling distribution of sample means will be a t-distribution with $N - 1$ degrees of freedom, with the sampling distribution mean equal to the population mean and variance equal to the variance of the population divided by the sample size. We also know that while the sample variance is not a good estimate of the population variance, the sum of squares of the sample divided by $N - 1$ is a good estimate.

All of this is the result of constructing a model of the process that generates sampling error in sample means and applying the results of probability theory to it. In the next chapter we will see how the sampling distribution can be put to practical use.

Advanced Topic 8.1   Sampling with Replacement and Sampling from Large Populations

Recall that simple random samples require *independence* – knowing some of the people who have been drawn into the sample provides no information on who else might be drawn. If we take a relatively large sample from a relatively small population, this property won't hold. If we think of drawing the sample as a sequence – draw the first person, then the second and so on – then we know if someone has been drawn as the first person they can't be drawn again. A very simple example will illustrate the point. Suppose you place five numbers – 1, 2, 3, 4, and 5 on cards in a box, shuffle them and then draw one at random. On the first draw, the probability of any one number coming up is one in five. Suppose you draw 2. Now if you do a second draw from the box, the odds of 2 coming up are zero and the odds of any of the four remaining numbers – 1, 3, 4, and 5 – are one in four. One way we can avoid this problem is by **sampling with replacement** – after drawing the 2, we put that card back in the box and shuffle again. Then the odds on the second draw are the same as on the first, each number has one chance in five of being selected. When we create populations on the computer for sampling experiments, we sample with replacement to have a simple random sample.

If the population from which you are drawing the sample is very large and the sample is relatively small, the loss of independence by not sampling with replacement doesn't matter. With five cards the probability of one of the remaining cards being on the second draw shifted from 0.2000 to 0.2500. If there had been a hundred cards in the box, the probability would shift from 0.0100 to 0.0101 – not really enough of a change to matter for most purposes. So when we draw samples for surveys of the US population, we don't worry that we are **sampling without replacement** – we don't allow anyone to appear in the sample twice.

Advanced Topic 8.2   Bootstrapping

There is a new technique called bootstrapping that takes the logic of using the computer to conduct sampling experiments one step further.[4] When we construct a hypothetical population for Monte Carlo simulation, we try to build that population so it looks very much like the population from which we're drawing the real sample and about which we want to generalize. We also make the sampling process on the computer mimic the process that we want to understand – the one by which the research sample was drawn. But if our computer generated population doesn't resemble the real population, then the Monte Carlo experiment is not a good guide to the true sampling distribution of samples from the real population. How can we be sure that the population on the computer resembles the real population we want to understand? If we have a probability sample, the best guide to the population is the sample. **Bootstrapping**

draws repeated random samples (with replacement – any case in the sample could appear more than once in a single bootstrapped sample) from the original sample. This mimics drawing repeated random samples from the population and allows estimation of the degree of uncertainty in many circumstances when analytical methods are not available.

Statisticians prefer an analytical solution to the problem of generating a sampling distribution, and many statisticians see either Monte Carlo simulation or bootstrapping as a second choice when the first choice of a mathematical solution is not available. This is because analytical solutions are more eloquent and not dependent on a single experiment. Indeed, when Bradley Efron began to publish papers on the logic of the bootstrap, he subtitled one of them "Thinking the Unthinkable" (Efron, 1979). But the bootstrap has become popular largely because Efron and others have done the analytic work that shows it produces useful results in many circumstances where no one has been able to find sampling distributions using traditional analytical tools.

Bootstrapping is an example of a resampling process in which new samples are drawn from the original sample. There are other related approaches, such as the jackknife and random sub-sample replication. In its simplest form, the jackknife performs an analysis on all the cases but one and repeats this process with each case left out, so there are the same number of analyses as there are cases, with each case left out of one analysis. In random sub-sample replications, we divide the sample (hopefully a large one) into two or more groups at random, and conduct our analysis in each group separately. This provides a sort of cross-check on the analysis. These resampling processes can be powerful tools for understanding the amount of error in the data.

Advanced Topic 8.3   Small Sample Versus Large Sample Estimators

The mathematics that allows statisticians to determine what the sampling distribution looks like (without actually having to do sampling experiments by hand or on the computer) work most easily when the sample being considered is very large. This is because a very large sample can be treated as if it were infinite in size, and in turn, the reciprocal of the sample size, $1/N$, can be treated as zero. This property sometimes allows statisticians to drop messy terms from key equations. If the term in the equation is divided by N, and if N is very large, the term can be treated as zero because we are dividing something by a very large number. Thus the messy terms essentially disappear, and the equations become much simpler. As a result, it can be easier to figure out sampling distributions for large samples than for small samples. Or, put differently, the fact that we have large samples makes unusual draws from the probability distribution of errors cancel each other out.

For example, if we draw a person for a sample who has a very low value for education, say 0 years of formal education, that outlier will be influential in a small sample. But in a large sample, that person will have less influence on the average and may well be balanced by someone with a very high level of education who is also drawn from the sample by chance. So large samples are "better behaved" in that they are less likely to produce results very different from what is typical of the population from which they are drawn. This is why large sample procedures for estimating the mean – the work based on the Law of Large Numbers – developed long before procedures based on small samples.

The results for large samples are called **asymptotic properties**, or large sample properties. Small sample properties are usually much harder to determine, and often statisticians have to rely on simulating the sampling experiment. Small sample properties are sometimes very different from asymptotic properties. Knowing what a sampling distribution looks like for large samples can be a poor guide to what will happen with a small sample. The fact that more is known about how large samples behave is one reason it is good to use large samples in research.

What is a large sample? It is important to remember that large means large enough to justify using statistical theory that assumes that the sample size is essentially infinite. If you use a statistical tool that depends on large sample theory with a small sample, you may get results that are very misleading. The sample size that will justify using asymptotic tools will vary. But certainly it is wise to have several hundred cases, and it is even better to have several thousand. When in doubt, it is a good idea to conduct the analysis using different statistical tools, some of which require large samples and some of which work with small samples and see if you draw the same conclusions.

Advanced Topic 8.4   More Properties of Estimators

Minimum variance – most efficient

It is sometimes possible to show analytically that an unbiased estimator will have a sampling variance as small as or smaller than that of any other unbiased estimator. Such an estimator is the most efficient unbiased estimator, and is called a **"best" unbiased estimator** (BUE). If it is formed through taking a linear function of the sample data, it is called a **best linear unbiased estimator** (BLUE) of the population parameter. The mean of a large sample or of a sample from a population that is normal is BLUE. So when statisticians refer to BLUE estimators, they do not refer to the color or mood of the estimator. They refer to the one estimator (of all of the unbiased linear estimators) that has the smallest variance. Recall that all of our samples are simple random samples unless we explicitly state otherwise.

Mean square error

The error that makes a sample statistic different from the population parameter comes from two sources. One is bias. But if the estimator is unbiased, the bias is zero and that source of error is eliminated. The other error comes from the sampling error that exists even for an unbiased estimator. It is described by the variance of the sampling distribution and the resulting efficiency of the estimator. These two can be summarized as the mean square error. For a sample statistic m used to estimate a population parameter M, the mean squared error (MSE) is

$$MSE = \frac{1}{K} \sum (m - M)^2 \qquad (8.9)$$

where K is the number of samples in the sampling experiment.

Remember that we are summing over all the samples we generated in the sampling experiment. Since the estimator is m, (m − M) is the error in using the sample m to estimate the population value M. The errors are squared (to eliminate negative signs) and averaged. The MSE is the variance of the sampling distribution of the estimator m when the estimator is unbiased, and the variance plus the bias when it is biased. Estimators can be evaluated based on their MSE, with BLUE estimators having the smallest MSE.

Consistency

It is sometimes possible to establish the behavior of an estimator when used with large samples but not when used with small samples. A consistent estimator is one whose sampling distribution "collapses" to the population parameter being estimated as the sample size approaches infinity. That is, if we conduct a series of sampling experiments and increase the size of the sample until it gets very large, the estimator may become unbiased and the sampling variance decreases towards zero even though in smaller samples the estimator was biased and not very efficient. Since we always work with finite samples, consistency is never a perfect guide to the actual behavior of an estimator in real samples, but when samples are "large," most statisticians are satisfied with consistency. Sometimes, "large" can be as small as 30. Usually several 100s can be treated as large, and several 1000s is nearly always considered large.

Robustness

We introduced the idea of robustness in the context of descriptive and exploratory analysis. The same idea applies to inferential and confirmatory analysis. An estimator is robust to the extent that its ability to provide unbiased and efficient estimates of population parameters are not degraded by unusual data points. Recall that the breakdown point of a statistic is the amount of outlying data it can tolerate before being pulled towards the outliers. For the mean family, the breakdown point is 0 percent, for the median family it is 50 percent. Estimators with breakdown points of 50 percent would be ideal, though that is not always possible. As it turns out, there is often a tradeoff between efficiency and robustness. This is because high efficiency estimators take account of all data points. But in guarding against outliers, robust estimators minimize the influence of some data points (those at the extremes of the distribution of data). For example, when sampling from a Normal population, the sample mean is a more efficient estimator of the population mean than the sample median. But if the population is not Normal, and outliers creep in, the mean's performance as an estimator (as measured by MSE) will be degraded, and the median will be preferable. This illustrates the importance of assumptions. We often assume a Normal population. If the assumption is true, the sample mean is the best possible estimator of the population mean. But for non-Normal populations, the mean may be a relatively poor estimator, and the median is preferred.

Maximum likelihood estimators

Maximum likelihood estimators are an important approach to constructing estimates. They are based on a clever but subtle line of logic. The exact values obtained for the mean or other characteristic of a particular sample are dependent on the characteristics of the population and random error. The probability of getting a particular sample will differ depending on the population from which is it drawn. The maximum likelihood estimation procedure assumes that the sample we draw is the most probable (maximally likely) sample for the population being studied. Thus it chooses an estimate of population parameters that make the actual sample as probable as possible.

Maximum likelihood estimators are consistent, asymptotically unbiased (though many of them have a bias inversely proportional to the sample size in finite samples), asymptotically efficient (minimum sampling variance), and asymptotically Normal (the sampling distribution for large samples approaches Normal shape). Thus they are "good" estimates by a variety of criteria. Unfortunately, these are asymptotic properties that may not hold up in finite samples. In addition, most maximum likelihood estimators assume that the population to which inferences are being made has a Normal distribution, or make some other rather restrictive assumptions, and are not robust if these assumptions are violated. In some situations bootstrapping provides maximum likelihood estimates that do not require any assumptions about the population distribution.

Applications

Two of our data sets, state homicide rates and environmental treaty participation are based on population data and the population size in each case is fairly small, so any sample we might draw to experiment with drawing samples would also be small. So for purposes of illustrating what we have learned in the chapter, it will be best to work with our larger data sets from which we can experiment with relatively large samples or small samples.

To further illustrate the Law of Large Numbers, Normal distribution, Law of Large Numbers, and Gosset's theorem with small samples, we will walk through another

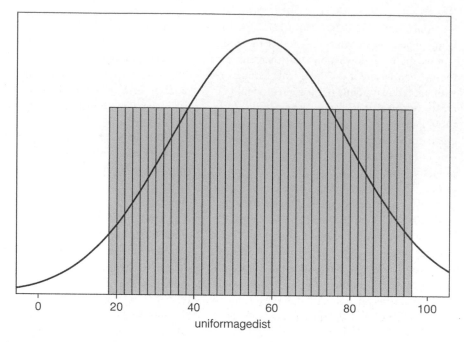

Figure 8A.1   Histogram of distribution for hypothetical uniform population with Normal distribution for comparison
Data source: Simulation.

set of sampling experiments that we have created. In this instance, we will use the age distribution in Great Britain as our example. In the animal concern data set (i.e., International Social Survey Programme data set), we had data on 971 British citizens, with one respondent who did not answer the age question. The mean age of the sample is 47.55, with an age range of 18 to 92 years and a standard deviation of 17.6. We will, therefore, start by selecting samples of size 971 for our sampling experiment.

We will begin by constructing a data set in which we assume there is an equal number of people of every age, beginning at age 18 and we will have an upper age limit of 95 for illustrative purposes. You will remember that this kind of flat distribution is called a uniform distribution. This hypothetical population is shown in Figure 8A.1. The mean age for this hypothetical population is 56.5.

We selected several hundred samples with a sample size of 971 and also selected several hundred samples with a sample size of 10 to make comparisons between the behavior of large and small samples. The histogram with the means for the samples of size 971 is shown in Figure 8A.2 and for the means for samples of size 10 is in Figure 8A.3. As Figure 8A.2 shows, the distribution of the sample means for the large samples approaches a Normal distribution. This is consistent with the Law of Large Numbers. The mean of the sample means for the large samples (N = 971) equals the mean of the hypothetical population to one decimal place – 56.5. This is consistent with the Law of Large Numbers, which states the mean of the means will be equal to the population mean. The range of the means is small, between

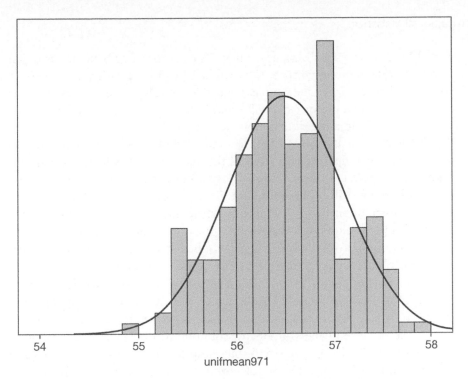

Figure 8A.2 Sampling distribution of means from a uniform population, N = 971

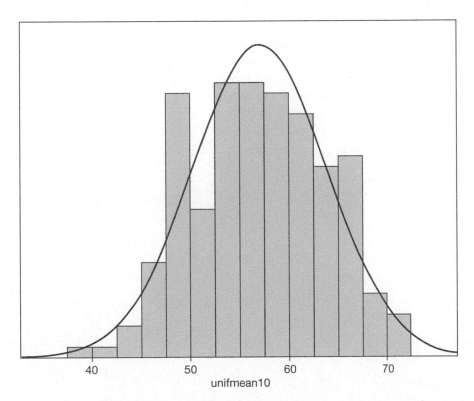

Figure 8A.3 Sampling distribution of means from a uniform population, N = 10

54.96 and 57.92, with a small variance of 0.35. The Law of Large Numbers says that the variance of the sampling distribution will be equal to the variance of the population divided by the size of the samples being selected. Therefore, the Law of Large Numbers proposes that the variance will equal 507.13/971 or 0.52, which is not too far from the actual sampling distribution variance of 0.35.

The distribution of the means for the small samples (N = 10) selected also approaches a Normal distribution. Gosset's theorem proposes when repeated random samples are selected from a Normally distributed population we get a t distribution, which you will recall looks a lot like a Normal distribution. Of course, the uniform distribution we have used to construct the hypothetical population is not Normal so we wouldn't necessarily expect the distribution of means to be bell-shaped, but in this case it is. While it also has a bell-shape, it has longer tails than a Normal distribution. Accordingly, the range of means is much larger than with the large samples, ranging between 39.10 and 72.50, with a large variance of 45.51. The mean of the sample means is 57.02, which is quite close to the population mean (56.5), which is also consistent with Gosset's theorem. The theorem also states that the variance of the sampling distribution will be equal to the variance of the population divided by the size of the samples. In this case, the variance of the population is 507.13 and the sample size is 10, so the sampling distirbution variance should be around 50.7. In fact, as shown in Table 8A.1 the actual variance of the sample means is 45.51, which is close to the value we calculated. So we see that for small samples and for large samples most sample means cluster around the population mean but when we work with small samples we are much more likely to get a sample mean distant from the population mean than when we work with large samples. And remember, Gosset's theorem works well only when the population from which we are drawing the samples has a Normal distribution, while

Table 8A.1   Means, variances, standard deviations, and range from four sampling experiments

Population and sample size	Mean of sample means	Variance of sample means	Standard deviation of sample means	Range
Uniform distribution, N = 971, Mean = 56.5	56.50	0.35	0.59	54.96 – 57.92
Uniform distribution, N = 10, Mean = 56.5	57.02	45.51	6.76	39.10 – 72.50
Bimodal distribution, N = 971, Mean = 52.13	52.11	0.25	0.49	50.72 – 53.51
Bimodal distribution, N = 10, Mean = 52.13	52.70	38.37	6.19	39.10 – 70.30

the law of large numbers works with samples from any population as long as the samples are large.

Of course the actual age distribution of Great Britain is not uniform. As in all populations, there are periods when the birth rate is higher and other periods when the birth rate is lower. In Great Britain, there were high birth rates during the periods between 1946 and 1951 and 1961 and 1966 (as well as 1986 to 1991, although these citizens would be too young to be included in our sample). Since the data from the ISSP are from 2000, individuals born during these periods would be ages 49–54 and ages 34–39 in 2000.

We will construct a second hypothetical population in which we redistribute the age distribution to reflect these "booms" in the population. This hypothetical population will have a small percent of cases in each age group from ages 18–95, with spikes at ages 34–39 and ages 49–54. This creates a bimodal distribution, as shown in Figure 8A.4. We have also selected two sets of samples. One is of size 971 and one of size 10, as in our previous analysis. We create several hundred samples of each of these sizes. The distribution of the sample means for the two sets of sampling experiments are shown in Figures 8A.5 and 8A.6. These sampling experiments are also summarized in Table 8A.1.

Consistent with the Law of Large Numbers and Gosset's Theorem for small samples, the mean of the sampling means in both bimodal sampling experiments is very close to the mean of the hypothetical population and variances close to the theorem's calculations. As we found with the uniform distribution, the spread of

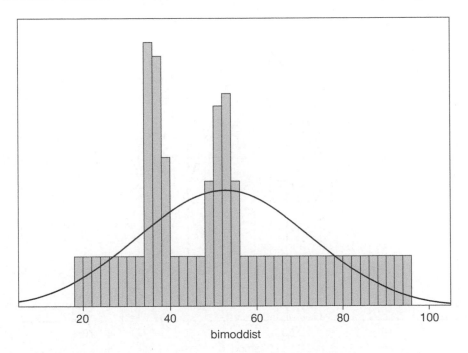

Figure 8A.4 Histogram of bimodal age distribution for hypothetical population with Normal distribution for comparison

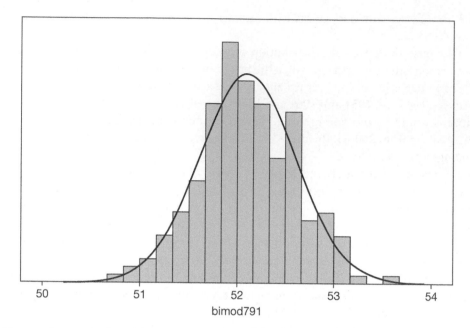

Figure 8A.5   Sampling distribution of means from a bimodal population, N = 971

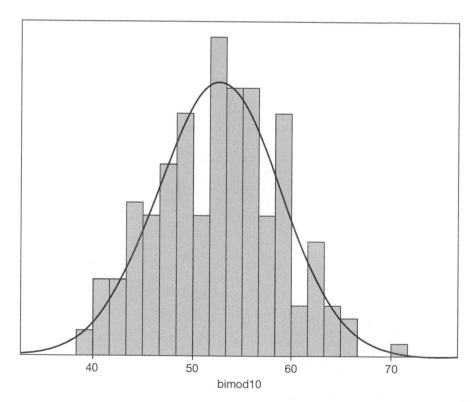

Figure 8A.6   Sampling distribution of means from a bimodal population, N = 10

means of the small samples is much larger than it is for the large samples. Both distributions are bell-shaped and resemble a Normal distribution. But again, we wouldn't expect Gosset's theorem to apply very exactly because it is meant for use when samples are drawn from Normally distributed populations.

Exercises

1. Table 8E.1 contains data from two sampling experiments from two hypothetical populations. Random samples were selected from the population data 800 times. For this question, we'll draw on the data from distribution no. 1 only.

a) What is the population mean for distribution no. 1? Are you confident in your response? Why or why not?

b) Calculate the variance of the population, if possible.

c) Can you determine the general shape of the distribution from the data you have been given? Why or why not?

d) What theorem have you been relying on in responding to questions about this distribution and how did you determine this?

Table 8E.1   Results of sampling experiments from two hypothetical populations

Population and sample size	Mean of sample means	Variance of sample means	Standard deviation of sample means	Range
Distribution no. 1 N = 824	24.01	0.68	0.82	23.64 – 25.03
Distribution no. 2 N = 25	24.02	57.23	7.57	16.50 – 31.32

2. Now let's turn to distribution no. 2.

a) What theorem will you draw on to learn more about this distribution? If the theorem is different from the one you used in Question 1, why are you using a different theorem? If the theorem is the same one you used in Question 1, why are you using the same theorem?

b) Can you determine the general shape of the distribution from the data you have been given? Why or why not?

c) What is the population mean for this distribution? How did you determine this value?

d) Calculate the variance of the population, if possible.

3. a)   What do the Law of Large Numbers and Gosset's Theorem tell us about why the variances and ranges of sample means of these two distributions differ so considerably?

    b)   What is the advantage of drawing large samples when carrying out a sampling experiment?

References

Dietz, T., Frey, R. S., and Kalof, L. 1987. Estimation with cross-national data: Robust and non-parametric methods. *American Sociological Review* 52, 380–90.

Dietz, T., Kalof, L., and Frey, R. S. 1991. On the utility of robust and resampling procedures. *Rural Sociology* 56, 461–74.

Efron, B. 1979. Computers and the theory of statistics: Thinking the unthinkable. *SIAM Review* 21, 460–80.

Efron, B. and Tibshirani, R. J. 1993. *An Introduction to the Bootstrap*. New York: Chapman & Hall.

Salsburg, D. 2001. *The Lady Tasting Tea: How Statistics Revolutionized Science in the Twentieth Century*. New York: W.H. Freeman and Company.

CHAPTER 9
USING SAMPLING DISTRIBUTIONS: CONFIDENCE INTERVALS

Outline

The last two chapters introduced the key concepts we use in making statements in the face of random error: the sampling experiment and the sampling distribution. In nearly every chapter from here on we will learn about tools that allow us to use the logic of the sampling experiment and sampling distribution to deal with sampling error in surveys, randomization error in experiments, measurement error or other sources of random error in data we are analyzing. While many tools have been developed by statisticians to deal with specific problems in data analysis, most of them are applications of these two basic concepts. In fact, most of the tools are really special cases of the two general tools we introduce in this chapter and the next: the **confidence interval** and the **hypothesis test**. Confidence intervals allow us to make estimates of values in the population, such as the population mean, while taking into account that our estimates are never certain. Hypothesis tests allow us to assess the degree to which some assertion about the population is reasonable to believe, given the information about the population we have in our sample. For example, a hypothesis test could tell us whether or not it's reasonable to believe that men and women have the same knowledge about how AIDS is transmitted, or have the same level of concern about animals.

In this chapter we focus on the confidence interval. In Chapter 8 we discussed the Law of Large Numbers and Gosset's work in describing sampling distributions, both of which help us make statements about the population based on sample data in two ways. (Remember that we can rely on the Law of Large Numbers when we have large simple random samples, and we can use Gosset's work when we have simple random samples of any size drawn from a Normally distributed population.)

First, they tell us that the mean of a sample is a good guess of the mean of the population. Below we discuss in detail what we mean by "good" but to preview: We know that for large samples, most samples will have means close to the population mean, few sample means will be far from the population mean, and if we use the sample mean as our estimate of the population mean we will be right on average. The same thing applies when we have simple random samples of any size drawn from a Normally distributed population – the mean of the sample means is a good guess of the population mean for the same reasons.

The use of the sample mean to estimate the population mean is called a **point estimate**. We are guessing at the population parameter with a single number. But sampling distributions let us do even better because the second way sampling distributions help us make statements about the population is by allowing us to develop an interval estimate. An **interval estimate** is a range that we can be quite certain includes the real population mean. This is important, since few sample means are exactly equal to the population mean, even though most are pretty close. We can have not only a single number that is a good guess of what's true in the population but a sense of how likely that guess is to miss (or accurately reflect) the true population value.

Confidence Intervals Using the Law of Large Numbers

While the sample mean may be the best single number to use as an estimate of the population mean, how can we take into account the uncertainty involved in making such a guess? To do this, we want to develop a high and low estimate around the sample mean so that this range will very likely include the population mean. That is, we can be pretty confident that for our research sample the confidence interval we build will include the real value of the population mean. We are confident because we know from sampling theory that we usually "catch" the value of the population mean in the confidence interval. So unless we are very unlucky, it's likely that the mean of the population we are studying is "captured" by the confidence interval we construct from data in our sample.

We can determine the percentage of samples in the sampling distribution for which the confidence interval "catches" the population mean. It is common to use 90 percent, 95 percent, and 99 percent. Thus, we can be 90, 95, or 99 percent certain that the confidence interval includes the population mean. When we use these values we are saying that the procedures we are using to construct the confidence intervals gives us intervals that cover the population mean for 90 percent or 95 percent or 99 percent of the samples in the sampling distribution. To be a bit more informal, we can say that we are 90 percent or 95 percent or 99 percent certain that the confidence interval we have constructed for our research sample "captures" (includes) the population mean.

Constructing the Confidence Interval for Large Samples

Let's start with the case of large simple random samples. If we have a large simple random sample, and we are trying to estimate the population mean, the confidence interval will be:

Upper limit: $\bar{x} + Z_{1-\alpha}\hat{\sigma}_{\bar{x}}$ (9.1)

Lower limit: $\bar{x} - Z_{1-\alpha}\hat{\sigma}_{\bar{x}}$ (9.2)

Or, in the form that statisticians use:

$$\bar{x} \pm Z_{1-\alpha}\hat{\sigma}_{\bar{x}}$$ (9.3)

These formulas are really rather simple if we take it one step at a time. First, we would read the equation as "x bar plus and minus Z sub one minus alpha times sigma hat sub x bar." We start with the sample mean, which is our best estimate of the population mean. To get the upper bound of the confidence interval we are going to add a number to the sample mean. To get the lower bound of the confidence interval, we will subtract the same number from the sample mean. We calculate the number to add and subtract by multiplying together two numbers. The first one, $Z_{1-\alpha}$, is found by looking in a "Z table" for the number corresponding to the confidence level we want. (The last row of Table A.1 in Appendix C displays Z values.) Let's say we want to be 95 percent certain that the confidence interval we build from our research sample will capture the population mean – that is, we want to use a procedure that will include the population mean in the interval for 95 percent of the samples in the sampling experiment. The Z table tells us the value of Z to use to make sure the confidence interval will include the mean of the population the percentage of times we want. The Z table is based on figuring out the areas under the Normal distribution, that is, how far we have to go to get a confidence interval big enough to work the specified percentage of times.

In a sense the Z table saves us the trouble of having to do a sampling experiment because the Law of Large Numbers tells us what would happen if we did the experiment. The Z table just records the results of the elaborate calculations that are required by the Law. Computer statistical packages do the calculations directly, so we don't use tables much these days except when we are learning how to build a confidence interval. In the next chapter it will become clear why the confidence level (95 percent in this case) is referred to as $1 - \alpha$. For the moment, accept it as an arbitrary label. Note that the right hand side of the expression:

$$Z_{1-\alpha}\hat{\sigma}_{\bar{x}} \tag{9.4}$$

is also referred to as the "**margin of error**." Thus you will sometimes see newspapers reporting that the results of a poll have a 5 percent margin of error. What they are saying is that a particular confidence interval (usually 95 percent, but often the newspapers don't say) is the mean plus and minus the reported margin of error.

The symbol

$$\hat{\sigma}_{\bar{x}} \tag{9.5}$$

is pronounced "sigma hat sub X bar" and is the symbol for the standard deviation of the sampling distribution of sample means. It is given a special name: the **standard error of the mean**, but it is just the standard deviation of the sampling distribution. We know from the Law of Large Numbers that

$$\sigma_{\bar{x}}^2 = \frac{\sigma^2}{N} \tag{9.6}$$

That is, the variance of the sampling distribution equals the variance of the population divided by the sample size. Then we can get the standard error by taking the square root. But since we don't know the variance of the population, what good does this do?

To actually build a confidence interval, we have to estimate the variance of the population. Gosset showed that we can get a good estimate of the population variance by dividing the sample sum of squares by N − 1. That is

$$\hat{\sigma}^2 = \frac{\sum(x - \bar{x})^2}{N - 1} \tag{9.7}$$

Or if you already have the variance of the sample s^2

$$\hat{\sigma}^2 = \left(\frac{N}{N - 1}\right)s^2 \tag{9.8}$$

Let's build a confidence interval for the animal concern question. Recall that we have a sample of 29,486 people and that the score can run from 1 to 5. The sample mean is 2.49 and the sample variance is 1.39.

1   Estimate the population variance: Given that the sample size is 29,486, we have:

$$\hat{\sigma}^2 = \left(\frac{N}{N - 1}\right)s^2$$

$$\hat{\sigma}^2 = \left(\frac{29,486}{29,485}\right)1.39 = 1.39 \tag{9.9}$$

In this case the sample size is so large that the correction does not change the variance, at least to the second decimal point.

2   Estimate the variance of the sampling distribution:

$$\hat{\sigma}_{\bar{x}}^2 = \frac{\sigma^2}{N} \tag{9.10}$$

That is, we take the estimate of the variance of the population and divide it by the size of our sample to get an estimate of the variance of the sampling distribution. This will be 1.39/29,846 = 0.000047.

3   Estimate the standard error: This is easy. Once we have an estimate of the variance of the sampling distribution, we just take the square root to get an estimate of the standard deviation of the sampling distribution – the standard error. This will be the square root of 0.000047, which equals 0.00686.

4   Find the Z value for the level of confidence we want: If we want a 95 percent confidence interval, the Z value will be 1.96 (see the Z values in the last row of Table A.1 in Appendix C).

5   Multiply the Z value by the standard error to get the "margin of error": This is 1.96 * 0.00686 = 0.013446.

6   Add the margin of error (in this case, 0.013) to the mean to get the upper bound: This gives 2.49 + 0.013446 = 2.503 or about 2.50.

7   Subtract the margin of error (in this case, 0.013) from the mean to get the lower bound: This gives 2.49 − 0.013 = 2.477 or about 2.48.

Thus, we can be 95 percent certain that the mean animal concern score for the adult population from which the sample was drawn is between 2.48 and 2.50. By 95 percent certain we mean that if we used this procedure in a sampling experiment, the confidence interval would cover the true population mean for 95 percent of all samples, and for 5 percent of all samples it would miss.

Sampling Experiments

It may be helpful to show how sampling experiments work using the two examples we explored in the last chapter. Recall that we created two populations from which to draw samples. One had a uniform distribution of years of education, with a mean of 10 years of education and every number of years of education from 0 to 20 having the same number of "people." The other was a "lumpy" distribution with a mean of 13.4 and with people "lumping up" in certain numbers representing the number of years of education that match categories, such as high school graduate or college graduate. We will again draw large samples of 901 "people."

For each sample in the sampling distribution, we will calculate a confidence interval using steps 1 to 7 above. We will see if the confidence interval for that sample includes the population mean. For the uniform distribution the mean was 10. For the non-uniform distribution the mean was 13.4. If the confidence interval includes the population mean, we have a success for that sample in that the confidence interval included the true population mean. The Law of Large Numbers says that a 95 percent confidence interval should be successful for 95 percent of the samples in the sampling distribution. Table 9.1 gives the results for both the uniform and non-uniform distributions.

It appears that the theory is pretty accurate – for about 95 percent of samples the confidence interval included the population mean, and it missed for about 5 percent.

Table 9.1   Ninety-five percent confidence interval for a sampling experiment

Distribution of population	Number of successes	Number of misses	Percentage successes	Percentage misses
Uniform	4,739	261	94.78	5.22
Non-uniform	4,747	253	94.94	5.06

Confidence Intervals Using the t Distribution

But what do we do if we have small samples? Before Gosset, researchers would calculate all confidence intervals, even for small samples, using the Law of Large Numbers. And before Gosset, they had to use the sample variances as the estimate of the population variation, because they hadn't figured out how to use the correction that yields an unbiased estimate of the population variance from the sample. We can try this with some sampling experiments using these older procedures with small samples to see how bad the problem can be. We've had the computer draw samples of size 7 from both the uniform and the non-uniform populations of education we've been using. Table 9.2 shows the performance of confidence intervals based on the Central Limit Theorem with a biased estimate of population variance.

It appears that we're in trouble. The 95 percent confidence interval is supposed to include the population mean for 95 percent of samples, and miss only 5 percent of the time. But it's working only around 90 percent of the time. That is, we're missing the true population mean nearly twice as often as we should be.

The problem is that using a Z value with small samples makes the size of the interval too small. Further, since we tend to underestimate the population variance when we just use the sample variance as our estimate, rather than using the formula that Gosset developed, that too will make the confidence interval too small. The result is the confidence interval misses the population mean more often than it should.

We can easily get around this by using Gosset's work to build the confidence interval. We proceed in exactly the same way as we did with the Z distribution. The only difference is that we now use a value for $t_{df, 1-\alpha}$ rather than for $z_{1-\alpha}$. Because we have a small sample from a Normally distributed population when we use Gosset's approach, the sampling distribution of sample means will have a t distribution. We use a t table (Table A.1 in Appendix C) to find these values of t we want that correspond to the accuracy of the confidence interval, just as we used a Z table to find the appropriate values of Z. But now we have to keep track of the number of degrees of freedom (N − 1 for estimating a confidence interval for the population mean) to find the right t value. Remember, we can only use Gosset's theorem and the

Table 9.2 Ninety-five percent confidence intervals based on applying the Law of Large Numbers to small samples

Distribution of population	Number of successes	Number of misses	Percentage successes	Percentage misses
Uniform	4,551	449	91.02	8.98
Non-uniform	4,446	554	88.92	11.08

t distribution when we have reason to believe that the population from which the sample was drawn is Normal in its distribution.

Suppose that instead of having 29,486 observations of the animal concern question scores, we only had 10 observations. This might be because we are studying something that is expensive to measure, or because we are doing a pilot study with limited data. To show how we would proceed, we have drawn a random sample of 10 cases from our sample of 29,486. A random sample of a random sample is still a random sample, so our 10 cases are a random sample of the countries included in the 2000 ISSP data set. Of course, this is not something we would do in research, but here it helps to show how we can build a confidence interval for the mean when we have a small sample.

The mean of our sample of 10 is 2.6 and the variance is 1.6. We will now build the confidence interval assuming that the population values of the animal concern question are Normally distributed. Of course, we don't know if that's true, and our confidence intervals will be accurate only if the population really is Normal. Figure 9.1 is a histogram of the sample of 10 cases imposed with a Normal curve.

The shape of the frequency distribution of the sample would make us cautious about the confidence intervals in that, while it has a slight tendency to peak near the middle, it's not very close to a Normal distribution. But of course, it's not the shape of the sample distribution we are concerned with but rather the shape of the *population* distribution. Still, we would want to be very cautious if an important decision rested on the confidence interval. But having noted that our confidence interval may not capture the population mean the right percentage of times if the population really isn't Normal, we will proceed through the steps of building the confidence interval.

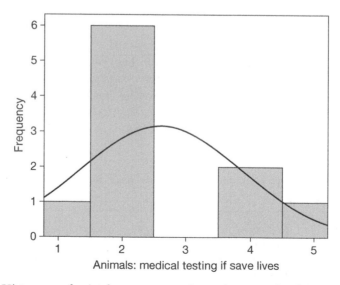

Figure 9.1 Histogram of animal concern score in random sample of size 10
Data source: 2000 ISSP data set, analyzed with Stata.

The steps we use in building the confidence interval are the same as before:

1   Estimate the population variance. Again, we take the sample variance and multiply by the correction factor.

$$\hat{\sigma}^2 = \left(\frac{N}{N-1}\right)s^2 = \left(\frac{10}{9}\right)1.6 = 1.78 \qquad (9.11)$$

2   Use that to estimate the variance of the sampling distribution.

$$\hat{\sigma}_{\bar{x}}^2 = \left(\frac{\hat{\sigma}^2}{N}\right) = \left(\frac{1.78}{10}\right) = 0.178 \qquad (9.12)$$

3   Estimate the standard error (take the square root of the variance of the sampling distribution found in step 2 to get the standard error of the mean).

$$\hat{\sigma}_{\bar{x}} = \sqrt{0.178} = 0.422 \qquad (9.13)$$

4   Find the t value (remember this is a small sample) that matches the desired level of confidence, 95 percent, and the number of degrees of freedom, which is N − 1 or 9. The t value is 2.262 (see the Table A.1 in Appendix C). Remember that the 95 percent Z value is 1.96. The t value takes the sample size into account and is larger than the Z value, thus making the confidence interval bigger than if we used a Z based on the Law of Large Numbers.
5   Multiply the t value by the standard error.

$$(2.262) * (0.422) = 0.955 \qquad (9.14)$$

6   Add this to the sample mean to get the upper bound.

$$2.6 + 0.955 = 3.555 \qquad (9.15)$$

7   Subtract from the sample mean to get the lower bound.

$$2.6 - 0.955 = 1.645 \qquad (9.16)$$

So we can be 95 percent certain that the true population mean for animal concern is between about 1.65 and 3.56. Remember that the 95 percent means that this confidence interval procedure would catch the real population mean in 95 percent of samples in a sampling experiment. Also remember that the 95 percent depends on the population distribution of education being Normal. If it's a bit different than Normal, then the confidence interval may hit less often.

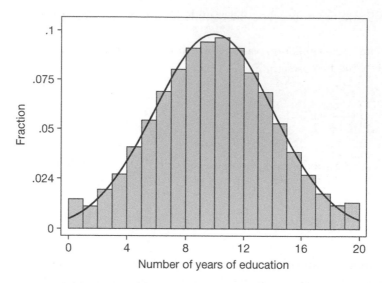

Figure 9.2   Population of years of education that is roughly Normally distributed
Data source: Simulation using Stata.

What if the population isn't really Normal? We never know for certain if the population is Normally distributed, though sometimes a body of previous research allows us to be fairly certain it's roughly Normal. We can see what happens when we apply Gosset's theorem to building confidence intervals with small samples when the population is roughly normal and when it is not by conducting another set of sampling experiments. We'll use the same two non-Normal distributions of education we used before, the uniform distribution and the lumpy distribution. We'll also add a third distribution – one that is a roughly Normal distribution of education with a mean of 9.96 and a variance of 16.43. Figure 9.2 shows this population. Note that it does deviate from the Normal a bit, in that there are too many cases in the "tails" and a bit too few in the middle. If we had a perfectly Normal distribution, the results based on Gosset's theorem would work perfectly. But here we want to see what happens if we are a little off.

Table 9.3 shows how often out of a thousand samples of size 10 in a sampling experiment the confidence intervals based on Gosset's theorem actually include the population mean. We can see that, in each case when we apply the confidence interval constructed on the assumption that the population is Normally distributed to non-Normal populations, we actually included the true population mean fewer times than the theory predicted. We should have had 950 out of 1,000 success and 50 misses according to theory. This should make us a bit cautious about working with small samples where we don't know the distribution of the population to be Normal. But even with the very non-Normal uniform and lumpy distributions, we don't go too far wrong. Applying Gosset's theorem to these non-Normal populations still created confidence intervals that "captured" the true population mean about 94 percent of the time.

Table 9.3 Accuracy of 95 percent confidence interval based on Gosset's theorem when the population is not Normally distributed, samples of size 10

Distribution of population	Number of successes out of 1,000	Number of misses out of 1,000	Percentage successes	Percentage misses
Normal population assumed by theory	950	50	95.0	5.0
Uniform	940	60	94.0	6.0
Non-uniform, lumpy	938	62	93.8	6.2
Nearly Normal	942	58	94.2	5.8

Size of Confidence Intervals

At the beginning of the chapter we said that you can build the confidence interval to be as certain about capturing the true population mean as you want. While 95 percent is a pretty good success rate, is it possible to be absolutely certain? Yes, we can be absolutely certain that the confidence interval for education runs from 0 (the lowest possible value) to infinity, or to some very large number. Of course, that's not very useful. In confidence intervals there is always a tradeoff between how wide the confidence interval is and how certain you are it catches the population mean. The wider the range, the more certain you are. Table 9.4 shows how the size of the confidence interval based on the Law of Large Numbers changes as we demand more certainty that we have captured the true population mean. Remember that the sample size for the animal concern variable we are using as an example is 29,486. As you can see, with such a large sample, the confidence interval is quite small and similar at all four probability levels. Table 9.5 shows how the size of the confidence intervals changes when we have a smaller sample size (a random sample of about 10 percent of the 2000 ISSP cases was selected or N = 2,945). Note though that almost 3,000 cases is still a very large sample. The more certain we want to be that the confidence interval captures the population mean, the broader the range of our estimate.

Table 9.4 Levels of confidence associated with interval estimates for mean animal concern score (N = 29,486)

Probability of hitting population mean	Upper bound	Lower bound	Size (upper limit minus lower limit)
0.75	2.50	2.49	0.01
0.90	2.50	2.48	0.02
0.95	2.50	2.48	0.02
0.99	2.51	2.48	0.03

Table 9.5 Levels of confidence associated with interval estimates for mean animal concern score for random sample of 10% of ISSP cases (N = 2,945)

Probability of hitting population mean	Upper bound	Lower bound	Size (upper limit minus lower limit)
0.75	2.49	2.44	0.05
0.9	2.50	2.43	0.07
0.95	2.50	2.42	0.08
0.99	2.52	2.41	0.11

In some applications we'd like to have the size of the confidence interval relatively small but the probability relatively high. How can we do that? Remember that there are several things that influence the size of the confidence interval:

1   *The confidence level.* The more certain we are of getting the population mean in the interval, the larger the interval.
2   *The sample size.* As we saw with the animal concern example above, the variance of the sampling distribution is inversely proportional to the sample size, so the larger the sample size the smaller the confidence interval. Since we actually multiply by the standard error, which is the square root of the variance of the sampling distribution, the size of the confidence interval changes proportionately with the square root of the sample size. If you want to cut the size of the confidence interval in half by increasing sample size, you have to quadruple the sample size. The sample size also plays a role in the t value in that the smaller the number of degrees of freedom, the larger the t.
3   *The variance of the population.* Usually, this is out of the researcher's control. But some variables may have smaller variance than others, so if you can work with variables known to have small variance, this could reduce the size of the confidence interval.

So, to reduce the size of the confidence interval without increasing the chances of missing the population mean, we must increase the sample size for a study. In fact, if we know how big the confidence interval should be, and we know how certain we must be and can make a guess at the population variance, we can calculate how large our sample needs to be.

The equation is

$$N = \frac{4\hat{\sigma}^2 t^2_{N-1,1-\alpha}}{c} \qquad (9.17)$$

Here c is the size of the confidence interval we want. Remember that $\hat{\sigma}^2$ is our estimate of the population variance and that $t_{N-1,\,1-\alpha}$ is the value of t from the table to get the probability of catching the population mean that we want. Of course the value for t depends on the confidence level but also the sample size. So you can start with a guess, look up a t value, calculate the sample size required, then use that sample size for a new t and repeat the process until the answer doesn't change. The tricky part can be guessing the value of the population variance. Usually we have to rely on prior research for this, using estimates of population variance from other studies.

Notice that the size of the population does not enter into the formula. As long as we are drawing a relatively small proportion of the population into the sample (so that we are approximating sampling with replacement), then the population size is irrelevant. For example, for a given variance, confidence level and size of confidence interval, you need the same size sample from one state as you do from the whole country.

Graphing Confidence Intervals

It is common to see graphs that display the mean and confidence interval of a variable conditional on some other variables. Figure 9.3 shows the mean animal

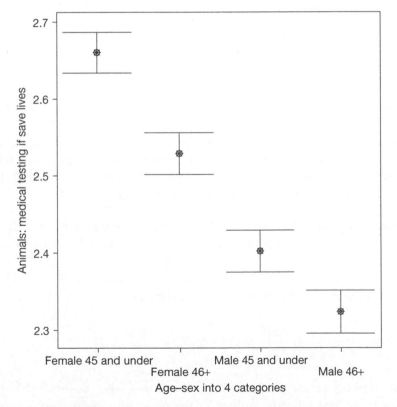

Figure 9.3   Mean animal concern score by gender and age group, Males ages 45 and under (N = 6,859); Males ages 46+ (N = 6,378); Females ages 45 and under (N = 8,379); Females ages 46+ (N = 7,460)
Data source: 2000 ISSP data set, analyzed with SPSS.

concern score by age group and gender. The small boxes represent the sample mean scores on the scale for each gender/age group (remember the sample mean is our best estimate of the population mean). The vertical bars extend to the 95 percent confidence intervals. (If you see an error bar graph in the literature, check to see the confidence level. A 95 percent confidence interval is the most common, but sometimes bars are constructed at two standard errors (this is because the Z value for 95 percent is 1.96, very close to 2) or at one standard error (with a large sample this would be about a 42 percent confidence interval).)

Note that because of the large sample size for each group, the confidence intervals are quite small. While the confidence intervals appear to be quite different, look closely at the Y axis. You will see that the smallest mean and confidence interval is for men ages 46+, ranging from about 2.30 to 2.36. The largest mean and confidence interval is for females ages 45 and under, ranging from about 2.64 to 2.67. The difference between these confidence intervals is small. None of the confidence intervals for the four groups overlap, which makes make us suspect that there are group differences. It appears that men are different from women of the same age group and that the two age groups are different for both men and women. The next chapter will examine how we determine whether differences across groups are likely to exist in the population, given evidence of differences in our sample.

Confidence Intervals for Dichotomous Variables

The Law of Large Numbers applies whenever we have a large simple random sample, whatever the distribution of the population. This means that we can use the Law of Large Numbers to construct a confidence interval for a dichotomous variable. The steps are just the same as those above for a continuous variable. As an example, we will construct a 99 percent confidence interval for the AIDS knowledge variable. Recall that the sample mean for this variable is 0.77.

1   Estimate the population variance. Here we use a simplification. For a dichotomous variable, if we label the proportion in the category labeled 1 as p, then the variance is just $p*(1 - p)$. We know that for this sample, p is 0.77, so then $p*(1 - p)$ will be $(0.77)*(1 - 0.77)$, which equates to $0.77*0.23 = 0.1771$.

We can now estimate the population variance using Gosset's formula:

$$\hat{\sigma}^2 = \left(\frac{N}{N-1} \right) s^2 = \left(\frac{8{,}310}{8{,}309} \right) 0.177 = 0.177 \tag{9.18}$$

Some statisticians use a different logic at this step. They note that the sample mean is the best estimate we have of the population mean. So we can take 0.77 as our estimate of the population mean, then apply the $p*(1 - p)$ formula to the population mean to get an estimate of the population variance. This gives the same result to three decimal places in this example because we have a reasonably large sample. The two approaches would differ with small samples, but we can't use this approach with small samples anyway.

2   Estimate the variance of the sampling distribution.

$$\hat{\sigma}_{\bar{x}}^2 = \frac{\hat{\sigma}^2}{N} \tag{9.19}$$

That is, we take the estimate of the variance of the population and divide it by the size of our sample to get an estimate of the variance of the sampling distribution. This will be 0.177/8310 = 0.000021.

3   Estimate the standard error: Once we have an estimate of the variance of the sampling distribution, we just take the square root to get an estimate of the standard deviation of the sampling distribution – the standard error. This will be the square root of 0.000021, which equals 0.005.

4   Find the Z value for the level of confidence we want. If we want a 99 percent confidence interval, the Z value will be 2.576 (See the Z table in the Appendix).

5   Multiply the Z value by the standard error. This is 2.576 * 0.005 = 0.013.

6   Add this to the mean to get the upper bound. This gives 0.77 + 0.013 = 0.783.

7   Subtract the number from step 5 from the mean to get the lower bound. This gives 0.77 − 0.013 = 0.757.

Thus, we can be 99 percent certain that the proportion of people in Uganda who know that condoms can help prevent the transmission of the AIDS virus is between about 76 percent and 78 percent. Saying that we are 99 percent certain means that if we used this procedure in a sampling experiment, the confidence interval would cover the true population mean for 99 percent of all samples, and for 1 percent of all samples it would miss.

Statisticians have developed a more precise approach based on what is called the binomial theorem. It's sometimes necessary to use this if the mean (the proportion in the category scored 1) is very close to one or zero. There is nothing in the calculations we have just described to keep the confidence interval from going below zero or over one, but such values don't mean anything when working with a proportion. In such cases the Law of Large Numbers doesn't work and the more appropriate binomial theorem does. But for most applications when the mean is not close to one or zero, the Law of Large Numbers works well. If we had used the binomial theorem for this problem the confidence interval would have ranged from 0.76 to 0.78 – no difference unless we go out more decimal places.

Rough Confidence Intervals

Remember that the choice of 90 percent, 95 percent and 99 percent rather than, for example, 85 percent, 97 percent or some other level of confidence is arbitrary. We want the confidence intervals to capture the true population mean most of the

Table 9.6   Certainty levels for confidence intervals with "nice" values of Z

Value of Z	Certainty level
1.0	0.6827
1.5	0.8664
2.0	0.9454
3.0	0.9973

time so we can be pretty certain we've captured it with the research sample in a particular study. But the common use of 90 percent, 95 percent and 99 percent instead of some other high level of confidence is just a conventional choice.[1]

It is common to see researchers conduct confidence intervals based on Z or t values of 1, 1.5, 2, or 3. They are choosing "nice" values for values of Z or t rather than "nice" values for the confidence level. There is nothing wrong with choosing "nice" values for Z or t rather than "nice" values for the level of confidence. Table 9.6 shows the confidence levels (the chances of catching the population mean) that correspond to Z values of 1, 1.5, 2, and 3. Both indicate a particular level of assurance that the confidence interval has captured the population mean. In management science, there is sometimes discussion of having quality control at the "six sigma" level, which means a Z value of six (six standard deviations from the mean), which corresponds to a confidence level of 0.9999, or one chance in ten thousand of missing the true value of the mean.

What Have We Learned?

The Law of Large Numbers and Gosset's work allow us to use the logic of the sampling experiment to estimate means of populations from sample data. The best estimate of the population mean is the sample mean. If we use just one number as the estimate, it is called a point estimate. But we know our sample mean may differ from the population mean because of sampling error. So we hedge our estimate by constructing a confidence interval that, for a designated large percentage of samples, will actually include the population mean. Several things influence the size of the confidence interval: the confidence level, the sample size, and the variance of the population. Constructing confidence intervals for large samples uses the Z distribution, and for small samples the t distribution is used. If we know how big the confidence interval should be, and we know how certain we must be and can make a guess at the population variance, we can calculate how large our sample needs to be.

Applications

Example 1: Why do homicide rates vary?

We can compare plots of means and confidence intervals for the homicide rates of different groupings of states, where the groups correspond to variables we think might cause variation in homicide rates. Before we do that, however, we need to stop and reflect on how to think about confidence intervals of state homicide rates. First, remember that the states differ substantially in their populations. So the average homicide rate across states is not the same as the average homicide rate for the US. This is because in taking the average across all states, which is 4.74, we give every state equal weight. But of course, states vary enormously in population, from Wyoming, which in 2000 had about 493,782 people, to California, which had about 33,871,648 people in 2000. So the homicide rate of 2.8 per 100,000 for Wyoming represents about 14 homicides while the rate of 6.8 for California represents about 2,303 homicides. If we want to get the homicide rate for the whole country, we would have to take the size of each state into account in what is called a **weighted average**.

Second, we have to remember that the confidence interval is trying to give us a good estimate of the value of the mean in the population, taking sampling error into account. But we have data on all 50 states. So what does the confidence interval mean? One way to think about it is to use the hypothetical logic that the set of states we actually have is a sort of sample from all the states that might exist with somewhat different configurations of homicide rates, histories, poverty levels, etc. Then the confidence interval is an estimate of what the mean of a group of states might be in that hypothetical population. As we've mentioned before, some researchers don't like this way of thinking; others do. But unless we have some random mechanism that underlies our data, then the confidence interval has no meaning. So if we are going to use confidence intervals in our research, we have to think in terms of a random process. If we are not comfortable with that logic for a data set, then we shouldn't use confidence intervals.

With those cautions in mind, we can now look at Figure 9A.1. It shows the 95 percent confidence interval around the mean homicide rate for states that were part of the Confederacy and those that were not. Clearly, the mean across the former Confederate states is higher. But we also know from previous chapters that the homicide rate is related to the amount of poverty in a state. So it would be helpful to take account of poverty in looking at the effect of region on homicide rates. It may be that many formerly Confederate states also have substantial amounts of poverty, and the poverty level is what is really driving homicide rates. In Figure 9.A.2 we plot the means for states that were formally in and not in the Confederacy, after first splitting the sample into two groups, those with higher and those with lower percentages of households in poverty. To split the states into these two groups, we used the median of the percentage of families in poverty

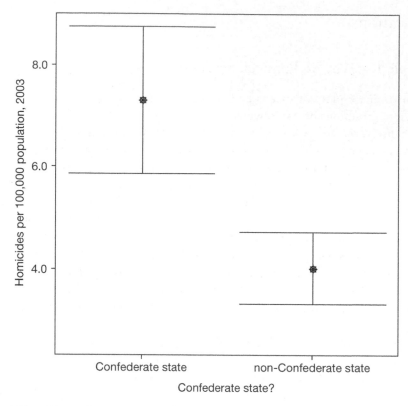

Figure 9A.1 Ninety-five percent confidence interval around the mean homicide rates for Confederate states (N = 11) and non-Confederate states (N = 38)
Data source: US Census Bureau, analyzed with SPSS.

across states in the Confederacy, 11.1. Those states above the median were considered to have more poverty; those below the median were considered to have less poverty.

Only one of the eleven confederate states has a low poverty rate (remember this is defined as poverty levels below the median value), which is why there is no confidence interval around the value. With only one state in this category, it is difficult to compare homicide levels by poverty level and region simultaneously. For states not in the Confederacy ("Northern" states), the mean for the low poverty group is lower than the mean for the high poverty group. However the 95 percent confidence intervals overlap quite a bit, so our interval estimates would not lead us to suspect big differences between the two groups. We also see the effects of small sample size (Confederate states with high poverty; N = 10) – the confidence intervals become rather large when we only have a handful of states in each group. And of course we don't have any reason to be sure that the "population" distribution of homicide rates is Normal, so the confidence intervals based on small samples may not be actually at 95 percent.

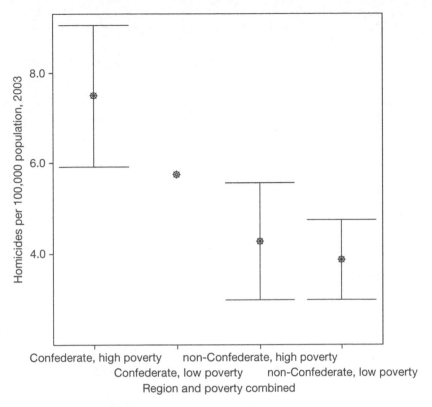

Figure 9A.2   A plot of means for Confederate states and non-Confederate states by low and high poverty level, non-Confederate state; low poverty (N = 24); non-Confederate state, high poverty (N = 15); Confederate state, low poverty (N = 1); Confederate state, high poverty (N = 10)
Data source: US Census Bureau, analyzed with SPSS.

Example 2:   Why do people vary in their concern with animals?

For an illustration using the animal concern question, please review the example in the text.

Example 3:   Why are some nations more likely to participate in environmental treaties than others?

Here we have the same issue as with the analysis of the state homicide data. The data set is not a random sample – it is all the data available. So when we build a confidence interval the "population" to which we are generalizing is a hypothetical one.

We can examine whether or not the number of NGOs in a country influences environmental treaty participation. It can be argued that NGOs exert pressure on governments to engage in internationally responsible behaviors, including partici-pating in environmental treaties. The median number of NGOs across countries is 494, so we will consider any country with 494 NGOs or less as having a small

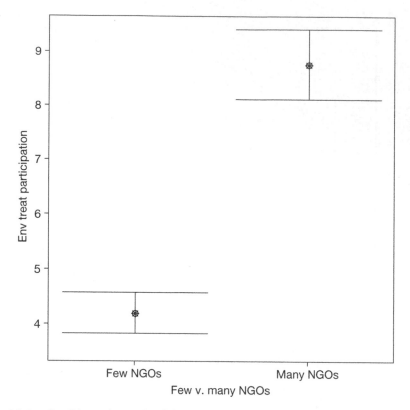

Figure 9A.3   Confidence intervals of the environmental treaty participation for countries with small and large numbers of NGOs; few NGOs (N = 85); many NGOs (N = 85)
Data source: Roberts et al., 2004, UNDP, 2002, analyzed with SPSS.

number of NGOs and any country with more than 494 NGOs as having a large number of NGOs.

Figure 9A.3 shows the confidence intervals on environmental treaty participation for the countries with large and small numbers of NGOs. The confidence intervals show a clear difference in extent of treaty participation among countries with a small number of NGOs compared to a large number of NGOs. The mean number of treaties signed for countries with a small number of NGOs is just above 4, and the mean number of treaties is around 9 for countries with a large number of NGOs. The confidence intervals do not overlap at all, suggesting number of NGOs influences environmental treaty participation.

Now let's also consider whether or not voice and accountability rates relate to environmental treaty participation. In later chapters we will consider tools that allow us to look at many variables at a time, but here we can certainly add one more variable. We will split the countries into low and higher rates of voice and accountability (at the median of −0.07). Figure 9A.4 shows the effects of both NGOs and voice and accountability.

Countries with few NGOs and low voice and accountability have the lowest mean number of treaties, while countries with many NGOs and high voice and account-

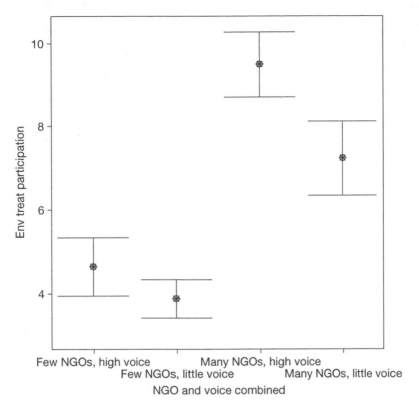

Figure 9A.4 Confidence intervals of environmental treaty participation for countries with small and large numbers of NGOs and low and high rates of voice and accountability; Countries with few NGOs and low voice (N = 52); few NGOs and high voice (N = 25); many NGOs and low voice (N = 27); many NGOs and high voice (N = 58)
Data source: Kaufmann et al., 2002, Roberts et al., 2004, UNDP, 2002, analyzed with SPSS.

ability have the highest mean number of treaties. Here we see that countries with few NGOs have participated in a fewer mean number of treaties compared to countries with many NGOs, and this relationship is true among both countries with lower and with higher voice and accountability rates. We also see that while the mean number of treaties is slightly higher among countries with few NGOs and high voice and accountability rates compared to countries with few NGOs and low voice and accountability rates, the confidence intervals overlap. None of the other confidence intervals overlaps. The graph suggests that voice and accountability influences treaty participation, but there is an interaction with NGOs.

Example 4: Why do people differ in their knowledge of how the AIDS virus is transmitted?

Remember that the mean of a zero-one categorical variable is just the proportion of people who fall into the category labeled one. Figure 9A.5 shows the confidence intervals around the proportion of men and women who had AIDS transmission knowledge.

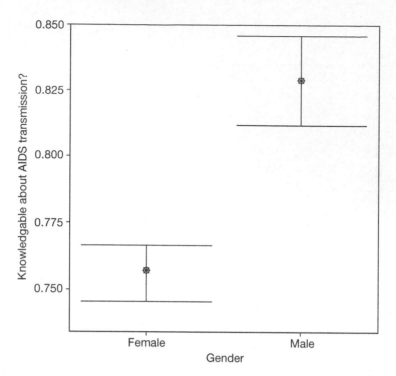

Figure 9A.5 Confidence intervals around the proportion of men and women who have AIDS transmission knowledge; Female (N = 6,424); Male (N = 1,886)
Data source: 2000 Ugandan DHS survey, analyzed with SPSS.

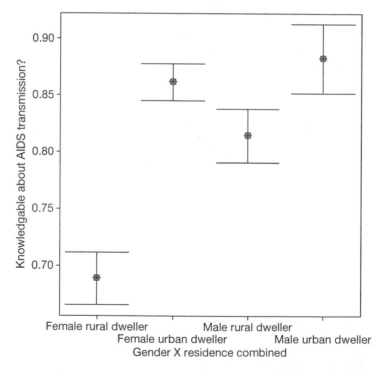

Figure 9A.6 Confidence intervals of AIDS transmission knowledge by gender and place of residence; Male urban dwellers (N = 588); Female urban dwellers (N = 2,269); Male rural dwellers (N = 1,298); Female rural dwellers (N = 4,155)
Data source: 2000 Ugandan-DHS survey, analyzed with SPSS.

Here the mean for men is higher and the confidence intervals do not overlap, so we would conclude that there may be a gender difference in AIDS knowledge. Now let's see how AIDS knowledge differs across both gender and place of residence (rural versus urban). The confidence intervals for this are displayed in Figure 9A.6.

Looking at the right side of the diagram, we see that male urban dwellers have the highest mean, followed closely by female urban dwellers. But the confidence intervals overlap so we wouldn't conclude that there's a difference between the two groups of urban dwellers in the population. Looking on the right, we see a different pattern, with rural dwelling men being much more likely to have AIDS knowledge than rural dwelling females, and the confidence intervals do not overlap at all. Furthermore, the means and 95 percent confidence intervals for rural dwelling men and women are below those of urban dwelling men and women. Therefore, we can conclude that there are differences by place of residence and there are gender differences but only among those living in rural areas.

Exercises

1. One thousand patients are surveyed nationwide about their views on the quality of care received from their physicians. The quality of care scale ranges from 0 to 10, with 10 reflecting the highest quality of care. The sample mean is 4.20, and the sample variance is 3.45. Based on this information, calculate the 95 percent confidence interval around the sample mean. What does this confidence interval tell us?

2. Is a 90 percent confidence interval for the data in Problem 1 going to have a smaller or larger range than a 95 percent confidence interval? What are the advantages and disadvantages of selecting a 95 percent versus a 90 percent confidence interval?

3. A researcher is interested in whether people, on average, work a full-time 40-hour work week. To examine this research question, he turned to the 2000 ISSP dataset. In a question asking respondents (N = 26,544) the average number of hours they worked in the prior week, the average hours worked was 40.53, with a range of 1 to 96 hours. The sample variance was 195.33. Based on this information, construct the 95 percent confidence interval and show your work. Do people tend to work 40 hours a week?

4. A group of researchers is interested in factors that relate to the number of children people have. One of the researchers suggests that, among other factors, educational attainment may relate to number of children, with the hypothesis that people with the highest degrees have fewer children (more focus on their professional lives) than those with less educational achievement. Data from 31,390 participants in the 2000 ISSP survey who are at least 35 years old are presented in Table 9E.1 by highest degree earned. (a) Graphically present the five 95 percent confidence intervals. (b) Does there appear to be a relationship between number of children and educational attainment?

Table 9E.1   Mean number of children by educational level

	N	Mean	Std. deviation	Std. error	95% confidence interval for mean	
					Lower bound	Upper bound
No formal qualification (i.e., schooling)	3,120	2.14	2.48	0.04	2.05	2.23
Lowest formal qualification	6,464	1.37	1.76	0.02	1.33	1.41
Above lowest qualification	6,906	1.37	1.46	0.02	1.34	1.41
Higher secondary schooling completed	6,651	1.07	1.33	0.02	1.03	1.10
Above secondary, not full university completed	3,808	1.22	1.44	0.02	1.18	1.27
University degree completed	4,441	0.96	1.29	0.02	0.92	1.00
Total	31,390	1.31	1.64	0.01	1.29	1.32

References

International Social Survey Programme (ISSP). 2000. 2000 Environment II data set. www.issp.org. Catalog no. ZA 3440. Cologne, Germany: GESIS-ZA Central Archive for Empirical Research.

Kaufmann, D., Kraay, A., and Zoido-Lobaton, P. (Jan. 2002). Governance Matters II: Updated Indicators for 2000/01. Policy Research Working Paper no. 2772. The World Bank Research Development Group and World Bank Institute; Governance, Regulation and Finance Division (http://hdr.undp.org/reports/global/2002/en/).

Roberts, J. T., Parks, B. C., and Vasquez, A. A. 2004. Who ratifies environmental treaties and why? Institutionalism, structuralism and participation of 192 nations in 22 treaties. Global Environmental Politics 4(3), 22–64.

Uganda Demographic and Health Surveys. 2001. Calverton, Maryland: UBOS and ORC Macro. (http://www.measuredhs.com/pubs/pdf/FR128/00FrontMatter.pdf).

United Nations Development Programme (UNDP). 2002. Human Development Report. Deepening Democracy in a Fragmented World. New York: Oxford University Press (http://hdr.undp.org/en/reports/global/hdr2002/).

US Census Bureau 2000. Table 33. Urban and rural population, and by state: 1990 and 2000 ("http://www.census.gov/prod/cen2000/index.html"\t"_blank"www.census.gov/prod/cen2000/index.html).

US Census Bureau 2002. Historical poverty tables: Table 21. Number of poor and poverty rate, by state: 1980 to 2006. Year 2002 ("http://www.census.gov/hhes/www/poverty/histpov/"\t"_blank"www.census.gov/hhes/www/poverty/histpov/hstpov21.html).

US Census Bureau 2003. Table 295. Crime rates by state, 2002 and 2003, and by type, 2003 (http://www.census.gov/prod/2005pubs/06statab/law.pdf).

CHAPTER 10
USING SAMPLING DISTRIBUTIONS: HYPOTHESIS TESTS

Outline

In the last chapter we learned how to think about and build confidence intervals in order to take account of uncertainty in estimating the population mean. Confidence intervals are one of the two most commonly used tools in statistics. The other is the hypothesis test. In this chapter we will examine the logic of hypothesis tests and learn how to conduct them, using tests about means as an example. Together hypothesis tests and confidence intervals carry us through the rest of the book. In every chapter we will examine statistical tools that allow us to build confidence intervals or test hypotheses.

Indeed, much of statistics can be thought of as the development of methods to build confidence intervals or test hypotheses. We start with confidence intervals for the mean and tests of hypotheses about the population mean. Then we look at other ways of analyzing data, for example looking at differences in means or proportions between groups. But as we learn new ways to examine the data, we will always return to the basic logic of the confidence interval and the hypothesis test. So if we develop a sound understanding of the basic logic of the confidence interval and hypothesis test, the rest of statistics is just a series of examples – the same basic logic applies.

Sometimes we are faced with an assertion about the world and we want to see how reasonable that assertion is, given the available data. We do this all the time in ordinary situations – if one of our friends is ten minutes late, we don't worry that something bad has happened because, based on past experience, being ten minutes late is typical of our friend. But if another friend is ten minutes late, we become concerned because in past experience this person is always on time. We use evidence to assess our understanding of the world – in this simple example trying to choose between a friend being late because something bad has happened and being late because they are just running a little behind.

Hypothesis testing is a way of thinking through the decisions we make about the state of the world and using a very systematic procedure, based on the sampling experiment, to assess what is likely to be true. The logic and tools of hypothesis testing are very powerful and are at the heart of most scientific research.

The simplest example of hypothesis testing is one where we have a strong statement – a hypothesis – about the mean of a population. For example, a government standard might be used to generate a hypothesis about a population mean. The US federal government wants drinking water systems to have less than 15 parts per billion of lead in the water. If chemical tests of drinking water show higher lead levels than that, then steps must be taken to reduce the amount of lead in the water. We could treat the standard as a hypothesis. Given some data, how likely is it that the amount of lead in a water supply is greater than the standard?[1]

Sometimes international agreements set standards. For example, in 1990 the World Summit for Children set the goal that in every country, 80 percent of all children should complete 4 years of schooling (US National Research Council, 1999, p. 39). Surveys might be used to calculate the percentage of children who have had four years of schooling in a country. We could then use the data from the survey to see how plausible it is to believe that in the population of children, the goal for four years of education has been reached.

In research, we often make assertions that involve more than one variable. Theory might suggest two groups will differ in some characteristics, or that two variables are related. For example, we might believe that there are differences between men and women in their concern about animals, or between Southern and non-Southern states in their homicide rates. Such assertions are very general – they don't say how different men are from women or how different Southern states are from states not in the South. It will be helpful to have precise hypotheses that make statements in terms of specific numbers, like 1 part per billion of lead in drinking water or 4 years of education for children. We can do this with hypotheses about differences or the relationships between variables by looking at the opposite of the statement. Instead of hypothesizing men are different from women, we can focus on the hypothesis that men and women are not different. This is a precise hypothesis in that the difference we are suggesting between men and women is a difference of zero. If we decide that a precise difference of "no difference" is not a reasonable thing to believe, then we must conclude that there is a difference. We begin by asserting that men and women are the same in their level of animal concern, or that Southern and non-Southern states have the same homicide rates so that our hypothesis is precise. Then if the data are inconsistent with the no-difference assertion we doubt the assertion and conclude that there is a difference between the two groups. In formal language, the hypothesis of no difference is rejected.

In general, the purpose of hypothesis testing is to assess how reasonable an assertion is in the face of sample evidence. We will begin with hypotheses about the value of the mean of the population as an example because it is the simplest case, though it is too simple to be interesting in most research problems. Then we turn to hypotheses about differences in means across two groups, which is often an interesting research problem.

The Logic of Hypothesis Tests

Suppose that it would be useful to know if some assertion about the value of the population mean is reasonable. For example, is it reasonable to believe that the level of lead in the drinking water is greater than 15 parts in a billion? Or is it reasonable to believe that all children have at least 4 years of education? A hypothesis test begins by converting the assertion into a precise quantitative hypothesis about the value of the population mean. This starting point is called the **null hypothesis**. We give it the benefit of the doubt – we assume it is true unless the evidence makes that belief implausible. Logically the null hypothesis must be true or false. If it is false something else must be true, which we will call an **alternative hypothesis**. That is, the population mean must either equal the specified value (the null hypothesis is true) or the mean must equal some other value (the null hypothesis is false and the alternative hypothesis true). For example, if the United States Environmental Protection Agency (EPA) standard for lead is 15 ppb, then the null hypothesis might be:

$$H_0: \mu = 15 \text{ ppb} \tag{10.1}$$

The population mean is 15 ppb. The null hypothesis, indicated by H_0, asserts that the mean level of lead in the drinking water system of a community is equal to the standard, 15 ppb. Then the alternative hypothesis, indicated by H_A, is just the logical complement:

$$H_A: \mu \text{ is not 15 ppb} \tag{10.2}$$

One or the other of these two assertions must be true.

Ideally, we'd like to see the lead level less than 15 ppb, so we'd be happy with rejecting the null hypothesis if we can then conclude that the lead level is well below the amount that causes health concerns. And we'd want to take action if the lead level is higher than the standard we have used as the null hypothesis. That is, we care about the "direction" in which we reject the null hypothesis. We will return to that issue below. For the moment, to keep things simple, we'll stay with just being concerned with whether the level of lead in the water is equal or not equal to the standard.

We can never disprove with absolute certainty the null hypothesis. But if it seems very unlikely that the null hypothesis is true (based on statistical procedures), then we would tend to believe the alternative hypothesis. Note that the statistical logic is rather like that in building a confidence interval. We don't know for certain that the confidence interval built from our sample has actually captured the population mean. But we know that the confidence interval procedure will work most of the time. So if we build a 95 percent confidence interval, we know that the procedure would produce a confidence interval that includes the population mean 95 percent of the time in a sampling experiment. In that sense we can be 95 percent certain that we've captured the population mean with the confidence interval we build with our research sample. We can make $1 - \alpha$, our level of confidence, as large as we want and thereby make the chances of missing the true population mean as small as we want.

The same thing happens with hypothesis tests. When we finish a hypothesis test, we don't know for certain that the null hypothesis is false. But if the hypothesis test procedure says we should reject the null hypothesis, we can be reasonably certain that the null hypothesis is wrong. And we can set the level of certainty to be as stringent as we want, as we will see below.

The example of a very simple coin tossing game may help to clarify the logic we are using. Suppose you play a coin tossing game with someone and they provide the coin. If the coin comes up heads, you win, if it comes up tails you lose. You would start the game with the assumption that your opponent is an honest player and that the coin is a fair one – it generates heads and tails with equal frequency. That means the coin acts like a random variable in a binomial distribution. You can predict the probability of getting a certain number of heads out of a certain number of tosses, using the logic of Figure 7.3 in Chapter 7 (on probability). For example, the chances of a head on the first toss is 0.5. The chance of two heads in a row is 0.25, and of three heads in a row is 0.125.

Suppose you lose the first toss. Are you being cheated? Most people would say "no." But you don't really know – it might be a fair coin that came up tails by chance, or it might be a loaded coin that will nearly always come up tails. What you really know is that the chances of that outcome with a fair coin are 0.5, and so losing on the first toss is a rather common thing, not something that will make you doubt that your opponent is honest. Suppose you lose the second toss as well. The chances of losing twice in a row are 0.25 (recall from Chapter 7 that is 0.5*0.5). That's a less likely event, and some cynical folks will become suspicious of their opponent. But most folks would let it ride. Suppose you lose three times in a row. Now the probability is 0.125 (0.5*0.5*0.5 = 0.125) that this is the result of a fair coin. Some people will now become very suspicious and accuse their opponent of being a cheat. Others might wait until heads came up four times in a row (probability = 0.0625) or five times in a row (probability = 0.03125) or six times (probability = 0.015625). At some point, when the results of the coin toss are sufficiently improbable if the coin is actually fair, all but the most optimistic would assume that they are being cheated.

But can you ever be certain you're being cheated? Not really. Remember the probability issues discussed in Chapter 7. We sometimes get six heads in a row. In fact, the probability of getting six heads in a row is 0.015625, which means that we would expect 6 heads to come up once or twice in a hundred tosses, and 15 or 16 times in 1,000 tosses. So it could be that even though your run of luck seems too bad to call luck, it might be the result of chance – the unfolding of a random event.

Let's think about the decisions you face. After a number of flips go against you, you can either call your opponent a cheat or let the results stand. And there are two possibilities for the truth. Either it's a fair coin, or it's not. If you call a cheat a cheat, or continue playing with an honest coin, you've made a correct decision. If you call an honest player a cheat, you've made a mistake. If you don't decide you're being cheated when you really are, you've also made a mistake. So there are two different ways to go wrong. Table 10.1 shows these choices, where one type of mistake is being taken in by a cheat and thus being a "sucker" and the other type of mistake is to call an honest opponent a cheat, thus "flying off the handle."

Before playing this game, you might want to decide how unlikely is too unlikely to be believed. This way you can assess the probability that seems too unlikely to be believed before you're caught up in the emotion of the game. You might pick some small number, such as 0.1 or 0.05 or 0.01. Then if the result of the flips is less likely than the value you've picked, you call your opponent a cheat. But notice

Table 10.1 Possible results from coin tosses

You decide to:	The coin is fair	The coin is loaded
Keep playing	Correct!	Mistake! (Sucker)
Call your opponent a cheat	Mistake! (Fly off the handle)	Correct!

that these unlikely runs of luck will happen sometimes with a fair coin, so you still might be making the mistake of calling an honest player a cheat – the "flying off the handle" mistake. When you pick the value you consider beyond belief – how improbable is too improbable to consider bad luck with a fair coin – you are also setting the chances you will make the mistake of calling an honest player a cheat. And if you want to avoid calling an honest player a cheat and set the chances of making that mistake very low, say 0.01, then you are easier to cheat than someone who has set the limit higher, say 0.1. So there is always a tradeoff between making one kind of mistake and the other.

We use the same kind of logic in courtrooms and medical testing. In the US in order to convict someone in a criminal case, the prosecution must convince the judge or the jury that the evidence is strong enough to indicate guilt "beyond a reasonable doubt." We want to avoid the error of convicting an honest person, which is logically equivalent to avoiding calling an honest player a cheat. In civil cases, because the consequences of this type of error are less, the standard of proof is less – the "preponderance of evidence" is sufficient to decide the case.

In setting up a medical test, we want to lessen the chance of assuming a person is well when in fact she has a disease and would benefit from treatment. Saying someone has the disease when she doesn't is like calling an honest player a cheat or convicting an innocent person. She would be treated unnecessarily. But the costs of a mistake here usually weigh more heavily on the other side. The costs of saying a healthy person has a disease may involve further testing, inconvenience, stress, expense and even some risks associated with further testing or treatment. But in most cases, we would rather accept those costs than the costs of missing the disease in someone who really has it and would benefit from treatment. Testing for the safety of a chemical that will be used in the environment or in food has the same logic as the medical example. We don't want to err by calling compounds safe that are actually toxic. In our hypothetical betting game and in the criminal court, we want to minimize one type of error – saying something is wrong when it's not. In medical testing and safety testing, we will accept that error in order to minimize the other error – missing the problem.

Of course, when we set up our decision rule for betting, rules of jurisprudence, or standards for interpreting the medical test or safety test, we try to make the chances of either type of error small. But you can see that there is always a tradeoff – if you push hard to eliminate one kind of error, you make the other kind of error more likely (though still rather unlikely in most situations). The statistical way of thinking about errors allows us to analyze the tradeoff logically and put our risks of a mistake where we want them.

The Formal Approach

Now let's return to the sample mean and the assertion about the population mean. The sample mean may differ from the population mean proposed under the null hypothesis because of sampling error *or* because the null hypothesis is not an accurate description of the population. The logic of the hypothesis test asks: "How likely is it that this sample would have been drawn from a population accurately described by the null hypothesis?" If it is reasonably likely, the null hypothesis seems plausible, and is allowed to stand (*not rejected* or *fail to reject* in the jargon). If the probability of getting the sample data from the hypothesized population is small, then the null hypothesis seems implausible and we doubt it (it is *rejected* in the jargon), and the alternative hypothesis is taken as more credible. While this sounds very technical, it is just the logic of the coin flip game restated – assume things are okay (the null hypothesis is true) unless the evidence really sides against it (the results are pretty unlikely) and then we must reject the null hypothesis and assume its alternative is true.

This framework was outlined by R.A. Fisher, one of the great scientists of the twentieth century.[2] But two other eminent statisticians, Jerzy Neyman and Egon Pearson, noted that worrying about rejecting the null hypothesis when it was true was not the only problem.[3] They emphasized the issue we have raised above in the context of the coin flip game, court decisions, medical testing, and safety assessment. There are two ways of making an error. Fisher built his interpretation around the idea of setting small chances of rejecting the null hypothesis when it is true really true. He was guarding against rejecting the null hypothesis when the sample mean differs from the hypothesized value of the population mean only because of chance. Neyman and Pearson urged that we always attend to the other type of error as well. Table 10.2 shows the errors and the labels that are applied to them. It is just Table 10.1 stated in the language of statistics. Again, rejecting the null hypothesis when it is false (calling a cheat a cheat) and letting it stand when it is true (not calling an honest player a cheat) are correct decisions. But there are two ways to make an error. We can reject the null hypothesis when it is true (calling an honest player a cheat). In the language of statistics this is called a **Type I** or **alpha error**. Or we can make a mistake by not rejecting the null hypothesis when it is false (not calling a cheat a cheat). This is a **Type II** or **beta error**. The terms Type I and Type II or alpha and beta are arbitrary labels statisticians working on these problems use to categorize these two ways of making an error.

As we noted before, there is a tradeoff between Type I and Type II errors. To make the chances of a Type I error very small raises the chances of a Type II error.

Table 10.2 Possible results from a hypothesis test

Decision from hypothesis test	Actual situation	
	Null hypothesis is true	Null hypothesis is false
Don't reject null hypothesis	Decision correct	Type II or β error
Reject null hypothesis	Type I or α error	Decision correct

Power is discussed at the end of this chapter as an Advanced Topic.

Steps in Testing a Hypothesis

Before performing calculations, the researcher should specify the chances she is willing to take in making a Type I error. This number, α, conventionally takes on values of 0.1, 0.05, and 0.01.[4] The α is the "critical value," or the point at which a sample is so improbable that we don't believe a rare event has occurred – we don't believe we've got an odd sample from the population described by the null hypothesis. Rather we conclude that the null hypothesis is not an accurate description of the population. Over the long haul, a researcher testing hypotheses at, say, the 0.05 level will be wrong in rejecting the null hypothesis (make a Type I error) 5 percent of the time. Or put in the core logic of statistics, if we use the same procedure in every sample in a sampling experiment, we will make a Type I error in 5 percent of all the samples.

How do we then assess the probability that our sample came from the population the null hypothesis is describing? Here the Law of Large Numbers and Gosset's work on the t-distribution can be used when we deal with sample means. Other methods for testing hypotheses also exist, and we'll come to those later. But a test about a mean is a simple application and a good starting point. The logic of hypothesis testing we've outlined is the same for any test, only the details of carrying it out differ.

To preview: After picking the chances of rejecting the null hypothesis when it's really true (the α level), we will construct a sampling distribution of samples drawn from the population described by the null hypothesis. In fact, for many problems the procedures are so well worked out that we don't need to actually construct a sampling distribution by drawing slips of paper, tossing dice or even having the computer do it. We can consult tables of the Normal distribution or the t-distribution. These tell us what the sampling distribution looks like.

The sampling distribution that assumes the null hypothesis is true tells us how likely, or unlikely, our sample is. If it is a sample that would come from the hypothesized population less often than the critical value we have set, we reject the null hypothesis and believe that the alternative hypothesis is a better description of the population. If our sample is more probable than the critical value, we let the null hypothesis stand, keeping in mind that we might have made a Type II error.

An example with a large sample

Suppose we want to test the hypothesis that the combined populations of the nations participating in the 2000 ISSP survey are neutral about animal concern in the sense that the mean of the animal concern scale for the population is at the mid-point of the scale. This is a hypothesis that the population mean is 3, since the scale runs from 1 to 5. Since we have a large sample, we will test the hypothesis based on the Law of Large Numbers. Next we will do an example using Gosset's theorem. The formal steps in testing this hypothesis are as follows:

Step 1 State the null and alternative hypotheses
In this case, the null hypothesis is that the mean score in the population is 3. Stated in a formal way:

$$H_0: \mu = 3$$
$$H_A: \mu \neq 3 \tag{10.3}$$

That is, the null hypothesis is that the population mean equals 3, and the alternative hypothesis is that the population mean does not equal 3. Rejecting the null means that it seems unlikely that the population mean is 3, since we are equally open to population means larger than 3 and smaller than 3. Below we will discuss the situation where only evidence in one direction matters, as would be the case if we were concerned about the concentration of lead in drinking water but would take action only if the level is too high, not if it is lower than the standard.

Step 2 Choose an α level
The alpha level is the chance we are willing to take of making a Type I error, rejecting the null hypothesis when it is in fact true. Typically, 0.05, 0.01, or 0.10 are used. Let's use 0.05.

Step 3 Determine the appropriate statistical procedure to examine the hypothesis
By this we mean that we have to think about what kind of hypothesis we have – one about the value of the mean, one about the difference between two groups or something else. We also have to think about what kind of sample we have: large or small and whether it came from a Normally-distributed population.

So far, we only have two ways to get at hypotheses about the population mean. One is for situations when we have large simple random samples (the Law of Large Numbers). Another is for any size sample, but requires that we have a simple random sample *and* that the population is Normally-distributed (Gosset's theorem). In the case of the survey data, we have a large enough sample ($N = 29,486$) that we can justify using the Law of Large Numbers. Later we will discuss methods for hypothesis about things other than just the population mean.

Step 4 Find the critical value that corresponds to the α level
Rather than actually conduct a sampling experiment, the Law of Large Numbers tells us what will happen if we did the sampling experiment. This simplifies our work

substantially. Instead of calculating probabilities, we can use the Z table to find a Z value that corresponds to the alpha value we have chosen. So we simply look in the Z table and find that a Z value of 1.96 corresponds to an α of 0.05. This value is called the critical value of Z because we use it as the basis for making our decision about the null hypothesis. Note that if we had a small sample, we would first have to calculate the number of degrees of freedom.

Step 5 Calculate the Z value for the sample
This is the second part of using the Law of Large Numbers rather than having to conduct a sampling experiment. We just picked the critical value of Z that matches our alpha probability level. Now we have to convert the sample mean into a Z value to compare to the critical value of Z from the table, following several substeps. The Z formula for the sample is:

$$Z_s = \frac{\bar{X} - \mu}{\hat{\sigma}_{\bar{X}}} \tag{10.4}$$

Z_s indicates the Z score for the sample. The sample mean we calculated to be 2.49. The mean of the population we are assuming, under the null hypothesis, to be 3. So the top of the equation for Z_s is just the sample mean minus 3, or $2.49 - 3.00 = -0.51$. When the sample mean is less than the hypothesized value, the numerator will be negative and the Z value for the sample will also be negative. When the sample mean is greater than the hypothesized value of the population mean, the difference will be positive and the Z value for the sample will be positive. When we compare the Z value for the sample with the critical value of Z from the Z table, we just worry about the absolute value of the sample Z.

The denominator in the equation is the standard error of the mean, or the standard deviation of the sampling distribution of sample means. This we already know how to estimate.

Substep a Estimate the population variance

$$\hat{\sigma}^2 = \frac{1}{N-1} \sum (X - \bar{X})^2 \tag{10.5}$$

or

$$\left(\frac{N}{N-1}\right) s^2 \tag{10.6}$$

Substep b Estimate the variance of the sampling distribution

$$\hat{\sigma}_{\bar{X}}^2 = \frac{\hat{\sigma}^2}{N} \tag{10.7}$$

The sample variance is 1.392. Using Gosset's formula to estimate the variance of the population we have:

$$1.392 * (29{,}486/29{,}485) = 1.39 \qquad (10.8)$$

Then the estimated variance of the sampling distribution is this estimate of the population variance divided by the sample size:

$$1.39/29{,}486 = 0.000047 \qquad (10.9)$$

Substep c Estimate the standard error by taking the square root of the estimated variance of the sampling distribution

$$\sqrt{0.000047} = 0.0069 \qquad (10.10)$$

Substep d Calculate the value of Z for the sample

$$Z_s = (2.49 - 3.00)/0.0069 = -0.51/0.0069 = -73.91 \qquad (10.11)$$

Step 6 Decide whether the null hypothesis can be rejected
The critical value was 1.96. We compare the critical value of Z with the absolute value of the sample Z. Here the absolute value of the sample Z (73.91) is much larger than the critical value (1.96). This indicates that our sample is way out in the tail of the sampling distribution that would result if the mean of the population really is 3. So there are two logical possibilities. We might have a sample that would come up one time in 20 or less, or the null hypothesis may be wrong. We have already decided to reject if the chances of the sample we have coming from the population described by the hypothesis are less than 0.05, so we follow our rule and reject the null hypothesis. Thus, we can conclude that the mean score on animal concern is not 3.

An example with a small sample

Suppose we wanted to see if African Americans are neutral with regard to animal concern. There are 96 African Americans who answered the animal concern scale in the 2000 ISSP data set. Since the sample size is relatively small, we will use Gosset's theorem and the t distribution. Of course, to do that we assume that the population distribution of animal concern among African Americans is Normal.

Step 1 State the null and alternative hypotheses
Again, the null hypothesis is that the mean score for African Americans is 3. Stated formally:

$$H_0: \mu = 3$$
$$H_A: \mu \neq 3 \tag{10.12}$$

Step 2 Choose an α level
We usually use 0.05. For variety, let's choose an α level of 0.01.

Step 3 Determine the appropriate statistical procedure to examine the hypothesis
We have a sample of 96, so we will use the t-test. The t-test requires that the animal concern variable have a Normal distribution in the population. Is that likely to be true? Figure 10.1 is the sample histogram.

The sample distribution does not depart radically from the Normal distribution drawn as the solid line. So we will proceed with the t-test but will keep in mind that since the population distribution may not be exactly Normal the results of the test may not be exactly correct.

Step 4 Find the critical value that corresponds to the α level
To do this in the t-test we have to calculate the number of degrees of freedom. For testing a hypothesis about the value of the mean of a single sample, we have N − 1 degrees of freedom. So the number of degrees of freedom for our test is:

$$df = N - 1 = 96 - 1 = 95 \tag{10.13}$$

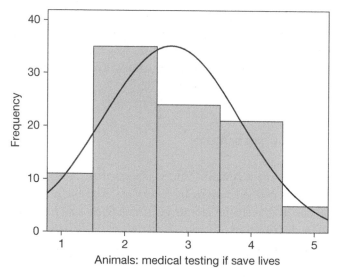

Figure 10.1   Histogram of animal concern for African–Americans, N = 96
Data source: 2000 ISSP data set, analyzed with Stata.

You can use the t table to find the critical value. Alternatively, you can use a website to get the exact critical value that matches your number of degrees of freedom and α level (for example: http://statpages.org/pdfs.html or http://www.cytel.com/Products/StaTable). Or if you are using a t table and your actual degrees of freedom is not available, select the next smallest number to find your critical value. Our table in Appendix C has values for degrees of freedom of 90 and 100, we would select 80 degrees of freedom. The critical value for the 0.01 level is 2.632. Using this critical value, we would reject the null hypothesis that African Americans have a mean of 3 if the sample value of t is greater than 2.632.

Step 5 Calculate the t value for the sample
Calculating a t value for the sample and comparing it to the critical value of t allows us to draw a conclusion about what would happen if we did a sampling experiment without having to do the sampling experiment. The formula for the sample t is the same as the formula for the sample Z:

$$t_s = \frac{\bar{X} - \mu}{\hat{\sigma}_{\bar{X}}} \qquad (10.14)$$

For this sample:

$$t_s = \frac{2.73 - 3.00}{\hat{\sigma}_{\bar{X}}} \qquad (10.15)$$

t_s indicates the t score for the sample. The sample mean for African Americans is 2.73. The mean of the population we are assuming, under the null hypothesis, to be 3. So the top of the equation for t_s is just the sample mean minus 3, or 2.73 − 3.00 = −0.27. As for the whole sample, the sample mean for African Americans is lower than the hypothesized value so the difference in the numerator is negative and the t value for the sample is negative.

We still need to calculate the denominator in the equation, which is the standard error of the mean, or the standard deviation of the sampling distribution of sample means. We calculate this is the same way as we did for the Z distribution, using a series of substeps.

Substep a Estimate the population variance

$$\hat{\sigma}^2 = \frac{1}{N-1}\sum (X - \bar{X})^2 \qquad (10.16)$$

or

$$\frac{N}{N-1}s^2 \qquad (10.17)$$

Substep b Estimate the variance of the sampling distribution

$$\hat{\sigma}_{\bar{X}}^2 = \frac{\hat{\sigma}^2}{N} \tag{10.18}$$

The sample variance is 1.189. Using Gosset's formula to estimate the variance of the population we have:

$$1.189 * (96/95) = 1.20 \tag{10.19}$$

Then the estimated variance of the sampling distribution is this estimate of the population variance divided by the sample size:

$$1.20/96 = 0.0125 \tag{10.20}$$

Substep c Estimate the standard error by taking the square root of the estimated variance of the sampling distribution

$$\sqrt{0.0125} = 0.112 \tag{10.21}$$

Substep d Calculate the value of t for the sample

$$t_s = \frac{2.73 - 3.00}{0.112} = -2.41 \tag{10.22}$$

Step 6 Decide whether the null hypothesis can be rejected
The critical value was 2.632. We compare the critical value of t with the absolute value of the sample t. The sample value is a bit samller than the critical value, so we will not reject the null hypothesis. This leads us to conclude that in the population, the mean animal concern score for African Americans may be equal to 3.

One-Sided and Two-Sided Tests

What stands as evidence against the null hypothesis? In some problems, not every improbable result would lead us to reject the null. Let's again consider the coin toss game, the problem of lead in drinking water, judicial deliberation, and medical testing. If you won every one of five or six tosses, while the probability is just as low as that of losing five or six tosses, you might not doubt the fairness of the coin. After all, you're a lucky person and deserve to win. As we noted above, we would be happy to find that the lead level in a drinking water supply is too low for us to believe that the concentration of lead is 15 ppb. In a court of law, if the evidence is beyond a reasonable doubt that someone is innocent, that person is acquitted. And if a medical test shows someone has a very low result on a test where only high values indicate a problem, that person will be considered disease free. That is, in some circumstances, evidence on one side automatically leads to non-rejection of the null hypothesis. It is only if there is considerable evidence on the other side that we reject. In hypothesis testing, we refer to this as a **one-sided test**. (Often it is called a **one-tailed test** because we are seeing if the sample falls in the tail of the sampling distribution developed by assuming the null hypothesis is true.) In practice, you decide to use a one-sided test when theory or practical application indicates that only evidence on one side counts against the null hypothesis. You then want all the chances of a Type I error in one tail of the sampling distribution, and none in the other tail. This means you use a different column in the t or Z table to find a critical value appropriate to get all α chances of a Type I error in one tail.

The more common circumstance is a **two-sided**, or **two-tailed**, test. Think of testing a new compound. We don't know if it will cure cancer, cause cancer, or have no effect. Or in the gambling situation, we suspect cheating whether we win or lose.[5] Evidence in either the tail above the hypothesized population mean or in the tail below it will make us doubt that the sample came from a population described by the null hypothesis. Then it is appropriate to use a two-tailed test. In social science research, we nearly always use a two-tailed test. This is because we most often test hypotheses about differences between groups, or associations between two variables. While theory might suggest the direction of the association or which group has the higher mean, the theory is never so strong as to justify using a one-tailed test.

The steps in carrying out a one-tailed and two-tailed test are identical except for two minor steps. In a two-tailed test, we compare the absolute value of the sample t or Z value to the critical value from the table. That is, we don't care whether the sample mean is above or below the hypothesized population mean. In a one-sided test, we only reject if the sample mean lies in the direction we are concerned with, for example, if the lead concentration in the sample is higher than in the standard. The only other difference is that there are different columns in a Z or t table for one-sided and two-sided tests. This is because if we choose a 0.05 level for alpha, we must decide the chances of making a Type I error between samples that by chance have too high a mean and those that by chance have to low a mean. We split the 0.05 in half, giving each tail of the sampling distribution 0.025. In a one-sided test, we only care about one tail of the sampling distribution, so the entire chances of a Type I error are present. This means a different value of t or Z is appropriate.

For example, suppose we wanted to know if a country has met the educational goal set by the World Summit for Children. The country has survey data on schooling levels of its children. We could code each child in the sample 1 if she or he has four years of schooling and 0 otherwise. Then we want to see if the population mean (for children of appropriate age) on this variable is 80 percent, which is the goal set by the Summit. Thus the null hypothesis is that the population mean is 0.80. We wouldn't want to reject the null hypothesis if the mean in the sample is greater than the standard. But we might accept that the country has met the standard even if the sample mean is less than the 80 percent goal because the sample of children might have come from a population of children where the true mean for having four years of schooling is 0.8. That is, we would say the country didn't meet the standard only if we can reject the hypothesis that the mean is 0.80 with a one-sided test.

Here only two steps differ from the procedure we've already used. First, we have to look in the Z or t table for a one-sided critical value corresponding to the alpha level we have chosen. Second, when we compare the Z value for the sample to the critical value, we have to keep track of the sign. That is, in this example we won't reject if the sample mean is greater than the standard. We will only reject if the sample mean is less than 0.80 (the sample Z is negative) and the sample Z value is larger in absolute value than the critical value.

Hypotheses about Differences in Means

So far we have discussed only the population mean. Hypothesis testing becomes much more interesting when we can test hypotheses about *differences* in group means. That's because we can look at differences and thus develop an understanding of why our dependent variables vary. We can test the plausibility of the assertion that men and women differ in their concern for animals, or that Southern states have higher homicide rates than do states that were not part of the Confederacy. The logic of hypothesis testing when we are interested in differences between groups is the same as when we are testing hypotheses about the mean. The steps are only slightly more complicated; there are five major steps and four sub-steps in the process. We provide an example below for the data in Table 10.3.

Versions of both the Law of Large Numbers and Gosset's theorem apply to differences between the means of two groups. For differences in means, the Law of Large Numbers becomes:

If large simple random samples with means $\bar{X}_1$ and $\bar{X}_2$ and sample sizes N_1 and N_2 are drawn from populations with means μ_1 and μ_2 and variances σ_1^2 and σ_2^2, then the

Table 10.3   Descriptive statistics on animal concern score for women and men

	Men	Women
Sample mean	2.37	2.60
Sample variance	1.287	1.454
Sample size	13,412	16,062

sampling distribution of differences in sample means will be Normal with mean $\mu_1 - \mu_2$ and variance $(\sigma_1^2/N_1) + (\sigma_2^2/N_2)$.

This seems complicated but if we look at it in steps, it will become familiar. We must have simple random samples from two populations. But if we draw a random sample from one population and divide it into two groups based on gender, or level of education, or any other variable, we can proceed as if the two sub-samples were simple random samples of the population of men and women, or people with more and less education, etc. This is one of the reasons that simple random samples are so useful.

Each sample has a mean, labeled 1 and 2 to keep them straight, and a sample size n_1 and n_2. The populations have means also, μ_1 and μ_2, and variances, σ_1^2 and σ_2^2. Let's think about the sampling experiment. We would draw a sample of size N_1 from population 1 and a sample of size N_2 from population 2. Then for each sample, we would calculate the mean ($\bar{X}_1$ and $\bar{X}_2$). Then for each pair of samples, we could then take the difference in means, which we might call d. We do this many times, each time taking a pair of samples (one for each group), calculating the means for each group and then calculating the difference in means. This is the sampling experiment. Then we plot these differences in means, one for each pair of samples in the sampling experiment. The sampling distribution of the differences (the ds) will be Normal with a mean equal to the difference in population means $\mu_1 - \mu_2$, which we might call D. The sampling distribution of the differences will have variance $((\sigma_1^2/N_1) + (\sigma_2^2/N_2))$. This says that the sampling distribution of differences in sample means, is centered at the difference in population means, is Normal, and has a variance equal to the variance of the first population divided by the size of the first sample plus the variance of the second population divided by the size of the second sample. It's a bit more complicated than the Law of Large Numbers for single samples, but only slightly so. We calculate a difference for each step in the sampling experiment, then we plot all those differences for the sampling distribution of differences. We find that the shape of the sampling distribution is the familiar Normal curve, that the mean of the sampling distribution is the difference between the two group means in the population and that the variance is a bit complicated but basically is the sum of the two population variances after each is divided by the sample size. As you will see later in the chapter, when we apply this and go through it step by step, it's not much more difficult than what you've already done.

We can use the Law of Large Numbers to construct confidence intervals on the differences in means. But it's more common to test hypotheses about differences in means. There is a special and interesting null hypothesis about differences in means. It's that the difference is zero. If we reject no difference (if we have the double negative "not no difference") we can conclude that there are differences between the two groups in the population. We'll come back to this below.

Gosset's theorem can be generalized in much the same way. For differences in means, Gosset's theorem becomes:

> If simple random samples with means $\bar{X}_1$ and $\bar{X}_2$ and sample sizes N_1 and N_2 are drawn from Normal populations with means μ_1 and μ_2 and variances σ_1^2 and σ_2^2 then the sampling distribution of differences in sample means will be a t-distribution with $(N_1 - 1) + (N_2 - 1)$ degrees of freedom, mean $\mu_1 - \mu_2$ and variance $(\sigma_1^2/N_1) + (\sigma_2^2/N_2)$.

Again, each sample has a mean, labeled 1 and 2 to keep them straight, and a sample size N_1 and N_2. The populations have means, μ_1 and μ_2, and variances, σ_1^2 and σ_2^2. The sampling distribution of differences in sample means is centered at the difference in population means. It has the shape of a t-distribution with degrees of freedom equal to the sum of the sample sizes minus two. The sampling distribution of differences in means has a variance equal to the variance of the first population divided by the size of the first sample plus the variance of the second population divided by the size of the second sample. That is, everything is the same as for the Law of Large Numbers applied to differences in means except that we end up with a t distribution rather than the Normal distribution.

We can use these versions of the Law of Large Numbers and Gosset's theorem to construct confidence intervals and test about differences in means. Remember, the Law of Large Numbers applies only when both samples are large. Gosset's theorem applies only when both the populations have Gaussian distributions. (Note that we are using the simplest versions of the formulas for estimating the sampling errors.)

The more correct, less simple formulas are discussed at the end of this chapter as an Advanced Topic.

Let's use this procedure to see if there are differences between men and women in the animal concern score. We have one survey but as noted above, if we consider it a simple random sample, we can use the theorems on differences in means as if we had a separate sample for men and for women.

Step 1 State the null and alternative hypotheses
In this case, the null hypothesis is that the mean for men is equal to the mean for women. Or, in other words we can hypothesize that in the population, the difference between the mean for men and the mean for women is zero, that is $D = 0$. Then the alternative hypothesis is that $D \neq 0$. If we reject the null we can conclude that there is a difference between the means for men and women.

Step 2 Choose an α level
An α of 0.05 is the conventional level. This means that we are willing to take one chance in 20 of concluding that in the population there is a difference between men and women when in fact there is no difference. Since we have no strong theory

Box 10.1 Caution: Rounding

You may notice that we included three decimal places in Table 10.3 for the sample variance, but only two for the sample mean. We are now doing calculations that build on prior calculations. We calculate the variance of the sample. Then we use that value in the calculations to estimate the variance of the population. Estimated population variances are then used to estimate standard errors that are the denominators for t and Z tests. Thus calculations involving the variance become the basis for more calculations that are the basis for still more calculations. This allows rounding error to build up across calculations so we include extra decimal places to keep rounding error small. Since the means are used only once for calculating the sample difference in means, rounding error is less of a problem. If you are doing calculations using a hand calculator or paper and pencil, it is important to carry a substantial number of decimal places (at least three more than in the data as reported) in order to prevent rounding error in each calculation to build up from one calculation to another. Nearly all important statistical calculations are done on computers where statistical packages use at least 16 digits in arithmetic to keep rounding error at a minimum. Three extra digits should suffice when you are doing calculations by hand. But remember that you may get results that are slightly different from those of your colleagues if a different number of digits you use in the calculations.

about the direction of the difference, just a rough argument that men may be less concerned than woman, we will use a two-tailed test.

Step 3 Determine the appropriate statistical method to examine the hypothesis
Since we have a large sample of men and a large sample of women, the Law of Large Numbers is selected and we will use the Z-test.

Step 4 Find the critical value that corresponds to the α level
In this step we find the critical value of Z. From the Z table we find that for an alpha of 0.05 for a two-tailed test, the critical value of Z = 1.96. (If we had small samples from Normally-distributed populations, we would use Gosset's theorem and look for a critical value of t, and we would need to calculate the appropriate number of degrees of freedom.)

Substep a (for hypotheses about differences in means): Calculate the sample differences in group means
This is just the mean for men minus the mean for women, or $2.37 - 2.60 = -0.23$. Since we subtracted the mean for women from the mean for men the negative sign

indicates that in the sample women have a higher score on the scale than do men. We could subtract men from women, which would just change the sign of the difference. When working with differences between groups, it doesn't matter which group we subtract from which – as long as we keep track of what we did.

Step 5 Calculate the Z value for the sample

Substep a Estimate the population variance
We must calculate the estimated population variance for each group. This is the same estimate we have used before, but now we must apply it twice, once for each group. We get the estimated population variance by applying the (N/N − 1) correction factor. (As is typical with large samples, the correction suggested by Gosset doesn't make any practical difference, but it is very important in small samples.)

$$\hat{\sigma}^2_{Men} = \left(\frac{N_{Men}}{N_{Men} - 1} \right) s^2_{Men} \tag{10.23}$$

and

$$\hat{\sigma}^2_{Women} = \left(\frac{N_{Women}}{N_{Women} - 1} \right) s^2_{Women} \tag{10.24}$$

Numerically, the estimate of the population variance for men is:

$$(13412/13411) * 1.29 = 1.29 \tag{10.25}$$

and for women:

$$(16062/16061) * 1.45 = 1.45 \tag{10.26}$$

Substep b Estimate the variance of the sampling distribution
The estimate of the variance of the sampling distribution of differences in means is:

$$\hat{\sigma}^2_d = \frac{\hat{\sigma}^2_{men}}{N_{men}} + \frac{\hat{\sigma}^2_{women}}{N_{women}} = \frac{1.29}{13,412} + \frac{1.45}{16,062}$$

$$= 0.000096 + 0.000090 = 0.000186 \tag{10.27}$$

Substep c Estimate the standard error of the differences

$$\hat{\sigma}^2_d = \sqrt{\left(\frac{\hat{\sigma}^2_{Men}}{N_{Men}} \right) + \left(\frac{\hat{\sigma}^2_{Women}}{N_{Women}} \right)} \tag{10.28}$$

Box 10.2 Caution: Order of Operations

When estimating the standard error of differences, the order in which you do the arithmetic matters. First divide the first estimated population variance by the size of the first sample, then divide the second population variance by the size of the second sample, *then* add the two, *then* take the square root.

or, for this data:

$$\hat{\sigma}_d^2 = \sqrt{\left(\frac{1.29}{13,412}\right) + \left(\frac{1.45}{16,062}\right)} = \sqrt{0.000096 + 0.000090}$$

$$= \sqrt{0.000186} = 0.014 \qquad\qquad (10.29)$$

Substep d Calculate the z value for the sample difference in means, d

$$t_{sample} = \frac{d - D}{\hat{\sigma}_d} = \left(\frac{-0.23 - 0}{0.014}\right) = -16.43 \qquad\qquad (10.30)$$

Step 6 Decide whether the null hypothesis can be rejected
The sample Z is far larger in absolute value than the critical value for Z for the α value we have chosen (0.05). So we reject the null hypothesis. This means that we

Box 10.3 Steps in a Hypothesis Test

- *Step 1* State the null and alternative hypotheses.
- *Step 2* Choose an α level.
- *Step 3* Determine the appropriate statistical method to examine the hypothesis.
- *Step 4* Find the critical value that corresponds to the α level (and the degrees of freedom, for the t-test on small samples).
  - *Substep a (for hypotheses about differences in means)*: Calculate the sample differences in group means.
- *Step 5* Calculate the Z or t value for the sample:
  - *Substep a* Estimate the population variance.
  - *Substep b* Estimate the variance of the sampling distribution.
  - *Substep c* Estimate the standard error by taking the square root of the variance of the sampling distribution.
  - *Substep d* Calculate the value for the sample.
- *Step 6* Decide whether the null hypothesis can be rejected.

can reject the notion that the sample came from a population in which the mean animal concern score for men is equal to the mean animal concern score for women. Therefore, it is reasonable to believe that there is a difference in animal concern between men and women. Remember that when we draw this conclusion, we have a 5 percent chance of being wrong – that the sample really did come from a population in which the mean score for men and women are equal. Finally, by looking at the sample means (which are our best estimates of the population means), women seem to score higher on this scale than do men.

A small sample difference in means test

We can use tests on differences in means with small samples but if and only if we believe that each of the two populations whose means we are comparing are normally distributed. As an example, let's see if states with the death penalty differ from other states in their homicide rates. In 2006 there were 12 US states with no death penalty (Alaska, Hawaii, Iowa, Maine, Massachusetts, Michigan, Minnesota, North Dakota, Rhode Island, Vermont, West Virginia, and Wisconsin). Table 10.4 has descriptive statistics on homicide rates for states with and without the death penalty.

The difference in means is $d = 2.88 - 5.33 = -2.45$. The negative sign indicates that the states without the death penalty have a lower mean homicide rate than those with the death penalty.

Step 1 State the null and alternative hypotheses
The null hypothesis is that there is no difference in homicide rates between states that have the death penalty and those that do not. The alternative hypothesis is that there is a difference. Of course, proponents of the death penalty might argue that the death penalty reduces homicide rates, which would suggest a one-sided test in which only a difference in means in which states with a death penalty have a lower homicide rate would be evidence in support of the alternative hypothesis. But opponents can argue that the use of executions by government legitimizes violence and so that states with a death penalty will have a higher homicide rate than those without. This would lead to a one-sided test in which only a difference in means in

Table 10.4   Descriptive statistics on homicide rate for states with and without the death penalty

	States without death penalty	States with death penalty
Sample mean	2.88	5.33
Sample variance	2.61	6.28
Sample size	12	38

which the states with a death penalty have a higher homicide rate would be supportive of the alternative hypothesis. It is often the case in science that we can construct plausible arguments for contrasting views, and this is why we usually use a two-sided hypothesis test that is neutral with regard to favoring or opposing the death penalty. Of course, the mechanics of the calculations don't differ.[6]

Step 2 Choose an α level
Since we have only 50 states, we might use the 0.10 level. It is often suggested that we should use relatively high α levels with small samples (accept greater chances of rejecting the null hypothesis when it is true).

Step 3 Determine the appropriate statistical method to examine the hypothesis
We have only 12 states in the no death penalty group and 38 states with a death penalty so the Law of Large Numbers and a Z test does not seem reasonable. We can use Gosset's theorem and the t test if we believe that the distribution of homicide rates is Normal for both populations (states with and without the death penalty). Figures 10.2 and 10.3 show the distributions.

The assumption that both populations are Normally distributed is a bit tenuous. We will proceed to calculate the t test for difference in means but given the distributions in our data set, we would be cautious about putting too much faith in the results – the p values may not really reflect what would happen in a sampling experiment because the data don't match the assumptions very well. Also remember that we have population data so the probabilities have to be interpreted by the logic of "all possible states," so we would again view the results as indicative but not definitive.

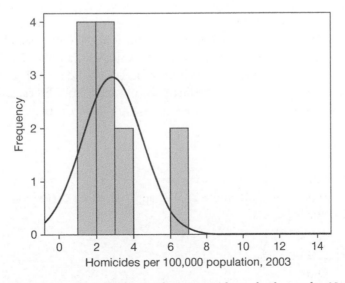

Figure 10.2 Histogram of homicide rate for states with no death penalty, N = 12
Data source: US Census Bureau, 2003, analyzed with Stata.

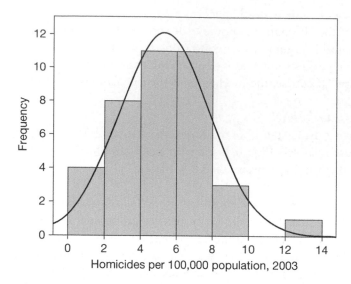

Figure 10.3   Histogram of homicide rate for states with the death penalty, N = 38
Data source: US Census Bureau, 2003, analyzed with Stata.

Step 4 Find the critical value that corresponds to the α level and number of degrees of freedom
For a difference in means test on small samples, the degrees of freedom is:

$$df = (N_1 + N_2) - 2 \tag{10.31}$$

For this sample, the number of degrees of freedom is $N_1 + N_2 - 2 = (12 + 38) - 2 = 48$ degrees of freedom. We have picked the 0.1 level – one chance in ten of a Type I error. The t-tables seldom have 48 degrees of freedom. The next smaller number of degrees of freedom is 40, for which the critical value of t is 1.684. The web-based calculators give an exact value for 48 degrees of freedom and the 0.1 level of 1.677. The difference between our rough estimate from a table and the calculated value for exactly the number of degrees of freedom we are using is small.

Step 5 Calculate the t value for the samples

Substep a Estimate the population variances
The sample variances are 2.61 and 6.28. We apply Gosset's correction to estimate the variance of each population.

$$\hat{\sigma}^2_{Without} = \left(\frac{N_{Without}}{N_{Without} - 1} \right) s^2_{Without} \tag{10.32}$$

and

$$\hat{\sigma}^2_{With} = \left(\frac{N_{With}}{N_{With} - 1}\right) s^2_{With} \tag{10.33}$$

Numerically, the estimate of the population variance for states with the death penalty is:

$$(12/11)*2.61 = 2.85 \tag{10.34}$$

and for states with the death penalty:

$$(38/37)*6.28 = 6.45 \tag{10.35}$$

Substep b Estimate the variance of the sampling distribution
The estimate of the variance of the sampling distribution of differences in means is:

$$\hat{\sigma}^2_d = \frac{\hat{\sigma}^2_{without}}{N_{without}} + \frac{\hat{\sigma}^2_{with}}{N_{with}} = \frac{2.85}{12} + \frac{6.45}{38} = 0.2375 + 0.1697 = 0.407 \tag{10.36}$$

Substep c Estimate the standard error of the differences as the square root of the estimated variance of the sampling distribution

$$\hat{\sigma}_d = \sqrt{\hat{\sigma}^2_d} = \sqrt{0.407} = 0.638 \tag{10.37}$$

Substep d Calculate the t value for the sample

$$t_{sample} = (-2.45 - 0)/0.638 = -3.84 \tag{10.38}$$

Step 6 Decide whether the null hypothesis can be rejected
The critical value is 1.677. The sample value is −3.84. So we would reject the null hypothesis and conclude that there is a difference in homicide rate between states with a death penalty and those without a death penalty.

Confidence intervals and hypothesis tests are discussed at the end of this chapter as an Advanced Topic.

Models for Differences in Means

Let's create a binary variable that takes on a value of 0 for one group in our study and 1 for the other group. For example, we might code gender as male, so women are scored 0 and men scored 1. We'll call this variable "male." Then we can think about what we expect an individual's animal concern score, which we'll call Y, to be as a function of her or his score on "male" and random error. The model is:

$$Y = (a + (b^*male)) + e \tag{10.39}$$

We are saying that each person's score on Y has three components: 1) the **intercept term**, or **a**, 2) the **slope**, or **b**, and 3) the **random error**, or **e**. Let's think about what this model predicts for each person. The prediction is just $a + (b^*male)$. Then e is the difference between what we predict for a person based on their gender (the variable "male") and their actual score.

For the moment, we want to focus on just the prediction. For women, male = 0, so $b^*male = 0$, and our prediction is a. For men, male = 1, so $b^*male = b$, and our prediction is $a + b$. The b coefficient is then the difference between men and women. Our hypothesis test is asking if it is reasonable to believe that in the population b = 0, and whatever value we find for b in the sample is just a result of random error that made women score differently from men – a difference that occurred purely by chance. Our prediction for each woman would be the mean score for women, and the prediction for each man would be the mean score for men. Then the intercept term, a, is just the mean for women. The b value would be the difference in means, or d in the notation we've been using above. Then $a + b$ is the mean score for men. This simple model is saying that everyone's score on Y can be explained by their gender and random error. The model predicts two values for Y, the population mean for men, and the population mean for women.

Using the example above, we see that the mean for women was 2.60. Thus, a = 2.60, since that is the value we predict for women. Then the difference between men and women (which we called d in the examples above) was −0.23. Thus, in the equation above, b = −0.23. The predicted score for men is $a + b = 2.60 + (−0.23) = 2.37$. So in the simple case of two groups, we can think about differences between the two groups in terms of a simple model in which the intercept is the mean for one group and the slope is the difference between the two groups.

We can state a model as we did the model for hypotheses about a single mean, but substitute d for the sample difference and D for the hypothesized population difference. Then the model is:

$$d = D \pm e \tag{10.40}$$

Differences in means as conditional distributions are discussed at the end of this chapter as an Advanced Topic.

Limits to Hypothesis Tests

The mean isn't everything. Suppose we test a hypothesis about differences in means and don't reject the null hypothesis. Does this imply the populations are the same? Of course, there's always a chance we've made a Type II error – that the means of the populations really are different. But let's put that aside for a minute and assume we didn't make such an error. We have only concluded that the two populations may have the same mean. We have concluded nothing about the shape of the distribution. Of course, in the case of Gosset's theorem, we can do the test only when the populations are Normal, but with the Law of Large Numbers that need not be true. It could be that the populations, while having equal means, differ in other ways. Both might be symmetric but with different variances. One or both might be asymmetric and different in shape from each other. It is important to remember that when we don't reject the hypothesis that populations have different means, we haven't investigated other aspects of the population distributions, and some of those might be quite different. While other sorts of hypothesis tests look at things other than the mean, each looks at only one thing at a time. Remember what hypothesis is being tested and see how well that captures the flavor of the research problem.

Repeated tests with the same data and variables erode the α level. It is common in exploratory studies to use a number of independent variables, one at a time, to predict the same dependent variable. We might want to predict animal concern based on gender (male/female), residence (rural/non-rural), and so on. If we use the same data set for each test, we are not really testing at the 0.05 level, even if we pick that as our α level in each test. In fact, the probability of a Type I error in repeated tests with the same data is

$$\text{Probability of Type I Error} = 1 - (1 - \alpha)^k \tag{10.41}$$

where k is the number of hypotheses tested. So in the above example, if we tested the three hypotheses, race, gender, and residence, each at the 0.05 α level, our real chances of a Type I error, accumulated across the three tests is

$$1 - (1 - 0.05)^3 = (1 - (0.95 * 0.95 * 0.95) = 1 - 0.857375 = 0.142625 \tag{10.42}$$

If we have a data set with many independent variables, this problem can be quite serious. Suppose we're trying to explain experiences of sexual coercion – trying to find out what groups in a population of college students more often experience sexual coercion. We might, for example, have as candidate independent variables, race (black/non-black), gender (male/female), residence (on/off campus), age (under 21/over 21), fraternity or sorority membership (yes/no), extra-mural athletic participation (yes/no), whether in a dating relationship (yes/no), prior experience with coercion (yes/no), frequent use of drugs or alcohol (yes/no), and class rank (first or second year/third year or later). That is, we're doing 10 tests, each at the

$\alpha = 0.05$ level. Then the overall chances, across all these tests, of a Type I error, is 0.5987. While 1 chance in 20 of a Type I error (0.05) is comforting, more than 1 chance in 2 is not.

The same problem occurs if we have many dependent variables and only one independent variable. While each single test has a α level of rejecting the null hypothesis when it is true, that is of finding a group difference when there is none, overall, the chances of rejecting a true null are much higher. The same formula applies.

There are three ways around this problem. One is to build a model with several independent variables. We will discuss this in detail later. A second strategy is to adjust the probability levels. The simplest way is to set α at a high level. In the 10 independent variable example on sexual coercion, if we had set α at 0.01, the overall chances of a Type I error would be 0.10. This is much more comforting than 0.6. But remember that in setting a high hurdle to avoid Type I errors, we also increase the chances of a Type II error. Statisticians have developed more subtle ways of making these adjustments in confidence levels for use with multiple hypothesis tests. The third approach is to split the sample into two groups at random. We then do our many hypothesis tests in one group only. When we think we have found the variables that matter, based on our hypothesis tests in the first randomly chosen group of the population, we then test those same hypotheses in the second group. If we find that in the second group we reject the null hypotheses that were also rejected in the first group, we can be optimistic that we avoided a Type I error. Unfortunately, unless the sample is very large, it may not be practical to split it into two random groups, and even if it is large, by decreasing the sample size we increase the chances of a Type II error. Again, there are techniques statisticians have developed to do this with more subtlety.[7]

Note that this problem arises when we have a single data set and test hypotheses using the same variables. The tests are not independent of one another. This is one of the reasons we have more confidence in research findings when they are based on analyses of several data sets. Splitting a single random sample into two groups at random is a way of creating multiple independent samples within a single data set.

Some cautions about reporting and interpreting hypothesis tests should also be mentioned. When we test hypotheses with computer statistical packages, the computer calculates the probability of our sample given that the null hypothesis is true. It is now common to report the p value when presenting research results. Then readers can pick their own α level – they can test the hypothesis at whatever level they choose. This is not unreasonable in that, following the logic of Neyman and Pearson, they have picked the chances they are willing to take of making a Type I error and with that comes some chance of making a Type II error. But sometimes the reporting of p values has led to sloppy thinking. It is not uncommon to see a comparison of p values of two different hypothesis tests and arguing that one value is "more significant" than the other. By the formal logic of hypothesis testing we have developed, a result is either significant at a given level or it is not. To compare two results, both of which lead to a rejection of the null hypothesis, one should look at the differences in the thing being studied (differences in attitudes or homicide rates, etc). In the procedures we have developed here and the applications section, we compare the means themselves for interpretation of what is a greater effect. Or one can use tests (which we will do later) of whether the effects of two independent variables are different. We note here that *p values should not be used to determine how important an effect is, but rather whether there is an effect.* If we start comparing p values, we lose track of the chances of a Type II error, and Type II errors can be as serious as Type I errors.

What Have We Learned?

The logic of sampling experiments and sampling distributions provides a way of using sample data to test assertions about the population. In particular, when they can be applied (large samples and a Normal population, respectively) the Law of Large Numbers and Gosset's approach provide a way to see how likely it was that the sample came from a population where the mean has the value asserted by the null hypothesis. They also can be used to determine if it's reasonable to believe that two groups defined by an independent variable have different population means on a dependent variable. This test for differences across groups in population means brings us back to the focus on models that describe how one variable influences another variable.

Advanced Topic 10.1   Power

In fact, three things influence the chances of making a Type II error in testing a hypothesis. First, as we have just said, the smaller the chances of a Type I error, the larger the chances of a Type II error. Second, the smaller the standard error, the smaller the chances of a Type II error. The standard error will be smaller when we have larger samples (this is yet another reason we like large samples) and when the population variance is smaller. Third, the bigger the difference between the sample mean and the hypothesized population mean, the smaller the chances of a Type II error. Because three factors are involved, we can't give you a sample table or graph that shows the chances of a Type II error in general. In particular, we have to know the difference between the sample mean and the population mean to calculate the chances.

To emphasize the importance of avoiding (or ignoring) Type II errors, statisticians have defined a term called the **power of a hypothesis test**. The power is 1 – beta, or the chances of NOT making a Type II error. Keep in mind that the chances of a Type II error are not 1 – α. For many common procedures for testing hypotheses, the power is above 0.8 when the α level is 0.05, so the chances of rejecting the null hypothesis when it is false (calling an honest player a cheat, a Type I error) is 0.05 and the chances of not rejecting the null hypothesis when it is false (not calling a cheater a cheat, a Type II error) is 0.2. But this is a broad generalization. In any study that conducts hypothesis tests, it is a good idea to think about how much power the test may have. If the sample size is small, you may be able to believe the tests that reject the null hypothesis, but you have to be cautious about the times the null was not rejected.

Advanced Topic 10.2   The More Correct, Less Simple Formulas

The approach we are using here for testing hypotheses about differences in means is a slight simplification to make things easier to understand. The formula for the variance of the sampling distribution of differences in means we are using assumes that the variances of the two populations are *not* equal. There is a more complicated formula to use if they are equal. The more complicated formula provides a better estimate of the standard error of differences when the variances of the two populations are equal because it is more efficient – it has a smaller variance for the sampling distribution. But in the text we use only the simpler formula that applies when the population variances are not equal.

The estimate of population variances when both population variances are assumed to have the same variance is:

$$\hat{\sigma}^2 = \hat{\sigma}_1^2 = \hat{\sigma}_2^2 = \frac{N_1 s_1^2 + N_2 s_2^2}{N_1 + N_2 - 2} \qquad (10.43)$$

Here we are labeling the two groups 1 and 2, which could stand for men and women or any two groups we want to compare. The left hand side says that if the two groups have the same population variance we can just call it the population variance without referring to a group number. We then estimate that population variance by weighting each sample variance by the size of the sample. Note that multiplying the sample variance by the sample size gives the sum of squares for that group, so we are just adding

the sum of squares for the two groups and then dividing by the number of degrees of freedom. When population variances are equal, this is a better approach than the one we have used in the text. But of course, we seldom know that the population variances are equal.

The formula for degrees of freedom for situations where the population variances are not equal is much more complicated than the formula we give here. The one we used is the proper formula for degrees of freedom in situations when the population variances are equal. The correct formula for degrees of freedom when population variances are not equal is:

$$df = \frac{\left(\dfrac{s_1^2}{N_1 - 1}\right) + \left(\dfrac{s_2^2}{N_2 - 1}\right)}{\left(\dfrac{s_1^2}{N_1 - 1}\right)^2 \left(\dfrac{1}{N_1 + 1}\right) + \left(\dfrac{s_2^2}{N_2 - 1}\right)^2 \left(\dfrac{1}{N_2 + 1}\right)} - 2$$

$$(10.44)$$

This is a weighted average of the two sample sizes where the weights depend on the variances.

Since these more proper formulas are complicated and hard to understand, and since most important calculations will be done with statistical software that can be told to use the correct formulas, we felt it was better to use the simpler approximations to aid understanding. For the more complex formulas, see any standard intermediate statistics text, such as Blalock (1979, pp. 227–32).

Advanced Topic 10.3   Confidence Intervals and Hypothesis Tests

You may have noticed that we use the symbol α to indicate the probability of a Type I error in a hypothesis test, and $1 - \alpha$ to indicate the level of confidence in a confidence interval. This is not a coincidence. Confidence intervals and hypothesis tests are really two different

ways of looking at the same problem. The Law of Large Numbers and Gosset's theorem both indicate that the sample mean is our best estimate of the population mean. But while we can't do better using a single number to tell us about the population mean, we know that we face random error. Therefore, we don't expect the sample mean to exactly equal the population mean. The confidence interval provides a range in which we can be fairly certain we will find the population mean. The hypothesis test tells us how likely it is that the sample we have came from the population described by the null hypothesis.

The relationship between confidence intervals and hypothesis tests is straightforward. If we construct a $1 - \alpha$ confidence interval using our sample data, any null hypothesis that proposes a population mean that falls inside the confidence interval (between its upper and lower value) will not be rejected at the α level. Any hypothesis that proposes a value for the population mean that falls outside the confidence interval will be rejected at the α level. So constructing a $1 - \alpha$ confidence interval on the mean or on differences in means allows you to test any hypothesis at the α level at a glance.

In some problems that are more complicated than the ones we present here, holding the null hypothesis to be true for the purpose of testing the hypothesis can lead to slightly different results than those obtained when simply estimating population values from the sample. For example, if we have a sample mean on a binary value of 0.2, our sample variance would be $s^2 = p(1 - p) = 0.2*0.8 = 0.16$. In a large sample, the Gosset correction wouldn't have much effect, so our estimate of the population variance would be about 0.16. But if we are testing a null hypothesis that the population mean (proportion in the category marked one) is 0.5, and we use that assumption about the population mean to construct the population variance, we get $0.5*0.5 = 0.25$. So if we strictly adhere to doing calculations based on the assumption that the null hypothesis is true, in some circumstances we get results a bit different from those that result from comparing the hypothesized value for the population to the confidence interval. But in most practical problems, using the confidence interval to test hypotheses will serve us well.

Advanced Topic 10.4 Differences in Means as Conditional Distributions

In chapter 5 we examined conditional distributions – the values taken on by the dependent variable at various values of the independent variable. The test of difference in means focused on one aspect of a conditional distribution. The independent variable has two values, or two categories. (Remember that a continuous variable or a variable with more than two categories can sometimes be collapsed into two categories, so the difference in means tests are more versatile than they first seem.) We focus not on the full conditional distribution of the dependent variable at each value of the independent variable, but on a single summary number, the mean. We then use the sample information to see if it is reasonable to believe that this summary measure is different in the population. But even if it isn't reasonable to believe that the means of the two groups on the dependent variable are different, there may be other differences in the population conditional distribution. The median might be different, or the variances or the inter-quartile range, or some other feature of the distribution. We focus on the mean for two parallel reasons. First, the mean family is the easiest place to learn how inference works. Second, we have a far more elaborate tool kit for working with the mean than for any other aspect of the conditional distribution. In fact, techniques for conducting inference on population differences in the median family and other summaries of data are still being researched. But remember that while focusing on the mean family gives important information, it is not the whole story.

Applications

Example 1:   Homicide rates

We can use a formal hypothesis test to assess the plausibility of the idea that states in the South have higher homicide rates than do those in other parts of the US. Recall that we are defining "south" as the Confederacy. The data for the test are in Table 10A.1.

We have two cautions before we can proceed. The first is that our data set is not a sample. So when we test a hypothesis, the "population" to which we are referring is one of all states that might exist, or some other rather hypothetical construct. Second, since we have small numbers of states in each group, we cannot rely on the Law of Large Numbers. Thus our results are dependent on the distribution of homicide rates in each of our hypothetical populations of Southern and non-Southern states being Normal. Graphs of the homicide rate for each of the two groups give some sense of the plausibility of that assumption.

These distributions are not exactly Normal, suggesting that the t values we use are not going to be completely accurate in reflecting the probabilities that the null hypothesis is true. But we will proceed because the use of t tests in situations like this – not exactly Normal samples and thus concern about the Normality of the population and data that is not exactly a random sample – is common practice. However, the more advanced techniques that might be better approaches to examining the differences in homicide rates across states follow the same basic logic, so this is a good practice.

First, we state the null and alternative hypotheses. Our null hypothesis is:

$$H_0: \mu_{\text{South}} = \mu_{\text{Not South}} \tag{10.45}$$

or equivalently

$$H_0: \mu_{\text{South}} - \mu_{\text{Not South}} = D = 0 \tag{10.46}$$

Then the alternative hypothesis is:

$$H_A: \mu_{\text{South}} \neq \mu_{\text{Not South}} \tag{10.47}$$

Table 10A.1 Descriptive statistics for Southern and non-Southern states, N = 50

	Mean	*Sample variance*	*Sample size*
South	7.31	4.69	11
Not South	4.02	4.63	39

Data source: US Census Bureau, 2002, 2003, analyzed with SPSS.

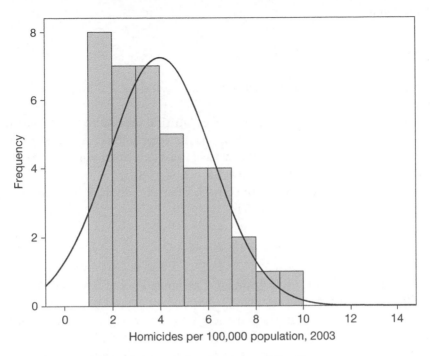

Figure 10A.1   Homicide rate for non-Southern states, N = 39
Data source: US Census Bureau, 2003, analyzed with SPSS.

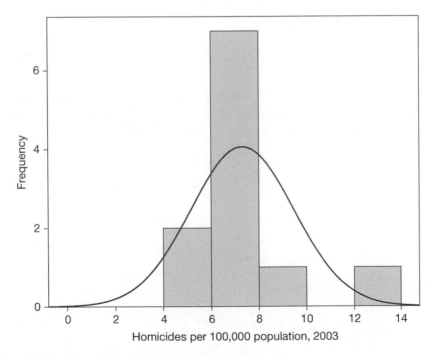

Figure 10A.2   Homicide rate for Southern states, N = 11
Data source: US Census Bureau, 2003, analyzed with SPSS.

or equivalently

$$H_A: \mu_{South} - \mu_{Not\ South} = D \neq 0 \tag{10.48}$$

Second, we choose the level at which we wish to test the hypothesis. We'll use the 0.05 level.

Third, we have to choose the appropriate statistical procedure from the two we have available. As we just noted, we don't have a large sample, so we will have to rely on Gosset's theorem and t values. We will use a two-sided test since we will be interested in a difference in homicide rates.

Fourth, we look to the table for the critical value of t. To do this we must know the number of degrees of freedom. This is:

$$(N_{South} + N_{Not\ South}) - 2 = (39 + 11) - 2 = 48 \tag{10.49}$$

Looking in the t table, the critical value for an alpha level of 0.05 and 48 degrees of freedom is not listed. But we have t values for 40 and 60 degrees of freedom. It is always better to be cautious and use a t for fewer degrees of freedom than we actually have. So we will use the t value for an alpha of 0.05 and 40 degrees of freedom. Thus the critical value of t is 2.021. If the t for the sample has an absolute value larger than this, we will reject the null hypothesis as being implausible.

Fifth, we calculate the value for the sample. Again, this has several sub-steps.

First, we calculate d, the sample difference in means. This is just the mean for South minus the mean for not South.

$$7.31 - 4.02 = 3.29 \tag{10.50}$$

Second, we calculate the estimated variance of each population and use that calculation to estimate the standard error of differences. We get the estimated population variance by applying the $(N/(N-1))$ correction factor.

$$\text{For Southern states: } \hat{\sigma}^2_{south} = \left(\frac{N_{South}}{N_{South} - 1}\right) s^2_{South} = \left(\frac{11}{11 - 1}\right) 4.69 = 5.16 \tag{10.51}$$

and

$$\text{For non-Southern states: } \hat{\sigma}^2_{not\ South} = \left(\frac{N_{not\ South}}{N_{not\ South} - 1}\right) s^2_{not\ South}$$

$$= \left(\frac{39}{39 - 1}\right) 4.63 = 4.75 \tag{10.52}$$

Then the estimate of the variance of the sampling distribution of differences in means is

$$\hat{\sigma}_d^2 = \frac{\hat{\sigma}_{South}^2}{N_{South}} + \frac{\hat{\sigma}_{not\ South}^2}{N_{not\ South}} = \frac{4.69}{11} + \frac{4.63}{39} = 0.43 + 0.12 = 0.55 \tag{10.53}$$

Finally, the standard error of differences, which is the denominator of the equation for the sample t, is just the square root of this:

$$\hat{\sigma}_d = \sqrt{\hat{\sigma}_d^2} = \sqrt{0.55} = 0.74 \tag{10.54}$$

Now we are ready to calculate the t for the sample:

$$t_{sample} = \frac{d - D}{\hat{\sigma}_d} = \left(\frac{3.29 - 0}{0.74}\right) = 4.45 \tag{10.55}$$

The absolute value of the t for the sample is greater than the critical value of t, so we will reject the null hypothesis. We conclude that the mean homicide rate for the Southern states is higher than the mean homicide rate for the non-Southern states. But we should also recall the non-Normality of the data and that the data are not a sample in the same sense as, for example, the Uganda data. So our conclusion, like all research conclusions, is tentative. The analysis seems to indicate a difference, but no analysis is perfect.

Example 2:   Animal concern

For an illustration using the animal concern variable, please review the example in the text of this chapter.

Example 3:   Environmental treaty participation

While our examples so far have used an independent variable that has two categories, we can also use the hypothesis testing procedures when we are willing to take a continuous independent variable and create a new variable that has two categories. For example, it has been suggested that countries in which citizens have more voice and whose governments are more accountable to citizens may be more likely to support international environmental treaties. We can test this idea by dividing the world into two groups of countries – those with little voice and accountability and those with greater voice and accountability. To do this we will define the little voice and accountability countries as those below the median value of −0.07. Table 10A.2 shows the basic statistics for the sample. On average, countries with greater voice and accountability have participated in three more treaties compared to lower voice and accountability countries.

Table 10A.2   Environmental treaty participation for lower and higher voice and accountability countries

	Mean	Sample variance	Sample size
Lower voice and accountability countries	4.98	6.26	85
Higher voice and accountability countries	8.02	12.12	84

Data source: Kaufmann et al., 2002, Roberts et al., 2004, analyzed with SPSS.

Stated formally, the null hypothesis of no difference between lower and higher voice and accountability nations is:

$$H_0: \mu_{\text{Lower voice}} = \mu_{\text{Higher voice}} \tag{10.56}$$

or equivalently

$$H_0: \mu_{\text{Lower voice}} - \mu_{\text{Higher voice}} = D = 0 \tag{10.57}$$

Then the alternative hypothesis is:

$$H_A: \mu_{\text{Lower voice}} \neq \mu_{\text{Higher voice}} \tag{10.58}$$

or equivalently

$$H_A: \mu_{\text{Lower voice}} - \mu_{\text{Higher voice}} = D \neq 0 \tag{10.59}$$

Let's again test the null hypothesis at the 0.05 level, and we will use a two-sided test. We do this even though our "theory" is that countries with more voice and accountability might participate in more environmental treaties, which is a directional hypothesis. In most theory driven research, we always use two-sided tests because we are rarely so certain that our theory is true that a one-sided test is justified. One-sided tests are usually reserved for research where our hypothesis is based on a standard, such as an expected performance level or a quality control goal.

These are still relatively small samples so we will need to invoke Gosset's theorem. Figures 10A.3 and 10A.4 show the sample distribution of treaty participation in each of the two groups of countries.

Here we find that both groups of countries have a relatively normal distribution, with lumps of cases for both groups between four and six treaties and a lump at fifteen for higher voice countries. So once again we can't expect that the t values we derive from our table will be precise in giving us a sense that a result is due purely to chance and the null hypothesis is true in the population. Further, as with the homicide data, our data set is not a simple random sample for a larger population.

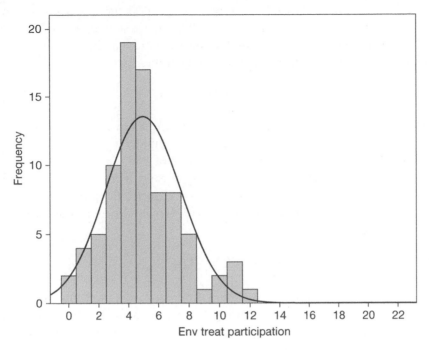

Figure 10A.3 Sample distribution of treaty participation in low voice and accountability countries, N = 85
Data source: Kaufmann et al., 2002, Roberts et al., 2004, analyzed with SPSS.

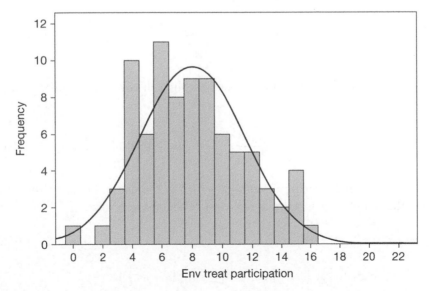

Figure 10A.4 Sample distribution of treaty participation in high voice and accountability countries, N = 84
Data source: Kaufmann et al., 2002, Roberts et al., 2004, analyzed with SPSS.

The "population" to which we refer is again a hypothetical one. But again, it is common to apply t tests in such circumstances, although we should keep the problem of non-Normality and the character of the data set in mind in drawing conclusions.

The number of degrees of freedom for the hypothesis test is

$$(N_{Lower\ Voice} + N_{Higher\ Voice}) - 2 = (85 + 84) - 2 = 167 \tag{10.60}$$

The critical value of t for 167 degrees of freedom and a one-sided test at the alpha level of 0.05 is not in the table. We have 120 and infinite degrees of freedom. We will choose 120 degrees of freedom and a t value of 1.98. Of course, we could go to a large t table or a calculator. And in most research a computer will calculate the exact value of the t for the data, that is, it calculates the probability that the samples came from populations in which there were no differences in means.

The sample difference in means is:

$$d = \bar{X}_{Lower\ Voice} - \bar{X}_{Higher\ Voice} = 4.98 - 8.02 = -3.04 \tag{10.61}$$

Now we must calculate the standard error of differences, first by estimating the variance of each population, then by estimating the variance of the sampling distribution of differences, and finally by taking its square root, which is the standard error.

For nations with little voice and accountability:

$$\hat{\sigma}^2_{Lower\ Voice} = \left(\frac{N_{Lower\ Voice}}{N_{Lower\ Voice} - 1}\right) s^2_{Lower\ Voice} = \frac{85}{(85 - 1)} * 6.26 = 6.33 \tag{10.62}$$

For nations with greater voice and accountability:

$$\hat{\sigma}^2_{Higher\ Voice} = \left(\frac{N_{Higher\ Voice}}{N_{Higher\ Voice} - 1}\right) s^2_{Higher\ Voice} = \frac{84}{(84 - 1)} * 12.12 = 12.27 \tag{10.63}$$

Then the estimate of the variance of the sampling distribution of differences in means is:

$$\hat{\sigma}^2_d = \frac{\hat{\sigma}^2_{Lower\ Voice}}{N_{Lower\ Voice}} + \frac{\hat{\sigma}^2_{Higher\ Voice}}{N_{Higher\ Voice}} = \frac{6.33}{85} + \frac{12.12}{84} = 0.07 + 0.14 = 0.21 \tag{10.64}$$

Finally, the standard error of differences, which is the denominator of the equation for the sample t, is just the square root of the above result:

$$\hat{\sigma}_d = \sqrt{\hat{\sigma}^2_d} = \sqrt{0.21} = 0.46 \tag{10.65}$$

Now we are ready to calculate the t for the sample:

$$t_{sample} = \frac{d - D}{\hat{\sigma}_d} = \left(\frac{-3.04 - 0}{0.46}\right) = -6.61 \tag{10.66}$$

The absolute value of the t value is higher than our critical value, so we reject the null hypothesis. Again, because this application of a hypothesis test based on Gosset's work doesn't exactly match the assumptions, we want to be cautious, but it does appear that the differences between lower and higher voice nations in treaty participation are worth further investigation.

Example 4:   AIDS transmission knowledge

We wonder if there are gender differences in the knowledge about how AIDS can be transmitted. We can examine such differences by proposing a null hypothesis that there are no differences between men and women and then seeing how plausible that hypothesis is given the data we have. The data are in Table 10A.3.

   We begin by stating the null hypothesis that there is no gender difference. In other words, in the population the mean for men is equal to the mean for women, or equivalently, that the difference between the mean for men in the population and the mean for women in the population is zero. Symbolically, this is:

$$H_0: \mu_{Men} = \mu_{Women} \tag{10.67}$$

or equivalently

$$H_0: \mu_{Men} - \mu_{Women} = D = 0 \tag{10.68}$$

Remember that if we score the AIDS transmission knowledge variable as 0 for those who did not know condoms can be used preventatively and 1 for those who did know, the mean is the same as the proportion in the category scored 1.

   Then the alternative hypothesis is:

$$H_A: \mu_{Men} \neq \mu_{Women} \tag{10.69}$$

or equivalently

$$H_0: \mu_{Men} - \mu_{Women} = D \neq 0 \tag{10.70}$$

Table 10A.3   Descriptive statistics on knowledge that condom use can reduce the likelihood of transmitting the AIDS virus for women and men

	Mean	Sample variance	Sample size
Men	0.83	0.142	1,886
Women	0.76	0.184	6,424

Data source: Uganda DHS data set, analyzed with SPSS.

Second, we choose the level at which we wish to test the hypothesis. Just for variation, let's test at the 0.01 level.

Third, we have to choose the appropriate statistical procedure from the two we have available. We will use a two-sided test since we will be interested in a difference whether it's one in which women are more likely or less likely to have knowledge of the efficacy of using condoms. We have a large sample of both men and women, so we are justified in using the Law of Large Numbers and thus the Z test.

Fourth, we look to the table for the critical value of Z. For a two-sided test with a 0.01 level of alpha, the critical value of Z is 2.576.

Fifth, we calculate the value for the sample. This has several sub-steps.

First, we calculate d, the sample difference in means. This is just the mean for men minus the mean for women:

$$d = \bar{X}_{Men} - \bar{X}_{Women} = 0.83 - 0.76 = 0.07 \tag{10.71}$$

That is, in the sample about 7 percent more men than women are knowledgeable.

Second, we calculate the estimated variance of each population and use that calculation to estimate the standard error of differences. We get the estimated population variance by applying the $(N/(N-1))$ correction factor, although for large samples this doesn't change things much.

For men:

$$\hat{\sigma}^2_{Men} = \left(\frac{N_{Men}}{N_{Men} - 1}\right) s^2_{Men} = \frac{1,886}{(1,886 - 1)} * 0.142 = 0.142 \tag{10.72}$$

For women:

$$\hat{\sigma}^2_{Women} = \left(\frac{N_{Women}}{N_{Women} - 1}\right) s^2_{Women} = \frac{6,424}{(6,424 - 1)} * 0.184 = 0.184 \tag{10.73}$$

Then the estimate of the variance of the sampling distribution of differences in means is

$$\begin{aligned}
\hat{\sigma}^2_d &= \frac{\hat{\sigma}^2_{Men}}{N_{Men}} + \frac{\hat{\sigma}^2_{Women}}{N_{Women}} = \frac{0.142}{1,886} + \frac{0.184}{6,424} \\
&= 0.000075 + 0.000029 \\
&= 0.000104
\end{aligned} \tag{10.74}$$

Finally, the standard error of differences, which is the denominator of the equation for the sample Z, is just the square root of the above result:

$$\hat{\sigma}_d = \sqrt{\hat{\sigma}^2_d} = \sqrt{0.000104} = 0.010 \tag{10.75}$$

Now we are ready to calculate the Z for the sample:

$$z_{sample} = \frac{d - D}{\hat{\sigma}_d} = \left(\frac{0.07 - 0}{0.010}\right) = 7.00 \qquad (10.76)$$

The absolute value of the Z for the sample is greater than the critical value of Z, so we will reject the null hypothesis. This means that, given our data, the proportion of men and the proportion of women who have AIDS transmission knowledge differs.

Exercises

1. We are interested in whether males and females differ in their satisfaction with life in general, their work life, and their family life as these were measured in the ISSP. These three questions are tested below using the t-test. (a) State the null hypotheses. (b) Summarize the statistical findings presented in Table 10E.1.

Table 10E.1 Hypothesis tests for gender differences in satisfaction with life, job, and family by gender

(a) Group Statistics

	Sex	N	Mean	Standard deviation	Standard error mean
Satisfaction with life in general	Male	20,257	2.69	0.99	0.1
	Female	25,507	2.76	1.03	0.01
Satisfaction with main job	Male	14,232	2.82	1.16	0.01
	Female	14,618	2.83	1.17	0.01
Satisfaction with family life	Male	19,848	2.45	1.08	0.01
	Female	25,053	2.55	1.10	0.01

(b) Independent samples test

	t	df	Sig.	Mean difference	Std error difference	95% confidence interval of the differences	
						Lower	Upper
Satisfaction with life in general	−7.1	45,762	0.000	−0.067	.01	−0.09	−0.05
Satisfaction with main job	−0.8	28,848	0.396	−0.012	.01	−0.04	0.02
Satisfaction with family life	−9.2	43,049	0.000	−0.096	.01	−0.12	−0.08

2. A researcher is interested in the relationship between the average number of hours spent watching television daily and academic performance among high school students. If she concludes, based on data she collects from a sample of two schools, that there is no relationship between television viewing and academic achievement when in fact in the population there is a relationship, what type of error did she make? Is there any way she could have increased the probability of making an accurate conclusion?

3. Let's return to the example of illiteracy rates. Data on male and female illiteracy rates from 25 random countries are presented in Table 10E.2. Based on this information, can you conclude that the mean difference between male and female illiteracy rates is statistically significant? Be sure to show your work.

Table 10E.2   Illiteracy rates for 25 random countries by gender

Country	Male illiteracy rate	Female illiteracy rate
Albania	7.9	23.0
Qatar	19.6	16.9
Argentina	3.2	3.2
Armenia	0.7	2.4
Bangladesh	50.6	69.8
Mexico	6.7	10.9
Belize	6.7	6.8
Guatemala	24.0	38.9
Iran	17.0	31.1
Brazil	13.0	13.2
Bulgaria	1.0	2.1
Rwanda	26.4	39.6
Chad	48.4	66.0
Liberia	29.8	63.3
Nigeria	27.8	43.9
China	7.9	22.1
Hong Kong	3.1	10.8
Uganda	22.5	43.2
Croatia	0.7	2.7
Cuba	3.2	3.4
Portugal	5.3	10.1
Dominican Republic	16.3	16.3
Ecuador	6.8	10.1
Spain	1.5	3.2
El Salvador	18.5	23.9

Data source: The United Nations Educational, Scientific, and Cultural Organization (http://unstats.un.org/), 2000 data.

4. Let's examine whether families are on average having two children. To test this, we will analyze data from the 2000 ISSP survey. Only respondents ages 35 and older were included. Use the information in Tables 10E.3 and 10E.4 to test this hypothesis. Show your calculations.

Table 10E.3 Frequency table of number of children

	Frequency	Percent	Cumulative percent
0	15,304	48.2	48.2
1	3,419	10.8	59.0
2	6,799	21.4	80.4
3	3,502	11.0	91.4
4	1,441	4.5	96.0
5	625	2.0	97.9
6	280	0.9	98.9
7	144	0.5	99.3
8 or more	232	0.7	100.0
Total	31,744	100.0	100.0

Data source: 2000 ISSP dataset.

Table 10E.4 Descriptive statistics for number of children

Category	Value
N	31,744
Mean	1.30
Standard error of mean	0.01
Standard deviation	1.63
Variance	2.67

Data source: 2000 ISSP dataset.

References

Blalock, H. M., Jr. 1979. *Social Statistics*. New York: McGraw-Hill Book Company.

Finifter, Bernard M. 1971. The generation of confidence: Evaluating research findings by random subsample replication. *Sociological Methodology* 3, 112–175.

International Social Survey Programme (ISSP). 2000. 2000 Environment II data set. www.issp.org. Catalog no. ZA 3440. Cologne, Germany: GESIS-ZA Central Archive for Empirical Research.

Kaufmann, D., Kraay, A., and Zoido-Lobaton, P. 2002, Jan. *Governance Matters II: Updated Indicators for 2000/01*. Policy Research Working Paper no 2772. The World Bank Research Development Group and World Bank Institute; Governance, Regulation and Finance Division (http://hdr.undp.org/reports/global/2002/en/).

Roberts, J. T., Parks, B. C., and Vásquez, A. A. 2004. Who ratifies environmental treaties and why? Institutionalism, structuralism and participation of 192 nations in 22 treaties. *Global Environmental Politics* 4, 22–64.

Uganda Demographic and Health Surveys. 2001. Calverton, Maryland: UBOS and ORC Macro. (http://www.measuredhs.com/pubs/pdf/FR128/00FrontMatter.pdf).

US Census Bureau 2000. Table 33. Urban and rural population, and by state: 1990 and 2000. ("http://www.census.gov/prod/cen2000/index.html"\t"_blank"www.census.gov/prod/cen2000/index.html).

US Census Bureau 2002. Historical poverty tables: Table 21. Number of poor and poverty rate, by state: 1980 to 2006. Year 2002 ("http://www.census.gov/hhes/www/poverty/histpov/"

\t"_blank"www.census.gov/hhes/www/poverty/histpov/hstpov21.html).

US Census Bureau 2003. Table 295. Crime rates by state, 2002 and 2003, and by type, 2003 (http://www.census.gov/prod/2005pubs/06statab/law.pdf).

US National Research Council. 1999. *Our Common Journey: A Transition Toward Sustainability*. Washington, DC: National Academy Press.

CHAPTER 11
THE SUBTLE LOGIC OF ANALYSIS OF VARIANCE

Outline

In Chapter 10, we introduced the t and Z tests to examine whether it was reasonable to believe that in the population, two groups had the same mean on a variable of interest. The Z and t test are sufficient as long as the independent variable can be treated as a dichotomy and the other assumptions regarding the use of these tests are plausible. To use the t and Z tests, we are concerned with differences in means, and for the Z, the sample must be large. For a t test, the sample must come from a population in which the dependent variable is Normally-distributed, and for both Z and t tests the sample must be a simple random sample. Of course we seldom will meet these assumptions exactly, but small deviations from the assumptions usually lead to only small problems in our analysis, so the tests are widely applicable.

But it is common to have an independent variable that has more than two categories and where collapsing it into only two categories does not seem appropriate. This is the greatest limit of the Z and t tests applied to differences in means. Gender is easily treated as a dichotomy. We might collapse a continuous variable like income or education into two categories by splitting into two groups at the median, although we lose information. But it doesn't make much sense to treat variables like race, ethnicity, religious affiliation or undergraduate major as dichotomies. They logically have three or more categories. In this chapter we will discuss **analysis of variance** (ANOVA), a method developed to compare means of a dependent variable across multiple categories of an independent variable.

A Review of the Two Groups Example

It may be helpful to review the logic of the Z and t tests. We will return to gender differences in the animal concern variable as an example. To refresh your memory, Table 11.1 has the key information on gender differences in means for the scale and the calculations we used in the last chapter to conduct the hypothesis test.

Table 11.1   Gender differences in means on animal concern

	Men	Women
Sample mean	2.37	2.60
Sample variance	1.29	1.45
Sample size	13,412	16,062
Estimated population variance	1.29	1.45
Difference in means		−0.23
Estimated standard error of differences		0.014
Z value for sample difference in means		16.43

Data source: 2000 ISSP data set, analyzed with SPSS.

More Than Two Groups Example

Suppose we wanted to know if there are age differences in animal concern. Figure 11.1 (a box and whisker diagram) and Table 11.2 describe the animal variable for six age groups. Figure 11.1 shows the conditional distribution of the animal variable for each age group.[1]

Because of the small range of values for animal concern, it is difficult to determine whether there are age differences with the boxplot. Table 11.2 presents the descriptive statistics for the six age groups. We see that there are some differences across age groups, with the youngest group in the sample being most concerned about animals and those aged 60 to 69 years the least concerned. Note that this might be an age effect, or a life cycle effect, or it could be a cohort effect.

How can we tell if the age differences we see in the sample are likely the result of real age differences in means in the population rather than sampling error?

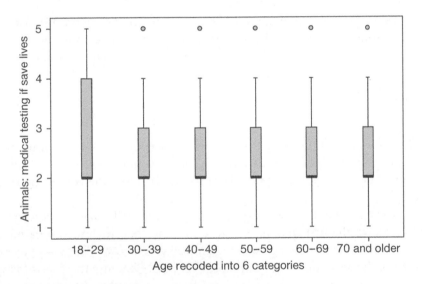

Figure 11.1   Boxplot of animal concern for six age groups
Data source: 2000 ISSP data set, analyzed with SPSS.

Table 11.2   Descriptive statistics on animal concern for six age groups

	18–29	*30–39*	*40–49*	*50–59*	*60–69*	*70 and over*
Sample mean	2.61	2.51	2.51	2.46	2.36	2.42
Sample variance	1.47	1.37	1.40	1.37	1.31	1.34
N	5,678	5,984	5,716	4,822	3,855	3,138

Data source: 2000 ISSP data set, analyzed with SPSS.

Table 11.3 Pairwise comparisons of age groups on animal concern

	18–29	30–39	40–49	50–59	60–69
30–39	4.55	–			
	(<0.001)				
40–49	4.64	0.17	–		
	(<0.001)	(0.863)			
50–59	6.36	2.13	3.44	–	
	(<0.001)	(0.033)	(0.001)		
60–69	10.00	6.13	1.93	3.94	–
	(<0.001)	(<0.001)	(0.054)	(<0.001)	
70+	7.11	3.44	3.24	1.51	−2.11
	(<0.001)	(0.001)	(0.001)	(0.132)	(0.035)

Data source: 2000 ISSP data set, analyzed with SPSS.

We could do Z tests comparing the means for each pair of age groups. Since we have 6 age groups, we will have 15 comparisons. The general formula for the number of pairwise comparisons which we can call c, when we have k groups, is:

$$c = \frac{(k)^{*}(k-1)}{2} \tag{11.1}$$

Table 11.3 shows all of the comparisons. The number in the first row in the cell is the Z value, and the number in parentheses (in the second row) is the p value for the null hypothesis that the population difference between the two group means is zero.

Here we can see that those who are 18–29 seem to differ from every other group. If we look down the first column of Z tests, we find that the tests comparing the 18–29 year olds to every other group are statistically significant. The 30–39 year old respondents also seem different from those over age 49 because the Z tests for ages 50+ in the second column of tests are significant, as are the test comparing this group to the 18–29 year olds. Among the older cohorts, the 40–49 year olds differ from the 50–59 year olds (see the third row and third column of tests) and those ages 70+.

However, we have just tested all of these 15 hypotheses with the same data, so they are not independent tests. If we intended to reject at the 0.05 level, then we have something like a 54 percent chance of making a Type I error across all the tests (we'll remind you how to do this calculation below). And we haven't addressed the question of whether age has an effect overall. We know that some categories differ from others and some do not, but we haven't thought about what it means to make a statement about the importance of age as a predictor of animal concern.

A Model

Recall the model we developed for a t or Z test on differences in means. Suppose Y is the animal concern score for an individual. Let $G = 1$ for women and 0 for men. The simple model is:

$$Y = (a + bG) + e \qquad (11.2)$$

For men, we expect the mean to be $(a + (b*0)) = a$ (so a is the mean for men). For women, we expect the mean to be $(a + (b*1)) = a + b$. The difference between the mean for men and the mean for women is b, which in the t test we called d. Note that in this model we are predicting the same value for all men and the same value for all women. If b (and therefore d since b and d are just two labels for the difference between men and women) is different from zero, we are saying that there are really two groups in the sample in the sense that men and women are different in their concern for animals. If we think about inference and use capital letters to indicate the population values of the parameters then we have:

$$Y = (A + BG) + E \qquad (11.3)$$

Then the inferential problem is to determine if it is reasonable to believe that in the population the value of B is zero. If we can reject the hypothesis that $B = 0$ (or in the notation used in the previous chapter, that $D = 0$) then we can believe that men and women differ in the mean score on the animal concern question.

We can test this hypothesis using Z (given that we have a large sample) or t (given that we believe that Y has a normal distribution in the population). But as we saw above, when we have more than two categories, we end up with many t tests on the same data set. Consequently, we have difficulty making a conceptual statement that the independent variable matters overall, not just that there are differences between some categories.

The model for many categories is just an elaboration of the two-category model. Suppose we create a series of 0–1 variables from our age categories, which we call binary or "dummy" variables. We would have:

A2 = 1 for those 18–29 and 0 for everyone else
A3 = 1 for those 30–39 and 0 for everyone else
A4 = 1 for those 40–49 and 0 for everyone else
A5 = 1 for those 50–59 and 0 for everyone else
A6 = 1 for those 60–69 and 0 for everyone else
A7 = 1 for those 70 or over and 0 for everyone else.

Notice that if we know someone's score on five of the six variables we can predict exactly where they are on the other variable. For instance, a respondent who is 71 would have the following scores: A2 = 0; A3 = 0; A4 = 0; A5 = 0; and A6 = 0. This person's score for A7 therefore must be 1. One of the variables is redundant. In the dichotomous case, we only needed one variable to represent two categories – if we know whether someone is a woman we know whether the person is a man. So one of the above variables is unnecessary. Then we can predict animal concern as:

$$Y = a + bA3 + cA4 + dA5 + fA6 + gA7 + e \qquad (11.4)$$

Again e is the error term. The intercept term is a, and it's what we'll predict when A3 through A7 are all zero. We exclude A2 because one variable is unnecessary or redundant. We could have selected any one of the age variables to exclude. The excluded variable is referred to as the reference group. When A3 through A7 are all zero, the person is 18–29, so a is the predicted value for that group. The other coefficients are the difference between that age group and the 18–29 group. The predicted values for the groups are:

18–29	a
30–39	a + b
40–49	a + c
50–59	a + d
60–69	a + f
70 and over	a + g

If we use the standard notation and let capital letters represent the population values of the parameters, saying that age has no effect is saying that B = C = D = F = G = 0. That is, in the population the age groups don't differ from one another.

Partitioning Variance

One way to think about the effect of a categorical independent variable is to ask if knowing someone's score on it helps to predict that person's value on the dependent variable. If we know your gender, does that improve our prediction of your concern for animals? We can get at this problem by considering the following logic.

The variation and variance in Y both start with looking at individual scores as deviations from the mean of Y, our dependent variable. That is:

$$Y - \bar{Y} \qquad (11.5)$$

We want to know why Y varies from observation to observation, that is, why all the deviations from the mean don't equal zero. One reason may be that the independent variable, usually called X, varies and that X causes Y. Thus variation in X is driving the variation in Y. We can examine that idea by taking the above expression and adding and subtracting a predicted value for $\hat{Y}$, called Y-hat. As we will see, this is what we predict for someone's score on Y based on the value the have on the independent variable X. Remember from algebra that adding and subtracting the same thing is not changing it. So we have:

$$Y - \bar{Y} = Y - \bar{Y} + (\hat{Y} - \hat{Y}) \tag{11.6}$$

Then rearranging:

$$Y - \bar{Y} = (Y - \hat{Y}) + (\hat{Y} - \bar{Y}) \tag{11.7}$$

Now we've divided the deviation of an individual's Y score from the mean of the Ys into two parts – the deviation of the Y score from the predicted value and the deviation of the predicted value from the mean of Y. We will use this partitioning of the variability in Y quite often in the rest of the book. Now let's use it to divide the variation in Y (the sum of squared deviations) into these two parts[2]:

$$\sum (Y - \bar{Y})^2 = \sum (Y - \hat{Y})^2 + \sum (\hat{Y} - \bar{Y})^2 \tag{11.8}$$

The first term, the one on the left of the equals sign, is just the sum of the squared deviations of the Y values from the mean of all the Y values. The first term after the equals sign is the sum of the squared deviations of each person's score from what was predicted for that person based on the independent variable. The last term is the sum of the squared deviations of the predicted values on Y from the mean of all the Ys.

Suppose men and women have the same mean value. That is, for men and women we predict the same thing. Then the second term becomes zero because $\hat{Y}$ (the prediction) is just the mean of Y, so there are no deviations of the predictions from the mean of Y. All the variation in Y is variation in the first term, variation of the individual Ys about the predictions. Remember that if the mean for men and the mean for women are the same, then both equal the overall mean. So if the mean for men and for women are the same, everyone has the same predicted score. But if men and women have different means, then we make one prediction for men and a different prediction for women. So some of the variation in Y will come from how the mean for men and the mean for women differ from each other and thus from the overall mean of all cases. The first term measures how much people differ from the mean for their gender, and the second term how much the means for each gender differ from the overall mean for everybody.

Conditional values and multiple categories in the independent variable are discussed at the end of this chapter as an Advanced Topic.

If much of the variability of Y is in the first term, it means that knowing group membership and using it to predict a score doesn't get us vary far (the means of the groups don't differ much from the overall mean and thus don't differ from each other). On the other hand, if the second term is large, it means that the means of the groups are different and thus knowing group membership helps predict the value of Y for individuals.

The first term on the right hand side of the equation (the term closest to the equals sign) is called the *within group sum of squares*, and the second term is called the *between group sum of squares*. The total variation in Y is called the overall sum of squares. If we have several groups and have, therefore, to keep track of several group means for the predicted values for Y based on X, things can get complex in terms of computation and notation. The logic, however, is the same – does knowing group membership help predict the dependent variable? If it does, that's saying the groups have different means. And if the independent variable predicts the dependent variable, then the second term will be relatively larger. If the independent variable doesn't predict the dependent variable then groups all have about the same mean and thus knowing group membership for a person doesn't aid prediction.

We then say we are *explaining* some of the variation (or variance) in Y by using X as an independent variable. If group membership doesn't help predict then the group means are essentially the same as the overall mean and the second term is mostly zeros or small numbers. Most of the variation in Y is *unexplained*. If we divide the "explained" or "between group" sum of squares (ESS) by the total sum of squares for Y (TSS), we get the *proportion of variation explained*, or R^2. In analysis of variance this used to be called eta-squared, but it is more common now to call it R^2. The equation is:

$$R^2 = \frac{\sum (\hat{Y} - \bar{Y})^2}{\sum (Y - \bar{Y})^2} = \frac{Explained\ Sum\ of\ Squares}{Total\ Sum\ of\ Squares}$$

$$= \frac{Between\ Group\ Sum\ of\ Squares}{Total\ Sum\ of\ Squares} = \frac{ESS}{TSS} \tag{11.9}$$

The square root of R^2 is R, and it's called **Pearson's correlation coefficient**, which we usually refer to with a lower case r rather than an upper case R. We'll talk more about the correlation coefficient later in the book. (Don't get confused here; using the capital letter for R is tradition, and we haven't yet talked about sample versus population even though we tend to use capital letters for the population parameters.)

An example is overdue. Consider the gender variable. We have already examined this problem using a Z test to look at differences in means. But there is often more than one way to test a hypothesis. Using the gender difference to try to explain variation in animal concern gives us another way to ask the same conceptual questions: Does gender influence animal concern? Table 11.4 gives the means (which will be the predicted values for y) and the within group (unexplained) and between group (explained) sum of squares.

Table 11.4   Means and sums of squares for men and women on animal concern

	Men	Women	Overall
Sample mean (predicted value for group)	2.37	2.60	2.49
N	13,412	16,062	29,474
Total SS			41,019.54
Within SS (unexplained)			40,618.74
Between SS (explained)			400.80
R^2			0.010

The R^2 = Between SS/Total SS, or R^2 = (400.80/41019.54) = 0.010
Data source: 2000 ISSP data set, analyzed with SPSS.

So gender explains about 1 percent of the variation in animal concern. This may seem trivial, but people's views are quite idiosyncratic, so when we work with survey data or other data for individuals, it's common to have a variable account for only a few percent of the variance and still be substantively important.

Now let's look at age effects, calculating the analysis of variance using a series of steps.

First we calculate the total sum of squares:

$$TSS = \sum (y - \bar{y})^2 \tag{11.10}$$

This is something we have been doing since the first time we calculated a variance. We take each individual's score on the dependent variable and subtract the overall mean of the dependent variable from that score. Then we square to eliminate the negative signs and add them up across all the people. This is the total sum of squares (TSS).

Next we calculate the explained (between group) sum of squares:

$$ESS = \sum (\hat{y} - \bar{y})^2 \tag{11.11}$$

In this calculation, the predicted values for people in each age group are just the means for those age groups. So those values become the $\hat{Y}$s for each person.

We can also calculate the unexplained (within group) sum of squares:

$$USS = \sum (y - \hat{y})^2 \tag{11.12}$$

Recall that the total sum of squares (TSS) equals the explained sum of squares (ESS) plus the unexplained sum of squares (USS), that is:

$$TSS = ESS + USS \tag{11.13}$$

Table 11.5 Means and sums of squares for six age groups on animal concern

	18–29	30–39	40–49	50–59	60–69	70+	Overall
Sample mean (predicted value for group)	2.61	2.51	2.51	2.46 ·	2.36	2.42	2.49
N	5,678	5,984	5,716	4,822	3,855	3,138	29,193
TSS (Total sum of squares)							40,546.05
USS (Unexplained or within sum of squares)							40,379.61
ESS (Explained or between sum of squares)							166.44
R^2							0.004

Data source: 2000 ISSP data set, analyzed with SPSS.

Then we can get the USS by subtracting the ESS from the TSS, rather than calculating it directly from the data:

$$USS = TSS - ESS \tag{11.14}$$

Then the ratio of the explained sum of squares to the total sum of squares is the R^2:

$$R^2 = \frac{ESS}{TSS} = \frac{\sum (\hat{y} - \bar{y})^2}{\sum (y - \bar{y})^2} \tag{11.15}$$

We see the results of these calculations in Table 11.5.

Age/cohort explains less than 1 percent of the variation in animal concern. That is, by predicting the animal concern scale score with the mean for a person's age group, we explain very little of the variation in the score.

So now we've seen how we go from an interest in whether the means of groups differ to looking at how much variation in the dependent variable is explained by using an independent variable to predict it, where the prediction is just the mean for the group to which a person in the data set belongs. Note that if we divide by the sample size, we'll have the sample variances (this is why it's called analysis of variance). If we divide by a correction factor, we can get an estimate of the population variance. This raises the issue of inference. Given that we see some mild difference across genders and across age groups in the sample, is there reason to believe that this sample came from a population in which the means for men and women and the means for the various age groups differ? Is it reasonable to believe that the sample differences in means are just a result of sampling error?

Inference in Analysis of Variance (the F-test)

The logic of inference in analysis of variance that was first developed by R.A. Fisher is rather subtle and can be hard to follow. So we'll try a simpler but equivalent approach. Above we argued that if the group means are different, then using the group means to predict the dependent variable Y will explain some of the variation in Y, and conversely, if the group means are the same, using them to predict does nothing. The same logic applies to the population. If the group means are equal in the population, then using the group means to predict will explain no variance. We would then expect any variance explained in the sample to be the result of sampling error that makes the means of the groups a bit different, even though in the population they are the same.

This suggests that we'd like to test the hypothesis that, in the population, R^2 (explained variance) is zero. It turns out that this is hard to do directly. R.A. Fisher found a way to test whether we are explaining variance with a model, that is, whether the R^2 in the population is zero. His approach requires three assumptions:

1   a simple random sample;
2   the populations are Normally-distributed or both sample sizes are large; and
3   the variances of the two populations are equal.

This last is a different assumption than we've used in the Z and t tests for differences in means, where the simple formulas don't assume equal populations variances. There is a technical term for equal population variances: **homoscedastic**[3] (same variances). When the population variances aren't equal, they are called **heteroscedastic** (different variances).

How do we know that the population variances are equal? We don't, but, as with population Normality, there are somewhat complicated statistical tools that help us get a sense of how reasonable that assumption may be. And there are alternative formulas to use when we have heteroscedasticity. But for the rest of the chapter, we will stay with the simplest case where we can assume that all groups have equal variances.

Fisher showed that if we construct a special ratio, it has a well-defined sampling distribution that he called the F distribution (we wonder why he picked that letter). The ratio is:

$$F_{df1,df2} = \frac{\left(\dfrac{R^2}{df1}\right)}{\left(\dfrac{1-R^2}{df2}\right)} \qquad (11.16)$$

The first thing to consider is the degrees of freedom, df1 and df2. In t-tests we had one degree of freedom to keep track of, now we have two – the numerator degrees of freedom and the denominator degrees of freedom. The numerator degrees of freedom, df1, is the number of parameters we have estimated to get the prediction minus one. We'll call the number of parameters used in the prediction equation k. For this type of problem it will be equal to the number of groups. So the equation for df1, the degrees of freedom associated with the variance explained is:

$$df1 = k - 1 \tag{11.17}$$

When the prediction equation uses one variable, as was the case for gender, then we are estimating two parameters, a and b, to get the prediction of the dependent variable. So the number of degrees of freedom is:

$$df1 = 2 - 1 = 1 \tag{11.18}$$

If we have six groups as in the age analysis, we will have five independent variables, plus the intercept a. This means that k = 6 so:

$$df1 = 6 - 1 = 5 \tag{11.19}$$

In the denominator of the equation we have the proportion of variance that wasn't explained which is just one minus the proportion that was explained. The associated degrees of freedom, df2, is the number of cases we have (N) minus the number of categories (k). That is:

$$df2 = N - k \tag{11.20}$$

For the gender example,

$$df2 = 29,474 - 2 = 29,472 \tag{11.21}$$

For the age analysis, we have:

$$df2 = 29,193 - 6 = 29,187 \tag{11.22}$$

What is Normally distributed? is discussed at the end of this chapter as an Advanced Topic.

Let's suppose we want to test the hypothesis that in the population gender explains no variance in animal concern. Again have two categories, two parameters in the prediction equation and 2 − 1 degrees of freedom in the numerator. Since the sample size was 29,474. Remember that when we use different variables the sample size

Box 11.1   Steps in the Hypothesis Test for Analysis of Variance

- *Step 1* State the null and alternative hypotheses.
- *Step 2* Choose an α level.
- *Step 3* Determine the appropriate statistical method to examine the hypothesis (in this chapter on analysis of variance, we will use the F test).
- *Step 4* Find the critical value that corresponds to the α level and the appropriate number of degrees of freedom.
  - *Sub-step a* Calculate the numerator degrees of freedom, df1. This is the number of categories in the independent variable (which is also the number of parameters in the model) minus one:

$$df1 = k - 1 \tag{11.23}$$

  - *Sub-step b* Calculate the denominator degrees of freedom, df2. This is the number of cases we have (N) minus the number of categories (k) in the independent variable:

$$df2 = N - k \tag{11.24}$$

  - *Sub-step c* Look up the critical value that corresponds to the α level and number of degrees of freedom.
- *Step 5* Find the F value for the sample
  - *Sub-step a* Calculate the total sum of squares

$$TSS = \sum (y - \bar{y})^2 \tag{11.25}$$

  - *Sub-step b* Calculate the explained sum of squares (remember that the predicted value is just the mean for the group into which the case falls):

$$ESS = \sum (\hat{y} - \bar{y})^2 \tag{11.26}$$

  - *Sub-step c* Calculate R-square

$$R^2 = \frac{ESS}{TSS} \tag{11.27}$$

  - *Sub-step d* Calculate the F value for the sample

$$F_{df1,df2} = \frac{\left(\dfrac{R^2}{df1}\right)}{\left(\dfrac{1 - R^2}{df2}\right)} \tag{11.28}$$

- *Step 6* Decide whether the null hypothesis can be rejected.

will shift depending on how many respondents answered each question. So, in the denominator we have:

$$N - k = 29{,}474 - 2 = 29{,}472 \text{ degrees of freedom} \tag{11.29}$$

If we want to work at alpha equals 0.05, an F table suggests that the **critical value** for alpha of 0.05 and 29,472 degrees of freedom is 3.89. (Actually, our F table only goes to 200 df, so to be conservative we used that value. In most research, a computer would give us a probability value for the F, and we wouldn't need to worry about tables.) Remember the explained variance (R^2) of animal concern with gender as the independent variable is .01. Then the F value for the gender analysis is:

$$F_{1,29472} = \frac{\left(\dfrac{.010}{1}\right)}{\left(\dfrac{1 - .010}{29472}\right)} = \frac{0.01}{0.000034} = 294.12 \tag{11.30}$$

Note that in these computations you have to carry plenty of decimal places, or better still, do the analysis with a good statistical package. This F value is much larger than the critical value of 3.92, so we reject the null hypothesis that knowing gender does not improve prediction. We conclude that this sample came from a population in which men and women had different scores on the animal concern scale. In fact, we could reject this hypothesis at the 0.0001 level if we wanted a stringent criterion.

This is a good place to note that **statistical significance** is driven, in part, by the size of an effect (the effect of gender on the animal concern scale in this case). It is also driven, in part, by sample size. The p value, therefore, cannot be interpreted as a measure of how "important" an effect is. Here we explain only a small amount of variance and have about one-quarter of a point difference between the genders in the attitude scale, but can conclude that there is an effect in the population. As to how important that effect is, we must look at the gender difference, the coefficients (equal to the d in the difference in means test), not the probability value.

A step-by-step example

Let's examine the relationship between animal concern and marital status. Marital status is divided into married, widowed, divorced, and never married.[4] Table 11.6 displays the mean animal concern score for each marital status. The differences are small, with never married persons having the highest mean animal concern. Let's test the hypothesis that there are no marital status differences.

Table 11.6   Mean animal concern score by marital status

	Married	*Widowed*	*Divorced*	*Never married*
Mean animal concern score	2.44	2.45	2.52	2.65
N	18,456	2,256	1,853	6,721

Data source: 2000 ISSP data set, analyzed with Stata.

Step 1 State the null (H_0) and alternative (H_A) hypotheses
  H_0: In the population, marital status explains no variance in animal concern.
  H_A: Marital status does explain some variance in animal concern
in the population.   (11.31)

Step 2 Choose an α level
We will use the 0.05 level.

Step 3 Determine the appropriate statistical method
This will be an F test. We assume we have a large simple random sample. The
sample size is large but the not all the countries in the ISSP used simple random
samples but are close enough for our purposes to proceed with the analysis.

Step 4 Find the critical value that corresponds to the α level and the
appropriate number of degrees of freedom

Substep a Calculate the numerator degrees of freedom, df1. We have 4 categories in
the independent variable, so:

$$df1 = k - 1 = 4 - 1 = 3$$ (11.32)

Substep b Calculate the denominator degrees of freedom, df2. We have 29,286
observations and 4 categories, so:

$$df2 = N - k = 29,286 - 4 = 29,282$$ (11.33)

Substep c Look up the critical value that corresponds to the α level. The critical value
for F at 3 and 29,282 degrees of freedom will be 2.65. That is as close as our table
comes to an F for 3 and 29,282 degrees of freedom.

Step 5 Calculate the F value for the sample

Substep a Calculate the total sum of squares

$$TSS = \sum (y - \bar{y})^2 = 40{,}708.54 \tag{11.34}$$

Substep b Calculate the explained sum of squares

$$ESS = \sum (\hat{y} - \bar{y})^2 = 220.75 \tag{11.35}$$

Substep c Calculate the R^2

$$R^2 = \frac{ESS}{TSS} = \frac{220.75}{40708.54} = 0.0054 \tag{11.36}$$

Substep d Calculate the F value for the sample

$$F_{df1,df2} = \frac{\left(\dfrac{R^2}{df1}\right)}{\left(\dfrac{1-R^2}{df2}\right)} = \frac{\left(\dfrac{0.0054}{3}\right)}{\left(\dfrac{1-0.0107}{29282}\right)} = \frac{0.0018}{0.000034} = 52.94 \tag{11.37}$$

Note that very large or very small values of R^2 with large sample sizes can lead to rounding error in calculating the F value for the sample.

Step 6 Decide whether the null hypothesis can be rejected
Since the F value for the sample, 52.94, is much greater than the critical value of F, 2.65, we would reject the null hypothesis. Thus we would conclude that marital status has an impact on animal concern.

F versus t and Z is discussed at the end of this chapter as an Advanced Topic.

For age groups we have:

$$F_{df1,df2} = \frac{\left(\dfrac{R^2}{df1}\right)}{\left(\dfrac{1-R^2}{df2}\right)} = \frac{\left(\dfrac{0.004}{5}\right)}{\left(\dfrac{1-0.004}{29187}\right)} = \frac{0.0008}{0.0000341} = 23.46 \tag{11.38}$$

Again, we reject the null hypothesis of no age differences in the population.

Feature 11.1   Analysis of Variance the Hard Way

The focus on the proportion of variance explained was not how Fisher originally conceptualized the F test in ANOVA, and it's not the way it's usually taught. Fisher's logic was subtle. If there is a difference between men and women, we can conceptualize them as being two distinct populations, each with its own mean. But if the null hypothesis is true, then we have one population. The division into men and women is arbitrary, at least with regard to the mean of the variable we are using as the dependent variable. Then came Fisher's clever insight. If gender has no effect on animal concern attitudes, then the sample of men and the sample of women can be treated as two separate independent samples for the population of people. The mean for men and the mean for women will differ in the sample just because of sampling error – the error that creates the variation in the sampling distribution. If the null hypothesis of no gender difference in population means is true, then the variability within the men and within the women are guides to the population variance, and can be used to construct an estimate of the population variance.

Fisher argued further that if the null hypothesis is true and there is no difference in population means across categories of the independent variable, then each group's sample mean (the mean for men and the mean for women in this example) is a sample mean drawn from the sampling distribution. The link between the variance of the sampling distribution and the variance of the population can be used to develop an estimate of the population mean. Recall that with large samples or with small samples from a Normally-distributed population, the variances of the sampling distribution of sample means is related to the population variance in the following way:

$$\sigma_{\bar{x}}^2 = \frac{\sigma^2}{N} \tag{11.39}$$

So if we have an estimate of the variance of the sampling distribution, we can use it to estimate the variance of the population:

$$\hat{\sigma}^2 = N\hat{\sigma}_{\bar{x}}^2 \tag{11.40}$$

We can apply this to each group. For group 1:

$$\hat{\sigma}_1^2 = N_1\hat{\sigma}_{\bar{x}_1}^2 \tag{11.41}$$

and parallel for each other group – the estimate of the population variance based on that group's deviation from the overall mean times the size of the sample for that group. Of course the "estimated sampling distribution variance" for each group is based on the deviation of that group's mean from the overall mean. To indicate the overall mean we will use two bars over the x, $\bar{\bar{x}}$, rather than just the one we use for a group mean. For group 1:

$$\frac{(\bar{x}_1 - \bar{\bar{x}})^2}{1} \tag{11.42}$$

Then the estimate of the population variance based on group 1 is:

$$N_1(\bar{x}_1 - \bar{\bar{x}})^2 \tag{11.43}$$

Then the estimated variance of the population based on the overall between group variation is just the sum of these between group estimates divided by k – 1 to account for the information used in the overall mean:

$$\hat{\sigma}^2_{between} = \frac{N_1(\bar{x}_1 - \bar{\bar{x}})^2 + N_2(\bar{x}_2 - \bar{\bar{x}})^2 + \ldots + N_k(\bar{x}_k - \bar{\bar{x}})^2}{k - 1} \qquad (11.44)$$

The numerator of this messy equation is called the *between group sum of squares*, which we'll label SSB. The SSB divided by the number of degrees of freedom is then an estimate of the population variance based on the assumption that, in the population, the groups all have the same mean, the overall population mean. This estimate is sometimes called the *mean square*.

If the null hypothesis of no differences in group means is true, then the estimate of the population mean developed by using the variation within each group and the one developed using the group means should be equal, except for differences that result from random sampling variation. However, if the null hypothesis is false and the groups have different means but equal variances (an assumption used in analysis of variances as noted above), then the variation within each group is still a way to get at the variance of the two groups in the population (which are equal by assumption and can be thought of as the population variance).

But if the two group means in the population are different, then the sample means are not two different means from the same sampling distribution. So if the two ways of estimating the population variance give the same result, it is consistent with the population means being equal. If the two ways of estimating the population variance (one based on variability within each group, the other based on the idea that the group means are drawn from the same sampling distribution) are different, then this casts doubt on the null hypothesis. Very subtle logic!

The estimate of the population variance based on the variation within groups is:

$$\hat{\sigma}^2_{within} = \frac{\sum_{all\ cases}(x_i - \bar{x}_{group})^2}{N - k} \qquad (11.45)$$

The numerator of this equation is called the *within group sum of squares* (SSW) and the estimated population variance is called the *within group mean square*.

Now if the null hypothesis is true, the two estimated variances should differ only because of sampling error. This can be tested using an F test:

$$F_{k-1, N-k} = \frac{\hat{\sigma}^2_{between}}{\hat{\sigma}^2_{within}} \qquad (11.46)$$

(In some texts, they use the mean square language: "the between mean square divided by the within mean square.")

This is a complicated logic but the language that comes from it (sums of squares and mean squares) is so commonly used that we felt you had to see it. It gives the same results as the R^2 approach we initially presented. The only real advantage over the R^2 approach is that there is less potential for rounding error in these calculations.

Feature 11.2   How to conduct many tests with the same data

The analysis of variance indicated that age made a difference in the level of animal concern. But how do the groups differ? Table 11.6 indicates the Z-tests comparing each group with every other group. But we are testing 15 hypotheses with the same data. Suppose we take an alpha level of 0.05 as our acceptable chance of a Type I error. Because we are applying it over and over we actually have a much higher chance of a Type I error. Using the formula we saw in the last chapter, with 15 tests at the 0.05 level, we actually have:

$$1 - (1 - .05)^{15} = 0.54 \text{ chances of making a Type I error} \tag{11.47}$$

Statisticians have addressed this problem by developing corrections to our formulas to account for the number of hypotheses being tested. One simple method, called the Bonferroni adjustment, argues that if we are conducting 15 tests and want an overall chance of a Type I error of 0.05, we should divide 0.05 by 15 and use that as the critical value for rejecting the null hypothesis. That is, we would reject only when the p value associated with a hypothesis test about differences between groups is less than 0.0033. In this example, we initially rejected the hypothesis on no difference 12 times (Table 11.3). If we used the Bonferroni correction, we would reject only 10 times (18–29 v. all five age groups; 30–39 v. 60–69; 30–39 v. 70+; 40–49 v. 50–59; 40–49 v. 70+; 50–59 v. 60–69). Most computer software actually adjusts the p values reported for a test. There are other somewhat better methods for making the correction, but the same basic logic holds – if you are testing many hypotheses with the same data, the alpha level for a single test is deceptive because the actual chances of making a Type I error are higher.

Table 11.7   Pairwise comparisons of age groups on animal concern with correction for multiple tests (comparisons in italic are significant after taking multiple tests into account)

	18–29	30–39	40–49	50–59	60–69
30–39	*4.55* (*<0.001*)	–			
40–49	*4.64* (*<0.001*)	0.17 (1.000)	–		
50–59	*6.36* (*<0.001*)	2.13 (0.517)	3.44 (0.805)	–	
60–69	*10.00* (*<0.001*)	*6.13* (*<0.001*)	*1.93* (*<0.001*)	*3.94* (*0.002*)	–
70+	*7.11* (*<0.001*)	*3.44* (*0.010*)	*3.24* (*0.018*)	1.51 (1.000)	–2.11 (0.588)

Data source: 2000 ISSP data set, analyzed with Stata.

What Have We Learned?

In the last chapter we learned how to test hypotheses about population differences in the mean value of a dependent variable across two values of a dichotomous independent variable. In this chapter we saw that this is a special case of testing whether the mean of the dependent variable changes across many categories of the independent variable. In developing the tools to test for changes across multiple categories, we also developed the concept of partitioning variance. We can examine how much variation there is between the predicted values for each group and also how much variation there is around the predicted value within each group. The ratio of the variance across predicted values to the overall variance tells us how well the model is doing – how much of the variation in the dependent variable is being explained by the independent variable. When the proportion of variance explained is high, the model predicts well, when it is low it appears that the independent variable doesn't tell us much about what is going on in the dependent variable.

Advanced Topic 11.1   Conditional Values and Multiple Categories in the Independent Variable

Using the language of conditional values, the first term measures how much variation there is about the conditional mean of the animal concern question for each gender, and the second term measures how much the conditional mean for each gender differs from the mean of the other gender.

The mean for everybody, ignoring the independent variable is sometimes called the "grand mean." It is equal to the mean of the means of each group only if the groups have equal size. This often happens in designed experiments but almost never in non-experimental data. If the groups we are looking at don't have equal size, then the overall (grand) mean is a weighted average of the group means where the weights depend on the proportion of the sample in each group.

Advanced Topic 11.2   What is Normally Distributed?

To use this test we have to have either a large sample or a Normally distributed population. But what needs to be Normally distributed? We can think about the residuals of the model, the e values, as being Normally distributed. Or we can think about the population values for the animal concern variable being Normally distributed in the population of men and Normally distributed in the population of women. When we get to regression later in the book, we will mostly think about Normality in the distribution of error terms. Given either a large sample or Normality, we also need to have data drawn from a simple random sample (the test can be applied to other kinds of probability samples with correction formulas that we're not going to burden you with). We also need to assume that the variances are equal for men and for women. Again, you can think about either the variance of the animal concern score in the population of men being equal to the variance of the animal concern score in the population of women, or you can think about the residual terms having the same variance in each group.

Thinking about the shape and variance of the population distribution of the variables is the traditional approach in most basic statistics courses, and it's what we've used in introducing t- and Z-based procedures. But thinking about the population distribution and variance of the residuals is the way we'll proceed in thinking about regression. There are both statistical tests and graphics that help us assess whether the assumptions about equal variance and Normality are appropriate, given the information we have in the sample. If we don't have Normality or equal variances, then there are other procedures to use or corrections to the basic procedures we're covering here.

Advanced Topic 11.3   F versus t and Z

Would the ANOVA approach to testing a hypothesis ever lead to a different answer than the t and Z tests conducted in Chapter 10? Not when both use the same assumptions. For small samples we must have population Normality, for large samples we don't need Normality. For large or small samples though, we need a simple random sample. But there is one difference. Recall that when we constructed the t test in the last chapter, we did not assume that the population variances of the two groups were equal. But in the F test, we do assume that the variances of the two groups are equal. So the difference in assumptions can lead to slightly different results.

Applications

Example 1:   Homicide rates

In previous chapters, we have made comparisons between Southern states, which we defined as those that had been members of the Confederacy, and all other states as a single group. It might be that lumping together of all other states misses regional variation in homicide rates. We can explore regional variation in more detail by developing four groups of states, presented in Table 11A.1. We retain our definition of the South as the former Confederate states for consistency. But this means that states that we sometimes might think of as southern, such as Kentucky and Missouri, have to be classified elsewhere, in this case in the Midwest. It would be interesting to work with even finer divisions, but with only 50 states, and as few as 11 in one of our regional categories (the South), we have to rely on Normality assumptions to use the F test. Finer subdivision, however interesting it might be, leaves us with too few states per group.

Table 11A.1   Boxplot of the state homocide rates by region

South	East	Midwest	West
Alabama	Connecticut	Illinois	Alaska
Arkansas	Delaware	Indiana	Arizona
Florida	Maine	Iowa	California
Georgia	Maryland	Kansas	Colorado
Louisiana	Massachusetts	Kentucky	Hawaii
Mississippi	New Hampshire	Michigan	Idaho
North Carolina	New Jersey	Minnesota	Montana
South Carolina	New York	Missouri	Nevada
Tennessee	Pennsylvania	Nebraska	New Mexico
Texas	Rhode Island	North Dakota	Oklahoma
Virginia	Vermont	Ohio	Oregon
	West Virginia	South Dakota	Utah
		Wisconsin	Washington
			Wyoming

Figure 11A.1 shows a boxplot diagram of the state homicide rates by region. In this version of the boxplot outliers outside the fences but not too far outside are indicated with a dot, those further out with a star.

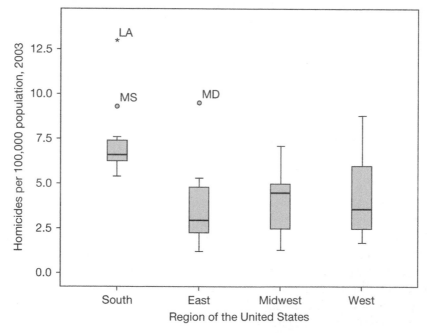

Figure 11A.1   Boxplot of the state homicide rates by region
South (N = 11), East (N = 12), Midwest (N = 13), West (N = 14)
Data source: US Census Bureau, 2003, analyzed with SPSS.

Table 11A.2   Mean homicide rates by region

Region	South	East	Midwest	West
Mean	7.31	3.60	3.94	4.45
N	11	12	13	14

Data source: US Census Bureau, 2003, analyzed with SPSS.

Here we see that the Southern rates are high compared to the other regions, especially the East. Maryland, a "border" state between the north and the south that the US Census classifies with the East, is an exception to the overall pattern for Eastern states. It has the second highest homicide rate in the data set, a rate that is higher than all the Western and Midwestern states and all the Southern states except Louisiana. Table 11A.2 shows the mean homicide rates by region.

Are these differences large enough to take seriously? One way to assess this is to ask if the region improves our ability to predict homicide rates. That is, if we use regional means to predict the homicide rates of individual states, do we explain much of the state to state variation? Put differently, if we think of the actual states as a sort of sample of all hypothetical states, is it reasonable to believe that the proportion of variance in homicide rates that can be explained using region to predict is zero? If we reject that hypothesis, then we can conclude that region is important.

Step 1 State the null and alternative hypotheses
The null hypothesis is that there are no differences across regions in the mean of state homicide rates. The alternative hypothesis is that there are differences across regions in state mean homicide rates.

Step 2 Choose an α level
Let us again use the conventional 0.05 α level.

Step 3 Determine the appropriate statistical method
We will use the F test.

Step 4 Find the critical value that corresponds to the α level and the appropriate number of degrees of freedom

Substep a Calculate the numerator degrees of freedom, df1. We have four regions so we'll have four categories in the independent variable, so the numerator degrees of freedom is

$$df1 = k - 1 = 4 - 1 = 3 \tag{11.48}$$

Substep b Calculate the denominator degrees of freedom, df2. We have 50 states, so the denominator degrees of freedom is

$$df2 = N - k = 50 - 4 = 46 \tag{11.49}$$

Substep c Look up the critical value that corresponds to the α level and number of degrees of freedom. The critical value of F for 3 and 46 degrees of freedom is 2.81. Note that our table does not include 46 degrees of freedom so we used a calculator.

Step 5 Calculate the F value for the sample

Substep a Calculate the total sum of squares

$$TSS = \sum (y - \bar{y})^2 = 315.64 \tag{11.50}$$

Substep b Calculate the explained sum of squares

$$ESS = \sum (\hat{y} - \bar{y})^2 = 97.73 \tag{11.51}$$

Substep c Calculate the R^2

$$R^2 = \frac{ESS}{TSS} = \frac{97.73}{315.64} = 0.310 \tag{11.52}$$

These calculations are reproduced in Table 11A.3.

Substep d Calculate the F value for the sample

$$F_{df1,df2} = \frac{\left(\dfrac{R^2}{df1}\right)}{\left(\dfrac{1 - R^2}{df2}\right)} = \frac{\left(\dfrac{0.310}{3}\right)}{\left(\dfrac{1 - 0.310}{46}\right)} = \frac{0.103}{0.015} = 6.87 \tag{11.53}$$

Table 11A.3   Sum of squares for homicide rate by region

Total SS	315.64
Within SS (unexplained, or USS)	217.92
Between SS (explained, or ESS)	97.73
R^2	0.310

South (N = 11), East (N = 12), Midwest (N = 13), West (N = 14)
Data source: US Census Bureau, 2003, analyzed with SPSS.

Step 6 Decide whether or not the null hypothesis can be rejected
The critical F value for an α of 0.05, and 3 and 46 degrees of freedom is 2.81. Since the F value for the sample, 6.87, is larger than the critical value of F, 2.81, we may reject the null hypothesis and conclude that region makes a difference. However note that we have modest size samples and can't be overly confident that that homicide rate is normally distributed in the population, so we want to view the α value at which we reject the hypothesis as approximate. We might still ask if only being in the South makes a difference. We will return to this hypothesis later.

Example 2:   Animal concern

For an illustration using the animal concern variable, please review the example in the main text of this chapter.

Example 3:   Environmental treaty participation

Many factors might influence the extent nations participate in environmental treaties. We may suspect that nations may feel particularly pressured to participate in environmental treaties if their neighboring nations are participating. We will measure this influence very roughly, by simply dividing nations of the world into continents. This analysis will focus on countries in South America, Europe, Asia, and Africa. North America and Australia have a small number of countries so they will be excluded. Figure 11A.2 shows treaty participation for these four continents.

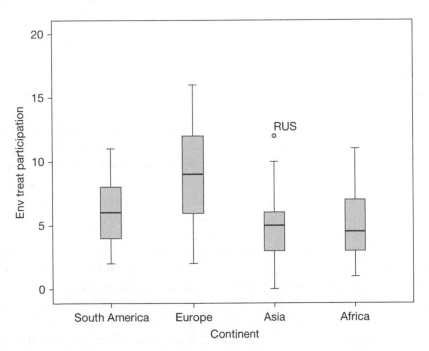

Figure 11A.2   Boxplot of environmental treaty participation by continent, N = 179
Data source: Roberts et al., 2004, analyzed with SPSS.

Table 11A.4   Mean environmental treaty participation by continent

Region	Africa	Asia	Europe	South America
Mean	5.10	4.71	9.24	6.03
N	58	51	41	29

Data source: Roberts et al., 2004, analyzed with SPSS.

Table 11A.5   Sum of squares of environmental treaty participation by continent

Total SS	1998.57
Within SS (unexplained, or USS)	1436.49
Between SS (explained, or ESS)	562.08
R^2	0.281

Africa (N = 58), Asia (N = 51), Europe (N = 41), South America (N = 29)
Data source: Roberts et al., 2004, analyzed with SPSS.

Looking at the median value, it appears that European countries are more likely than South American, Asian, and African countries to participate in environmental treaties. While African countries have the lowest median number of treaties ratified, the continent's IQR overlaps with the IQRs of Asian and South American countries, and to some extent Europe's IQR. Therefore, we will need to conduct further tests to determine whether there truly are differences in treaty participation between the four continents.

We can again use an analysis of variance, with four categories in the independent variable. Table 11A.4 shows the mean number of environmental treaties ratified for the countries in each of the four continents. Table 11A.5 shows the sum of squares that results by using these four means to predict the treaty participation score for each country within the zone. We can explain about 28.1 percent of the variance in environmental treaty participation scores by using continent as a predictor variable. You will have noticed that we can usually get higher R^2 values when we work with aggregate data (such as states or countries) than we get when we work with individual level data. This is not too surprising; individuals are hard to predict, but their idiosyncrasies tend to average out at a more aggregate level.

Step 1 State the null and alternative hypotheses
Our null hypothesis is again that, in the population, continent explains no variance in environmental treaty participation, or, numerically, that in the population the R^2 value is zero.

Step 2 Choose an α level
We can again test at the 0.05 α level, allowing ourselves one chance in twenty of making a Type I error.

Step 3 Determine the appropriate statistical method
We will use the F test.

Step 4 Find the critical value that corresponds to the α level and number of degrees of freedom

Substep a Calculate the numerator degrees of freedom, df1. We have four categories in the independent variable, so

$$df1 = k - 1 = 4 - 1 = 3 \tag{11.54}$$

Substep b Calculate the denominator degrees of freedom, df2. We have 179 data points, so

$$df2 = N - k = 179 - 4 = 175 \tag{11.55}$$

Substep c Look up the critical value that corresponds to the α level and number of degrees of freedom. The critical value for F at 3 and 175 degrees of freedom is 2.66. Again we used a statistical calculator rather than the tables in Appendix C.

Step 5 Find the F value for the sample

Substep a Calculate the total sum of squares

$$TSS = \sum (y - \bar{y})^2 = 1998.57 \tag{11.56}$$

Substep b Calculate the explained sum of squares

$$ESS = \sum (\hat{y} - \bar{y})^2 = 562.08 \tag{11.57}$$

Substep c Calculate the R^2. These calculations are reproduced in Table 11A.5.

$$R^2 = \frac{ESS}{TSS} = \frac{562.08}{1998.57} = 0.281 \tag{11.58}$$

Substep d Calculate the F value for the sample

$$F_{3,179} = \frac{\left(\dfrac{0.281}{3}\right)}{\left(\dfrac{1 - 0.281}{175}\right)} = \frac{0.094}{0.004} = 23.1 \tag{11.59}$$

Step 6 Decide whether or not the null hypothesis can be rejected
Since the F value calculated for the sample, 23.1, is larger than the critical value for our chosen α level and the number of degrees of freedom we have, 2.66, we can reject the null hypothesis. This means it is reasonable to believe, given this data, that continent predicts treaty participation. Good researchers would now elaborate this analysis by trying to find out exactly what is going on, but that will require more data and more statistical tools than we will deal with now.

Example 4:   AIDS transmission knowledge

We have hypothesized that AIDS knowledge differs by marital status. Table 11A.6 give the proportion of people within each marital status who know that condoms can help prevent AIDS transmission. We don't show a boxplot because the dependent variable has only two values: respondents who understood the value of condom use and those who don't.

These are pretty small differences. Is there reason to believe that marital status has an effect in the population?

Step 1 State the null and alternative hypotheses
We can hypothesize no effect as a null hypothesis, which implies that the means for each marital group are equal to each other and equal to the mean of everyone, ignoring marital status. This is the same as hypothesizing that if we predict each person's score on the AIDS knowledge variable with the mean score for their marital status, we will not explain any variance.

Step 2 Choose an α level
We will select an α level of 0.05, allowing one chance in twenty of a Type I error or rejecting the null hypothesis when it is true.

Step 3 Determine the appropriate statistical method
Since the AIDS knowledge variable is a 0–1 dichotomy, we have to rely on the Law of Large Numbers to justify applying the F test.

Step 4 Find the critical value that corresponds to the α level and number of degrees of freedom

Substep a Calculate the numerator degrees of freedom, df1. The numerator degrees of freedom is the number of categories in the independent variable (3) minus 1 $(3 - 1 = 2)$, so

$$df1 = k - 1 = 3 - 1 = 2 \tag{11.60}$$

Substep b Calculate the denominator degrees of freedom, df2. The denominator degrees of freedom is the sample size (8,310) minus the number of categories in the independent variable (3), so

$$df2 = N - k = 8,310 - 3 = 8,307 \tag{11.61}$$

Table 11A.6 Proportion of people with AIDS knowledge by marital status

Marital Status	Mean	Total N
Never married	0.82	2,100
Currently married	0.76	5,250
Formerly married	0.76	960

Data source: 2000 Uganda DHS data set, analyzed with SPSS.

Substep c Look up the critical value that corresponds to the α level and number of degrees of freedom. The critical value for F at 4 and 8,307 degrees of freedom is 3.07.

Step 5 Find the F value for the sample

Substep a Calculate the total sum of squares.

$$TSS = \sum (y - \bar{y})^2 = 1,460.14 \tag{11.62}$$

Substep b Calculate the explained sum of squares.

$$ESS = \sum (\hat{y} - \bar{y})^2 = 5.26 \tag{11.63}$$

Substep c Calculate the R^2. These calculations are reproduced in Table 11A.7.

$$R^2 = \frac{ESS}{TSS} = \frac{5.26}{1460.14} = 0.0036 \tag{11.64}$$

Substep d Calculate the F value for the sample.

$$F_{df1, df2} = \frac{\left(\dfrac{R^2}{df1}\right)}{\left(\dfrac{1-R^2}{df2}\right)} = \frac{\left(\dfrac{0.0036}{2}\right)}{\left(\dfrac{1-0.0036}{8307}\right)} = \frac{0.0018}{0.00012} = 15.0 \tag{11.65}$$

Table 11A.7 Sum of squares for AIDS knowledge by marital status

Total SS (TSS)	1460.14
Within SS (unexplained, or USS)	1454.89
Between SS (explained, or ESS)	5.26
R^2	0.0036

Never married (N = 2,100), Currently married (N = 5,250), Formerly married (N = 960)
Data source: 2000 Uganda DHS data set, analyzed with SPSS.

Step 6 Decide whether or not the null hypothesis can be rejected
The critical value of F for 2 and 8,307 degrees of freedom is 3.07, and the calculated F value for the sample is 15.0. Since the F value for the sample, 15.0, is larger than the critical value of F, 3.07, we can reject the null hypothesis. Thus, given this data, it is reasonable to assume marital status has an effect on AIDS knowledge. But we also know that while we can believe the effect is not zero, it is also not very large. The "currently married" and "previously married" have the same percentage of correct respondents, and are only 6 percent lower than the "never married." Because we have a very large sample compared to many studies, small difference can be statistically significant. The judgment of the researcher has to determine if statistically significant differences are also important in the context of the research being conducted.

Exercises

1. We are going to examine whether there are differences in the percent of the population under age 15 for three continents, using analysis of variance. The data are presented in Table 11E.1. To help you, the within sum of squares equals 181.63 and the total sum of squares equals 3607.96.

   a) State the null hypothesis.
   b) Calculate the degrees of freedom needed to compute the critical value.
   c) What is the critical F value for an α of 0.05?
   d) Calculate the R^2.
   e) Calculate the F value.
   f) Can we reject the null hypothesis? What do you conclude?

Table 11E.1   Percent of the population under age 15 for 24 countries in Africa, Europe, and South America

	Percentage of population under age 15
Africa	
Cote d'Ivoire	41
Ethiopia	46
Kenya	41
Liberia	47
Rwanda	45
Somalia	48
Uganda	50
Zimbabwe	43
South America	
Argentina	27
Brazil	28
Chile	27
Columbia	32
Ecuador	33
Peru	33
Uruguay	24
Venezuela	33
Europe	
Austria	16
France	19
Germany	15
Hungary	16
Italy	14
Romania	17
Spain	14
Switzerland	16

Data source: The Statistics Division and Population Division of the United Nations Secretariat, 2003, website: http://unstats.un.org/

2. We are interested in whether work status affects satisfaction with: 1) family life; 2) work life; and 3) life in general. The F-test was used to test these research questions. Summarize the statistics presented in Tables 11E.2 and 11E.3.

Table 11E.2 Descriptive statistics on work status and life, job and family satisfaction

		N	Mean	Std deviation	Std error	95% confidence interval for mean	
						Lower bound	Upper bound
Life satisfaction	1 WORKING FULLTIME	20,923	2.68	0.94	0.01	2.67	2.69
	2 WORKING PARTTIME	3,963	2.63	0.96	0.02	2.60	2.66
	3 TEMP NOT WORKING	721	2.69	0.98	0.04	2.62	2.74
	Total	25,607	2.67	0.95	0.01	2.66	2.68
Job satisfaction	1 WORKING FULLTIME	20,695	2.83	1.15	0.01	2.81	2.84
	2 WORKING PARTTIME	3,890	2.85	1.16	0.02	2.81	2.88
	3 TEMP NOT WORKING	612	2.96	1.24	0.05	2.86	3.05
	Total	25,197	2.83	1.15	0.01	2.82	2.85
Family satisfaction	1 WORKING FULLTIME	20,722	2.47	1.05	0.01	2.45	2.48
	2 WORKING PARTTIME	3,931	2.40	1.02	0.02	2.37	2.44
	3 TEMP NOT WORKING	706	2.47	1.05	0.04	2.39	2.55
	Total	25,359	2.46	1.05	0.01	2.44	2.47

Data source: 2000 ISSP data set, analyzed with SPSS.

Table 11E.3 Analysis of variance of work status and life, job, and family satisfaction

		Sum of squares	df	Mean square	F	Sig.
Satisfied with life in general	Between groups	8.44	2	4.22	4.71	.009
	Within groups	22,928.59	25,604	.90		
	Total	22,937.02	25,606			
Satisfied with main job	Between groups	10.80	2	5.40	4.07	.017
	Within groups	33,434.44	25,194	1.33		
	Total	33,445.25	25,196			
Satisfied with family life	Between groups	13.38	2	6.69	6.11	.002
	Within groups	27,746.76	25,356	1.09		
	Total	27,760.14	25,358			

Data source: 2000 ISSP data set, analyzed with SPSS.

3. Think of two research hypotheses that can be tested using analysis of variance (do not use previous examples). For both, write the null and alternative hypotheses.

---

4. A researcher wants to test whether ethnicity relates to level of animal concern. Table 11E.4 is an ANOVA table for testing this. What conclusion can the researcher draw from this output?

Table 11E.4 Analysis of variance of relationship between animal concern and ethnicity

Animal Concern Scale	Sum of squares	ANOVA		
		df	F	Sig.
Between groups	41.691	2	5.757	0.003
Within groups	5,007.537	1,383		
Total	5,049.227	1,385		

Data source: 2000 ISSP data set, analyzed with SPSS.

---

5. You ran an ANOVA test to see whether former confederacy states differed with other states in the legitimate violence index (an index measuring factors contributing to a "culture of violence" in communities). Unfortunately, you spilled coffee on the output and cannot recognize all the numbers. Given the remaining numbers on Table 11E.5, your ANOVA table, can you reconstruct the output?

a) What is the Between groups sum of squares?
b) What is the total degrees of freedom?
c) Can you calculate the F-ratio?
d) What conclusion can you draw from the results you have?

Table 11E.5 Analysis of variance of relationship between state membership in the Confederacy and the Legitimate Violence Index

Legitimate Violence Index	Sum of Squares	ANOVA		
		df	Mean Square	Sig.
Between groups	–	1	2173.500	.018
Within groups	17,346.420	48	–	
Total	19,519.920	–		

Data source: US Census Bureau, 2003, Baron and Strauss 1988, analyzed with SPSS.

---

6. To understand better the relationship between work status and satisfaction with work (analyzed in question 2), the Bonferroni test is presented in Table 11E.6. What does this test tell us?

Table 11E.6 Bonferroni test of relationship between work status and satisfaction with work

a) Dependent Variable: Work satisfaction
Descriptives

		N	Mean	Standard deviation	Standard error
rjobsat	1 Employed-full time	20,695	5.1727	1.14727	0.00798
	2 Employed-part time	3,890	5.1517	1.16194	0.01863
	3 Employed-< part-time	612	5.0441	1.24417	0.05029
	Total	25,197	5.1663	1.15213	0.00726

b) Multiple Comparisons – Bonferonni

(I) wrkstat R: Current employment status	(J) wrkstat R: Current employment status	Mean difference (I-J)	Standard error	Sig.	95% confidence interval	
1 Employed-full time	2 Employed-part time	0.02098	0.02013	0.892	−0.0272	0.0692
	3 Employed-< part-time	0.12853*	0.04725	0.020	0.0154	0.2417
2 Employed-part time	1 Employed-full time	−0.02098	0.02013	0.892	−0.0692	0.0272
	3 Employed-< part-time	0.10755	0.05010	0.095	−0.0124	0.2275
3 Employed-< part-time	1 Employed-full time	−0.12853*	0.04725	0.020	−0.2417	−0.0154
	2 Employed-part time	−0.10755	0.05010	0.095	−0.2275	0.0124

* The mean difference is significant at the 0.05 level
Data source: 2000 ISSP data set, analyzed with PSSS.

References

Baron, L. and Straus, M. A. 1988. Cultural and economic sources of homicide in the United States. The Sociological Quarterly 29, 371–90.

International Social Survey Programme (ISSP). 2000. 2000 Environment II data set. www.issp.org. Catalog no. ZA 3440. Cologne, Germany: GESIS-ZA Central Archive for Empirical Research.

Roberts, J. T., Parks, B. C., and Vasquez, A. A. 2004. Who ratifies environmental treaties and why? Institutionalism, structuralism and participation of 192 nations in 22 treaties. Global Environmental Politics 4(3), 22–64.

Uganda Demographic and Health Surveys. 2001. Calverton, Maryland: UBOS and ORC Macro. (http://www.measuredhs.com/pubs/pdf/FR128/00FrontMatter.pdf).

US Census Bureau 2000. Table 33. Urban and rural population, and by state: 1990 and 2000. ("http://www.census.gov/prod/cen2000/index.html"\t"_blank"www.census.gov/prod/cen2000/index.html).

US Census Bureau 2002. Historical poverty tables: Table 21. Number of poor and poverty rate, by state: 1980 to 2006. Year 2002 ("http://www.census.gov/hhes/www/poverty/histpov/"\t"_blank"www.census.gov/hhes/www/poverty/histpov/hstpov21.html).

US Census Bureau 2003. Table 295. Crime rates by state, 2002 and 2003, and by type, 2003 (http://www.census.gov/prod/2005pubs/06statab/law.pdf).

CHAPTER 12
GOODNESS OF FIT AND MODELS OF FREQUENCY TABLES

Outline

The Normal distribution was first investigated because scientists noted the bell-shaped pattern in their measurements. This is why it came to be called the "normal" distribution – it was the typical, or normal, pattern for errors in measurement. Later the Normal distribution was found in the theory of sampling distributions – it is the shape of the sampling distribution of sample means if the samples are large, and it is the shape of population distributions that generate t-shaped sampling distributions of sample means. But in between its birth in the physical sciences and its current role in sampling theory, during the late nineteenth and early twentieth centuries many social and biological scientists used the Normal and other distributions to describe the distribution of variables in their data. Scientists tried to understand the distribution of human height, weight, strength, length of arms, legs, and fingers, and a myriad of other measurements.

This work started with the idea that there was an "average" person and all other persons were somehow deviations from that average.[1] The Belgian scientist Adolphe Quetelet emphasized this approach to research. In contrast, Charles Darwin's cousin, Francis Galton, emphasized not the mean but the distribution of variables itself. (Galton also developed the ideas of correlation and regression that we'll cover in Chapter 13.) Quetelet's view of the world goes back to Plato and argues for an ideal type or form and deviations from that form as in some sense flawed. Modern scientific theory, particularly in the historical and evolutionary sciences, takes a much different view, and sees the average (the mean) as simply the center of a distribution. Variance is the stuff of life and the object of science.[2] This is a major advance in scientific thinking. From the late nineteenth to the early twentieth centuries many scientists focused on measuring this variability for all sorts of variables: size of plants and animals, the number of Prussian cavalry officers killed by horsekicks, the number of children various groups of people were having, and so on. There was a strong interest on the part of Galton in seeing what the shape of the distributions looked like.

An important turn away from focusing on ideal types and averages came when Karl Pearson developed a statistical method for testing the hypothesis that a batch of data comes from a population described by a hypothesized distribution.[3] That is, instead of just graphing a distribution of data and visually comparing it to a theoretical distribution like the Normal distribution, Pearson developed a method of testing the hypothesis that the data came from a population in which the variable had the hypothesized distribution. In 1892, he analyzed the results of the roulette wheels in Monte Carlo. He discovered that it was very improbable that the outcomes were the result of a uniform distribution. That is, he rejected the hypothesis (though this language had not yet been developed – it waited for R. A. Fisher) that the wheel was fair with every number having an equal chance of coming up on each spin.

The test Pearson invented is called the χ^2 **test** (pronounced **chi-square**). It is what statisticians call a **goodness of fit** test. We compare the data we have in the sample to what we would expect if the null hypothesis were true. We calculate a number that represents the difference between what we have in the sample and what the null hypothesis tells us to expect. The resulting number has a sampling distribution. If the number we have for our sample is improbable, we reject the null hypothesis.

The idea of goodness of fit comes up repeatedly in statistics. The variance and the median absolute deviation (MAD) from the median are both measures of goodness of fit in the sense that they describe how well the data fit a model that says every score should be equal to the mean or the median. The larger the variance, the less well we do predicting everyone's score with the mean. The larger the MAD, the less well we do predicting everyone's score with the median. We can even think of the t or Z value in a hypothesis test on a single mean or a difference in means as a very simple goodness of fit test. The Z or t tells us how close the sample mean or difference in means is to the hypothesized population mean or hypothesized population difference in means. And as we have seen, large values for Z or t tell us that our sample is unlikely to have come from a population that the null hypothesis describes accurately. If the probability is large enough, we reject the null as an inadequate description of the population from which our sample was drawn. And the more intuitive form of the F test for analysis of variance is based on the amount of variance in the dependent variable that can be predicted by using the independent variable in a model. R^2 is a measure of goodness of fit, and we can test the hypothesis that in the population it has the value 0 – no variance is explained by the model. If we have large values of F, we reject the hypothesis of no variance explained. Chi-square works the same way – large values mean bad fit to the hypothesis. This means a low probability that the null hypothesis is an adequate description of the population.

But chi-square differs from the statistics discussed in previous chapters in an important way. In previous chapters we have dealt with values of variables for individual cases, and means, variances, medians, MADS, and so on, as summaries of those. The fit we are interested in is between a prediction from a model and an individual score. *Chi-square works only with categorical variables.* Instead of predicting the score, we predict counts – the number of observations that fall in a particular cell in a contingency table or a bin in a histogram or frequency distribution.

Chi-square Applied to a Frequency Table

In Chapter 8 we explored how education might be distributed in the US population. We offered the unlikely but simple hypothesis that education might be described by a uniform distribution, with every number of years of education having equal representation in the population. We can now test that hypothesis, which provides an easy introduction to a procedure we can then use to test more interesting hypotheses.

Table 12.1 shows the actual data on education in a national telephone interview survey we conducted. It also shows what we would expect if every level of education was an equally likely stopping point – the uniform distribution. The second and third columns show the number of cases and the proportion of cases that actually fall in each category. The second column is the frequency table – the count of the

---

Box 12.1   Caution: Rounding Error

Rounding error matters in working with chi-square. You will notice that the sample size in Table 12.1 is 420 and both columns 2 and 5 add up to 420. Columns 3 and 4 should add up to 1.00 but both are a little off because of rounding error.

---

number of people who fell in the category. This is the table we will work with in actually testing hypotheses, although we need the other columns for steps in the process. The sample frequency in column 2 has 21 cells, one for each number of years of education from 0 to 20. The fourth column shows the proportion of cases that would fall in that category if the null hypothesis of a uniform distribution is true. It is $1/21 = 0.0476$. That is, if the distribution is uniform, we expect an equal proportion in each category. The last column simply multiplies the expected proportion by the sample size to give us the expected frequency in each cell in a sample of size 420, if the uniform distribution is the right description of the

Table 12.1   Observed and expected values for education

Years of education	Sample frequency	Sample proportion	Expected proportion	Expected frequency
0	0	0	0.0476	20.00
1	0	0	0.0476	20.00
2	0	0	0.0476	20.00
3	2	0.0048	0.0476	20.00
4	0	0	0.0476	20.00
5	0	0	0.0476	20.00
6	0	0	0.0476	20.00
7	3	0.0071	0.0476	20.00
8	10	0.0238	0.0476	20.00
9	2	0.0048	0.0476	20.00
10	5	0.0119	0.0476	20.00
11	10	0.0238	0.0476	20.00
12	103	0.2452	0.0476	20.00
13	39	0.0929	0.0476	20.00
14	51	0.1214	0.0476	20.00
15	31	0.0738	0.0476	20.00
16	84	0.2	0.0476	20.00
17	18	0.0429	0.0476	20.00
18	22	0.0524	0.0476	20.00
19	12	0.0286	0.0476	20.00
20	28	0.0667	0.0476	20.00
Total	420	1.0001	0.9996	420

Data source: Collected by the authors.

population. We would expect 20 people per cell. This is called the "expected value" for the cell under the null hypothesis of a uniform distribution.

We can think of the fourth and fifth columns as displaying a model. The model predicts the frequency of a cell (the number of observations in that cell) as:

Cell frequency = (sample size) *
(proportion expected under null hypothesis) + e   (12.1)

Under the very simple hypothesis we are using, we expect equal numbers in each cell. In other situations, we might have a different set of expected proportions representing a different hypothesis. The e term is again random error. It is what makes the observed cell frequency differ from the expected cell frequency. That is, if the null hypothesis is true, the cell frequency we observe in the sample is the same as what the expected table predicts except that it has been shifted a bit by random error (because of sampling error or measurement error or randomization error).

To actually calculate the goodness of fit, we need labels to keep track of things. The label for the **observed cell frequency** (what we have in the sample) is O. The label for an **expected cell frequency** is M, what is predicted by the null hypothesis model. (In most texts this is called E for "expected" but since we use E for error, using E for expected would be confusing.) Since we have lots of cells, we need to keep track of them with subscripts. We can simply label the cells 1–21 (for years of education 0–20) and use that as the subscript. Thus we have, for zero years of education, $O_1 = 0$, $M_1 = 20$. For 12 years of education we have, $O_{13} = 103$, $M_{13} = 20$, and so on.

The formula for the chi-square goodness of fit statistic is:

$$X^2 = \sum \frac{(O - M)^2}{M}$$ (12.2)

First, we have to calculate the expected values based on the null hypothesis. Then we subtract the expected value from the observed value. This tells us, for each category, how far off our sample frequency is from what is expected under the null hypothesis. This is the e term in the model above. It is shown in column 4 of Table 12.2. Note that some of the differences are positive and some are negative – the same problem that occurred in looking at deviations from the mean and median. We again solve that problem by squaring, and these results are in column 5. Then we divide each squared deviation by M. This is shown in column 6. Finally, to get the summary measure of goodness of fit, the sample chi-square under the null hypothesis, we add the numbers in column 6. For this sample and this null hypothesis, the chi-square value is 778.429.

This is the number that has a sampling distribution that Karl Pearson discovered. We use the sampling distribution for chi-square in the same way that we used the sampling distributions to test hypotheses about the sample mean, differences in means and R^2 values. If our sample chi-square is one that would occur only rarely from the hypothesized population, instead of thinking that by chance we have an unusual

Table 12.2   Calculation of the chi-square statistic

Years of education	Sample frequency	Expected frequency	0−M	$(0-M)^2$	$(0-M)^2/M$
0	0	21	−21	441	21
1	0	21	−21	441	21
2	0	21	−21	441	21
3	2	21	−19	361	17.19048
4	0	21	−21	441	21
5	0	21	−21	441	21
6	0	21	−21	441	21
7	3	21	−18	324	15.42857
8	10	21	−11	121	5.761905
9	2	21	−19	361	17.19048
10	5	21	−16	256	12.19048
11	10	21	−11	121	5.761905
12	103	21	82	6,724	320.1905
13	39	21	18	324	15.42857
14	51	21	30	900	42.85714
15	31	21	10	100	4.761905
16	84	21	63	3,969	189
17	18	21	−3	9	0.428571
18	22	21	1	1	0.047619
19	12	21	−9	81	3.857143
20	28	21	7	49	2.333333

sample that was drawn from a population well described by the null hypothesis, we reject the null hypothesis. Again, there are tables of critical values of chi-square, just as there are tables of critical values of t and Z that correspond to our α level for the hypothesis. Appendix C has a table of critical chi-square values, or you can use a statistical calculator.

As with t, we have to keep track of degrees of freedom for the chi-square test. If we know the sample size, then once we have multiplied the expected proportion under the null hypothesis times the sample size for all but one of the cells, we could just subtract to get the last cell. That is, the last cell must make the sum of all the expected value cells add up to the sample size. In the statistical jargon, this is sometimes discussed as the model under the null hypothesis "preserving" (that is, reproducing in the expected table) the sample size. We will use the sample size in calculating the expected frequencies. So, as in the t procedure we have "used up" one piece of information. For a chi-square analysis of a one variable frequency table, the number of degrees of freedom is the number of cells minus 1 (that is, k − 1).[4] For our table this is 21 − 1 or 20.

So the null hypothesis is that in the US population, number of years of education has a uniform distribution. The alternative hypothesis is that it has some other pattern.

Let's use the 0.01 α level, that is, we will take one chance in 100 of making a Type I error. Then we would look up the critical value for chi-square with an α of 0.01 and 20 degrees of freedom. The table indicates that this critical value is 37.56. The value indicating goodness of fit to the null hypothesis' predicted values for our table is 778.43. This is larger than the critical value, so we reject the null hypothesis and conclude that in the US population, education does not have a uniform distribution.

Assumptions for Chi-square

Chi-square can be used whenever two conditions are met: 1) the error terms for the cells in the table are independent of each other, and 2) the smallest expected cell frequency is greater than five. The idea that the error terms are independent means that we have a random sample from a population, or that subjects were assigned to experimental and control groups at random, or that measurement error is random (here measurement error means assigning someone to the wrong cell in the table). We justify making this assumption by understanding how the data were collected: Was it a random sample? A randomized experiment? Is it plausible that misclassification was random? The assumption about smallest expected frequencies is easy to check – skim the expected table and make sure no cell value is smaller than five. But note that the number five is a rule of thumb. If you have only one expected cell frequency below five the theoretical critical probability from the table will be pretty close to the proper critical value for your table, so it's probably safe to proceed. But if you have lots of cells near five, you may want to be cautious in interpreting the probability levels of the chi-square test.

Chi-square and the Association between Two Qualitative Variables

We seldom use chi-square to test hypotheses about the distribution of a single variable in a frequency distribution because better measures have been developed to deal with this problem. But the single variable problem serves as a simple introduction to the chi-square method just as the tests of hypotheses for a single mean were a good prelude to hypotheses about differences between two means. Like tests of hypotheses about means, chi-square is most interesting when applied to differences between groups. The difference in means test is useful only when we have two categories in the independent variable and the dependent variable is either continuous or binary. In contrast, the chi-square can be used in any two-variable table, no matter how many categories are in each variable.[5] Of course, with many categories in each variable, we have numerous cells in the table and thus run the risk of getting expected cells less than five, which violates one of the assumptions of the chi-square test.

Table 12.3   AIDS knowledge by gender

Knowledgeable about AIDS	Men	Women
Percent "Yes"	82.9	75.6
N	1,563	4,857

Data source: 2000 Uganda DHS data set, analyzed with Stata.

In the applications section of Chapter 10 we examined gender differences in AIDS knowledge and found there was reason to believe that there are gender differences in the population. We tested the hypothesis of no gender differences by invoking the Law of Large Numbers and thus a Z test. But we can also test the hypothesis by using chi-square for a table that cross-tabulates gender and the answer to AIDS knowledge question. The percentage table is Table 12.3.

This is an easy table to read. But for working with a chi-square test, it will be useful to have a fuller table (Table 12.4). The numbers in parentheses are the percentages based on the total for each column, while the numbers without parentheses are the frequencies in each category. Recall that the bottom row and the column furthest to the right are called the **marginals** of the table. The second column, labeled "Men" indicates that 323 out of 1,886 men incorrectly answered the AIDS question "No" and 1,563 correctly answered the question. That is, 82.9 percent of the men correctly answered the AIDS transmission question. In contrast, 75.6 pecent of the women did. Is it reasonable to believe that these data came from a population in which there is no difference between men and women in AIDS knowledge?

First we'll sketch the model used to test for no association. With two variables in a table, we need to add some notation to handle the general case. Table 12.5 shows the notation we need for the observations, using a 2 × 2 table (called a two-by-two table because there are two categories of two variables) as a simple case. The table deals only with the frequencies, so in the example you can ignore the numbers in parentheses.

Table 12.4   AIDS knowledge by gender with marginals included

Correctly answered AIDS transmission question	Men	Women	Total
No	323	1,567	1,890
	(17.1)	(24.4)	(22.7)
Yes	1,563	4,857	6,420
	(82.9)	(75.6)	(77.3)
Total	1,886	6,424	8,310
	(100.0)	(100.0)	(100.0)

Data source: 2000 Uganda DHS data set, analyzed with Stata.

Table 12.5   Notation for 2×2 chi-square table

	Column 1	*Column 2*	*Row marginal*
Row 1	O_{11}	O_{12}	$O_{1.}$
Row 2	O_{21}	O_{22}	$O_{2.}$
Column marginal	$O_{.1}$	$O_{.2}$	$O_{..} = N$

The two rows are labeled Row 1 and Row 2, and the two columns are labeled Column 1 and Column 2.[6] Then the observed value for the first row and the first column is O_{11} where the 11 in the subscript indicates the first row and first column. Then for the first row and second column, the observed frequency is O_{12}; for the second row and first column, the observed frequency is O_{21}; and for the second row and second column, the observed frequency is O_{22}. So for our example, O_{11} is 323, and O_{12} is 1,567. We also have to keep track of the marginal values. The marginal for the first row (this is the frequency for the first category of the row variable) is $O_{1.}$ and for the second row it is $O_{2.}$. The marginal for the first column is $O_{.1}$ and the marginal for the second column is $O_{.2}$. Finally, we can continue to call the sample size N, but in this notation it could be called $O_{..}$. Notice that there are some arithmetic relations in the table. The row margins must sum to N, and the column margins must sum to N. The cell values in a row or a column must sum to the marginal for that row or column. And the sum of all the cells must be equal to the sample size, N.

The deterministic part of the model included the sample size and a set of proportions determined by the null hypothesis. In the 2×2 table, we also reproduce the sample size. That is, we have no hypothesis about the sample size, and accept it as what it is. We also accept the row and column marginals as what they are, having no hypothesis about how the variables are split on the column variable or the row variable. That is, we want the expected table to reproduce the row and column marginals as well as the sample size. But what else is there to have a hypothesis about? We hypothesize that the association between the two variables is zero – that knowing what category someone falls into on one variable does not change our prediction of what category they fall into on the other variable. That means we need to include a term for the association between the row variable and the column variable. The full model then looks like this

$$M = [(N)^*(\text{Row effect})^*(\text{Column effect})^* (\text{Row column association effect})] + e \tag{12.3}$$

The null hypothesis is that there is no row column association, so the row column association effect is 1, which is no effect when we multiply. Then the null hypothesis is

$$O = [(N)^*(\text{Row effect})^*(\text{Column effect})] + e \tag{12.4}$$

The null hypothesis says that we can accurately predict the cell frequencies using information about the sample size, the row marginals and the column marginals. There is no need for information about the association between the two variables because the null hypothesis says there is no association. To the extent that the observed table differs from the expected table, it is only because of the random error e.

To calculate, we have:

$$M = (N)^*(\text{Row proportion})^*(\text{Column proportion}) \tag{12.5}$$

We don't use the error term in making the predictions – it's what makes the observed values differ from what the null hypothesis predicts, and thus is not part of the null hypothesis calculation. In fact, we estimate the values of e for each cell in the chi-square calculation.

Remember that the row proportion is the row marginal divided by the sample size. Then the expected cell formula is:

$$M = (N)^*(\text{Row marginal}/n)^*(\text{Column marginal}/N) \tag{12.6}$$

In the notation, the cell in the first row and the first column will have expected value:

$$M_{11} = (N)^*(O_{1.}/N)^*(O_{.1}/N) = (O_{1.}{}^*O_{.1})/N \tag{12.7}$$

Notice how the equation simplifies. We multiply by the sample size once, then working through the equation, we divide by it twice, so one multiplication and one division cancel each other out, leaving just one division. So the formula for an expected cell simplifies to "row marginal times column marginal divided by sample size." But the logic is still that we are predicting cell frequencies based on column proportions, row proportions and the sample size.

For example, for the cell for men who responded "No" we would have:

$$M_{11} = M_{\text{Men,no}} = (8,310)^*(1,890/8,310)^*(1,886/8,310) = 429 \tag{12.8}$$

That is, if the null hypothesis is true and in the population there is no relationship between gender and response to the AIDS knowledge question, we would expect to see about 429 cases in that cell.

The expected values for the other cells are:

$$M_{22} = (N)^*(O_{2.}/N)^*(O_{.2}/N) = (O_{2.}{}^*O_{.2})/N \tag{12.9}$$

$$M_{12} = (N)^*(O_{1.}/N)^*(O_{.2}/N) = (O_{1.}{}^*O_{.2})/N \tag{12.10}$$

$$M_{21} = (N)^*(O_{2.}/N)^*(O_{.1}/N) = (O_{2.}{}^*O_{.1})/N \tag{12.11}$$

The chi-square statistic for the sample is then calculated the same way as in the one-dimensional example (see Equation 12.2), and we repeat that equation below:

$$X^2 = \sum \frac{(O - M)^2}{M} \tag{12.12}$$

Here we are testing a null hypothesis of no association between the two variables, so the logic and the formula for calculating the expected values differs from the one-dimensional example, but the calculation of goodness of fit does not.

In a two-dimensional table, the number of degrees of freedom is:

$$df = (\text{number of rows} - 1)*(\text{number of columns} - 1) \tag{12.13}$$

For a 2 × 2 table this is:

$$df = (2 - 1)*(2 - 1) = 1, \text{ or one degree of freedom} \tag{12.14}$$

For a 3 × 3 table, this would be:

$$df = (3 - 1)*(3 - 1) = 4, \text{ or four degrees of freedom} \tag{12.15}$$

For a 4 × 5 table, it would be:

$$df = (4 - 1)*(5 - 1) = 12, \text{ or 12 degrees of freedom} \tag{12.16}$$

We will complete the example of AIDS knowledge and gender, using our standard steps in a hypothesis test.

Step 1 State the null (H_0) and alternative (H_A) hypotheses

H_0: There is no gender difference in AIDS knowledge.
H_A: There is a gender difference in AIDS knowledge. (12.17)

Step 2 Choose an α level
Let's let α be 0.05. As always, this is the probability of rejecting the null hypothesis when in fact it's true.

Step 3 Determine the appropriate statistical procedure to examine the hypothesis
The variables in the hypothesis (gender and AIDS knowledge) are categorical, so we will use the chi-square test.

Step 4 Find the critical value that corresponds to the α level and the appropriate number of degrees of freedom

Substep a Calculate the degrees of freedom.

$$df = (\text{\# of rows} - 1)*(\text{\# of columns} - 1) \tag{12.18}$$

or

$$(2 - 1)*(2 - 1) = (1)*(1) = 1 \tag{12.19}$$

Substep b Look up the critical value that corresponds to the a level and degrees of freedom. The critical value of chi-square for a 0.05 α level and 1 degree of freedom is 3.84.

If our observed chi-square value (the value calculated for our data in the table) is larger than this, we reject the null hypothesis and conclude that there is a gender difference in AIDS knowledge.

Step 5 Calculate the chi-square value for the sample

$$X^2 = \sum \frac{(O - M)^2}{M} \tag{12.20}$$

We will solve the above chi-square formula in a series of substeps:

Substep a Calculate the expected values, M. For each cell, the expected value, M, is the row proportion times the column proportion divided by the sample size. The expected values are shown in Table 12.6.

We've placed the original marginal from the real table in the bottom row and far right column. The cells should add to these values within rounding error. This is a good way to check your arithmetic – use the "(row marginal)*(column marginal)/sample size" to calculate each cell, then add them to get the marginals of the expected table. If the expected marginals don't come very close to the marginals in the sample table, there is an arithmetic error.

Table 12.6   Expected values for crosstabulation of gender and AIDS knowledge

Has AIDS knowledge	Men	Women	Original marginal
No	429	1,461	1,890
Yes	1,457	4,963	6,420
Original marginal	1,886	6,424	8,310

Table 12.7 Differences between observed and expected values for crosstabulation of gender and AIDS knowledge

Has AIDS knowledge	Men	Women
No	$323 - 429 = -106$	$1{,}567 - 1{,}461 = 106$
Yes	$1{,}563 - 1{,}457 = 106$	$4{,}857 - 4{,}963 = -106$

Substep b Calculate the differences between the observed and expected values (O − M)
The differences between the observed and expected values are shown in Table 12.7.

Substep c Square the differences between the observed and expected values (O − M)²
The squared differences are shown in Table 12.8.

Substep d Divide the squared differences by the expected value to get each cell's contribution to the chi-square value, $\dfrac{(O - M)^2}{M}$
The results are shown in Table 12.9.

Substep e Sum the quotients $\sum \dfrac{(O - M)^2}{M}$
Adding across all the cells ($26.19 + 7.69 + 7.71 + 2.26$), we get a sample (or observed) chi-square value of 43.85.

Step 6 Decide whether or not the null hypothesis can be rejected
The sample chi-square measuring how well the null hypothesis fits the data (43.85) is larger than the critical value of 3.84. Therefore we can reject the null hypothesis and conclude that we have strong evidence of a gender difference in AIDS knowledge. It is comforting that this is the same result that we had when we used a Z test on the gender differences.

Table 12.8 Squared differences between observed and expected values for crosstabulation of gender and AIDS knowledge

Has AIDS knowledge	Men	Women
No	11,236	11,236
Yes	11,236	11,236

Table 12.9   Division of the squared differences between observed and expected values by the expected value for crosstabulation of gender and AIDS knowledge

Has AIDS knowledge	Men	Women
No	26.19	7.69
Yes	7.71	2.26

Box 12.2   Steps in the Hypothesis Test for Association between Two Categorical Variables

- *Step 1* State the null and alternative hypotheses.
- *Step 2* Choose an α level.
- *Step 3* Determine the appropriate statistical method to examine the hypothesis.
- *Step 4* Find the critical value that corresponds to the α level and the appropriate number of degrees of freedom.
  - *Substep a* Calculate the degrees of freedom.

$$\text{df} = (\text{\# rows} - 1)^*(\text{\# columns} - 1) \tag{12.21}$$

  - *Substep b* Look up the critical value that corresponds to the α level and the degrees of freedom.
- *Step 5* Calculate the chi-square value for the sample.

$$X^2 = \sum \frac{(O - M)^2}{M} \tag{12.22}$$

  - *Substep a* Calculate the expected values, M

$$M = [(\text{row proportion})^*(\text{column proportion})]/\text{sample size} \tag{12.23}$$

  - *Substep b* Calculate the differences between the observed and expected values $(O - M)$.
  - *Substep c* Square the differences between the observed and expected values $(O - M)^2$.
  - *Substep d* Divide the squared differences by the expected value to get each cell's contribution to the chi-square value, $\frac{(O - M)^2}{M}$.
  - *Substep e* Sum the quotients $\sum \frac{(O - M)^2}{M}$
- *Step 6* Decide whether or not the null hypothesis can be rejected.

Table 12.10   Crosstabulation of 3 levels of animal concern by gender and age group

Animal concern	Young men	Middle-aged men	Older men	Young women	Middle-aged women	Older women
Percent with "low" score	64.5	67.5	70.9	54.6	59.1	63.0
Percent with "medium" score	15.1	14.0	12.7	18.6	15.7	15.3
Percent with "high" score	20.4	18.5	16.3	26.7	25.1	21.7
N	5,274	4,822	3,189	6,387	5,715	3,794

Data source: 2000 ISSP data set, analyzed with Stata.

When faced with so many cells, we may want to "collapse" a variable by combining categories. We can do this with the animal concern scale so that it has only 3 categories (high for scores of 4 and 5; medium for a score of 3; and low for scores of 1 and 2). We would further collapse age into young (ages 18–39), middle (ages 40–59) and older (ages 60+). Then the animal concern by gender by age table would be Table 12.10. Note that one of the percentage rows is unnecessary. If for a given gender/age group we know the percent with a low and a medium score, we can calculate the percent with a high score, for example.

If we treat the intersection of age and gender as a single variable (with values "young men," "middle-aged men," "older men," "young women," "middle-aged women," "older women") we have a 3 × 6 table that has 18 cells.

We could test the hypothesis that there is no overall relationship between gender/age group and animal concern by applying a chi-square test to the table. We would have (3 − 1)*(6 − 1) = 2*5 = 10 degrees of freedom. If we test at the 0.05 level, the critical value for chi-square is 18.37. We asked the computer to calculate the chi-square value for the table and the result was a chi-square value of 364.76. This means we can reject the null hypothesis of no effect and conclude that age group/gender has an effect on animal concern.

However, this seems a rather ineloquent way of approaching the problem as we haven't learned much about what the effects are. Looking at the percentage table and focusing on the "high" row, we find that young and middle-aged women have the highest percentages in that group, followed by older women, young men, and then middle aged and older men. If we look at "low" scores we have a pattern that is consistent with what we found when we looked at the "high" row. This row indicates those with the least concern. Young and middle aged women are least likely to be in this row, while middle-aged and older men are the most likely to have low levels of animal concern.

One way to move forward is to consider the 3 × 2 × 3 table as three 3 × 2 tables, one for each age group. This is just a different way of thinking about the problem

Table 12.11   Crosstabulation of 3 levels of animal concern by gender for young people

Animal concern	Men	Women	Total
Low	3,403	3,490	6,893
Medium	795	1,191	1,986
High	1,076	1,706	2,782
Total	5,274	6,387	11,661

Data source: 2000 ISSP data set, analyzed with Stata.

that makes it easier to handle. We now have three null hypotheses. One is that there is no association between gender and animal concern among young people. The second is that there is no association between gender and animal concern among middle-aged people. And the third is that there is no association between gender and animal concern among older people. We can do a chi-square test for each of the two tables separately. Each of the three tables has $(3 - 1)*(2 - 1) = 2$ degrees of freedom. If we test at the 0.05 level, the critical value of chi-square is 5.99.

Tables 12.11, 12.16, and 12.21 are the frequency tables (one table for each age group) corresponding to the percentage table above (Table 12.10).

Step 1 State the null and alternative hypotheses
   H_0: There is no gender difference in levels of animal concern among young people.
   H_A: There is a gender difference in levels of animal concern among young
         people.   (12.24)

Step 2 Choose an α level
Let's let α be 0.05. As always, this is the probability of rejecting the null hypothesis when in fact it's true.

Step 3 Determine the appropriate statistical procedure to examine the hypothesis
The variables in the hypothesis (gender and level of animal concern) are categorical, so we will use the chi-square test.

Step 4 Find the critical value that corresponds to the α level and the appropriate number of degrees of freedom

Substep a Calculate the degrees of freedom.

   df = (# of rows − 1)*(# of columns − 1)   (12.25)

or

   $(3 - 1)*(2 - 1) = (2)*(1) = 2$ (12.26)

Substep b Look up the critical value that corresponds to the α level and degrees of freedom. The critical value of chi-square for a 0.05 α level and 2 degrees of freedom is 5.99. If our observed chi-square value (the value calculated for our data in the table) is larger than this, we reject the null hypothesis and conclude that there is a gender difference in level of animal concern.

Step 5 Find the chi-square value for the sample

$$X^2 = \sum \frac{(O - M)^2}{M} \tag{12.27}$$

We will solve the above chi-square formula in a series of substeps.

Substep a Calculate the expected values, M. Let's begin by calculating the expected values under the null hypothesis for young people. The expected table will be formed by using the "row marginal times column marginal divided by sample size" formula. The expected values are shown in Table 12.12.

Substep b Calculate the differences between the observed and expected values $(O - M)$. The differences between the observed and expected values are shown in Table 12.13.

In addition to being a step in the calculation of chi-square for the sample, this table can be used to see where the discrepancies between the data and the null

Table 12.12   Expected values for crosstabulation of gender and level of animal concern for young people

Animal concern	Men	Women	Row marginal
Low	3,117.54	3,775.46	6,893
Medium	898.22	1,087.78	1,986
High	1,258.23	1,523.77	2,782
Column marginal	5,274	6,387	11,661

Table 12.13   Differences between observed and expected values for crosstabulation of gender and level of animal concern for young people

Animal concern	Men	Women
Low	3,403 − 3,117.54 = 285.46	3,490 − 3,775.46 = −285.46
Medium	795 − 898.22 = −103.22	1,191 − 1,087.78 = 103.22
High	1,076 − 1,258.23 = −182.23	1,706 − 1,523.77 = 182.23

Table 12.14   Squared differences between observed and expected values for crosstabulation of gender and level of animal concern for young people

Animal concern	Men	Women
Low	81,487.41	81,487.41
Medium	10,654.37	10,654.37
High	33,207.77	33,207.77

hypothesis appear. There are more men and fewer women in the low category than should be the case if there is no association between gender and animal concern among young people. Compared to what we expect from the null hypothesis, there are too few men in the high category and too many women there, and too few men and too many women in the medium category.

Substep c Square the differences between the observed and expected values $(O - M)^2$. The squared differences are shown in Table 12.14.

Substep d Divide the squared differences by the expected value to get each cell's contribution to the chi-square value

$$\frac{(O - M)^2}{M}$$

The results are shown in Table 12.15.

Some statisticians suggest that this table of "contributions to chi-square" is the best place to look for where the observed table differs from what is expected under the null hypothesis. Here we find the biggest discrepancies for men in the low category and in the high category. Looking back to the deviations table we know that this is because fewer men than expected have high scores and quite a few men had low scores.

Table 12.15   Division of the squared differences between observed and expected values by the expected value for crosstabulation of gender and level of animal concern for young people

Animal concern	Men	Women
Low	26.14	21.58
Medium	11.86	9.79
High	26.39	21.79

Table 12.16 Crosstabulation of 3 levels of animal concern by gender for middle-aged people

Animal concern	Men	Women	Row marginal
Low	3,255	3,380	6,635
Medium	673	900	1,573
High	894	1,435	2,329
Column marginal	4,822	5,715	10,537

Data source: 2000 ISSP data set, analyzed with Stata.

Substep e Sum the quotients.

$$\sum \frac{(O-M)^2}{M}$$

If we add the quotients we have the chi-square value for the sample, which is:

$$26.14 + 11.86 + 26.39 + 21.58 + 9.79 + 21.79 = 117.55. \tag{12.28}$$

Step 6 Decide whether or not the null hypothesis can be rejected
Since the sample chi-square value, 117.55, is larger than the critical value of 5.99, we reject the null hypothesis of no association and conclude that for young people there is a relationship between gender and level of animal concern.

So, we have rejected the null hypothesis of no association between gender and level of animal concern for young people. But what is the situation for middle-aged people? Table 12.16 is the frequency table for level of animal concern among middle-aged people.

Step 1 State the null and alternative hypotheses
    H_0: There is no gender difference in levels of animal concern among middle-aged people.
    H_A: There is a gender difference in levels of animal concern among middle-aged people.   (12.29)

Step 2 Choose an α level
Let's let α be 0.05. As always, this is the probability of rejecting the null hypothesis when in fact it's true.

Step 3 Determine the appropriate statistical procedure to examine the hypothesis
The variables in the hypothesis (gender and level of animal concern) are categorical, so we will use the chi-square test.

Step 4 Find the critical value that corresponds to the α level and the appropriate number of degrees of freedom

Substep a Calculate the degrees of freedom.

$$df = (\text{\# of rows} - 1)*(\text{\# of columns} - 1) \tag{12.30}$$

or

$$(3 - 1)*(2 - 1) = (2)*(1) = 2 \tag{12.31}$$

Substep b Look up the critical value that corresponds to the α level. The critical value of chi-square for a 0.05 α level and 2 degrees of freedom is 5.99. If our observed chi-square value (the value calculated for our data in the table) is larger than this, we reject the null hypothesis and conclude that there is a gender difference in level of animal concern among middle-aged people.

Step 5 Find the chi-square value for the sample

$$X^2 = \sum \frac{(O - M)^2}{M} \tag{12.32}$$

We will solve the above chi-square formula in a series of substeps.

Substep a Calculate the expected values, M. Table 12.17 shows the expected values under the null hypothesis. All the cells have expected values greater than 5.

Substep b Calculate the differences between the observed and expected values $(O - M)$. The differences between the observed and expected values are shown in Table 12.18.

Table 12.17 Expected values for crosstabulation of gender and level of animal concern for middle-aged people

Animal concern	Men	Women	Row marginal
Low	3,036.35	3,598.65	6,635
Medium	719.84	853.16	1,573
High	1,065.81	1,263.19	2,329
Column marginal	4,822	5,715	10,537

Table 12.18 Differences between observed and expected values for crosstabulation of gender and level of animal concern for middle-aged people

Animal concern	Men	Women
Low	3,255 − 3,036.35 = 218.65	3,380 − 3,598.65 = −218.65
Medium	673 − 719.84 = −46.84	900 − 853.16 = 46.84
High	894 − 1,065.81 = −171.81	1,435 − 1,263.19 = 171.81

Table 12.19   Squared differences between observed and expected values for crosstabulation of gender and level of animal concern for middle-aged people

Animal concern	Men	Women
Low	47,807.82	47,807.82
Medium	2,193.99	2,193.99
High	29,518.68	29,518.68

Substep c Square the differences between the observed and expected values $(O - M)^2$. The squared differences are shown in Table 12.19.

Substep d Divide the squared differences by the expected value to get each cell's contribution to the chi-square value

$$\frac{(O - M)^2}{M}$$

The results are shown in Table 12.20.

Substep e Sum the quotients.

$$\sum \frac{(O - M)^2}{M}$$

Adding the quotients gives us the chi-square value for the sample:

$$15.75 + 3.05 + 27.70 + 13.28 + 2.57 + 23.37 = 85.72 \tag{12.33}$$

Step 6 Decide whether or not the null hypothesis can be rejected
Since the sample chi-square value, 85.72, is larger than the critical value of 5.99, we can reject the null hypothesis and conclude that for middle-aged people, there is a relationship between gender and level of animal concern.

Table 12.20   Division of the squared differences between observed and expected values by the expected value for crosstabulation of gender and level of animal concern for middle-aged people

Animal concern	Men	Women
Low	15.75	13.28
Medium	3.05	2.57
High	27.70	23.37

Table 12.21   Crosstabulation of 3 levels of animal concern by gender for older people

Animal concern	Men	Women	Row marginal
Low	2,262	2,390	4,652
Medium	406	582	988
High	521	822	1,343
Column marginal	3,189	3,794	6,983

Data source: 2000 ISSP data set, analyzed wtih Stata.

We have rejected the null hypothesis of no association between gender and level of animal concern for young and middle-aged people. But what is the situation for older people? Table 12.21 is the frequency table for level of animal concern among older people.

Step 1 State the null and alternative hypotheses

H_0: There is no gender difference in levels of animal concern among older people.

H_A: There is a gender difference in levels of animal concern among older people.   (12.34)

Step 2 Choose an α level
Let's let α be 0.05. As always, this is the probability of rejecting the null hypothesis when in fact it's true.

Step 3 Determine the appropriate statistical procedure to examine the hypothesis
The variables in the hypothesis (gender and level of animal concern) are categorical, so we will use the chi-square test.

Step 4 Find the critical value that corresponds to the α level and the appropriate number of degrees of freedom

Substep a Calculate the degrees of freedom.

$$df = (\# \text{ of rows} - 1)^*(\# \text{ of columns} - 1) \tag{12.35}$$

or

$$(3 - 1)^*(2 - 1) = (2)^*(1) = 2 \tag{12.36}$$

Substep b Look up the critical value that corresponds to the α level. The critical value of chi-square for a 0.05 α level and 2 degrees of freedom is 5.99. If our observed chi-square value (the value calculated for our data in the table) is larger than this, we reject the null hypothesis and conclude that there is a gender difference in level of animal concern among older people.

Step 5 Find the chi-square value for the sample

$$X^2 = \sum \frac{(O - M)^2}{M} \tag{12.37}$$

We will solve the above chi-square formula in a series of substeps.

Substep a Calculate the expected values, M. Table 12.22 shows the expected values under the null hypothesis. All the cells have expected values greater than 5.

Substep b Calculate the differences between the observed and expected values (O − M). The differences between the observed and expected values are shown in Table 12.23.

Substep c Square the differences between the observed and expected values (O − M)². The squared differences are shown in Table 12.24.

Table 12.22 Expected values for crosstabulation of gender and level of animal concern for older people

Animal concern	Men	Women	Row marginal
Low	2,124.48	2,527.52	4,652
Medium	451.20	536.80	988
High	613.32	729.68	1,343
Column marginal	3,189	3,794	6,983

Table 12.23 Differences between observed and expected values for crosstabulation of gender and level of animal concern for older people

Animal concern	Men	Women
Low	2,262 − 2,124.48 = 137.52	2,390 − 2,527.52 = −137.52
Medium	406 − 451.20 = −45.20	582 − 536.80 = 45.20
High	521 − 613.32 = −92.32	822 − 729.68 = 92.32

Table 12.24   Squared differences between observed and expected values for crosstabulation of gender and level of animal concern for older people

Animal concern	Men	Women
Low	18,911.75	18,911.75
Medium	2,043.04	2,043.04
High	8,522.98	8,522.98

Table 12.25   Division of the squared differences between observed and expected values by the expected value for crosstabulation of gender and level of animal concern for older people

Animal concern	Men	Women
Low	8.90	7.48
Medium	4.53	3.81
High	13.90	11.68

Substep d Divide the squared differences by the expected value to get each cell's contribution to the chi-square value,

$$\frac{(O - M)^2}{M}$$

The results are shown in Table 12.25.

Substep e Sum the quotients

$$\sum \frac{(O - M)^2}{M}$$

Adding the quotients gives us the chi-square value for the sample:

$$8.90 + 4.53 + 13.90 + 7.48 + 3.81 + 11.68 = 50.30 \tag{12.38}$$

Step 6 Decide whether or not the null hypothesis can be rejected
Since the sample chi-square value, 50.3, is larger than the critical value of 5.99, we can reject the null hypothesis and conclude that for older people, there is a relationship between gender and level of animal concern.

Our application of the three chi-square tests leads us to conclude that there are gender differences among young, middle-aged, and older people. But there does not appear to be an interaction effect, meaning women in all three age groups were more likely to have a high level of animal concern and men were more likely to have a low level of animal concern.

What Have We Learned?

Here we have shifted to a different sort of model, one that predicts, not individual scores on a variable from another variable, but counts of people that have various combinations of values on two variables. The null hypothesis is that knowing someone's score on one variable doesn't help predict the score on the other variable. We see how well this null model fits the actual data, and then decide, using a sampling distribution, how likely it was that this sample came from a population in which the null hypothesis was true – there is no association between the two variables. If we reject the null hypothesis we are rejecting the notion of no association and concluding that in the population the two variables must be associated.

Applications

In the text we have examined the application of the chi-squared test to both the animal concern data and the AIDS knowledge data, so the applications will focus on homicide and environmental treaty participation and an additional animal concern example.

Example 1:   Homicide rates

Most of the homicide data we have on states is continuous and our sample size is only 50, so chi-square tests on contingency tables are not the best way to analyze the variation in homicide rates. To use chi-square for this example we will have to "collapse" continuous variables into categories, and we will frequently encounter small expected cells. But we can proceed to demonstrate how the procedure would work. We can convert homicide rate into a dichotomy by comparing states with homicides rates less than the median (which is 4.65) to those with homicide rates above the median (see Table 12A.1). This neatly divides our sample in half.

All of the former Confederate states are above the median homicide rate. Table 12A.2 shows the frequencies.

We have a cell with no cases (Confederate states below the median homicide rate). Some statistical procedures that can be used with contingency tables cannot be used when there are zero cells, but for chi-square the critical issue is whether or not expected values in the cells fall below five, a rule of thumb for the smallest cell that we can have and still use chi-square.

Using the six steps in a hypothesis test, we examine the association between high and low homicide rates and state membership in the Confederacy.

Table 12A.1 Percent former Confederate and non-Confederate states with homicides rates above the median

	Former Confederate state	Former non-Confederate state
Percent with homicide rates above median	100.00	35.9
N	11	39

Data source: US Census Bureau, 2003, analyzed with SPSS.

Table 12A.2 Observed values of high and low homicide rates for former Confederate and non-Confederate states

	Former Confederate state	Former non-Confederate state	Total
Low homicide rate	0	25	25
High homicide rate	11	14	25
Total	11	39	50

Data source: US Census Bureau, 2003, analyzed with SPSS.

Step 1 State the null and alternative hypotheses
The null hypothesis is that there is no association between state membership in the Confederacy and homicide rate. The alternative hypothesis is that there is an association between these variables.

Step 2 Choose an α level
Given the small sample, we might want to test the null hypothesis at an α of 0.10 rather than the more common 0.05. This gives a higher chance of a Type I error than the 0.05 level but lowers our chances of a Type II error.

Step 3 Determine the appropriate statistical method to examine the hypothesis
Both of the variables are categorical, so we will use the chi-square test.

Step 4 Find the critical value that corresponds to the α level and the appropriate number of degrees of freedom

Substep a Calculate the degrees of freedom.

$$\text{df} = (\text{\# rows} - 1)*(\text{\# columns} - 1) \tag{12.39}$$

We have a 2×2 table so the number of degrees of freedom is $(2 - 1)*(2 - 1) = 1$.

Substep b Look up the critical value that corresponds to the α level and the degrees of freedom. The critical value for chi-square with one degree of freedom and an α of 0.1 is 2.70.

Step 5 Calculate the chi-square value for the sample

$$X^2 = \sum \frac{(O - M)^2}{M} \tag{12.40}$$

Substep a Calculate the expected values, M.

$$M = [(\text{row proportion})^*(\text{column proportion})]/\text{sample size} \tag{12.41}$$

The expected values are shown in Table 12A.3.

Two cells have expected values just above 5. We can proceed with a test of the hypothesis based on the chi-square test but should note that the p values may be a bit different than what the table says. That is, if we reject or don't reject at what we believe is the 0.1 level, we should be aware we might actually be making a decision with a slightly different chance of a Type I error.

Substep b Calculate the differences between the observed and expected values (O − M). The results are shown in Table 12A.4.

Substep c Square the differences between the observed and expected values (O − M)². The results are shown in Table 12A.5.

Table 12A.3   Expected values of high and low homicide rates for former Confederate and non-Confederate states

	Former Confederate State	Former non-Confederate state	Total
Low homicide rate	(25*11)/50 = 5.50	(25*39)/50 = 19.50	25
High homicide rate	(25*11)/50 = 5.50	(25*39)/50 = 19.50	25
Total	11	39	50

Data source: US Census Bureau, 2003, analyzed with SPSS.

Table 12A.4   Differences between observed and expected values for crosstabulation of state membership in Confederacy and high and low homicide rates

	Former Confederate State	Former non-Confederate state
Low homicide rate	(0 − 5.50) = −5.50	(25 − 19.50) = 5.50
High homicide rate	(11 − 5.50) = 5.50	(14 − 19.50) = −5.50

Data source: US Census Bureau, 2003, analyzed with SPSS.

Table 12A.5 Squared differences between observed and expected values for crosstabulation of state membership in Confederacy and high and low homicide rates

	Former Confederate State	Former non-Confederate state
Low homicide rate	$(-5.50)^2 = 30.25$	$(5.50)^2 = 30.25$
High homicide rate	$(5.50)^2 = 30.25$	$(-5.50)^2 = 30.25$

Data source: US Census Bureau, 2003, analyzed with SPSS.

Substep d Divide the squared differences by the expected value to get each cell's contribution to the chi-square value,

$$\frac{(O - M)^2}{M}$$

The results are shown in Table 12A.6.

Substep e Sum the quotients

$$\sum \frac{(O - M)^2}{M}$$

Adding the quotients $(5.50 + 5.50 + 1.55 + 1.55)$ gives us a sample chi-square of 14.10.

Step 6 Decide whether or not the null hypothesis can be rejected
The calculated chi-square for the sample, 14.10, is greater than the critical value of 2.70, so we reject the null hypothesis. We conclude that there is an association between state membership in the Confederacy (in this case South versus non-South) and high and low homicide rates. Of course, whenever we work with state data, we have to be cautious in interpreting the results because we don't have a sample or an experiment. Again, we can think about these results as if we had a sample of hypothetical states.

Table 12A.6 Division of the squared differences between observed and expected values by the expected value for crosstabulation of state membership in Confederacy and high and low homicide rates

	Former Confederate State	Not former Confederate state
Low homicide rate	$30.25/5.50 = 5.50$	$30.25/19.50 = 1.55$
High homicide rate	$30.25/5.50 = 5.50$	$30.25/19.50 = 1.55$

Data source: US Census Bureau, 2003, analyzed with SPSS.

Example 2:   Animal concern

In the chapter we examined whether animal concern varied by gender and three age groups. Now let's ask if age alone has an effect on animal concern. For the animal concern application, we again have a continuous dependent variable. We will again split the dependent variable at the median to create a dichotomy. This is not how we would usually analyze these data for research, since we are throwing away a great deal of information by treating everyone who scored 2 and below and equal as everyone who scored 3 and above as equal. But by dichotomizing we can show how we would use the chi-square procedure, so we will proceed with the dichotomy.

We can ask if age has an effect on animal concern, using the dichotomized version of animal concern. We will again use the strategy of looking at two two-way tables.

We can see in Table 12A.7 that the youngest age group is most likely to express high concern. The 30–39 and 40–49 year olds are next most likely to have high concern, followed by the 50–59 year olds and lastly the 70 and over and 60–69 year old age groups.[7] This might mean that animal concern is highest among youth and drops off as people age. But remember that if we have a single survey taken at one point in time, age is also a cohort. It may be that newer cohorts are developing a high level of concern when they are young and this will persist over time as they age. We would only be able to tell this if we also had data on animal concern from a more recent survey.

Before we spend much thought on interpreting these results, we should see if there is reason to believe that there are age differences in the population or if these differences might be just random error from a population in which there are no age differences. We can apply the chi-square test to assess this possibility.

To calculate the sample chi-square, we need the frequency table. Table 12A.8 is the frequency table.

Table 12A.7   High animal concern levels by age group

	18–29	30–39	40–49	50–59	60–69	70 and over	Overall
Proportion with high animal concern	0.432	0.387	0.383	0.355	0.331	0.337	0.377
N	5,678	5,984	5,716	4,822	3,855	3,138	29,193

Data source: 2000 ISSP data set, analyzed with SPSS.

Table 12A.8   Crosstabulation of levels of animal concern and age group

	18–29	30–39	40–49	50–59	60–69	70 and over	Overall
Low concern	3,225	3,668	3,525	3,110	2,580	2,079	18,187
High concern	2,453	2,316	2,191	1,712	1,275	1,059	11,006
N	5,678	5,984	5,716	4,822	3,855	3,138	29,193

Data source: 2000 ISSP data set, analyzed with SPSS.

Step 1 State the null and alternative hypotheses
The null hypothesis is that the data are from a population in which there are no age differences in animal concern, and the alternative hypothesis is that there are age differences.

Step 2 Choose an α level
We will test the hypothesis using a 5 percent chance of a Type I error.

Step 3 Determine the appropriate statistical method to examine the hypothesis
We will use the chi-square test.

Step 4 Find the critical value that corresponds to the α level and the appropriate number of degrees of freedom

Substep a Calculate the degrees of freedom.

$$\text{df} = (\# \text{ rows} - 1)*(\# \text{ columns} - 1) \tag{12.42}$$

We have $(2 - 1)*(6 - 1) = 5$ degrees of freedom

Substep b Look up the critical value that corresponds to the α level and the degrees of freedom. Checking the chi-square table we find the critical value for 5 degrees of freedom and the 0.05 level is 11.07. So if the sample chi-square is larger than this, we will reject the null hypothesis; otherwise we will let it stand.

Step 5 Calculate the chi-square value for the sample

$$X^2 = \sum \frac{(O - M)^2}{M} \tag{12.43}$$

Substep a Calculate the expected values, M.

$$M = [(\text{row proportion})*(\text{column proportion})]/\text{sample size} \tag{12.44}$$

We need to calculate the expected values for each cell in the table. The expected value for a cell is the row marginal times the column marginal divided by the sample size. Remember that this is an algebraic simplification from the column percentage times the row percentage times the sample size.

The expected table is Table 12A.9. We can see that every expected cell is well over 5.

Substep b Calculate the differences between the observed and expected values (O − M). The results are shown in Table 12A.10.

Substep c Square the differences between the observed and expected values, (O − M)². The results are shown in Table 12A.11.

Table 12A.9   Expected values for crosstabulation of levels of animal concern and age group

	18–29	30–39	40–49	50–59	60–69	70 and over	Overall
Low concern	3,537.35	3,727.98	3,561.02	3,004.07	2,401.63	1,954.95	18,187
High concern	2,140.65	2,256.02	2,154.98	1,817.93	1,453.37	1,183.05	11,006
N	5,678	5,984	5,716	4,822	3,855	3,138	29,193

Data source: 2000 ISSP data set, analyzed with SPSS.

Table 12A.10   Differences between observed and expected values for crosstabulation of levels of animal concern and age group

	18–29	30–39	40–49	50–59	60–69	70 and over	Overall
Low concern	3,225 – 3,537.35 = –312.35	3,668 – 3,727.98 = –59.98	3,525 – 3,561.02 = –36.02	3,110 – 3,004.07 = 105.93	2,580 – 2,401.63 = 172.37	2,079 – 1,954.95 = 124.05	18,187
High concern	2,453 – 2,140.65 = 312.35	2,316 – 2,256.02 = 59.98	2,191 – 2,154.98 = 36.02	1,712 – 1,817.93 = –105.93	1,275 – 1,453.37 = –172.37	1,059 – 1,183.05 = –124.05	11,006
N	5,678	5,984	5,716	4,822	3,855	3,138	29,193

Data source: 2000 ISSP data set, analyzed with SPSS.

Table 12A.11   Squared differences between observed and expected values for crosstabulation of levels of animal concern and age group

	18–29	30–39	40–49	50–59	60–69	70 and over
Low concern	$(-312.35)^2$ = 97,562.52	$(-59.98)^2$ = 3,597.60	$(-36.02)^2$ = 1,297.44	$(105.93)^2$ = 11,221.16	$(172.37)^2$ = 29,711.42	$(124.05)^2$ = 15,388.40
High concern	$(312.35)^2$ = 97,562.52	$(59.98)^2$ = 3,597.60	$(36.02)^2$ = 1,297.44	$(-105.93)^2$ = 11,221.16	$(-172.37)^2$ = 29,711.42	$(-124.05)^2$ = 15,388.40

Data source: 2000 ISSP data set, analyzed with SPSS.

Substep d Divide the squared differences by the expected value to get each cell's contribution to the chi-square value

$$\frac{(O - M)^2}{M}$$

The results are shown in Table 12A.12.

Substep e Sum the quotients

$$\sum \frac{(O - M)^2}{M}$$

27.58 + 0.97 + 0.36 + 3.74 + 12.37 + 7.87 + 45.58 + 1.59 + 0.60
+ 6.17 + 20.44 + 13.01 = 140.28

(12.45)

Table 12A.12   Division of squared differences between observed and expected values by expected value for crosstabulation of levels of animal concern and age group

	18–29	30–39	40–49	50–59	60–69	70 and over
Low concern	27.58	0.97	0.36	3.74	12.37	7.87
High concern	45.58	1.59	0.60	6.17	20.44	13.01

Data source: 2000 ISSP data set, analyzed with SPSS.

Step 6 Decide whether or not the null hypothesis can be rejected

Since the calculated chi-square for the sample, 140.28, is larger than the critical value of 11.07, we reject the null hypothesis of no difference across age groups in the population. Thus we conclude that there are age differences in animal concern. It appears that younger people are more concerned than older ones.

Example 3:   Environmental treaty participation

To examine environmental treaty participation we will again have to collapse continuous variables into discrete categories. We will do this by categorizing countries as high on treaty participation if they are at or above the median (6) and low on treaty participation if they below the median. We will again consider two factors that might influence treaty participation. One is number of NGOs in a nation, with the expectation that more NGOs will exert greater pressure on governments to participate in treaties. We will divide countries into low and high NGO presence at 494 – the median. In addition, we have hypothesized that the extent citizens have freedom of speech and a role in selecting government officials may relate to treaty participation. Governments that have little accountability and give their citizens few liberties may be more likely to ignore pressures to participate in environmental treaties. The degree of voice and accountability will be characterized as low and high (split at the median score on the index of −0.07). Table 12A.13 shows the percentage of nations that have high levels of treaty participation by NGO presence and by voice and accountability.

Table 12A.13   Percentage of nations with high levels of treaty participation by NGO presence and voice and accountability

	Few NGOs Less voice and accountability	Few NGOs More voice and accountability	Many NGOs Less voice and accountability	Many NGOs More voice and accountability
Percent higher treaty participation	11.5	36.0	77.8	91.4
N	52	25	27	58

Data source: Kaufmann et al., 2000, Roberts et al., 2004, analyzed with SPSS.

Table 12A.14 Crosstabulation of levels of environmental treaty participation and voice and accountability for nations with few NGOs

	Low voice and accountability	High voice and accountability	Total
Low treaty participation	46	16	62
High treaty participation	6	9	15
Total	52	25	77

Data source: Kaufmann et al., 2002, Roberts et al., 2004, analyzed with SPSS.

Table 12A.15 Crosstabulation of levels of environmental treaty participation and voice and accountability for nations with many NGOs

	Low voice and accountability	High voice and accountability	Total
Low treaty participation	6	5	11
High treaty participation	21	53	74
Total	27	58	85

Data source: Kaufmann et al., 2002, Roberts et al., 2004, analyzed with SPSS.

Countries with many NGOs have high levels of treaty participation regardless of the level of voice and accountability. But that pattern might be the result of random error. To see if it is plausible to believe that the pattern is greater than would be expected from chance, we can test the hypothesis that there is no relationship between NGO presence and treaty participation for countries with little and more voice and accountability. The data are presented in Tables 12A.14 and 12A.15.

Step 1 State the null and alternative hypotheses
The null hypothesis is that there is no relationship between NGO presence and environmental treaty participation for the little voice and accountability and the higher voice and accountability groups of nations. The alternative hypothesis is that there is a relationship.

Step 2 Choose an α level
We will test at the 0.05 alpha level.

Step 3 Determine the appropriate statistical method to examine the hypothesis
For a test of association between categorical variables, we use the chi-square test.

Step 4 Find the critical value that corresponds to the α level and the appropriate number of degrees of freedom

Substep a Calculate the degrees of freedom.

$$df = (\# \text{ rows} - 1)^*(\# \text{ columns} - 1) \tag{12.46}$$

For a 2×2 table, $df = (2 - 1)^*(2 - 1) = 1$

Substep b Look up the critical value that corresponds to the α level and the degrees of freedom. The critical value of chi-square at an α of .05 with one degree of freedom is 3.84.

Step 5 Calculate the chi-square value for the sample

$$X^2 = \sum \frac{(O - M)^2}{M} \tag{12.47}$$

Substep a Calculate the expected values, M.

$$M = [(\text{row proportion})^*(\text{column proportion})]/\text{sample size} \tag{12.48}$$

The results are shown in Table 12A.16 and 12A.17.

Table 12A.16   Expected values for crosstabulation of levels of environmental treaty participation and voice and accountability for nations with few NGOs

	Low voice and accountability	High voice and accountability	Total
Low treaty participation	(62*52)/77 = 41.87	(62*25)/77 = 20.13	62
High treaty participation	(15*52)/77 = 10.13	(15*25)/77 = 4.87	15
Total	52	25	77

Data source: Kaufmann et al., 2002, Roberts et al., 2004, analyzed with SPSS.

Table 12A.17   Expected values for crosstabulation of levels of environmental treaty participation and voice and accountability for nations with many NGOs

	Low voice and accountability	High voice and accountability	Total
Low treaty participation	(11*27)/85 = 3.49	(11*58)/85 = 7.51	11
High treaty participation	(74*27)/85 = 23.51	(74*58)/85 = 50.49	74
Total	27	58	85

Data source: Kaufmann et al., 2002, Roberts et al., 2004, analyzed with SPSS.

Substep b Calculate the differences between the observed and expected values (O − M). The results are shown in Table 12A.18 and 12A.19.

Substep c Square the differences between the observed and expected values (O − M)². The results are shown in Table 12A.20 and 12A.21.

Substep d Divide the squared differences by the expected value to get each cell's contribution to the chi-square value,

$$\frac{(O - M)^2}{M}$$

Table 12A.18   Differences between observed and expected values for crosstabulation of levels of environmental treaty participation and voice and accountability for nations with few NGOs

	Low voice and accountability	*High voice and accountability*
Low treaty participation	46 − 41.87 = 4.13	16 − 20.13 = −4.13
High treaty participation	6 − 10.13 = −4.13	9 − 4.87 = 4.13

Data source: Kaufmann et al., 2002, Roberts et al., 2004, analyzed with SPSS.

Table 12A.19   Differences between observed and expected values for crosstabulation of levels of environmental treaty participation and voice and accountability for nations with many NGOs

	Low voice and accountability	*High voice and accountability*
Low treaty participation	6 − 3.49 = 2.51	5 − 7.51 = −2.51
High treaty participation	21 − 23.51 = −2.51	53 − 50.49 = 2.51

Data source: Kaufmann et al., 2002, Roberts et al., 2004, analyzed with SPSS.

Table 12A.20   Squared differences between observed and expected values for crosstabulation of levels of environmental treaty participation and voice and accountability for nations with few NGOs

	Low voice and accountability	*High voice and accountability*
Low treaty participation	$(4.13)^2 = 17.06$	$(-4.13)^2 = 17.06$
High treaty participation	$(-4.13)^2 = 17.06$	$(4.13)^2 = 17.06$

Data source: Kaufmann et al., 2002, Roberts et al., 2004, analyzed with SPSS.

Table 12A.21 Squared differences between observed and expected values for crosstabulation of levels of environmental treaty participation and voice and accountability for nations with many NGOs

	Low voice and accountability	High voice and accountability
Low treaty participation	$(2.51)^2 = 6.30$	$(-2.51)^2 = 6.30$
High treaty participation	$(-2.51)^2 = 6.30$	$(2.51)^2 = 6.30$

Data source: Kaufmann et al., 2002, Roberts et al., 2004, analyzed with SPSS.

The results are shown in Table 12A.22 and 12A.23.

Substep e Sum the quotients.

$$\sum \frac{(O - M)^2}{M}$$

The chi-square for nations with few NGOs is:
$(0.41 + 0.85 + 1.68 + 3.50) = 6.44$ (12.49)

Table 12A.22 Division of squared differences between observed and expected values by the expected value for crosstabulation of levels of environmental treaty participation and voice and accountability for nations with few NGOs

	Low voice and accountability	High voice and accountability
Low treaty participation	$17.06/41.87 = 0.41$	$17.06/20.13 = 0.85$
High treaty participation	$17.06/10.13 = 1.68$	$17.06/ 4.87 = 3.50$

Data source: Kaufmann et al., 2002, Roberts et al., 2004, analyzed with SPSS.

Table 12A.23 Division of squared differences between observed and expected values by expected value for crosstabulation of levels of environmental treaty participation and voice and accountability for nations with many NGOs

	Low voice and accountability	High voice and accountability
Low treaty participation	$6.30/ 3.49 = 1.81$	$6.30/ 7.51 = 0.84$
High treaty participation	$6.30/23.51 = 0.27$	$6.30/50.49 = 0.12$

Data source: Kaufmann et al., 2002, Roberts et al., 2004, analyzed with SPSS.

The chi-square for nations with many NGOs is:
$$(1.81 + 0.84 + 0.27 + 0.12) = 3.04 \tag{12.50}$$

Step 6 Decide whether or not the null hypothesis can be rejected
The chi-square value for the data on nations with few NGOs is 6.44, and the critical value is 3.84, so we can reject the null hypothesis and conclude that among nations with few NGOs, voice and accountability influences treaty participation. For the nations with more NGOs, the sample chi-square is 3.04 (the critical value is again 3.84), so we cannot reject the null hypothesis. We conclude that voice and accountability does not influence treaty participation among nations with many NGOs. So it appears that the relationship between voice and accountability and environmental treaty participation is conditioned by number of NGOs in the nation. In other words, there appears to be an interaction between NGOs and voice and accountability on environmental treaty participation. Voice and accountability seem to matter only when there are few NGOs.

Exercises

1. A researcher is interested in whether there are gender differences in beliefs about whether it is a good idea for a couple to live together before marriage. Data on cohabitation for 45,228 respondents from the 2000 ISSP dataset are presented, by gender, in Table 12E.1.

a) State the null hypothesis.

b) Calculate the number of degrees of freedom.
c) Construct the table of expected counts.
d) Calculate the Chi-square value, and show your work.
e) Using a p value of 0.05, what is the critical value? Can you reject the null hypothesis? What do you conclude?

Table 12E.1   Crosstabulation by gender of view of couples living together without marrying

				Sex		Total
				1 Male	2 Female	
Couple living together without marriage	1	Strongly agree	Count	4,806	6,511	11,317
			% within sex	24.0	25.8	25.0
	2	Agree	Count	8,195	9,640	17,8354
			% within sex	40.9	38.3	39.4
	3	Neither agree nor disagree	Count	2,485	3,125	5,610
			% within sex	12.4	12.4	12.4
	4	Disagree	Count	3,110	4,036	7,146
			% within sex	15.5	16.0	15.8
	5	Strongly disagree	Count	1,438	1,882	3,320
			% within sex	7.2	7.5	7.3
Total			Count	20,034	25,194	45,228
			% within sex	100.0	100.0	100.0

Data source: 2000 ISSP data set, analyzed with SPSS.

2. The relationship between gender and beliefs about whether "what women really want is a home and children" is shown in Table 12E.2. Interpret the cross-tabulation and Chi-square, and summarize your findings. Include in your response whether a relationship exists between the variables, and describe the relationship you find (if any).

Table 12E.2 Crosstabulation by gender of view of what women really want

				Sex		Total
				1 Male	2 Female	1 Male
What women	1	Strongly agree	Count	2,608	3,577	6,185
really want is			% within sex	13.6	14.3	14.0
home and kids	2	Agree	Count	6,581	7,589	14,170
			% within sex	34.4	30.4	32.2
	3	Neither agree nor disagree	Count	4,134	4,783	8,917
			% within sex	21.6	19.2	20.2
	4	Disagree	Count	4,080	5,754	9,834
			% within sex	21.3	23.1	22.3
	5	Strongly disagree	Count	1,708	3,234	4,942
			% within sex	8.9	13.0	11.2
Total			Count	19,111	24,937	44,048
			% within sex	100.0	100.0	100.0

Chi-Square Test

Pearson Chi-square value	df	p value
260.90	4	<0.001

0 cells (.0%) have expected count less than 5. The minimum expected count is 2144.17
Data source: 2000 ISSP data set, analyzed with SPSS.

3. A researcher is interested in the relationship between marital status (married versus unmarried) and the belief that a "bad marriage is better than no marriage." Interpret the cross-tabulation and Chi-square in Table 12E.3, and summarize your findings. Include in your response whether a relationship exists between marital status and belief about bad marriages, and describe the relationship you find (if any).

Table 12E.3 Crosstabulation by marital status of view that a bad marriage is better than no marriage

				Married?		Total
				1 Unmarried	2 Married	
Bad marriage better than no marriage	1	Strongly agree	Count	561	741	1,302
			% within married	3.0	2.9	2.9
	2	Agree	Count	1,128	1,900	3,028
			% within married	6.1	7.3	6.8
	3	Neither agree nor disagree	Count	1,169	2,042	3,211
			% within married	6.3	7.9	7.2
	4	Disagree	Count	6,421	10,438	16,859
			% within married	34.6	40.4	37.9
	5	Strongly disagree	Count	9,305	10,734	20,039
			% within married	50.1	41.5	45.1
Total			Count	18,584	25,855	44,439
			% within married	100.0	100.0	100.0

Chi-Square Tests

Pearson Chi-square value	df	p value
337.47	4	<0.001

0 cells (0%) have expected count less than 5. The minimum expected count is 544.48
Data source: 2000 ISSP data set, analyzed with SPSS.

References

International Social Survey Programme (ISSP). 2000. 2000 Environment II data set. www.issp.org. Catalog no. ZA 3440. Cologne, Germany: GESIS-ZA Central Archive for Empirical Research.

Kaufmann, D., Kraay, A., and Zoido-Lobaton, P. (Jan. 2002). Governance Matters II: Updated Indicators for 2000/01. Policy Research Working Paper no. 2772. The World Bank Research Development Group and World Bank Institute; Governance, Regulation and Finance Division (http://hdr.undp.org/reports/global/2002/en/).

Roberts, J. T., Parks, B. C., and Vasquez, A. A. 2004. Who ratifies environmental treaties and why? Institutionalism, structuralism and participation of 192 nations in 22 treaties. *Global Environmental Politics* 4(3), 22–64.

Uganda Demographic and Health Surveys. 2001. Calverton, Maryland: UBOS and ORC Macro. (http://www.measuredhs.com/pubs/pdf/FR128/00FrontMatter.pdf).

US Census Bureau 2000. Table 33. Urban and rural population, and by state: 1990 and 2000. ("http://www.census.gov/prod/cen2000/index.html"\t"_blank"www.census.gov/prod/cen2000/index.html).

US Census Bureau 2002. Historical poverty tables: Table 21. Number of poor and poverty rate, by state: 1980 to 2006. Year 2002 ("http://www.census.gov/hhes/www/poverty/histpov/"\t"_blank"www.census.gov/hhes/www/poverty/histpov/hstpov21.html).

US Census Bureau 2003. Table 295. Crime rates by state, 2002 and 2003, and by type, 2003 ("http://www.census.gov/prod/2005pubs/06stata"www.census.gov/prod/2005pubs/06statab/law.pdf).

CHAPTER 13
BIVARIATE REGRESSION AND CORRELATION

Outline

In Chapter 5 we examined conditional distributions – how the values of one variable relate to the values of another variable, and we used graphics to examine relationships between two variables. Graphics are powerful tools, but we don't have methods that allow us to make inferences from sample graphics to the graphics we would obtain if we had population data, or otherwise deal with stochastic error in a systematic way. Regression analysis is a way of summarizing the conditional distribution of the dependent variable given values of the independent variable. Just as the mean or median is a method of summarizing the typical value of a set of data, regression has the same strengths and limits, and it is not the whole story. There can be more to the conditional distribution than can be captured by a regression analysis, just as there is more to a single variable of data than is captured by the mean. But regression can be used to make inferences and thus deal with the random error present in almost all data.

As long as we are attentive to its limitations, regression analysis is one of the most powerful tools we have for understanding the relationships among variables and assessing the plausibility of theories. We will see that most of the techniques

Feature 13.1   The Invention of Regression

The basic ideas of regression and correlation were invented by Sir Francis Galton.[1] Galton was very interested in heredity. In 1875 he began an analysis of the relationship between the weight of sweet pea seeds in parents and in their offspring. (He chose sweet peas because they were easy to grow and because they self-fertilize so each offspring has only one parent.) In graphing the weight of parent seeds versus the weight of offspring peas (a scatterplot) he began to develop the idea of regression and correlation. In fact, the term regression comes from his thinking about the degree to which the offspring "regressed" (moved towards) the characteristics of their parent. However, Galton worked mostly with statistics in the median family – the median and the interquartile range. As we have mentioned, even though the median family has a lot of intuitive appeal, it is hard to use mathematically. It was not until 1896 when Karl Pearson, who had worked as Galton's assistant, developed the regression approach we use today, which is based on the mean family (Pearson, 1896). Today we call this the correlation coefficient, and we almost always use the term, Pearson correlation coefficient, acknowledging Pearson's work and ignoring Galton's.

It is typical in most statistics books to discuss correlation before discussing regression. In fact, regression was developed first. We point this out to emphasize that correlation is really a special way of looking at regression. In both regression and correlation we are looking at a way to summarize a scatterplot of variables. In both we summarize the relationship with a straight line that fits all the data points as closely as possible. The only real difference is that when we do a regression, we use the original units of measurement of the variables, while when we do a correlation we first standardize the variables by subtracting the mean and dividing by the standard deviation, creating a Z score. When we run a regression with Z scores we get a "Pearson" correlation coefficient.

for statistical inference we have learned so far – t tests and Z tests on differences in means and F tests for analysis of variance – can be thought of as special applications of regression. In addition regression can handle many kinds of data analysis problems that we haven't covered yet, and many that are beyond the scope of this book. Regression is a tool that allows us to understand numerous types of statistical problems, using a similar statistical framework for all of them.

In this chapter we discuss bivariate regression and correlation. **Bivariate regression** is the simplest form of regression, and correlation is just another way of looking at a bivariate regression. Bivariate regression and correlation involve only two variables. One is the dependent variable whose variability we want to understand. The other is the independent variable whose variability we think may account for some of the differences across observations in the dependent variable. Bivariate regression and correlation are sometimes useful themselves, but they are also a simple beginning for understanding multiple regression, the topic of the next chapter. In **multiple regression**, we have a number of independent variables that are being used to predict the dependent variable and thus explain its variability.

Straight Lines

The fundamental concept in regression is the straight line. As you will recall, a straight line relating an independent variable to a dependent variable can be described by a simple equation:

$$Y = A + BX \tag{13.1}$$

where Y is the dependent variable, X is the independent variable, A is called the **intercept** or **constant**, and B the **slope**.[2] This is a model of the sort we've been discussing since Chapter 1, but for the moment we will leave out the random error to focus on the prediction line that shows how we think X is related to Y. Remember there are a number of values for X and Y, one for each different observation or data point, but A and B are the same for all data points.

In a straight line, the value of Y corresponding to a particular value of X can be found by multiplying B times X and adding A to the product. B tells the amount of change in Y that accompanies each change in X. A tells us the value of Y when X is equal to zero. If X increases by 10, then Y increases by 10B. Note that A is measured in the same units as Y, while B is in the units of Y divided by the units of X, so that BX is (units of Y divided by units of X) times (units of X), resulting in units of Y.[3]

Two examples, one hypothetical and one real, may help us better understand the straight line. For the hypothetical example, imagine that a dairy cooperative has agreed to donate milk to low-income families, and they would like to determine how much milk should be given per family. It is reasonable to assume that milk consumption is a function of the number of children in a family. (It might be more

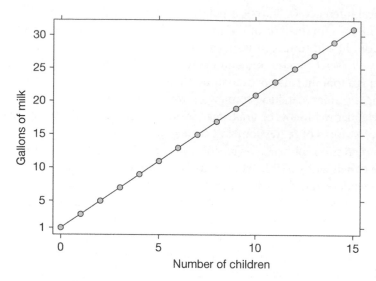

Figure 13.1   Straight line showing the relationship between number of children and gallons of milk needed

realistic to take the age of the children into account, but for explication we will keep it simple. We'll also assume that there are the same number of adults in all low income families, even though that is not realistic.) A straight line could describe the amount of milk that should be provided.

Figure 13.1 is a graph of such a line. The units of Y are gallons of milk consumed per week and X is the number of children. B is the gallons of milk per child and A is the gallons of milk one would give to a family with no children. If a family has no children, then X = 0 and B*X = 0 so the line suggests that milk consumption should be equal to A. Perhaps we believe that each child consumes 2 gallons of milk per week, and that milk consumption in a family with no children is 1 gallon per week. Then for a family with 1 child, the line predicts milk consumption to be A + (B*1) which is 1 + (2*1) = 3 gallons, for four children 1 + (2*4) = 9 gallons, and so on. Table 13.1 shows this relationship.

Table 13.1   Gallons of milk by number of children in a household

Number of children (X)	Gallons of milk (Y)
0	1
1	3
2	5
3	7
4	9
5	11
6	13

This hypothetical example has no error term. The equation has no relationship to empirical data. But to develop values for A and B, we could use information from a sample of families on how much milk they consume and how many children are in the family. Then we would use the methods of this chapter to find the A and B that will describe the milk consumption of those families as a function of number of children. If we did this with a group of families whose milk consumption is not constrained by low incomes, then we could use the results of the analysis to calculate how much milk to give the families that need assistance.

As another example, let's consider the relationship between state homicide rates and the percent of the population in poverty. Figure 13.2 displays the now familiar scatterplot of the data with a straight line imposed as a summary. We will discuss how the values of A and B were determined later. For these data the straight line we will use to summarize the relationship between X and Y has A = 0.59 homicides per 1,000 and B = 0.36 homicides per 1,000 per percent change in poverty. This means that for every 1 point increase in the poverty rate (which is a 1 percent increase in the percentage of families in poverty) the homicide rate is expected to increase by just over a one-third of a point. We would expect that a state with no one in poverty would have almost a homicide rate of about 0.6. Of course, we do not have any states with zero poverty. The lowest poverty rate is for New Hampshire (NH), with a rate of 5.8 percent. Often, if the data do not include a value of zero, then the intercept does not have a meaningful interpretation.

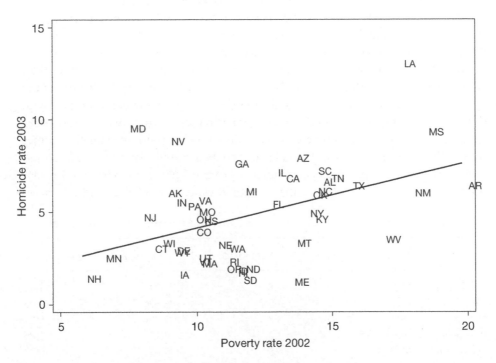

Figure 13.2   Scatterplot of poverty rate by homicide rate, N = 50
Data source: US Census Bureau, 2002, 2003, analyzed with Stata.

The line comes close to the actual values for some states, such as Colorado (CO), Florida (FL), Minnesota (MN), and Texas (TX), but is far from others like Louisiana (LA) , Maine (ME), Maryland (MD), and Nevada (NV). For example, for California (CA), with a poverty rate of 13.1, we would predict that the homicide rate is:

$$0.59 + (0.36*13.1) = 5.31 \text{ homicides per } 100,000 \tag{13.2}$$

The actual homicide rate is 6.8 homicides per 100,000, so the line predicts 1.49, too low a homicide rate. For Nevada, with a poverty rate of 8.9, the line predicts a homicide rate of:

$$0.59 + (0.36*8.9) = 3.79 \tag{13.3}$$

But the actual rate is 8.8, so the line underpredicts by 5.01. The discrepancy between the line and the actual data is at the heart of regression analysis. Table 13.2 shows these relationships for a few states (these data are also shown in Table 1.1). Again, for some states the line predicts homicide rather well. For some states it misses substantially. Sometimes the errors are in predicting a homicide rate that is too high and sometimes in predicting a homicide rate that is too low.

As we work with regression, remember that rounding error can creep in. Here we are doing examples by using only two decimal places, one more decimal place that was reported in the data on homicide and poverty rate. But if we compare our results with those we get from a computer statistical package, rounding error will lead to some discrepancies. For Stata, A, the intercept term isn't 0.59. Rather, Stata displays it as 0.5879896. Stata displays B, the regression coefficient, as 0.3553473 rather than 0.36. If we have Stata calculate the predicted value for California, it tells us that the prediction is 5.24304 rather than 5.31 and that the residual is 1.55696 rather than 1.49. For Nevada, Stata gives us 3.750581 for the prediction and 5.049419 for the residual. So if you are comparing calculations done by different people, or done with software rather than a calculator, or even comparing results from two software packages, remember that one source of small differences is rounding error that comes from differences in how many decimal places are being used. For research, we would always use a software package that carries many decimal places. In most of its calculations, Stata uses 7 digits, far more that we would want to do with a hand calculator.

Table 13.2 Poverty rate, homicide rate, predicted value and error for selected states

State	X_i (Percent in poverty)	Y_i (Homicide rate)	$F(X_i) = A + (B*X_i)$	E_i
California	13.1	6.8	5.31	1.49
New York	14.0	4.9	5.63	−0.73
Nevada	8.9	8.8	3.79	5.01
Vermont	9.9	2.3	4.15	−1.85
Oklahoma	14.1	5.9	5.67	0.23

Data source: US Census Bureau, 2002, 2003.

Fitting a Straight Line to Data

Lines as summaries

Straight lines that relate one variable to another without error occur only in mathematics, or in the case where the relationship between two variables is definitional, as in the relationship between temperature in degrees Fahrenheit and temperature in degrees Centigrade. But the straight line is often an excellent way to summarize the relationship between two variables. In the case of the relationship between the homicide rate and the poverty level, how did we pick a line? The line is called the **regression line**, and it's based on what it is we want the line to do – summarize the link between X and Y. We will discuss how to calculate the regression line, but first we should think about what we mean by a good summary of the relationship between X and Y.

There are many ways we could draw a line and many different lines we could draw. The simplest approach is an "eyeball" fit. We take a ruler and move it around until is seems to run through the middle of most of the data points. That is, we move the ruler around until the line defined by its edge is as close as possible to as many points as possible. This is one way to choose a line, but the results will vary from person to person. If we plotted the line by "eyeballing" it, then used algebra to calculate the slope and the intercept, while someone else did the same thing, each of use would get different values for A and B.

The problem is rather like that of summarizing the location (central tendency, typical value) of a set of data. It's a matter of choosing a summary measure, which in this case is a line defined by the values for A and for B. The eyeball method is analogous to looking at a frequency distribution and picking a number that is typical. If all the points fall exactly on a straight line, then the eyeball fit will be easy to apply. It would be like the problem of choosing the best number to summarize a set of data when all the cases have the same value on the variable. The challenge arises with real data, where there may be a general relationship between X and Y that can be summarized by a straight line, but the summary will not perfectly describe each X, Y pair (that is, each data point). With real data the data points may fall near the summary line but we never expect to see them all fall exactly on the line. As a result, we have to think about how well the line summarizes the data.

Error in fit

To move beyond the eyeball method, we have to consider how to judge the fit of the line to the data points. By convention we measure the vertical distance between the line and a particular data point.[4] That vertical distance between the line and the data point is how far off the line is in describing that data point. Algebraically, we must distinguish between the Ys' actual value and $\hat{Y}$ (pronounced "Y-hat"), which is the Y value predicted by the line. Equation 13.4 shows how E, the error, is just the observed value of Y minus the predicted value of Y ($\hat{Y}$) and that the predicted value is just the line (A + (BX)). On a graph, this will be the vertical distance between Y and the line.

$$E = Y - \hat{Y} = Y - (A + BX) \qquad (13.4)$$

If every data point fell on the line, then Y would equal $\hat{Y}$ for every data point, and for every data point, E would equal zero. But the actual Ys will not fall on the line; thus there is a difference between our prediction, $\hat{Y}$, and the actual value of Y. This is the error we have been discussing since Chapter 1. In regression it's also called the disturbance or the residual.

For each data point we have an X, a Y, a $\hat{Y}$, and an E. For all data points in a given analysis (with one line), we have one A and one B. E is the measure of error, or how far the line is from a data point. We want a line that fits well in the sense that the line makes the Es small. Picking a line is thus a matter of minimizing some summary of the Es. The technique for fitting will depend on what summary of the Es we choose.

Ordinary least squares

The most common method of fitting a line to a point is called ordinary least squares (OLS). It is logical, easy to compute and has properties that allow us to make generalizations about population OLS lines by using sample data.

Some of the Es will be positive, and some will be negative. If we just add them, the negatives and positives will cancel. OLS finds the line that minimizes the sum of the squared errors, which we will abbreviate USS for *Unexplained Sum of Squares*:

$$USS = \sum (E^2) = \sum (Y - \hat{Y})^2 = \sum (Y - (A + BX))^2 \qquad (13.5)$$

This equation looks formidable, but as usual, we will examine it one step at a time. Let's start at the left. The USS is the sum of the squared errors. Each error is the actual value of Y minus the predicted value, so the next term shows us that the sum of the squared errors is just the sum of the squared differences between the actual value of Y and the value the line predicts ($\hat{Y}$). Finally, the last term on the right reminds us that the errors we are going to square and then add are just the actual value of Y minus (A + (BX)).

The logic of squaring to eliminate negatives is something we discussed in earlier chapters. Recall that the arithmetic mean (average) minimizes the sum of the squared deviations from the mean:

$$\sum (X - \bar{X})^2 \qquad (13.6)$$

The OLS (ordinary least squares) regression line is the two-variable analogy to the mean. Both use a least squares criterion in summarizing data – they minimize the sum of the squared deviations around the summary. In fact, the OLS regression line is part of the mean family of statistics and in a sense is a generalization of the mean to two dimensions. It was first developed by the mathematician Karl Frederich Gauss, who also did the key work on the Normal distribution.

Breakdown point is discussed at the end of this chapter as an Advanced Topic.

Box 13.1   A Note about Notation

Not all statistics books use the same abbreviations for the sums of squares, although all use the same logic. We have three kinds of sums of squares in regression.

There is the sum of squares for the dependent variable which is the sum of the squared deviations of the dependent variable from its mean. Since we usually call the dependent variable Y, this can be called SSY. It is also called the Total Sum of Squares or TSS.

There is the variation of the predicted value for the dependent variable around the mean of the dependent variable. We are calling this Explained Sum of Squares or ESS. Because it is predicted by the regression equation, some books refer to it as the Regression Sum of Squares or RSS.

There is the variation of the actual values of Y around the predicted values. Since the difference between the dependent variable and its predicted value is E, the error term, this is the variation in E. We call this the Unexplained Sum of Squares, or USS. But some books call this the Error Sum of Squares, or ESS. Unfortunately, that's the same term we are using for the Explained Sum of Squares.

So if you are looking at another statistics book or talking with a friend who took a different course, be careful to check how the abbreviation ESS is being used.

OLS and conditional distribution

If we had a large amount of data, we could imagine having many values of Y for each value of X. We thus have a conditional distribution of Y for every value of X. Or we can think of taking very small slices along the X axis and having many values of Y for that small set of values of X. We could then calculate the conditional mean of Y for every value of X. If the conditional means have a straight line relationship to the values of X, then the conditional means of Y will fall exactly on the OLS regression line. Put another way, the OLS regression line predicts the conditional means of Y for every value of X, assuming that the relationship between the X values and the conditional means are a straight line.

Calculating the OLS Line

Given that we want to minimize USS, how do we find A and B so that USS will be small? Calculus allows us to solve such minimization problems. It turns out that the following formulas will give least squares values for A and B.

Calculating B

First we need to define some terms. The following formulas define SSX (sum of squares of X), SSY (sum of squares of Y), and SXY (sum of cross products of X and Y). These are also referred to as the variation in X, variation in Y and covariation in X and Y. The SSX and SSY should look familiar because they are the numerator for the variances of X and Y. Notice also that the SSY is the total sum of squares since Y is the dependent variable.

$$SSX = \sum (X - \bar{X})^2 \tag{13.7}$$

$$SSY = \sum (Y - \bar{Y})^2 \tag{13.8}$$

$$SXY = \sum (Y - \bar{Y})(X - \bar{X}) \tag{13.9}$$

For the homicide rate–poverty example these are:

$$SSX = 478.60 \tag{13.10}$$

$$SSY = 315.64 \tag{13.11}$$

$$SXY = 170.07 \tag{13.12}$$

Note that if we divide each of these by the number of data points, we would have the variance of X, variance of Y and covariance of X and Y. If we are working with sample data, we could divide by the number of degrees of freedom to estimate the population variances and covariances. All the calculations that follow to find A, B and so on can be done with sums of squares (variations), with sample variances and covariances or with estimated population variances and covariances. We present the equations with sums of squares because most readers find it clearer than the other versions of the equations.

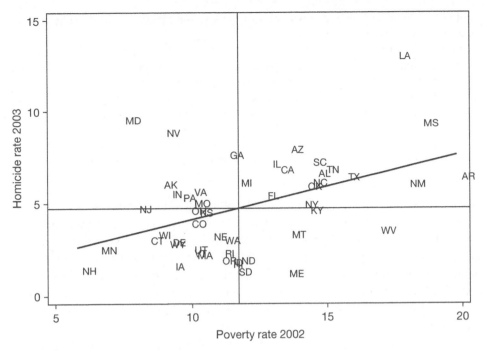

Figure 13.3   Graph of poverty rate by homicide rate with lines drawn at the mean of X and the mean of Y, N = 50
Data source: US Census Bureau, 2002, 2003, analyzed with Stata.

The sum of squares for a single variable must be positive, since negative deviations from the mean become positive squared deviations from the mean, and thus all the items in the sum are positive. The covariation (SXY) can be positive or negative depending on the mix of numbers in the data. To understand how this works, look at Figure 13.3. We have redrawn the scatterplot and inserted a vertical line at the mean for X (11.69) and a horizontal line at the mean for Y (4.74). This divides the data into four quadrants. If we call the upper right hand corner the first quadrant, then each state in that quadrant (for example, Tennessee (TN)) has an X value above the mean of X and a Y value above the mean of Y. For each of these states, the deviation from X and from Y are both positive. The covariation is calculated by taking the deviation from the mean of X for each state, and multiplying it by the deviation from the mean of Y for that state, then adding them together. When we multiply the two positive deviations for each state in this quadrant, we get a positive number. So for data points in this quadrant, the contribution to the SXY is a positive number. Table 13.3 provides some examples.

Now consider the lower right quadrant, which we'll call quadrant 2. Every state here is above the mean on X but below the mean on Y. Here the deviation for X values is positive but the deviation for Y values is negative, so the product of the deviations will be negative. Data points in this quadrant make a negative con- tribution to the SXY. If we move to the lower left quadrant (quadrant 3), we have

Table 13.3 Table of contributions of various states to the covariation between homicide rates and poverty rates

State	Quadrant	X	Y	X deviation	Y deviation	Product of deviations
TN	1 (upper right)	14.8	6.8	+3.11	+2.06	+6.41
MT	2 (lower right)	13.5	3.3	+1.81	−1.44	−2.61
VT	3 (lower left)	9.9	2.3	−1.79	−2.44	+4.37
VA	4 (upper left)	9.9	5.6	−1.79	0.86	−1.54

Data source: US Census Bureau, 2002, 2003.

values below the mean for both X and Y. Thus deviations from mean of X and deviations from mean of Y are both negative, and the contributions to the SXY will be based on multiplying a negative times a negative, which produces a positive. Finally, quadrant 4 in the upper left hand corner has data points above the mean for Y but below the mean for X. Like the second quadrant, one contribution to the SXY (the contribution from the deviation from Y) is positive, the other (from the deviation from X) is negative, so the product will be negative. Thus for data points in this quadrant, we will be adding in negative numbers in creating the sum.

If most data points are in the first and third quadrants (the upper right and lower left) most of the numbers added to make the SXY will be positive. If most data points are in the upper left and lower right, then the SXY will be composed of mostly negative numbers. Thus SXY will be negative when data points are mostly in the upper left and lower right and positive when data points are mostly in the lower left and upper right. This matches a sense of pattern in the data. A line running from upper left to lower right would have a negative slope indicating that as X goes up Y goes down. One running from lower left to upper right has a positive slope indicating that as X goes up, Y goes up. Of course, the actual value of the SXY will depend on the size of the contributions from each quadrant, as well as the number of cases there.

Once we have the sums of squares, then the OLS value for B can be calculated quite simply as:

$$B = \frac{SXY}{SSX} \tag{13.13}$$

As noted above, for the effect of the poverty rate on homicide,[5]

$$B = (170.07/478.60) = +0.36 \tag{13.14}$$

Calculating A

Once I have B, the formula for A is:

$$A = \bar{Y} - (B\bar{X}) \tag{13.15}$$

Table 13.4 Table of predicted value and error in the regression of homicide rates on poverty rates

State	X	Y	$\hat{Y}$	E
TN	14.8	6.8	5.92	+0.88
MT	13.5	3.3	5.45	−2.15
VT	9.9	2.3	4.15	−1.85
VA	9.9	5.6	4.15	+1.45

Data source: US Census Bureau, 2002, 2003.

The OLS regression line always passes through the point $(\bar{X}, \bar{Y})$, that is the point of the means of the two variables. You can see this in Figure 13.3. The formula for A is based on this property. For the homicide data:

$$A = (4.74) - (0.355*11.69) = 4.74 - 4.15 = 0.59 \tag{13.16}$$

Calculating $\hat{Y}$

Now, given A and B, for each value of X in the data we can calculate $\hat{Y}$, the predicted value of Y. This is the algebraic equivalent of drawing the line.

$$\hat{Y} = A + (BX) \tag{13.17}$$

Calculating E

Given Y and $\hat{Y}$ we can calculate E for each data point:

$$E = Y - \hat{Y} = Y - (A + (BX)) = Y - A - BX \tag{13.18}$$

Table 13.4 gives predicted values for the homicide rate and for the error in prediction for the states we examined in Table 13.3.

Goodness of Fit

Fit and error

Statistical theory tells us that the OLS line fits the data using the ESS criterion better than any other line that could be drawn. But this doesn't tell us how *well* it fits. The logic we developed with analysis of variance applies here. We will calculate a measure that tells us how well our predicted values match the actual values. But first we have to look at the equation for error (E) a bit more closely. There are a number of equations here, but you'll see that they are the same equations we dealt with in Chapter 11 when we were discussing analysis of variance. Let's start with the definition of E:

$$E = Y - \hat{Y} \tag{13.19}$$

Going back to algebra, we know we can add $\hat{Y}$ to each side of the equation and still have both sides equal. Since the positive and negative $\hat{Y}$ on the right hand side will cancel each other out, we can re-arrange and get:

$$Y = \hat{Y} + E \tag{13.20}$$

Since $\hat{Y} = A + BX$, we can substitute this for $\hat{Y}$ and get back to our original equation for the model:

$$Y = (A + BX) + E \tag{13.21}$$

This is our model. $A + BX$ is the deterministic part – the part of Y that can be explained by X. E is the random error – the part of Y that can't be explained by X. The equation partitions the value of Y into two parts, the part predicted by the regression equation, which we will call "explained" (corresponding to $\hat{Y}$) and the part that is error (E), or misprediction, which can be called "unexplained." The diagram in Figure 13.4 shows this. The line is the predicted value, $A + BX$, and E is the discrepancy between the line and the actual value of Y.

Partitioning sums of squares or variances

We've partitioned Y into two parts, the part predicted by the model (the fit of the model) and error. Our concern is explaining variation in Y in terms of X. For example, we want to be able to say that some part of the variation from state to state in the homicide rate is a result of variation from state to state in poverty level.

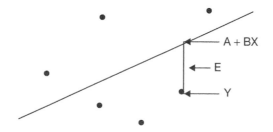

Figure 13.4　Line with error for Y indicated

Remember that we are defining variation as variation about the mean. Thus we are trying to explain why the homicide rate in various states differs from the mean homicide rate of all states. Put differently, we want to know the degree to which the conditional means of Y vary across values of X.

We can understand how well we are doing in explaining the variation in Y by breaking Y into two parts, but different parts than we did above. For each value of Y, the difference between that value and the mean is

$$Y - \bar{Y} \tag{13.22}$$

the deviation of Y from the mean of all the Ys. This is, for a single value of Y, the variation we want to explain – the fact that not all states have the same value on Y.

By doing a little algebra, we can get at a way of understanding which part of the variation in Y is explained by the independent variable and which part is not. We can add and subtract $\hat{Y}$ to the deviation (recall that adding and subtracting the same number to a number is effectively adding zero):

$$(Y - \bar{Y}) + (\hat{Y} - \hat{Y}) \tag{13.23}$$

Rearranging this term gives:

$$(Y - \hat{Y}) + (\hat{Y} - \bar{Y}) \tag{13.24}$$

The deviation from the mean for a particular value of Y can be broken into two parts: $(\hat{Y} - \bar{Y})$ the deviation of the predicted value from the mean, and $(Y - \hat{Y})$ the deviation of Y from its predicted value $\hat{Y}$, that is E.

Table 13.5 shows this for the four states we have been using as examples. If we add together the deviation of the predicted value from the mean and the deviation of the predicted value from the actual value, we get the deviation of the actual value from the mean. So we have broken the variation in Y into two components, one explained by our model and the other unexplained.

We defined the sum of squares of Y, or variation in Y, above as:

$$SSY = \sum (Y - \bar{Y})^2 \tag{13.25}$$

Table 13.5 Table of homicide rates decomposed into deviations

State	Y	$Y - \bar{Y}$	$\hat{Y}$	$\hat{Y} - \bar{Y}$	$Y - \hat{Y}$ (E)
TN	6.8	+2.06	5.92	1.18	+0.88
MT	3.3	−1.44	5.45	0.71	−2.15
VT	2.3	−2.44	4.15	0.59	−1.85
VA	5.6	0.86	4.15	0.59	+1.45

Data source: US Census Bureau, 2003.

That is, we take the deviations of Y from the mean of Y, square and add them. We have just broken the deviation of Y from the mean into two parts, and we can substitute each of those two parts into the equations for the sum of squares of Y. The first part, which is the deviation of the predicted values from the mean of Y, is called the explained sum of squares or regression sum of squares. (We label it ESS.) This is the same as the explained or "between" sum of squares in the analysis of variance. The equation for it is:

$$ESS = \sum (\hat{Y} - \bar{Y})^2 \tag{13.26}$$

The second part of our partition was the deviation of the actual value from the predicted value, $(Y - \hat{Y})$. We can develop a sum of squares for this as well. It is called the *unexplained sum of squares*. It is the same sum of squares we developed in Chapter 11 where it was called USS or the "within" sum of squares. The equation for it is:

$$USS = \sum (Y - \hat{Y})^2 \tag{13.27}$$

SSY can also be called the *total sum of squares*, TSS. Using the term TSS reminds us that there is a total variability in Y that we have now allocated to two sources: the variability of our predicated value around the mean of Y and the variability of the actual value of Y around the predicted value – the error. The ESS is the variation in Y that is explained by X. The USS is the variation in Y that cannot be attributed to X and is thus attributed to E, the error term.

It is important to note that there are multiple ways to make these computations. For example, instead of using sums of squares, we could have done the analysis after dividing the sums of squares by the number of cases. Then instead of working with variations, we would be working with sample variances. Or the sums of squares could be divided by degrees of freedom to estimate population variances. Everything else still follows the same pattern.

R^2

Finally we are now ready to measure how well the line fits the data. We can think about the proportion of all the variation in Y (TSS) that is explained by X (ESS). Then the measure of fit is:

$$R^2 = \frac{ESS}{TSS} \tag{13.28}$$

R^2 is the proportion of variation in Y explained by X. It is the ratio of the ESS to the TSS. Since R^2 is a proportion, it ranges from 0 to 1. An R^2 of zero indicates that X explains no variation in Y. An R^2 of 1 indicates that X perfectly predicts

Table 13.6   Sums of squares for model predicting homicide rate with poverty rate, N = 50

Category	Value
Total Sum of Squares (TSS)	315.642
Explained Sum of Squares (ESS)	60.434
Unexplained Sum of Squares (USS)	255.208
R^2 (ESS/TSS)	0.191

Data source: US Census Bureau, 2002, 2003.

Y – there is no error, and all E terms are zero. For the model predicting the homicide rate with the poverty rate, Table 13.6 gives us the sums of squares and the R^2.

How large an R^2 should we expect? What is a "good" R^2? The answers depend on the kind of data you are using and the type of phenomena you are studying. Some things are intrinsically more predictable than others. And data based on the aggregation of many individuals (e.g., census tracts, states, nations) tends to produce higher R^2 values than data on individuals. So there are no universal rules about how large an R^2 is "good." As seen in Table 13.6, poverty rate explains 19.1 percent of the variance in homicide rate, but, as you will see in the applications, we get smaller R^2 values when we use survey data on individuals.

Pearson's correlation coefficient

Of course, the square root of R^2 is R. But while we usually use the uppercase letters for R^2 we usually use the lowercase r for the square root of R^2. r has a special name, the **Pearson correlation coefficient**. It was developed by Karl Pearson as a measure to show the relationship between two variables. It is one of the most commonly used measures in statistics.

While the Pearson correlation coefficient is defined only for the relationship between two variables, R^2 can be used when there are more than two variables. The lower case r is usually used to indicate the bivariate correlation, while uppercase R and R^2 are usually used for the coefficient of determination in either bivariate or multivariate regression. Of course, for bivariate regression, r and R are the same thing.

The Pearson's correlation coefficient ranges from −1 to +1. The sign for r matches the sign of the B coefficient in the regression equation. If increasing X leads to increasing Y, then r will be positive. If increasing X leads to decreasing Y then r will be negative. By taking the square root of R^2 for the model predicting homicide with poverty, we get an r value of +0.437. We know it is positive because the B value in the regression of homicide on poverty is positive. We can use R^2 with more than one independent variable but then its relationship to Pearson's bivariate r is more complex.

Box 13.2 Steps in calculating an OLS regression

Step 1 Calculate the sums of squares

Substep a Calculate the sums of squares of X.

$$SSX = \sum (X - \bar{X})^2 \tag{13.29}$$

Substep b Calculate the sums of squares of Y.

$$SSY = \sum (Y - \bar{Y})^2 \tag{13.30}$$

Substep c Calculate the sum of cross products of X and Y, XY.

$$SXY = \sum (Y - \bar{Y})(X - \bar{X}) \tag{13.31}$$

Step 2 Calculate B

$$B = \frac{SXY}{SSX} \tag{13.32}$$

Step 3 Calculate A

$$A = \bar{Y} - (B\bar{X}) \tag{13.33}$$

Step 4 Calculate R^2

Substep a Calculate the total sum of squares, TSS.

$$TSS = SSY = \sum (Y - \bar{Y})^2 \tag{13.34}$$

Substep b Calculate the explained sum of squares, ESS.

$$ESS = \sum (\hat{Y} - \bar{Y}) \tag{13.35}$$

Substep c Calculate R^2.

$$R^2 = \frac{ESS}{TSS} \tag{13.36}$$

The formula for calculating Pearson's correlation coefficient directly from the data is:

$$r = \frac{SXY}{\sqrt{SSX * SSY}} = \frac{\sum_{1}^{N}((X_i - \bar{X})(Y_i - \bar{Y}))}{\sqrt{((\sum_{1}^{N}(X_i - \bar{X})^2(\sum_{1}^{N}(Y_i - \bar{Y})^2))}} \tag{13.37}$$

However, if you already have the regression equation, there's a third way to calculate r – it is B times the ratio of the standard deviation of X to the standard deviation of Y. That is:

$$r = B * (s_x/s_y) \tag{13.38}$$

This last equation for r hints at one of the reasons r is so popular in research. Suppose we standardize both X and Y by subtracting the mean and dividing by the standard deviation. Recall that these standardized values are also called Z scores. If we calculated the OLS regression not with the original values of X and Y but with the standardized values, the intercept term A will be zero and the B coefficient will be r, the Pearson correlation coefficient. Thus the Pearson correlation coefficient is telling us how many standard deviation changes in Y we can expect when X changes by one standard deviation. In the homicide rate example, if we compare two states that differ by one standard deviation in their poverty rates, the Pearson's r value of 0.437 tells us that we should expect that the state with the higher poverty rate should be about 0.44 standard deviations higher in homicide rate than the state with the lower poverty rate.

Both B and r in bivariate regressions are telling us about the relationship between the two variables. Each has advantages and disadvantages. B helps us keep track of what we have actually measured since it tells us how much change in Y we expect when X changes by one unit, while the Pearson correlation coefficient puts everything into standardized scores, which are a bit abstract. However, using B means that we have to decide which variable is the independent (the predictor) variable and which is the dependent (the predicted variable). B will have a different numerical value depending on which variable is taken as dependent and which as independent.[6] Pearson's r does not require us to pick dependent and independent variables. Usually we want to think about what is causing what, so we will prefer B coefficients. But sometimes we are only interested in a correlation and so Pearson's r is preferable because it is agnostic about causation. Recall that Pearson's chi-square also requires no consideration of what is causing what.

Pearson's r and B coefficients are discussed at the end of this chapter as an Advanced Topic.

Interpreting Regression Lines

Interpreting A

We usually do not spend much effort interpreting A. It is simply the value for Y expected when X equals zero. Most social science theory and many policy applications don't assign an interesting role to such a situation, so we often ignore A. There will be exceptions to that rule in some special situations and when some more advanced techniques are used. In the homicide/poverty example, the predicted value for a state with a poverty level of zero is 0.59.

Interpreting B

B is the reason for doing the analysis. B tells how, in the data at hand, values of X are related to values of Y. If B equals zero, then for every value of X, the expected value of Y, $\hat{Y}$, is equal to A. That is equivalent to saying the line is flat, and that there is no relationship between X and Y. B can be positive or negative, depending on whether large values of X are associated with large or small values of Y. B can be thought of as predicting a typical value of Y for each value of X. The numeric values of B that can be considered large or small depend on the application. When we compare states that differ by 1 percentage point in the percentage of families in poverty we expect the homicide rate to be higher by B, that is by 0.36 per 100,000. If we compared two states that differed by 10 points in their poverty rates, we would expect that the homicide rates would be 3.6 points higher in the state with the higher poverty level.

Interpreting R^2

This is the measure of goodness of fit. The larger it is, the better the fit. If B equals 0 then R^2 equals 0. What numeric values of R^2 can be considered large depends on the applications. With data on individuals, very small values (0.05–0.10) can be important; with aggregate data larger values are common (0.1–0.5). But the interpretation of R^2, like that of B, depends on the application, not arbitrary rules of thumb. Sometimes people become too concerned with R^2 and think that the higher the R^2 the better, and that the study with the highest R^2 is best. It is true when we move to multiple regression that R^2 has some important uses in testing hypotheses, but, like chi-square, it is probably better used for that purpose than deciding the value of a study. In the case of the relationship between poverty rate and homicide rate, the R^2 is 0.191. In other words, 19.1 percent of the variance in homicide rate is explained by poverty rate.

Interpreting r

Pearson's correlation coefficient for the poverty/homicide rate example is 0.437. The positive sign for r tells us that higher values of poverty are associated with higher homicide rates – the same thing we learned from the positive value on B. This is because B and r are telling us the same information in different ways. The value of r means that if we compare two states that differ by one standard deviation in poverty we expect them to differ by 0.44 standard deviations in the homicide rate.

Interpreting E

For each data point, the regression line generates a value for E, the discrepancy between the predicted value $\hat{Y}$ and the actual value Y. E tells us how much we mispredict for each observation. The patterns of error can be of interest. For policy purposes, it might be interesting to identify people, organizations, or governments that do better or worse than might be expected and follow up with further study. For example, if some school districts are doing much better in training students than would be expected from their resources, it would be nice to know why. Or if some hospitals have a higher mortality rate than expected, they probably should be given close scrutiny.

E is the stochastic or random part of the model. As described in Chapter 1, we expect that E is generated by sampling error, measurement error, other variables or permutation error. That is, the reason a state or person doesn't fall on the line may be because of random sampling error, because of error in measuring Y, because variables other than X effect Y, or just random factors of the sort captured in the permutation error logic.

β is discussed at the end of this chapter as an Advanced Topic.

Inference in Regression: A Basic Approach

E and the processes that generate it are critical for making inferences in regression analysis. Recall our two ways to make inferences about the mean. One is when we have a large simple random sample from a population. Then we can use the Law of Large Numbers to construct confidence intervals and test hypotheses. When we have small simple random samples from a population that is Normally-distributed, we can use Gosset's theorem for inference.

The situation with regression can get more complicated. Let's assume we have a large simple random sample from a population. Then statistical theory quite similar to the Law of Large Numbers allows us to create confidence intervals for and test hypotheses about regression coefficients. In fact, we'll use the Z procedure and the

F test just as we have in previous chapters. But when we don't have large simple random samples, the problem of inference in regression is harder to describe. We will leave discussion of it to an advanced topics section at the end of this chapter. For the rest of this discussion we will assume we have a large simple random sample. When that's the case, hypothesis testing and confidence interval construction will look familiar, although the formulas for standard errors for regression are a bit more complicated than for tests on means.

Working with b

We are now going to deal with inference from samples to populations so we will refer to the sample regression statistics as a and b and the corresponding population parameters as A and B. If we have a large simple random sample, the sampling distribution for the sample regression coefficient b is a Normal distribution. If we have a small sample but the conditions 1–7 described in Advanced Topic 13.5 are true, the sampling distribution is a t distribution with $N - 2$ degrees of freedom. We will use the t distribution formulation because when the sample size gets large the t distribution and the Z distribution are the same. So the sampling distribution of sample b coefficients is a t distribution with $N - 2$ degrees of freedom, has a mean equal to B (the regression coefficient in the population), and has a variance we will label σ_b^2 (sigma squared sub b). For large samples or when the conditions in the advanced topic are met, the ordinary least squares b calculated for the sample is a good estimate of the population regression coefficient B. And when we have a relatively small sample and the conditions of the advanced topic section are not met exactly, the probability values we are using will be off. A more advanced course in regression analysis can give you a sense of how far off our inferences about the population are likely to be in various circumstances.

As in testing hypotheses about the population mean, we don't actually know σ_b^2 but it can be estimated as:

$$\hat{\sigma}_b = \frac{\sqrt{\dfrac{USS}{N-2}}}{\sqrt{SSX}} \tag{13.39}$$

where N is the sample size and $N - 2$ is the number of degrees of freedom.

We can construct a confidence interval around b just as we did for the mean. It will be:

$$b \pm t_{1-\alpha, N-2}\hat{\sigma}_b \tag{13.40}$$

where b is the sample estimate, and t is the t value corresponding to $N - 2$ degrees of freedom at the desired confidence level (90 percent, 95 percent, 99 percent, etc.).

We can also test hypotheses about B using t. The test value of t for a test that B in the population equals a hypothesized value (B_{Hyp}) is:

$$t = \frac{b - B_{Hyp}}{\hat{\sigma}_b} \tag{13.41}$$

In most applications we will test the hypothesis that B equals zero, which means that X has no direct effect on Y. Rejecting this hypothesis leads to the conclusion that X does have an effect on Y. The arguments about spuriousness and indirect effects in Chapter 6 can be conceptualized in terms of situations where B values might be zero. Thus tests about B being equal to zero can be used to make arguments about causality. This is one of the reasons regression is so powerful in testing theory.

Working with a

If the above assumptions are true, then the sampling distribution of a is a t distribution with N − 2 degrees of freedom, mean A and variance σ_a^2 (sigma squared sub a). The standard error can be estimated as:

$$\hat{\sigma}_a = \sqrt{\frac{1}{N} + \frac{\bar{x}^2}{SSX}} \tag{13.42}$$

Again, confidence intervals can be constructed and hypothesis tests may be conducted. In most applications, A is of little interest, so it is unusual to test hypotheses about its value or construct confidence intervals around sample estimates for A, though there are some exceptions when we work with categorical independent variables.

Working with R^2

Just as in analysis of variance, we can use the following F test to test the hypothesis that in the population R^2 equals zero:

$$F_{2,N-2} = \frac{R^2/1}{(1 - R^2)/N - 2} \tag{13.43}$$

This is equivalent to testing the hypothesis that B equals zero, and is thus equivalent to the t test for B above. The numerical value of t, when squared, equals the numerical value of F.

Table 13.7   Regression coefficient table for the effects of poverty on homicide rates

	Coefficient	Standard error	t	P
a	0.588	1.275	0.461	0.647
Poverty	0.355	0.105	3.371	0.001
R^2	0.191	–	$F_{1,48} = 11.37$	0.001

N = 50
Data source: US Census Bureau 2002, 2003.

Technically, the F test is a test that the correlation between Y and $\hat{Y}$ is zero, or that the unexplained sum of squares (USS) equals the total sum of squares (TSS) and thus that the explained sum of squares (ESS) is zero (no variance explained).

> *Confidence intervals for predictions* are discussed at the end of this chapter as an Advanced Topic.

Table 13.7 is the kind of regression table often presented in the literature. We will use it to examine our model of the effects of poverty on homicide rates.

The t test on poverty indicates that we can reject the hypothesis that the population B is equal to zero at the 0.001 level. This means that, at any conventional level of hypothesis testing, we can reject the notion that poverty rate has no relationship to the homicide rate. The t test on the intercept term, a, shows we cannot reject the hypothesis that in the population, A = 0. However, given that the values of the independent variable don't include zero, this is of little interest since the intercept itself is not very meaningful in this data set. The F test lets us reject the hypothesis that in the population there is no correlation between $\hat{Y}$ and Y ($R^2 = 0$) or, equivalently, that X explains no variance. If you look carefully you will notice that the t value for our independent variable is the square root of the F value for the test on R^2. This is no coincidence but a mathematical relationship between the F test and the t test on the regression coefficient in bivariate regression. The idea that the B value is zero is logically equivalent to the idea that X explains no variance in Y, so it's not surprising that the F test and the t test always give the same result in bivariate regression. In the next chapter we will see that things get a bit more complicated when we have more than one independent variable.

> *Thinking about inference in regression* and *Binary dependent variables* are discussed at the end of this chapter as Advanced Topics.

What Have We Learned?

This chapter showed the calculations that underpin the basic bivariate regression model. While the calculations can become complicated, they are based on a simple logic. We pick a straight line that shows how values of Y change with values of X. We do this by making the predicted value of Y (based on X) come as close to the actual values of X as possible. We use the criterion of ordinary least squares, that is we make the sum of the squared errors as small as possible. This is predicting the mean values of Y at each value of X, given that these means all fall along a straight line. While there are other ways to pick a straight line, ordinary least squares is the most commonly used, provides a good tool for inference and is a starting place for understanding more complicated approaches.

Advanced Topic 13.1   Breakdown point

Like the mean, the OLS regression line is heavily influenced by points far from the line. Adding one or two data points that are very different from the overall pattern of the data (outliers) can produce dramatically different values for A and B and thus a very different line. In statistical terminology, the OLS regression line is not robust with regard to outliers. This notion can be made more precise by recalling the concept of the breakdown point. Suppose we are looking at the mean of a data set. How many outliers can be in the data set before the mean loses its power as a summary of the typical values in that batch of data? The answer is none. One really bad data point will throw the mean off so that it is not representative of the data. Since the mean can't tolerate any outliers before losing its value as a summary, it has a breakdown point of zero – it can tolerate 0 percent outliers. In contrast the median won't become a poor descriptor of the data until fully half the data are outliers, at which point it's hard to justify focusing on the 50 percent considered "good" rather than the 50 percent considered "bad." This implies the median has a breakdown point of 50 percent, which is as good as can be

done with any descriptor. The OLS line, like the mean, has a breakdown point of zero, one very different data point (outlier) distorts it totally.

The high breakdown point of the median is one of its advantages as a summary of a set of data, and the low breakdown point of the mean is one of its disadvantages. Just as the median is a robust measure of location, there are robust procedures to fit lines to a scatterplot. Minimizing the sum of the absolute deviations, rather than the sum of the squared deviations, will produce a robust line that is an analogy to the median. That is, the line could be thought of as the conditional medians of Y given differing values of X.

Nor do we have to fit a line through the center of the data. We could choose another summary of the conditional distribution of Y given X. For example, we could fit a line through the 75th percentile (the upper quartile) or the 25th percentile (the lower quartile) or any other percentile we like. For the moment the key point to keep in mind is that the OLS regression line is trying to predict the mean of the conditional distribution of Y at each value of X by putting a line through the mean of those distributions.

Advanced Topic 13.2   Pearson's r and B coefficients

Pearson's r is related to the B coefficients in an interesting way. We have calculated B here by predicting Y with X because that makes theoretical sense to us. We could also run the regression the other way, and predict X with Y. This might not make theoretical sense, but it can be done to explore the mathematical relationships involved. The regression coefficient for predicting X with Y, which we will call B*, would be calculated as:

$$B^\star = \frac{SXY}{SSY} \qquad (13.44)$$

Now, if we multiply B times B*

$$BB^\star = \left(\frac{SXY}{SSY}\right)\left(\frac{SXY}{SSY}\right) = \frac{(SXY)^2}{(SSX)(SSY)} \qquad (13.45)$$

Then if we take the square root

$$\sqrt{\frac{(SXY)^2}{(SSX)(SSY)}} = \frac{SXY}{\sqrt{(SSX)(SSY)}} = r \qquad (13.46)$$

This produces the formula for the Pearson correlation coefficient. That is, the B value for predicting Y with X, multiplied by the B* for predicting X with Y produces a number that is the square of the Pearson correlation coefficient. The value of B in regression depends on which variable is dependent and which is independent. The value of r (and R^2) does not. So using B requires that we have some sense of causal direction.

Advanced Topic 13.3   β

The lower case Greek β (**beta**) is used in three different ways in statistics texts. Sometimes it is used to indicate the chances of a Type II error in testing a hypothesis. Often it is used to indicate the population regression (slope) coefficient when distinguishing the population regression coefficient from the sample regression coefficient. (We are using B for that purpose.) β is also used to represent the standardized regression coefficient. A standardized variable is one that has been converted to a Z-score by subtracting the mean and dividing by the standard deviation. As we saw above, this is just the Pearson correlation coefficient, r. Thus each person's (or country's) value on a particular variable is expressed not in the units in which it was measured, but as how many standard deviations that person or country is from all persons or countries in the data set. Z scoring is done to be able to compare variables that are measured on different metrics (or scales). If we use standardized, rather than

"regular" values in our regression analysis, we end up with a standardized slope coefficient, rather than B. The computational formulas are the same; the only difference is that instead of using X and Y in raw form, we enter Z-scores for X and for Y.

If we conduct an analysis using Z-scores, A will always equal zero. This happens because the mean of the standardized version of X is zero, as is the mean of the standardized version of Y. The OLS regression line goes through the mean of X and the mean of Y. With the mean of the standardized X being zero, the right hand side of the equation will be A + (B*0), which then equals A. This must equal the mean of standardized Ys, which is zero, so A = 0.

In bivariate regression, the beta value is equal to the Pearson's correlation coefficient. This gives us another way to interpret the correlation coefficient – it is the regression coefficient when both variables are standardized. For the homicide/poverty model, β = +0.55, which is just the value of Pearson's r. Note that when we move to multiple regression, the correlation coefficient and the standardized regression coefficient are no longer equivalent.

Some social scientists believe that standardized regression coefficients are easier to interpret than unstandardized, and there is a small literature on the subject. Our view is that it is usually better to work in the units in which the data are measured because this forces us to think about those units, and thus about what is measured and how to model it. It seems more meaningful to us to use unstandardized regression coefficients (Bs). When we compare states that differ by 1 percentage point in their percentage of families in poverty, we expect the homicide rate to increase 0.36 per 100,000. This seems clearer and more faithful to the data than to say that if we compare states that differ by one standard deviation in their poverty rates, we expect them to differ by

0.55 standard deviations in their homicide rate. Of course, this is clearly a matter of preference. And there are times when standardized units seem quite natural, as when we construct attitude scales that have no natural units and that are constructed to have a mean of 0 and a variance of 1.0 – they are already standardized.

However, it is usually not a good idea to compare βs across samples. Remember that β depends not only on the slopes but also on the standard deviations that are used to create the Z scores. In fact, we can calculate β from B as:

$\beta = B * (s_x/s_y)$

This means that the value of β depends on the amount of variability in the sample, and that can differ from variable to variable and sample to sample. So even if you choose to use β for making interpretations within a data set, it is probably not a good idea to use β when comparing across samples.

Advanced Topic 13.4   Confidence intervals for predictions

In applied work it is sometimes necessary to forecast a value for Y given a value of X. There is a subtle complication in making such forecasts. The prediction itself (the point estimate) is:

$$\hat{Y} = a + (bX) \tag{13.47}$$

But there are two kinds of confidence intervals for the forecast of Y. One assumes that we are simply predicting the mean value of all the Ys at the value of X used in the equation above. This is called simply the confidence interval for the mean of Ys at a particular X and can be calculated as:

$$\hat{Y} \pm t_{N-2,1-\alpha}\hat{\sigma}_{\hat{Y}} \tag{13.48}$$

where the t distribution has N − 2 degrees of freedom. The standard deviation of the $\hat{Y}$ can be estimated as:

$$\hat{\sigma}_{\hat{y}} = \sqrt{\frac{1}{N} + \frac{(X - \bar{X})^2}{SSX}} \tag{13.49}$$

Notice that the value of the standard error, and thus the size of the confidence interval for the mean of Y is dependent on how far X is from the mean of X. The farther X is from the mean, the larger the standard error and thus the larger the confidence interval.

If the problem is viewed as predicting a single value of Y, then there is more variability to be considered, since the individual Ys will vary about each value of X. The confidence interval is called a prediction interval or confidence interval for the individual case; the predicted value for Y is the same ($\hat{Y}$), and the formula for the confidence interval is the same. But the appropriate formula for the standard error becomes:

$$\hat{\sigma}_{\hat{y}} = \sqrt{1 + \frac{1}{N} + \frac{(X - \bar{X})^2}{SSX}} \tag{13.50}$$

The +1 under the radical expands the size of the confidence interval substantially.

Advanced Topic 13.5   Thinking about inference in regression

Background

Given sample data, generally we will want to make statements about population values of regression coefficients. Recall that we have generally used uppercase A and B to refer to the intercept and slope without discussing the difference between the population and the sample. As we consider statistical inference, we will have to make that distinction. Now A and B will be the population intercept and slope, and lower case a and b will refer to the sample estimate of the population intercept and slope. R^2 is, by convention, always uppercase. R^2 refers to the percent of variance explained in the sample. There is no commonly used symbol to indicate the population variance explained. Before getting into the procedures for inference, we have to consider what is meant by population and sample in this setting, and understand the assumptions used in making inferences about regression coefficients in the population.

The sampling experiment for regression

Earlier we discussed sampling experiments as a process of drawing repeated random samples of a single variable. With two variables, X and Y, the conceptual framework gets a bit morecomplicated. The simplest approach to the sampling experiment for regression comes from the logic of experimental research. From this logic we can develop an understanding of the sampling process that will guide analysis in non-experimental data.

Suppose we are conducting a study in which we can manipulate the values of the variable X by experimental control. Then we can think of the sampling experiment as a situation in which the values of x are always the same in each sample we draw in the sampling experiment, because we have "fixed" them by how the research is conducted. (This is often referred to as the "fixed x" approach. Note that in this case the x values in any sample in the sampling experiment are the same as the X values in the population – we are not really sampling x; we are fixing them by the design of the study.)

For example, we might conduct a study of how class size affects performance on a statistics test. We conduct one study in which 200 students are in classes of size 20, 200 in classes of 50, and 200 in classes of 100. We use the score on the test as a dependent variable (y) and class size (x) as the independent variable. For each student we have a class size and a test score. We think of the sampling experiment as the situation in which we would do this study over and over, each time sampling a different class from each size category. Note that the x values (class size) don't change from sample to sample in the sampling experiment.

If x doesn't vary from sample to sample in the sampling experiment, $\hat{y}$, the predicted value based on x, won't vary from sample to sample either. We have classes of size 20, 50 and 100, and those are the only values we will have for x, and so the only values we'll have for $\hat{y}$ are a + bx. What will vary from sample to sample is e. In fact, we can think of e as a random variable drawn from some probability distribution. If e varies from sample to sample, y will also vary across samples. Because of e we will get many different values of y, even though we only have three values for x. This is because y = a + bx + e. The values of y in a sample are generated by both a + bx (or $\hat{y}$) and by e. So while (a + bx) won't vary across samples because of the way we've designed the experiment, e will vary and so does y. That is, the e values and the y values in a sample are random subsets (samples) of the E and Y values in the population. The value of e and y we get will vary from sample to sample, while the values of x are always 20, 50, and 100. The random element in this approach is the error term.

Essentially, we are sampling values of e. Then, because of the fixed (same across all samples) values of x, y will differ from sample to sample because e differs from sample to sample.

A simple diagram may help. Notice in Figure 13.5 that there are only three values for X. This is the fixed X process we are thinking about. (If X were equal to 20, 50, and 100 as in the class size effect on testing example, the X values would be further apart but that would make the graph harder to read. So for now we have simplified the graph.) In the population we have a straight line relationship between X and the average value of Y at any particular value of X. That is, the plot of the conditional means of Y is a straight line. We can think of the line as predicting the average of the Y values that occur at a given value of X. Since multiple values of Y are associated with each value of X, there are multiple values of E associated with each value of X. We assume (below) that the sample values of x are without error, so they are in essence X, the population values. But in the sample we get values of y, and thus e's that are a subset of all possible values of Y and E. From sample to sample we get different values of y and e, representing different values of Y and E in the population. For each value of X, there are several values of Y and E, and only some of those possible values turn up as y and e in any particular sample. The logic of the "fixed x"

way of thinking is that we always take the same x values (thus the X values) in each sample. But we get different values for e and y (subsets of E and Y) in each sample because we think of the process as sampling e values that are added to the predicted y values to yield the actual y values. For example, we might get two data points at each value of x. Then at each value of x (or X – the same thing in this case), we are getting a random sample of the E and Y values that exist for that value of X.

Why does e vary from sample to sample?

As noted in Chapter 1, there are several interpretations of why e exists – why test score isn't perfectly predicted by class size, or in general, why x doesn't perfectly predict y. All of these may be operating in any study, yet each is a slightly different way of thinking about the random error in data.

1 *Sampling error*: As in the classic application of inferential statistics with a simple random sample from a larger population, each sample will differ a little from the next due to luck of the draw, or sampling error. Since x is fixed, it is y and e that differ. We are sampling values of e and thus of y.

2 *Randomization error*: Suppose students are assigned at random to classes. Then from experiment to experiment the quality of students in each class size will vary by luck of the draw. By chance, some classes will include students with more ability or more time to study; other classes will, by chance, have students who will do less well. This will cause e to vary from sample to sample.

3 *Measurement error on y*: We can assume reasonably that we measure x, class size, without error, but we know that no test measures true ability perfectly. That is, if we think of Y as the true understanding for each student, then y is a student's understanding

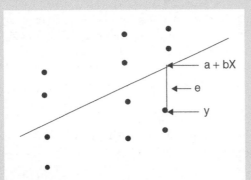

Figure 13.5 Decomposing y into fit and error

as measured by a fallible test. Thus even though there may be a perfect relationship between X and Y in the population (all Ys fall perfectly on the line – class size perfectly predicts learning with no variance in Y for any value of X), errors in measuring Y by using y lead to error in the relationship between x and y observed in the sample. That is, if X and Y are linked by a perfect straight line, each y we observe is "contaminated" by measurement error e, and thus the y observed will differ from the true Y. Then e will vary from sample to sample in the sampling experiment.

4 *Left out variables*: The residual e measures the effects of all variables not included in the regression equation. This is essentially the same as randomization error for this example, in that we have not considered the effects of student ability in statistics, motivation or other factors. Thus all these things are "left out" variables whose effects are captured by e.

5 *Superpopulations and permutations*: Suppose we have data on all nations in the world for a given year. We can imagine nations that don't actually exist, but might, and that have different combinations of values for Y at a value of X than those we observe. The actual nations of the world that constitute our data are in some sense a sample from this "superpopulation." Then e represents a kind of sampling error. By a parallel logic, we can think of the hypothesis tests as comparing the effects of the actual variables to random numbers, so that when we reject a null hypothesis that states a regression coefficient is zero, we are saying that the variable whose effect is described by that coefficient has more effect than would a variable that is created by a random number generator. We will reiterate this point in a moment.

At this level of analysis, and in most practical applications, these five approaches are identical in the logic of the fixed X.[7] It is very important to think about what generates the stochastic error that causes sample data to differ from the population data. To carry out statistical inference (build confidence intervals or test hypotheses), it is essential to make assumptions about that error. Our ability to use standard inferential techniques depends on the validity of those assumptions. Indeed, hypothesis tests and confidence intervals are meaningful only as statements about a stochastic model. If we don't know what is generating the stochastic error (e), then we don't know what a confidence interval or hypothesis test means.

If we use the "fixed x" conceptual framework, we will only have to worry about the properties of e in order to do statistical inference. For example, in order to do inference in small samples, we will need to assume that the E values scattered around each particular value of X are Normally-distributed. That is, if we took all the Es associated with one value of X and plotted a stem and leaf of the Es, they would form a Normal distribution. There are other assumptions as well, which we will discuss below.

Another approach that will lead to essentially the same result is to assume that X and Y have a "joint" normal distribution. That means that a plot of X and Y would each be Normal. Imagine a three dimensional scatterplot, with X and Y plotted on the flat surface and the frequency (proportion of cases) plotted on the vertical axes that comes up of the plane, would produce a three dimensional histogram that looks like a normal distribution that has been spun around its center (this is easier to see than describe.) Usually joint normality leads to the same conclusions as normally distributed Es.

Using regression with population data

Many data sets of great interest in social research seem to represent the population, rather than a sample, in that no other observations are possible. In looking at the relationship between homicide rate and poverty using states as observations, we have included all states. Why would we use inferential statistics in this situation? There are several rationales that parallel the sources of error just discussed.

1   It is a strong (overwhelming) tradition in the literature. This seems to be changing as researchers develop a better understanding of what an inferential procedure means. But it is still true that it is nearly impossible to present quantitative material to a professional audience without reporting standard errors and/or hypothesis tests, even when it is not obvious what kind of process generated the errors and thus how the hypothesis tests and confidence intervals should be interpreted.

2   Standard errors provide a means of assessing uncertainty. This is certainly true; in fact it is why standard errors are used in traditional sampling and experimental situations. But unless there is an understanding of the process that generated the errors, and thus led to the standard error calculations, it's not clear how the errors can be interpreted. That is, it's not clear why we would use the standard error as calculated rather than some other measure of uncertainty. The standard error has meaning only in relationship to a stochastic process that generated the E values.

3   The observed values of e are a sample from the population of all possible values of E. If E represents measurement error or the effects of variables not included in the model (variables other than X in the bivariate case), this is a reasonable way of thinking about the issue. But we must be careful to think about the character of that process. Would it, for example, produce Es that are normally-distributed?

4   The observed values are a sample of all the possible configurations of X, Y, and E.

When we do not have an experiment, we have no control over the observed values of variables, but we can think about other values as equally reasonable. The sample is thought of as a sample of "all possible (or at least plausible) worlds," the superpopulation.

The real problem in using inferential statistics with population data comes when it is necessary to go from them to a precise description of the process that generated the error – a description precise enough to indicate the population distribution of E and thus Y. If we cannot make these assumptions, and thus generate a conceptual model for what generated E and Y, then the common procedures for inference are not appropriate. In many circumstances the assumptions are not justified if we have "population" data, such as state data. There are other regression methods that are more appropriate but they are beyond what we can cover in this book.

Assumptions of regression needed for inference

We must make a number of assumptions to perform inferences in OLS regression. Here we will simply introduce the assumptions. There are several ways to test for these assumptions and even ways to correct for some of the violations of the assumptions, but these are beyond the scope of this book.

1   *Correct model (linearity)*. In the population, $Y = A + BX + E$, which means a straight line describes the relationship between X and Y and thus that the conditional means of Y for every value of X all fall on the straight line defined by A and B (**linear relationship**). This implies that the relationship is not a more complicated line (a curve), and that any other major influences on Y can be

considered part of the effects of E. When we get to multiple regression there can be more than one independent variable, in which case we assume that we have not included variables that are not a cause of Y, nor left out any variables that are a cause of Y. The consequences of the former mistake are trivial, the consequences of the latter can be very important. If the relationship is not linear, the effects won't be accurately represented by the correlations, and the strength of the relationship will likely be underestimated. In bivariate regression, a scatterplot can give a sense of whether or not the relationship between the two variables is adequately summarized by a straight line. In multiple regression, more complex diagrams are required to assess linearity. If X and Y are related by a curve more complicated than a straight line, alternatives to linear regression can be used.

2   *No measurement error in X, measurement error in Y is described by E.* This is sometimes referred to as the assumption of fixed X, described above. Some authors assume no measurement error in Y, in which case E cannot represent measurement error. They then must assume joint normality of X and Y. We prefer allowing the possibility of measurement error in the conceptual model because the interpretation of E becomes broader. But note that we assume x is measured without error, that is, that the sample values of x are perfect representations of the population values of X. If this is not true, our inferences will be somewhat incorrect – the p values for hypothesis tests will not be quite right, and confidence intervals will not have the right probability of capturing the population mean.

3   *E has a mean of zero.* If we could take the average all the Es for a particular value of X, we would find their average is zero. This is equivalent to saying that the mean of all the Ys for a particular value of X will fall exactly on the regression line. In a sense, this is restating assumption 1 above. Indeed, one way to interpret the OLS regression line is that it gives a line that for each value of X goes through the mean of all the Y values for that value of X. We mentioned criteria other than least squares for fitting regression lines. These simply substitute some other summary measure, such as the median, a trimmed mean or even a quartile, for the mean. Thus the least absolute deviation lines fits a line through the data that passes through the median of the Ys at each value of X.

4   *Homoscedasticity.* Homoscedasticity means constant error variance (from the Greek). It means that the variance of the Es at any particular value of X is equal to the variance of the Es at any other particular value of X. That is, the spread in the plot of the Ys around X doesn't change with the value of X. In other words, Y is not more variable around X at high values of X than at low values of X, or vice versa. Suppose we take all the Es for one value of X and calculate its variance. Then we take a different value of X, and there too calculate the variance of the Es. The two variances should be the same. This can be seen in a two dimensional scatterplot. If the y values take on a funnel shape, or in general aren't scattered with equal variability around the line at all values of x, then we may have heteroscedasticity, differing variances of the E values across values of X. There are ways to correct for this and ways of detecting it when the scatterplots are hard to read.

5   *No correlation in E across values of X.* Sometimes called no autocorrelation. Knowing the E for one observation does not give any information useful in predicting the E for some other observation. If we have observations over time, this means that high values for E do not tend to be followed by either high or low E values in any systematic way. If we have observations over space, adjacent

units of observation do not have systematically similar, or systematically dissimilar, E values. Again, there are ways to detect this. A simple one is to look at the correlation of the E values across time or space. This is the same as saying that the E values were drawn with independence.

6   *E and X are uncorrelated.* This is not the same as heteroscedasticity, which says there is no relationship between the variance of E and the value of X (that the value of X is not correlated with the value of the variance of E). It's also not the same as not having values of E be correlated with each other. This says there is no relationship between the value of X and the value of E. If you think of E as measuring the effects of all left out variables, all other variables influencing Y, then those variables, at least in their combined effect (E) are not correlated with X. If they are correlated with X, then the calculation of b will attribute the effects of E, in part, to X, not to E. This is a critical problem. And there is no simple way to detect it other than using data on a candidate independent variable that is suspected of being correlated with the X variable already being considered.

Suppose some variable that is a cause of Y is correlated with X, but is not included in a regression and thus will be part of the error term e. In fact, regression will attribute the effects of that left out variable to x, not e. The only way to detect and correct this is to gather data on the left out variable and include it in a multiple regression model or a related diagnostic procedure. If it is left out, then x will pick up its effects to the extent it is correlated with x. Thus the estimate of the effects of x will include some effects that should be attributed to the other variable. Of course, if we have data on the other variable, the problem can be solved by including it in the model. This is one reason multiple regression is so popular and powerful.

7   *Normality of the residuals/error (Normally-distributed error variance).* So far we assumed that at each value of X, the Es are Normally-distributed, with mean zero and variance σ_E^2. The heteroscedasticity assumption says that for each X, the value of σ_E^2 is the same as for every other X – the variance of E does not change across values of X. Finally, in parallel to our theorems on the sampling distribution of sample means, if we are working with a small sample, we must assume that the conditional distributions of the E terms are Normal at every value of X. In large samples the normality assumption is not critical, but in small samples making this assumption when it is not warranted can lead the analysis astray. The Law of Large Numbers allows us to do inference (make estimates and test hypotheses) with large samples no matter what the population distribution. But if we have small samples, we need to assume that the distribution of the population is Normal. In this case, it is the residuals that must be Normal.

The combination of homoscedasticity and independence from above is sometimes called the "identically, independently distributed" assumption, labeled i.i.d. If the Es are also Normally-distributed, then we can say they are Normal i.i.d.

What we get from these assumptions

These are many conditions that need to be met, but OLS works very well even if only some of them turn out to be correct. If all of them hold for our population, then OLS is a very powerful way to construct a regression line. If the population values of E are Normal i.i.d., have mean 0, all relevant independent variables are included in the model (for the bivariate case this just means that X and E are uncorrelated) and we can consider X fixed (no measurement error in x), then the OLS estimates of coefficients and their standard errors are best unbiased estimators. That is, no other method for combining the data to estimate the line will be more efficient, and the OLS estimates will be unbiased (the mean of a and the mean of b across repeated random samples will be equal to A and B, respectively), as will the standard errors. If we drop the assumption of Normally-distributed E values, then the OLS estimator is BLUE (the most efficient unbiased estimate of the population parameters that can be formed from a linear combination of the data). But there can be biased estimators that are more efficient. If the population distribution of E values is non-Normal, some non-linear ways of combining the data to form a and b, the estimates of A and B, may produce results that are unbiased and more efficient than OLS. If the situation meets all the assumptions of OLS, there is no better method for estimating A and B. How well it does when the assumptions aren't met depends on which assumption is violated.

As long as the X is fixed and the errors have zero mean and are uncorrelated with the X values, then OLS will be consistent. That is, as the sample size gets infinite, the estimates a and b will "collapse" toward the true values A and B. If X and E are really uncorrelated, then we can even relax the assumption that x is fixed (measured without error and thus the same in all samples in the sampling experiment), and OLS will still be consistent.

What all this means is that OLS works very well for large samples (though there are some problems that occur even in large samples). When samples are small or moderate, the use of OLS for inference depends more on the assumptions. But if there is a straight line relationship between X and the conditional means of Y in the population, OLS almost always gives us a good estimate of A and B, even when the estimates of standard errors that go with the coefficient estimates may be unreliable.

How can we make all these assumptions?

While it is traditional to call these the assumptions that underpin OLS regression, another way to think of them is as "the conditions under which OLS performs really well." When we move away from those conditions, OLS performs less well. But small departures from some of the ideal conditions for using OLS often mean only small declines in how well OLS does for us. For example, if we have a population in which the conditional distributions of the Ys are not really Normal, but close, then our hypothesis tests will not actually be at the α level we specify, but they'll be close. In more advanced texts you'll learn how to check the assumptions and what to do if they are not met by a particular data set and model.

Advanced Topic 13.6   Binary dependent variables

The appropriate techniques for examining the effect of a continuous independent variable on a dichotomous dependent variable look much like regression.[8] We predict the dependent variable as a function of the independent variable and error.

$$Y = f(X) + E \qquad (13.51)$$

The complication comes in because in this case Y must be interpreted as the probability of being in the category scored one rather than in the category scored zero. Using a regression model assumes that a straight line links education to the probability of having AIDS transmission knowledge. There are problems with this straight-line model. For example, we could get predicted values of Y that are greater than 1 or less than 0, even though probabilities can only range from 0 to 1. Further, the estimates from OLS regression are not efficient in small samples. More advanced techniques use an f(X) for linking a continuous independent variable to a categorical dependent variable that are not straight lines but rather curves that make sense in terms of probabilities.

However, the AIDS knowledge question is from a large sample, and ordinary least squares regression, while not the best tool for analyzing the effects of education on AIDS knowledge, will still give us a reasonable idea of the pattern in the data. Table 13.8 shows the regression of the AIDS knowledge variable on years of education.

The b coefficient is positive and indicates that for each additional year of education, people are about 0.02 more likely to have AIDS knowledge. The t and F tests indicated that we can reject the hypothesis that in the population, education has an effect (p is less than 0.001). The intercept tells us what to expect for someone with no education – that they have about a 64 percent chance of knowing that condom use reduces the risk of AIDS.[9]

Table 13.8 Regression coefficients for the relationship between education and AIDS knowledge, $N = 8,306$

	Coefficient	Standard error	t	P
Education	0.022	0.001	20.33	<0.001
Intercept	0.645	0.008	83.39	<0.001
R^2	0.047	–	$F_{(1,8304)} = 413.48$	<0.001

Data source: 2000 Uganda DHS data set, analyzed with SPSS.

Applications

We have used the homicide example throughout the chapter, so we will examine the other three research questions in these applications.

Example 2:   Animal concern

We will treat the animal concern question as a continuous level variable (the traditional approach to bivariate regression requires that both the independent and dependent variables are continuous level measurements). We have explored the possibility that concern for other species varies with gender and age. In Chapter 11 we treated gender and age as categorical independent variables and analyzed them with analysis of variance. We will discuss the relationship between regression and analysis of variance below.

Number of years of education is a continuous variable that could have a linear effect on animal concern. It may be that concern either increases or decreases with increased education. Figure 13A.1 shows the scatterplot of years of education and the animal concern score. It's not very helpful because there are only five values

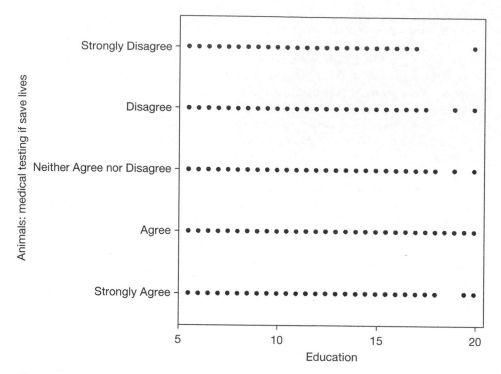

Figure 13A.1　Scatterplot of animal concern by years of education, N = 26,476
Data source: 2000 ISSP data set, analyzed with SPSS.

for the animal concern question and education runs from zero to 30 years. We have 26,476 observations. With a large number of observations and a relatively small number of values for the dependent and independent variables, many data points fall on top of each other. Thus a dot in this graph might represent one person or many people.

The regression of animal concern on education is in Table 13A.1.

We cannot reject the null hypothesis that education has no effect on animal concern based on the t test and the F test. Remember, in bivariate regression, these tests always give identical results. We see the limited ability of education to predict

Table 13A.1　Regression coefficients for the relationship between education and animal concern, N = 26,476

	Coefficient	*Standard error*	*t*	*P*
Education	−0.003	0.002	−1.74	0.082
Intercept	2.524	0.023	107.78	<0.001
R^2	0.000	–	$F_{(1,26475)} = 3.02$	0.082

Data source: 2000 ISSP data set, analyzed with SPSS.

animal concern in the model. The R^2 value of 0.000 also shows that education explains 0 percent (or at least less than 0.01 percent) of the variance in animal concern. If we are interested in explaining why animal concern varies from person to person, we will have to keep developing and testing hypotheses.

We saw in Chapter 11 that we can use gender (coded as a binary variable) as an independent variable in an analysis of variance. When we do that analysis of variance, with the animal concern item as dependent and the binary gender variable as independent, we are doing the same thing as running a bivariate regression. The F value for the ANOVA is the same as the F value for testing that R^2 in the bivariate regression is zero. The intercept term (a) in the regression is the mean for the group coded zero on the binary variable, the regression coefficient (b) is the difference between men and women, and a + b is the mean on the dependent variable for the group coded 1. Analysis of variance and regression are just two different ways of talking about the same kind of analysis.

Example 3: Environmental treaty participation

We suggested earlier that having more NGOs in a country may relate to a greater number of environmental treaties being ratified because of the pressures put on by the organizations. In Figure 13A.2 we show a scatterplot of the relationship between number of NGOs and number of environmental treaties ratified with

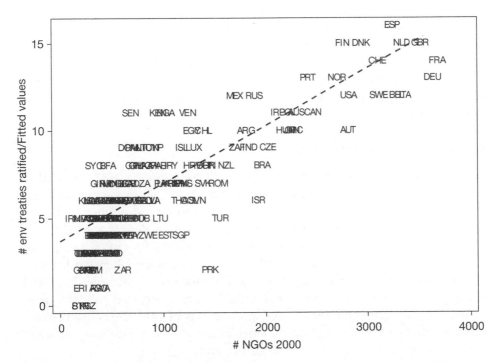

Figure 13A.2 Scatterplot of environmental treaties by number of NGOs, N = 170
Data source: Roberts et al., 2004, analyzed with SPSS.

Table 13A.2   Regression coefficients for the relationship between number of NGOs and environmental treaty participation, N = 170

	Coefficient	Standard error	t	P
Number of NGOs	0.003	0.00015	20.62	<0.001
Intercept	3.65	0.194	18.83	<0.001
R^2	0.717	–	$F_{(1,168)} = 425.05$	<0.001

Data source: Roberts et al., 2004, analyzed with Stata.

the OLS regression line plotted. There is a very strong pattern evident from the graph; the fewer the NGOs, the fewer the number of treaties ratified and the greater the number of NGOs, the more treaties ratified. And unlike other scatterplots we have looked at, like homicide and poverty rates, there are few cases that fall far from the pattern (all nations are close to the regression line), suggesting a strong relationship.

Table 13A.2. shows the regression of treaty participation on number of NGOs. The effect of NGOs on treaty participation is statistically significant – we can believe that the effect we are seeing is bigger than would have occurred if we had used a random number to predict treaty participation. The slope is positive as predicted by theory – greater participation in environmental treaties is associated with a greater presence of NGOs in a country. The effects are strong in that we have explained 72 percent of the variance in treaty participation. With each additional NGO, treaty participation increases by 0.003. This doesn't mean much by itself but gives an indication that the presence of many NGOs in a country are needed to significantly influence treaty participation. Remember that in complex calculations like these rounding error can matter.

Example 4:   AIDS knowledge.

In this application, we will consider the effects of education on AIDS knowledge. The AIDS knowledge variable is a dichotomy, with people who know condoms can prevent AIDS transmission having a score of 1 and those not knowing that having a score of 0. Bivariate regression is not the best way to analyze this relationship. We have already seen how we can use tests on differences in means across groups and chi-square tests for contingency tables to examine the relationship between dichotomous dependent variables and categorical independent variables. We could also break a continuous independent variable into categories and use those techniques. But techniques for examining the effects of continuous independent variables on dichotomous dependent variables are a bit complex, and we won't cover them other than in Advanced Topic 13.6 above.

Exercises

1. Let's examine how happiness in general, happiness with one's job and happiness with family life are related using Pearson correlations. Higher scores on each of the three variables reflect greater satisfaction. Using Table 13E.1, summarize the relationships, and be sure to include the strength of the relationships in your response.

Table 13E.1 Correlations of happiness in general, happiness with job, and happiness with family life

		Happy in general	Happy with job	Happy with family life
Happy in general	Pearson correlation	1.000	0.381	0.637
	Sig.	.	0.000	<0.001
	N		28,599	44,503
Happy with job	Pearson correlation	0.381	1.000	.336
	Sig.	0.000	.	<0.001
	N	28,599		28,335
Happy with family life	Pearson correlation	0.637	0.336	1.000
	Sig.	0.000	0.000	.
	N	44,503	28,335	

Data source: 2000 ISSP data set.

---

2. Let's also return to the relationship between the proportion of doctors in each state and state infant mortality rate. The bivariate regression output is presented in Tables 13E.2 and 13E.3.

a) Write the regression as an equation.
b) What is the R value? What does this value tell us?

Table 13E.2 Descriptive statistics on number of doctors and infant mortality rate

	Mean	Std. Deviation	N
INFMORT	7.0860	1.3271	50
DOCTRATE	236.02	57.35	50

c) Is there a statistically significant relationship between doctor and infant mortality rates? How did you conclude this?

Table 13E.3 Regression coefficient table for the effects of doctor rate on infant mortality rate in the 50 US states

		Coefficient	Standard error	T		P
R^2	.075	–		F = −1.976	.054
A	8.584	.780		11.009	<0.001
B	−.006	.003		−1.976	.054

Data source: US National Center for Health Statistics; 2000 data. http://www.cdc.gov/nchs/default.htm.

---

3. We will now try to identify some factors that may affect satisfaction with one's job (the original question was recoded so that responses now range from 1 = completely dissatisfied to 7 = completely satisfied). The bivariate relationships between the following six variables and job satisfaction: (1) respondent age (Table 13E.4); (2) hours worked weekly (Table 13E.5);

(3) hours spent weekly doing housework (Table 13E.6); (4) view of job as stressful (1 = strongly disagree to 5 = strongly agree) (Table 13E.7); (5) how difficult it is to fulfill family responsibilities (1 = never to 4 = several times a week) (Table 13E.8); and (6) number of young children (under age 18) in the household (Table 13E.9).

Table 13E.4 Bivariate relationship between age and job satisfaction

	Coefficient	Standard error	T	P
R^2	.004	–	F = 130.730	<0.001
Intercept	4.93	.023	218.87	<0.001
Age	.006	.001	11.43	<0.001

Table 13E.5 Bivariate relationship between hours worked and job satisfaction

	Coefficient	Standard error	T	P
R^2	R	–	F = 4.846	.028
Intercept	5.121	.023	223.457	<0.001
Work hrs	.001	.001	2.201	.028

Table 13E.6 Bivariate relationship between housework hours and job satisfaction

	Coefficient	Standard error	T	P
R^2	.000	–	F = 0.062	0.804
Intercept	5.202	.012	427.023	<0.001
Housework hr	.000	.001	−0.249	0.804

Table 13E.7 Bivariate relationship between view of job as stressful and job satisfaction

	Coefficient	Standard error	T	P
R^2	.018	–	F = 499.977	<0.001
Intercept	4.814	0.017	279.379	<0.001
Job stressful	.129	0.006	22.360	<0.001

Table 13E.8 Bivariate relationship between difficulty fulfilling family responsibilities and job satisfaction

	Coefficient	Standard error	T	P
R^2	0.015	–	F = 395.717	<0.001
Intercept	5.449	0.016	340.739	<0.001
Family respon	−1.40	0.007	−0.122	<0.001

Table 13E.9 Bivariate relationship between number of young children and job satisfaction

	Coefficient	Standard error	T	P
R^2	0.000	–	F = 0.903	0.342
Intercept	5.140	0.062	83.487	<0.001
Young children	0.019	0.019	0.950	0.342

Data source: 2000 ISSP.

a) Briefly summarize the relationships between these six variables and job satisfaction.

b) Were any of these findings contrary to your expectations? Explain.

c) Interpret the intercept value of 4.814 for the relationship between view of job as stressful and job satisfaction. Is this value meaningful?

d) Interpret the R^2 value of .015 for the relationship between difficulty fulfilling family responsibilities and job satisfaction. Is this value statistically significant?

e) What is the direction of the slope for the relationship between age and job satisfaction, and what does this mean?

f) Interpret the A value of 5.140 for the relationship between number of young children and job satisfaction. What does the beta value of .020 (not presented in table) mean? How does the beta value differ from the B value?

References

International Social Survey Programme (ISSP). 2000. 2000 Environment II data set. www.issp.org. Catalog no. ZA 3440. Cologne, Germany: GESIS-ZA Central Archive for Empirical Research.

Long, J. S. 1997. *Regression Models for Categorical and Limited Dependent Variables*. Thousand Oaks, CA: Sage.

Menard, S. 2002. *Applied Logistic Regression Analysis. 2nd edn.* Thousand Oaks, CA: Sage.

Pampel, F. C. 2000. *Logistic Regression. A Primer.* Thousand Oaks, CA: Sage.

Pearson, K. 1896. Mathematical contributions to the theory of evolution: III. Regression, heredity and panmixia. *Philosophical Transactions of the Royal Society of London* 187, 253–318.

Press, S. J. 1972. *Applied Multivariate Analysis.* New York: Holt, Rinehart and Winston.

Roberts, J. T., Parks, B. C., and Vasquez, A. A. 2004. Who ratifies environmental treaties and why? Institutionalism, structuralism and participation of 192 nations in 22 treaties. *Global Environmental Politics* 4(3), 22–64.

Stanton, J. M. 2001. Galton, Pearson, and the peas: A brief history of linear regression for statistics instructors. *Journal of Statistics Education* 9(3).

Uganda Demographic and Health Surveys. 2001. Calverton, Maryland: UBOS and ORC Macro. (http://www.measuredhs.com/pubs/pdf/FR128/00FrontMatter.pdf).

United Nations Development Programme (UNDP). 2002. Human Development Report. Deepening Democracy in a Fragmented World. New York: Oxford University Press (http://hdr.undp.org/en/reports/global/hdr2002/).

US Census Bureau 2000. Table 33. Urban and rural population, and by state: 1990 and 2000. ("http://www.census.gov/prod/cen2000/index.html"\t"_blank"www.census.gov/prod/cen2000/index.html).

US Census Bureau 2002. Historical poverty tables: Table 21. Number of poor and poverty rate, by state: 1980 to 2006. Year 2002 ("http://www.census.gov/hhes/www/poverty/histpov/"\t"_blank"www.census.gov/hhes/www/overty/histpov/hstpov21.html).

US Census Bureau 2003. Table 295. Crime rates by state, 2002 and 2003, and by type, 2003 ("http://www.census.gov/prod/2005pubs/06stata"www.census.gov/prod/2005pubs/06statab/law.pdf).

CHAPTER 14
BASICS OF MULTIPLE REGRESSION

Outline

While bivariate regression is a useful research tool in some circumstances, the complex character of the world usually requires that realistic models use more than two independent variables. That is, most things worth studying in the social sciences have more than one cause, and these causal variables are usually correlated with each other. In an experiment, we control for the effects of causes other than the one on which we are focusing, either by including them explicitly in the experimental manipulation or by the randomization process. This is called *experimental control*. In regression (and cross-tabulations with more than two variables) we control by incorporating all relevant variables into the statistical model. This is called *statistical control*.

For example, we have offered a number of ideas about what influences environmental treaty participation, and in some cases we have found independent variables that seem to be able to explain differences in environmental treaty participation across nations. But those independent variables are themselves correlated, so we may have spurious effects. For example, we might want to consider the influence of voice and accountability (again, meaning extent of citizen participation in government and personal freedoms) on environmental treaty participation controlling for the presence of NGOs. We obviously can't do an experiment with manipulating level of voice and accountability or number of NGOs, so we have to use statistical procedures to control for the effects of one independent variable while looking at another. This is why multiple regression is so useful in the social sciences where many important issues cannot be addressed by an experiment.

We use multiple regression to estimate the net effects of any independent variable on the dependent variable, controlling for other variables we believe to also cause changes in the dependent variable. In principle, experimental control is a better way to assess causal relationships, but it cannot be applied to many problems of interest in the social sciences. And in most experimental situations, there is concern over how well the experimental situation mimics the rest of the world – the problem of external validity (recall that external validity refers to the degree to which the results of a study done in one situation generalizes to other situations). Thus regression is one of the most valuable tools for understanding causal relationships.[1]

Two Independent Variables

Terminology and notation

We will use the term multiple regression when more than one independent variable is used to predict a single dependent variable.[2] We will continue to call the dependent variable Y but will label one independent variable X2 (voice and accountability in our example) and the other X3 (number of NGOs). We start

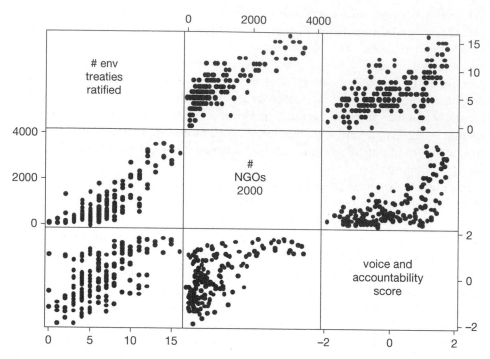

Figure 14.1   Scatterplot matrix of environmental treaty participation, number of NGOs and voice and accountability, N = 162
Data source: Kaufmany et al., 2002, Roberts et al., 2004, analyzed with Stata.

with X2 rather than labeling the first variable X1 because we want to save the label "1" for Y, the dependent variable. The reason for this will become clear below. Our argument about NGOs is that they will exert pressure on governments to participate in treaties. Our argument about voice and accountability is that governments will be more likely to participate in environmental treaties when citizens have more voice (participate in electing officials and freedom of expression) and governments are more accountable to their citizens for their actions. A scatterplot matrix of these three variables is displayed in Figure 14.1. With a scatterplot matrix including the country symbols makes the graph very messy so we have not used them. (We only use countries that have data on all three variables so the sample size is a bit smaller than when we look at one variable at a time.)

The reason we have to move from bivariate regression to multiple regression is because of the spuriousness problem we discussed in Chapter 6. In the scatterplot matrix there seems to be some relationship between number of NGOs and environmental treaty participation. And there also appears to be a relationship between treaty participation and voice and accountability. But if you look at the bottom middle graph, with voice and accountability on the vertical axis and number of NGOs on the horizontal axis, there also seems to be a link between NGOs and voice and accountability. So either NGOs or voice may be influencing treaty participa-

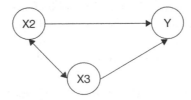

Figure 14.2   A hypothetical causal model allowing both X2 and X3 to have a causal effect on Y

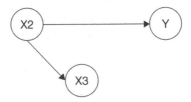

Figure 14.3   A hypothetical model in which X3 is spurious and X2 is causal

tion, with the other independent variable showing some correlation but not a causal relationship.

A hypothetical causal model allowing both X2 and X3 to have a causal effect could be diagrammed as in Figure 14.2. We use a double-headed arrow to indicate that there may be a relationship between X2 and X3, but that we are not going to analyze that causal relationship – we are interested for the moment only in what effects X2 and X3 have on Y, not on how they may influence each other. The alternative, which has X3 as spurious and X2 as causal, could be diagrammed as Figure 14.3. The first model implies an effect of X3 on Y when X2 is controlled (included in the multiple regression equation). The second model implies that when X3 and X2 are both used to predict Y, the effect of X3 will not differ from zero. Of course, if neither X2 nor X3 have regression coefficients different from zero, then they should not be considered causes of Y.

When looking at a regression coefficient, we need to know the following: 1) What is the dependent variable? 2) What is the independent variable? 3) What other variables are controlled for in estimating the effect of that independent variable on the dependent variable? That is, what other variables are in the regression equation? Because we need to keep track of these things, our notation gets more complicated.[3] The basic notation in its most elaborate form will refer to several different X variables, each with a different subscript.

$$Y = a_{1 \cdot 23} + b_{12 \cdot 3}X2 + b_{13 \cdot 2}X3 + e_{1 \cdot 23} \tag{14.1}$$

(We have left the variables in capital letters even though we will be dealing with sample data. This is because, with all the subscripts, it is easier to read the capital letters.) Again, we have started with X2 so that the subscript "1" is reserved for the dependent variable, in this case Y. To read the coefficients, for example, the regression coefficient for X2, we would say "b one two dot three." For the error term we would say "e one dot two three." To be perfectly proper we should say "b sub one two dot three" and so on, but even without the "sub" to indicate lower case, this is quite a lot to manage.

The notation for the constants (the a and the bs) indicate that the dependent variable is variable 1, while the numbers to the right of the dot are those variables that are controlled in the equation. If we used this notation for bivariate regression, there would be nothing to the right of the dot for the b term, and the number of the single X variable would be to the right of the dot for the intercept. In many applications we will be able to simplify the regression coefficients to a single number with each b (e.g., b_2, b_3, etc.) representing the independent variable. But it's best to learn the full notation at first to avoid confusion, and we'll need it to develop some ideas that are important.

Geometric view

We examined the bivariate regression line as a means of summarizing the relationship between two variables expressed in a scatterplot. This was an extension to two dimensions of the use of a mean to summarize a set of data. The mean is the point on the one-dimensional number line such that the sum of the squared deviations about that line are minimized. The OLS regression line through a scatterplot is the line such that the sum of the squared deviations about the line are smaller than the sum of squares around any other line (that was how we picked a and b, the OLS criterion). For two independent variables, think of a cube in which the two independent variables form a scale along the bottom of the cube (the X2 by X3 scatterplot is laying flat on the bottom), and the Y variable is scaled vertically. Then each data point fits somewhere in the cube, just as each data point in the scatterplot fits somewhere on a piece of paper. Figure 14.4 shows this, with lines from the data point drawn onto the plane of the independent variables to help us see where things are. If we had enough data, there would be a conditional distribution of Y and each combination of values of X2 and X3 (each point on the plane indicated by the values of the two independent variables). Even if we don't have enough data to see this, we can still imagine a conditional distribution. We are then trying to find a set of regression coefficients that summarize the conditional means.

To find the appropriate a and bs for the two independent variable problem, we are actually trying to find a plane that cuts as close as possible to all points in the cube. Again, we will employ the least squared errors (OLS) criterion, which says that the sum of the squared deviations from the plane (the predicted values) should be as small as possible. Just as we used a point to summarize the one-dimensional array of data, and a line to summarize the two-dimensional array, now we use a plane to summarize the three-dimensional array. If we are correct that the relationship between the dependent variable and the independent variables is linear (that is, can

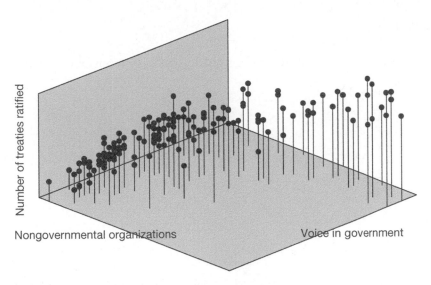

Figure 14.4 Diagram of multivariate relationship of two independent variables to one dependent variable, N = 162
Data source: Kaufmann et al., 2002, Roberts et al., 2004, analyzed with Stata.

be described as a plane rather than a curved surface), then the conditional means for every combination of values of X2 and X3 will all fall on the regression plane. There also is a geometry of regression when we have more than two independent variables, but the hyper-cubes and hyper-planes (cubes in planes in many dimensions) that result, while mathematically tractable, are quite hard to visualize.

Algebraic view

The algebraic view is simple. We wish to find an a and two bs so that a predictive equation for Y minimizes the sum of the squared residuals. The equation for the prediction of Y is:

$$\hat{Y}_{1 \bullet 23} = a_{1 \bullet 23} + b_{12 \bullet 3}X2 + b_{13 \bullet 2}X3 \tag{14.2}$$

Then we can calculate the e term for each point:

$$e_{1 \bullet 23} = Y - \hat{Y}_{1 \bullet 23} \tag{14.3}$$

The goal in finding a and the bs is, again, to make the sum of the squared errors as small as possible. That is, we still wish to minimize:

$$\sum (Y - \hat{Y})^2 \tag{14.4}$$

Again, we could use trial and error, but one advantage to the OLS criterion for finding the line is that the solutions for a and the bs are exact. No guessing is needed.

Calculations

First, we need to define a variety of sums of squares (variations) and covariations:

$$SSY = \sum (Y - \bar{Y})^2 \tag{14.5}$$

$$SSX2 = \sum (X2 - \overline{X2})^2 \tag{14.6}$$

$$SSX3 = \sum (X3 - \overline{X3})^2 \tag{14.7}$$

$$SX2X3 = \sum (X2 - \overline{X2})(X3 - \overline{X3}) \tag{14.8}$$

$$SX2Y = \sum (X2 - \overline{X2})(Y - \bar{Y}) \tag{14.9}$$

$$SX3Y = \sum (X3 - \overline{X3})(Y - \bar{Y}) \tag{14.10}$$

This is a place where you can easily get "equation shock." But if you walk through these one at a time you will see that they are actually old acquaintances that you have been calculating for several chapters. The equations for SSY, SSX2 and SSX3 are just the sums of squares for each of the variables that are the starting point for calculating the variances. We take each value for a variable, subtract the mean for the variable, square to eliminate the negative signs and then add them all up.

SX2X3, SX2Y and SX3Y are called the **covariations**, sometimes called the **cross-products**. If we divided by N we would have the sample covariances; if we divided the sum of squares by N we would have the sample variances. These are also the denominators for one way of calculating the correlation coefficient. Recall from our discussion of bivariate regression that the sign on the covariations depends on whether the pattern between the two variables is best described by a line with a positive slope or a negative slope, or equivalently with a positive or a negative correlation.

Having defined the variations (sums of squares) and covariations (cross products), we can use them to calculate the slopes (bs), the intercept (a) and the coefficient of determination (R^2). The calculations are composed of simple pieces. For the slopes we have:

$$b_{12 \bullet 3} = \frac{(SSX3)(SX2Y) - (SX2X3)(SX3Y)}{(SSX2)(SSX3) - (SX2X3)^2} \tag{14.11}$$

$$b_{13 \bullet 2} = \frac{(SSX2)(SX3Y) - (SX2X3)(SX2Y)}{(SSX2)(SSX3) - (SX2X3)^2} \tag{14.12}$$

Note that the denominators for both equations are the same. Note also that the effect of X2 on Y depends on both the covariation (or the correlation) between X2 and X3 and the covariation (correlation) between Y and X3. The effect we are estimating for each variable depends critically on what other variables are included in the regression equation.

The formula for a is simple:

$$a_{1 \cdot 23} = \bar{Y} - (b_{12 \cdot} \overline{X2} + b_{13 \cdot} \overline{X3}) \tag{14.13}$$

As in bivariate regression, the regression equation will include the point defined by the means of all the variables.

Finally, we can define the total sum of squares (TSS), Explained sum of squares (ESS) and the Unexplained sum of squares (USS) and from them define the coefficient of determination, R^2:

$$TSS = SSY = \sum (Y - \bar{Y})^2 \tag{14.14}$$

$$ESS = \sum (\hat{Y}_{1 \cdot 23} - \bar{Y})^2 \tag{14.15}$$

$$USS = \sum (Y - \hat{Y}_{1 \cdot 23})^2 = \sum e^2_{1 \cdot 23} \tag{14.16}$$

$$R^2 = \frac{ESS}{TSS} \tag{14.17}$$

Again, these formulas may look complex, but they are more or less the same as those we have seen in previous chapters. We have a total variation in the dependent variable, Y, which is the SSY and also labeled TSS. This is what we want to understand – why does Y vary from country to country? Then we have the amount of variation in Y that is "explained" by the regression equation (ESS). That is, the regression or explained variation in Y is the variation of the predicted value around the mean of Y. The unexplained sum of squares is the variation in the actual value of Y around the predicted value of Y – the squared errors (USS). The R^2 is the proportion of the total variation in Y explained by the regression equation – the explained sum of squares divided by the total sum of squares.

Table 14.1 shows the results of using these formulas to estimate the regression for the environmental treaty participation data. For comparison, Table 14.2 shows the two bivariate regressions with treaty participation as a dependent variable and also the **auxiliary regression** in which the independent variables are used to predict each other. The dependent variable is in the column. We will interpret the regression in the next section. These auxiliary regressions will be helpful in thinking about some aspects of regression later in the chapter. Because there are more variables, we have simplified the table from those used in the last chapter. We display only the coefficients and the t values, not the standard errors or the p values. We show statistical significance using asterisks (*), a very common approach used in published studies. One asterisk indicates that the p value for the test of the hypothesis that the population value is zero is less than 0.05, two asterisks indicates that p is less than 0.01, and three asterisks indicates that p is less than 0.001. Note that some authors will use one asterisk for p less than 0.10. Always check the footnotes of the table to determine what the asterisks mean.

Box 14.1 Steps in Calculating Two Independent Variable Multiple Regression

Step 1 Calculate the means, sums of squares and covariations

$$SSY = \sum (Y - \bar{Y})^2 \tag{14.18}$$

$$SSX2 = \sum (X2 - \overline{X2})^2 \tag{14.19}$$

$$SSX3 = \sum (X3 - \overline{X3})^2 \tag{14.20}$$

$$SX2X3 = \sum (X2 - \overline{X2})(X3 - \overline{X3}) \tag{14.21}$$

$$SX2Y = \sum (X2 - \overline{X2})(Y - \bar{Y}) \tag{14.22}$$

$$SX3Y = \sum (X3 - \overline{X3})(Y - \bar{Y}) \tag{14.23}$$

Step 2 Calculate the regression coefficients

$$b_{12 \bullet 3} = \frac{(SSX3)(SX2Y) - (SX2X3)(SX3Y)}{(SSX2)(SSX3) - (SX2X3)^2} \tag{14.24}$$

$$b_{13 \bullet 2} = \frac{(SSX2)(SX3Y) - (SX2X3)(SX2Y)}{(SSX2)(SSX3) - (SX2X3)^2} \tag{14.25}$$

Step 3 Calculate the intercept term

$$a_{1 \bullet 23} = \bar{Y} - (b_{12 \bullet 3}\overline{X2} + b_{13 \bullet 2}\overline{X3}) \tag{14.26}$$

Step 4 Calculate the Explained and Unexplained Sum of Squares

$$TSS = SSY = \sum (Y - \bar{Y})^2 \tag{14.27}$$

$$ESS = \sum (\hat{Y}_{1 \bullet 23} - \bar{Y})^2 \tag{14.28}$$

$$USS = \sum (Y - \hat{Y}_{1 \bullet 23})^2 = \sum e_{1.23}^2 \tag{14.29}$$

Step 4 Calculate the R^2

$$R^2 = \frac{ESS}{TSS} \tag{14.30}$$

Table 14.1 Regression of environmental treaty participation on number of NGOs and voice and accountability, N = 162

	Coefficient	T
Number of NGOs	0.003	14.57***
Voice and accountability	0.365	1.88
Intercept	3.811	16.68***
R^2	0.731	$F_{(2,160)} = 215.67$

*p<0.05, **p<0.01, ***p<0.001
Data source: Kaufmann et al., 2002, Roberts et al., 2004, analyzed with SPSS.

Box 14.2   Caution: Shifting Number of Cases

Statistical packages and the number of cases

When calculating regressions, most statistical packages will include every observation for which all the variables being used in the regression are available. If we are also running other regressions with different sets of variables, the sample size may change from regression to regression. For example, if we run the bivariate regression predicting treaty participation with just number of NGOs as an independent variable and then another bivariate regression with just voice and accountability as the independent variable, the sample size for these two regressions will be 170 and 169 because there are some countries for which we have data on one of the variables but not the other. In Table 14.2 where we are comparing the two independent variable regressions, the two bivariate regressions, and the two auxiliary regressions, we have restricted every analysis to the 162 countries for which we have data on all three variables. Thus, every regression is based on the same set of countries. It is important to be careful about how the observations in a data set may change from analysis to analysis because of missing data. Here the change is significant – we had treaty participation data for 191 countries but 29 of these countries are excluded from analyses due to missing data on NGOs and/or voice.

In technical terminology, cases are omitted when there is missing data on any one of the variables examined. This is referred to as **listwise deletion**. In contrast, **pairwise deletion** is when all cases are used that have complete data for pairs of variables. When there are more than two variables, there is a potential for the analyses to be based on different sets of cases, depending on patterns of missing data. Listwise deletion is typically preferred because we know how many cases we have. The exception to this is when there is a substantial amount of missing data, and there is a sophisticated statistical literature on how to estimate regressions in those cases.

Table 14.2 Bivariate and auxiliary regressions for model predicting treaty participation with number of NGOs and voice and accountability, $N = 162$

	Environmental treaty participation	Environmental treaty participation	Number of NGOs	Voice and accountability
Number of NGOs	0.003*** (20.52)	–	–	0.001 (10.711)***
Voice and accountability	–	2.187 (9.719)***	588.504 (10.711)***	–
Intercept	3.601 (17.91)***	6.451 (30.457)***	852.632 (16.49)***	−0.575 (−7.066)***
R^2	0.725	0.371	0.418	0.418
	$F_{(1,160)} = 421.07$***	$F_{(1,160)} = 94.46$***	$F_{(1,160)} = 114.72$***	$F_{(1,160)} = 114.72$***

*$p<0.05$, **$p<0.01$, ***$p<0.001$
Data source: Kaufmann et al., 2002, Roberts et al., 2004, analyzed with SPSS.

Interpretation

Interpreting a

The intercept term, a, has the same interpretation as before, and again, it's usually of minimal interest. It is the value predicted for Y when both Xs are equal to zero. In this case that means a country with no NGOs and about a mean score (0) on the voice and accountability scale (since it runs from negative to positive values). In this case, it is possible for a country to have no NGOs and a score of zero on the voice and accountability scale so the intercept term is meaningful.

Interpreting b

Each b represents the change in the dependent variable that can be expected when the independent variable corresponding to the b changes by one unit and there is no change in the other independent variable. As the notation makes clear, the value of each b depends on the other variable(s) being controlled. In Table 14.1, we find that for each additional NGO, the number of environmental treaties ratified increases by 0.003. Of course, this increase is very small, but it tells us that treaty participation will increase by one treaty for about every 335 NGOs. When we consider the t-test for statistical significance, we find that the t value is 14.57. This is statistically significant at 0.001, so we would conclude that we can reject the hypothesis and conclude that controlling for voice and accountability, NGOs have an effect on treaty participation.

The coefficient for voice and accountability is not statistically significant at the 0.05 level. This means that controlling for NGOs, extent of voice and accountability in a nation does not affect treaty participation. To interpret the b coefficient (for illustrative purposes only), for each one point increase in voice and accountability, environmental treaty participation increases by about 0.365. Since the calculations of scores on the voice and accountability scale are complex, this is difficult to interpret, but the positive coefficient indicates the relationship between voice and treaty participation is positive. Overall, these results partially support the first model – number of NGOs has an effect on environmental treaty participation, but voice and accountability has no effect when controlling for the effects of NGO presence. As Table 14.2 shows, the two independent variables are correlated.

Interpreting R^2

R^2 is still the percentage of the variation in the dependent variable that can be predicted using the $\hat{Y}$ generated by the regression equation. It reflects the predictive ability of the entire equation, of both variables acting together. In this case we explain about 73 percent of all variation across nations in environmental treaty participation using these two independent variables. The F test is a test that R^2 in the population equals zero – that the equation has explained no variance and therefore there is no correlation between Y and $\hat{Y}$. We can reject the hypothesis of no correlation between the actual Y and the predicted Y, or no variance explained in Y by either X2 or X3, at the 0.001 level. Note that in bivariate regression the F test on $R^2 = 0$ is equivalent to the t test that the population value of the single independent variable is zero. This link between the F test for the R^2 for the whole regression and the t test for one variable does not hold in multiple regression. As we will see below, there is an F test in multiple regression that is linked to the t test for a single variable, but it is not the F test on the overall R^2, which tests that *all* population B values are zero.

Adjusted R^2 is discussed at the end of this chapter as an Advanced Topic.

Relationship between Bivariate and Multiple Regression

In addition to the regression equation with two independent variables (Equation 14.31), we have also estimated two bivariate regression equations, each using one of the independent variables as a predictor of the dependent variable (Equations 14.32 and 14.33). There are also two bivariate equations relating one independent variable to the other (Equations 14.34 and 14.35). As we mentioned above, these equations relating the independent variables to each other are called the "auxiliary regressions." All four of these regressions are displayed in Table 14.2. What are the relationships between the bivariate regression and the multiple regression?

If there is no correlation between the two independent variables, that is, if the bs and R^2 (and R) in the auxiliary regressions are zero, then the b coefficients in the multiple regression will be equal to the b coefficients in the two bivariate regressions. Then the R^2 for the multiple regression will be the sum of the R^2s for the bivariate regressions.

To look at this more formally, we first must look at the five equations involved:

$$Y = a_{1 \cdot 23} + b_{12 \cdot 3}X2 + b_{13 \cdot 2}X3 + e_{1 \cdot 23} \tag{14.31}$$

$$Y = a_{1 \cdot 2} + b_{12}X2 + e_{1 \cdot 2} \tag{14.32}$$

$$Y = a_{1 \cdot 3} + b_{13}X3 + e_{1 \cdot 3} \tag{14.33}$$

$$X2 = a_{2 \cdot 3} + b_{23}X3 + e_{2 \cdot 3} \tag{14.34}$$

$$X3 = a_{3 \cdot 2} + b_{32}X2 + e_{3 \cdot 2} \tag{14.35}$$

Thus, if $b_{23} = 0$ (and therefore $b_{32} = 0$), it follows that:

$$b_{12 \cdot 3} = b_{12} \tag{14.36}$$

$$b_{13 \cdot 2} = b_{13} \tag{14.37}$$

$$R^2_{1 \cdot 23} = R^2_{1 \cdot 2} + R^2_{1 \cdot 3} \tag{14.38}$$

That is not the case here – NGOs and voice and accountability are related to one another – as indicated by the non-zero regression coefficients and non-zero R^2s in the auxiliary regressions. (In this case the relationship is quite strong – the independent variables explain about 42 percent of the variance in each other.) Note that it doesn't matter which of the two independent variables we take as dependent in the auxiliary regressions because these are not being used to describe causal relationships – they are just tools to help think through what a regression does. The t, F and R^2 values, as seen in Table 14.2, are the same.

This change in the size and significance of the effect of one independent variable on the dependent variable when a second independent variable is included in the equation is why multiple regression is such an important research tool. It allows us to enter control variables into the equation – variables that may also have an influence on the dependent variable. Of course which variable is the focus of your research and which variable is the control variable is a matter of your research question – the regression procedure treats all independent variables the same way. Multiple regression allows us to test for spurious relationships – bivariate relationships between an independent and the dependent variable that disappear when another independent variable is added. And sometimes the opposite can happen – by controlling for one independent variable a second independent variable that is not significant in a bivariate regression becomes significant.

Inference

Tests regarding values for a and b

As in bivariate regression, we can use the t test to examine the plausibility that the population A and Bs take on specific values. The form of these tests is the same as the one presented in Chapter 13 for bivariate regression.

$$t_b = \frac{b - B_{Hyp}}{\sigma_b} \tag{14.39}$$

$$t_a = \frac{a - A_{Hyp}}{\sigma_a} \tag{14.40}$$

The major difference between inference in bivariate regression and multiple regression is that in multiple regression, the formula for the standard errors of the bs are somewhat more complex. For example, the standard error for the $b_{12.3}$ (the coefficient estimating the effect of X2 on Y controlling for X3) is:

$$\hat{\sigma}_{b_{123}} = \sqrt{\frac{ESS/_{N-k-1}}{(SSX2)(1 - R^2_{2\bullet3})}} \tag{14.41}$$

It is necessary to keep track of the number of degrees of freedom involved in estimating the population coefficients. This will be N − k, where N is the number of cases in the sample and k is the number of parameters in the prediction equation. This will be equal to the number of independent variables in the equation plus one to account of the estimation of the constant. Note that we are subtracting the number of parameters in the regression equation from the sample size to obtain the number of degrees of freedom. Some texts don't include the constant in their definition of k, so their k will be the number of independent variables. Then the equation above will use N − k − 1 instead of N − k. Since we are now moving to models with very complex computations, we will nearly always use a computer to do the calculations and so won't burden you with the rest of the standard error formulas.

Test for R^2

In bivariate regression, testing the hypothesis that b equals zero is equivalent to testing the hypothesis that R^2 equals zero. In multiple regression, one independent variable may have a substantial effect on the dependent variable, while another independent variable may have no effect. The prediction equation may still do a good job of explaining variance in the dependent variable. In other words, the R^2

test tells us whether the overall model is predictive of the variance in the dependent variable, not whether *each* variable is predictive. We can perform a test of the hypothesis that the population value of R^2 equals zero using the F test:

$$F_{k,N-k} = \frac{R^2_{1\bullet 23}\Big/2}{(1 - R^2_{1\bullet 23})\Big/N - 2 - 1} \tag{14.42}$$

This F test is equivalent to testing the hypothesis that both b coefficients are both equal to zero:

$$b_{12\bullet 3} = b_{13\bullet 2} = 0 \tag{14.43}$$

Partitioning Variance and Statistical Control

In a multivariate experiment, it is possible to control the relationship between the independent variables in such a way that there will be no correlation between the independent variables. In this situation the multiple regression will yield the same results as the two bivariate regressions – the correlation between the independent variables will be zero. In using regression with non-experimental data (and with some experimental data where absolute control is not practical), the independent variables are not controlled by manipulation and so must be controlled by analysis. This is an imperfect method of control, and it is necessary to think carefully about its meaning.

Venn diagrams are a popular way to visually depict the interrelationships among independent and dependent variables.[4] The Venn diagram, as in Figure 14.5, provides a geometric representation of the process of partitioning variance.

The upper circle represents the variation in the dependent variable Y. The lower two circles represent the variation in the independent variables X2 and X3. The area at the very top of the circle for Y (labeled 1) is the portion of the variation in Y that cannot be explained by X2 or X3. The area on the bottom left side of the Y circle (labeled 2) is the variance in Y that is explained by X2 over and above what is explained by X3. The area on the bottom right side of Y (labeled 3) is the variability in Y explained by X3 over and above what is explained by X2. The area in the bottom center of the Y circle (labeled 4) is the area explained by X2 and X3 that cannot be uniquely attributed to either independent variable. It is real variance explained in Y but we cannot give "credit" to either X2 or X3 because they are correlated. The total variance explained in Y consists of the areas labeled 2, 3 and 4. The correlation or covariation between X2 and X3 are the areas labeled 4 and 5.

We can have a situation in which we explain a great deal of variance in Y but cannot attribute much of the variance to either X2 or X3. This will occur when the

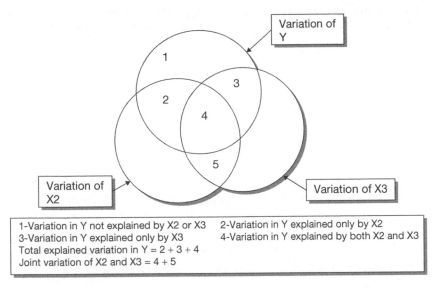

Figure 14.5 Venn diagram showing the interrelationships among independent and dependent variables and the partitioning of variance

area 4 is large compared to areas 2 and 3. Since the correlation between X2 and X3 is represented by the areas 4 and 5, having area 4 large means that the correlation between the two independent variables is large. This is called **multicollinearity**.

Using multiple regression, we can calculate the magnitude of the variance uniquely explained by each independent variable and the amount of variance explained that cannot be uniquely attributed to one independent variable or the other. This is more precise than the Venn diagrams, which are useful when we are first thinking about how much variance each variable explains, but don't take us very far in analysis. To find out what each variable contributes, we calculate the R^2 for the two independent variable regression, and the R^2s for each bivariate regression. By subtracting the R^2 for the bivariate regression for X2 from the R^2 for the two variable regression, we have an estimate of the portion of the variance in the dependent variable explained by X3 independent of X2. A parallel operation will assess the magnitude of the variance explained by the first X independent of the second.

Changes in R^2 by adding X2 to a model that already includes X3 can be represented in an equation in two ways. In Equation 14.44 we use a set of parentheses and a dot to indicate the change. In Equation 14.45 we use the Greek letter delta (Δ) to represent change.

$$R^2_{1(2\bullet3)} = R^2_{1\bullet23} - R^2_{1\bullet3} \tag{14.44}$$

or

$$\Delta R^2_{1\bullet3\to1\bullet23} = R^2_{1\bullet23} - R^2_{1\bullet3} \tag{14.45}$$

We also can represent changes in R^2 from adding X3 to a model that already includes X2 by either of two equations, using parentheses and dots (Equation 14.46) or the Δ (Equation 14.47).

$$R^2_{1(3\bullet2)} = R^2_{1\bullet23} - R^2_{1\bullet2} \tag{14.46}$$

or

$$\Delta R^2_{1\bullet2 \to 1\bullet23} = R^2_{1\bullet23} - R^2_{1\bullet2} \tag{14.47}$$

Equation 14.44 (or 14.45) shows us how much variance X2 can explain over and above what can be explained by X3. It is the equation to find the area in the section of the Venn diagram labeled 2. This is the difference between the variance explained by an equation that includes X2 and X3 (in the Venn diagram areas 2, 3 and 4) and one that includes only X3 (area 3 and 4 in the Venn diagram). For the model of treaty participation, if we consider voice and accountability as X2 and NGOs as X3, then:

$$R^2_{1(2\bullet3)} = R^2_{1\bullet23} - R^2_{1\bullet3} = 0.731 - 0.371 = 0.36 \tag{14.48}$$

That is, NGOs can uniquely explain about 36 percent of the variation in treaty participation over and above what is explained by voice and accountability uniquely and the variance in treaty participation explained by both in a shared way.

Next, consider the equation for how much adding variable X3 increases variance explained over and above that explained by X2.

$$R^2_{1(3\bullet2)} = R^2_{1\bullet23} - R^2_{1\bullet2} = 0.731 - 0.725 = 0.006 \tag{14.49}$$

Here we see that number of NGOs can explain almost none of the variation in treaty participation over what is explained uniquely by voice and accountability and what is explained jointly by the two variables.

Statisticians refer to the square roots of these R^2 subtractions as **part correlations**.[5] The part correlations have the same sign as the b coefficient. So the part correlation of voice with treaty participation controlling for NGOs is 0.60, and the part correlation of NGOs with treaty participation controlling for voice is 0.08.

We can test the hypothesis that the **part R^2** equals zero, which is the hypothesis that a particular variable explained no variance when the other variable was controlled. That is, we are testing the hypothesis that adding a variable to the model did not increase our ability to predict the dependent variable. To do that, we use the following F test:

$$F_{1,N-k} = \frac{\Delta R^2 / 1}{(1 - R^2_{1\bullet23}) / N - k} \tag{14.50}$$

The numerical value of this F test is the square of the numerical value of the t associated with the test that the regression coefficient for the X variable under consideration is zero. The two tests are equivalent, as we would hope since logically no variance explained means b equals zero – if a variable doesn't explain variance then changes in that variable don't lead to changes in the dependent variable.

Multiple Regression

The move from two independent variables to an unlimited number of independent variables is straightforward, although the computations involved are cumbersome if done by hand. The regression equation is:

$$Y = a_{1 \bullet 23...} + b_{12 \bullet 3...}X2 + b_{13 \bullet 2...}X3 + ... + e_{1 \bullet 23...} \tag{14.51}$$

where by mathematical convention, the "..." means more stuff that looks like what came before, in this case more independent variables. Suppose that we want to also consider a nation's ability to implement and carry out policies and programs as a predictor of treaty participation, and that we label it X4. Then the regression with three independent variables is:

$$Y = a_{1.234} + b_{12.34}X2 + b_{13.24}X3 + b_{14.23}X4 + e_{1.234} \tag{14.52}$$

We have to use all the subscripts to keep track of what variables are being included in the equation since the value of the regression coefficient for a variable depends on which other variables are controlled. Writing all this down is a bit messy, but the logic is straightforward.

Again, we can estimate standard errors and perform the test of the hypothesis that any coefficient has a particular population value using a t test:

$$t_b = \frac{b - B_{Hyp}}{\sigma_b} \tag{14.53}$$

This is a test that the variable has no effect on the dependent variable, given that the effects of the other variables are statistically controlled by being included in the regression. So we can conduct a t-test of the hypothesis that in the population the regression coefficient equals zero for each independent variable.

An F-test can be used to test the hypothesis that the population R^2 is zero, which means that all Bs are simultaneously zero:

$$F_{k,N-k} = \frac{R^2_{1 \bullet 23...} \Big/ k}{(1 - R^2_{1 \bullet 23...}) \Big/ N - k} \tag{14.54}$$

While some researchers find that this isn't very useful, most seem to feel that if this test can't reject the hypothesis that no variance is explained by the model overall, then one shouldn't go on to tests on individual variables. Note that this is the same kind of F test we developed in analysis of variance. Analysis of variance is just a special case of regression where all the independent variables are categorical.

The increase in variance explained by a particular variable net of all other variables in the equation can be calculated, and the hypothesis that in the population this increment is zero can be tested using an F test.

$$F_{1,N-k} = \frac{\Delta R^2 / 1}{1 - R^2_{1 \bullet 23...} / N - k} \tag{14.55}$$

We are testing the hypothesis that a particular independent variable does not add anything to the ability of the regression to predict the dependent variable. This is logically equivalent to the t test that the population regression coefficient for a particular variable equals zero. In fact, if you square the value of that t test, you get the value of the F test. Again, saying a variable explains no variance over and above the other variables in the equation is equivalent to saying that the b value for the variable is zero.

In addition, we can test the hypothesis that a set of variables adds no predictive power to the equation (a so-called multiple part correlation) using the following calculation:

$$F_{k1-k2,N-k1} = \frac{\Delta R^2_{k2 \to k1} / k1 - k2}{1 - R^2_{k1.} / N - k1} \tag{14.56}$$

Remember that when we are looking at changes in R^2, we are comparing two regressions. One regression is larger – it has more variables. The other regression has only a subset of the variables in the large regression. In the example we have been using, the larger regression has two independent variables, number of NGOs and voice and accountability. Then to calculate part R^2 – the proportion of variance in the dependent variable uniquely explained by the independent variable, we also looked at a smaller regression that had only one of the independent variables. The difference in R^2 between, for example, the regression using both NGOs and voice and accountability to predict treaty participation and one using only NGOs is the amount of variance explained in treaty participation that can be uniquely attributed to voice and accountability. That is the part R^2 we are working with here.

There are k1 parameters in the larger regression, k2 in the smaller, so k1 − k2 is the number of variables added going from the smaller to the larger. So with three independent variables in the smaller equation and four in the larger (along with an intercept in each) we have k1 − k2 = 5 − 4 = 1. The F test in Equation 14.56 is to see if we can believe that the added variables explain some variance over and above that explained by the variables in the smaller equations. If we have only two independent variables in the large equation, this is the same as testing whether or not a single variable explains variance in the population (equivalent to testing whether or not the b value is zero in the population) over and above the variance explained by the other independent variable. But we can have many variables and compare the large equation with all variables with a smaller equation in which several variables, not just one, are omitted to see of the set of omitted variables matters.

The numerator of the equation is the ratio of the change in variance explained by adding the block divided by the number in the block. The denominator is the variance unexplained by the larger regression divided by the degrees of freedom in that regression. This is equivalent to testing the hypothesis that all variables in the block have Bs equal to zero.

Other Features of Multiple Regression

β coefficients

Standardized regression coefficients, also known as beta weights, can be calculated to estimate effects in multiple regression. The simplest interpretation of standardized regression coefficients is that they are the regression coefficients that result from running the multiple regression after all variables have been transformed into Z scores. We generally prefer **unstandardized coefficients** that relate variables to each other in the units of measurement. An exception is when all variables are scores on scales that are standardized so that all scales have equal standard deviations (many attitude scores are in this form). Then **standardized coefficients** seem appropriate. Often variables are measured on different scales and therefore unstandardized regression coefficients are difficult to compare. Thus, beta weights are commonly reported in the literature. But betas are confounded by variable range so that a variable with greater variability will have a larger beta weight.

Collinearity

As we noted above, if there is no relationship between the independent variables (the R^2 in a regression predicting one independent variable with all the other independent variables equal to zero) then the multiple regression results are the same as the results of the bivariate regressions. In an experiment, we can design the study so that the independent variables are not correlated. This is one of the great advantages of experiments. But when we don't have an experiment that has created data with no correlation among independent variables, and whenever we have data that does not come from a designed experiment, we will have correlations among the independent variables. As a result the b coefficients and tests of significance in a multiple regression will not be the same as what we get with the bivariate regressions using one independent variable at a time. This is exactly why we use multiple regression. We want to control for other independent variables – possible causes of the dependent variable – when we estimate the effects of a particular dependent variable.

Unfortunately, if independent variables are too highly correlated, the ability of regression to estimate the effects of each variable is much weakened – sometimes to the point where we really can't draw any conclusions. This problem is called multicollinearity. Texts in regression analysis deal with it in some detail. Informally, we can see that if independent variables are too highly correlated, they are behaving as if they are the same thing in the data we have, and thus regression cannot differentiate the effects of one from the other. Regression analysis gives "credit" to an independent variable only for the variance that it explains in the dependent variable over and above the other independent variables. Recall the Venn diagrams for two independent variables. The larger the areas of shared variance explained (area 4) and the smaller the area of unique variance explained (area 2 for X2 and area 3 for X3) then the less the effect that can be clearly attributed to each of the independent variables. The amount of area explained by the two independent variables that cannot be attributed uniquely to one or the other (area 4) depends on the correlation of the two independent variables. When it is high (area 4 and 5) then less is left for unique variance explanation (areas 2 and 3). We will return to this issue later, when we discuss the meaning of statistical control.

Auxiliary regressions are a good way to detect colinearity. Recall that an auxiliary regression for an independent variable uses all the other independent variables in a regression to predict that variable. There is one auxiliary regression for each independent variable. As a rough guide, many researchers suggest that one should be concerned about multicollinearity degrading the ability of a regression to give useful results when the R^2 values of the auxiliary regressions are above 0.4 and that above 0.6 serious problems may be occurring. There are more elaborate diagnostics for multicollinearity described in regression texts.

Sample size

Technically, one can run a multiple regression whenever we have one more observation than we have independent variables. So for a bivariate regression, we can calculate the regression with two cases, and if we have two independent variables we can run the regression with just three cases. But this is very bad practice – there's not really enough data to have believable results from the regression unless we have substantially more cases than independent variables. A rough guide would be that we should have at least five times as many cases as independent variables and it's better to have 10 or even 20 cases per independent variable. Thus for a bivariate regression we should have at least 5 and ideally 10 or 20 cases, for two independent variables 10 cases is minimal and 20–40 is better, for three independent variables, at least 15 cases and 30–60 is better. These are not absolutes of course. If we want to use the Law of Large Numbers to justify statistical inference, it would be good to have 100 or more cases. And in small samples, the problem of collinearity is often worse than it is in larger samples. We can also have a problem with very large samples. Since the size of standard errors for the b coefficients is inversely proportional to the square root of sample size, the larger the sample the smaller the standard error and thus the larger the t value for a particular b value. So with very large samples (some surveys have thousands or even tens of thousands of cases) nearly every b coefficient (even those indicating very small substantive effects on the dependent variable) is statistically significant. It is with large samples that we have to be very careful to distinguish between statistical significance (we can be confident that in the population the coefficient is not zero) and substantive importance (we can be sure the effect is not zero and the effect we see is large enough to matter in the context of what we are studying.)

Statistical Control

In an experiment, control for the effects of a variable occurs in one of two ways: the variable is either one of those included in the experimental manipulation, or it is one of the variables whose effects should be distributed at random across observations because of the random assignment of individuals to various experimental and control conditions. With non-experimental data, control for the effects of variables is accomplished by including the variables in the model. What does control mean in this context?

Think back to the Venn diagrams. We found that X3, NGOs, has a significant effect on Y, net of (controlling for) X3. This means that the variance in Y *that cannot be explained by X3* can be explained by the variance in X2 *that cannot be explained by X3*. An independent variable has an effect in a regression if and only if it provides predictive ability beyond what the other independent variables predict *and* that predictive power comes after the other independent variables' effect on the independent variable in question has been "removed." This gives a sense of why collinearity among independent variables is such a problem. If one independent variable is well predicted by the others, then there is no variance left to explain variance in the dependent variable. This is the problem of multicollinearity we mentioned above.

Residuals and statistical control are discussed at the end of this chapter as an Advanced Topic.

A Three Independent Variable Example

We have already examined the effects of number of NGOs and voice and accountability on treaty participation. Let's now add a nation's ability to implement and carry out policies and programs as a predictor of treaty participation. To measure this, we will use the "Government Effectiveness Index" created by the same researchers who developed the voice and accountability index (Kaufmann et al., 2002). This measure is also on a −2.5 to 2.5 scale. The scale measures a government's ability to design and implement policies and programs.[6] It is hypothesized that countries that are more "effective" will be more likely to participate in environmental treaties. We can add the government effectiveness variable to our two independent variable model and see what variables influence treaty participation.[7]

Figure 14.6 is the scatterplot matrix for the data. We have restricted the plot to the 148 countries for which all four variables are available. We note that there appear to be positive relationships between government effectiveness and the other three variables. It also appears that the scatterplot between government effectiveness and treaty participation has quite a bit of "scatter," meaning several data points fall far from the overall pattern of the data. Now we can use multiple regression to examine whether treaty participation is predicted by government effectiveness, controlling for the effects of NGOs and voice.

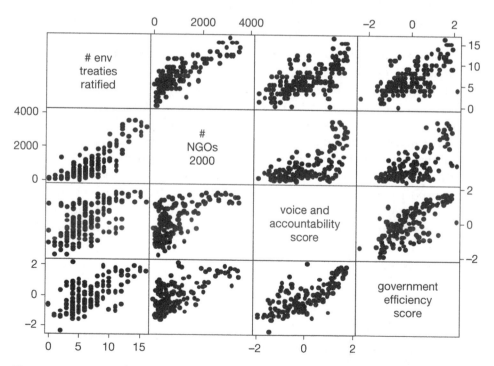

Figure 14.6   Scatterplot of treaty participation, NGOs, voice and accountability, and government effectiveness, N = 148
Data source: Kaufmann et al., 2002, Roberts et al., 2004, analyzed with SPSS.

Table 14.3 shows the multiple regression results. Consistent with the previous multiple regression model, number of NGOs is a statistically significantly predictive of treaty participation, while voice and accountability and government efficiency are not predictive of treaty participation when the effects of the other variables are controlled.

Table 14.3   Regression of treaty participation on NGOs, voice and accountability and government effectiveness, N = 148

Independent variable	Coefficient (t value)
Number of NGOs	.003
	(12.45)***
Voice and accountability	.302
	(1.27)
Government effectiveness	.185
	(.78)
Intercept	4.17
	(16.49)***
R^2	0.729
	$F_{(3,144)} = 129.44$***

*p<0.05, **p<0.01, ***p<0.001
Data source: Kaufmann et al., 2002, Roberts et al., 2004, analyzed with SPSS.

Regression and analysis of variance are discussed at the end of this chapter as an Advanced Topic.

What Have We Learned?

Multiple regression is similar to bivariate regression except that we are modeling the effects of several independent variables on a dependent variable (remember bivariate regression includes one independent variable). As with bivariate regression we can also do inference from a sample to a population when the sample size is large or when the sample size is small and the residuals have the properties described in the last chapter. The power of multiple regression is that it allows us to compare the effects of several variables at once in non-experimental data. But in doing so, it is important to keep in mind how regression attributes effects to variables. Each variable gets credit for what variance it can predict in the dependent variable over and above the variance predicted by all the other independent variables in the same model.

Advanced Topic 14.1   Adjusted R²

While R^2 is a useful way to look at how well we are doing in predicting the dependent variables with the independents, it has a flaw. The more variables we add, the better prediction we will get, even if those variables are not important. For example, to illustrate, we generated a random number with a uniform distribution and added it to the equation predicting treaty participation with NGOs and voice and accountability. While the random number is not statistically significant, as we would hope, adding it increases the R^2 from about 0.73 to about 0.74. Statisticians sometimes refer to this effect as the regression "capitalizing on chance." By this they mean that the OLS regression will do the best job it can to fit the Y values with the data, and even a random number allows for some small adjustments that slightly improve fit. So to compare fit across models, we use the "adjusted R^2" that corrects for how many predictor variables are being used. The equation for the adjusted R^2 (which we label R_A^2) is:

$$R_A^2 = 1 - \left(\left(\frac{N-1}{N-k} \right) (1 - R^2) \right) \qquad (14.57)$$

In the equation, N is the number of data points, k is the number of independent variables and R^2 is the proportion variance explained. So for the treaty participation model, we have:

$$R_A^2 = 1 - \left(\left(\frac{162-1}{162-3} \right) (1 - 0.731) \right)$$
$$= 0.728 \qquad (14.58)$$

This is a bit smaller than R^2. We can use the adjusted R^2 as a diagnostic when we are adding new variables to a model. If the adjusted R^2 does not increase, it means that the new variable does little to explain variance in the dependent variable.

It is important to use R^2 and adjusted R^2 only to compare across models of a particular

dependent variable within a data set. It is tempting to look at the percentage variance explained in a field of research and to think that this gives some sense of how "advanced" that field is. But in fact, R^2 depends not just on the quality of the model, but on several other factors, such as the variability in the dependent variable. The variability in the dependent variable in turn depends on the variability in the independent variables. But it is also a result of variability in the error terms, E, in the population. The e values in our equation are samples of the E values in the population. If E is measurement error, then better measurements may improve the variance explained. If E is the result of variables left out of the model, then including them will increase the variance explained. But to the extent that E is caused by factors like randomization or sampling error, it is not clear how better science will lead to smaller typical values for E and therefore larger values for R^2. As an example, it is nearly always the case that studies that use individuals for the units of analysis yield smaller R^2 than studies that use nation states or other aggregate units of analysis.

One of our statistical mentors, Tim Tardiff, performed an interesting analysis. He built a model using individual level data on how people traveled to work, and the R^2 was small, about 0.10. Then he used data for Census tracts and got a much larger R^2. Why? Tim argued that the errors in the individual level model were truly random factors and so when you average them out across all the individuals in a Census tract, they in fact average out to zero. So despite the smaller R^2 for the individual level data, the individual level model was a more powerful predictor – something you couldn't see if you just compared the R^2 for the individual regression with the R^2 for the regression based on Census tract data. Thus, we should only compare the R^2 values within one data set and for models with the same dependent variables.

Advanced Topic 14.2   Residuals and Statistical Control

We have discussed statistical control by looking at Venn diagrams and part R^2 values. We can also think about statistical control in terms of residuals. We will look at this with just two independent variables. In addition to the multiple regression, we have four bivariate regressions, one with each independent variable predicting the dependent variable and one with each independent variable predicting the other independent variable. Results for such a set of regressions for our example are in Table 14.2. For each of these bivariate regressions, we have a set of residuals that can be indicated using the standard notation as follows,

$$e_{1 \bullet 2} = (Y - \hat{Y}_{1 \bullet 2}) \qquad (14.59)$$

This is the residual from using X2 alone to predict Y. In the example it's the residual we get when we predict treaty participation using only voice and accountability.

$$e_{1 \bullet 3} = (Y - \hat{Y}_{1 \bullet 3}) \qquad (14.60)$$

This is the residual from the regression using just X3 to predict Y, in the example using only number of NGOs to predict treaty participation.

$$e_{2 \bullet 3} = (X2 - \hat{X}_{2 \bullet 3}) \qquad (14.61)$$

This is the residual from one of the auxiliary regression where we have used X3 to predict X2, in the example using NGOs to predict voice and accountability.

$$e_{3 \bullet 2} = (X3 - \hat{X}_{3 \bullet 2}) \qquad (14.62)$$

This is the residual from using voice and accountability (X2) to predict NGOs (X3).

Remember that the auxiliary regression among the independent variables doesn't have to make theoretical sense, they are just tools for working with the statistics. But the regression predicting the dependent variable should always make sense theoretically.

If we run regression using these residuals, we can obtain the b coefficients from the multiple regression in the following way:

$$e_{1 \bullet 2} = b_{13 \bullet 2} e_{3 \bullet 2} \qquad (14.63)$$

$$e_{1 \bullet 3} = b_{12 \bullet 3} e_{2 \bullet 3} \qquad (14.64)$$

That is, the b coefficient that links voice and accountability to treaty participation in the multiple regression of voice and accountability and NGOs on treaty participation can be thought of as the effect of voice on treaty participation after the effects of NGOs have been "removed" from both voice and treaty participation. (Note that the intercept terms disappear from the equations because the residuals have a mean of zero, so the intercept will be zero.) So too, the b coefficient for predicting treaty with NGOs while controlling for voice can be obtained by regressing the residual of environmental treaty participation predicted by voice on the residual of NGOs predicted by voice. Again, the effect of NGOs on treaty participation, as it is calculated in the multiple regression, is the effect after the predictive power of voice has been eliminated from *both* the dependent variable treaty participation and the independent variable we are examining, NGOs. Of course we don't do the calculations this way, but it gives a different way to think about what a statistical control in a regression means.

Advanced Topic 14.3   Regression and Analysis of Variance

You will have noticed that we are again calculating explained and unexplained sums of squares, just as we did in the chapter on analysis of variance. Further, we use an F test to see if independent variables help predict the dependent variable. Regression analysis and analysis of variance can be thought of as the same statistical technique. We use the term analysis of variance when the independent variables are categorical and we build a model using binary independent variables. We use the term multiple regression when the independent variables are continuous. But these are conventional distinctions. An independent variable in an analysis of variance can be continuous and an independent variable in a regression can be a binary variable or can be represented by a set of binary variables.

Applications

In this final set of applications we will bring together many of the variables we have used throughout the previous chapters and see what we can conclude about what causes variation in each of the dependent variables we have been studying.

Example 1:   Homicide rate

In previous chapters we have considered a number of possible causes of state homicide rate. Among these have been: the percentage of people with less than a high school education; the poverty rate; the percentage of people who are single; the percent urban; and whether or not the state was in the South. Figure 14A.1 is the scatterplot matrix for these variables. Note that because the South variable is a dichotomy, the scatterplot is not very helpful. All the variables seem to have some relationship to the homicide rate, which we see by looking across the first row in the scatterplot matrix. However, we also see strong associations among the independent variables.

Table 14A.1 provides the regression of the homicide rate on all these independent variables. We have added a column that includes the R^2 from the auxiliary regressions because we suspect that there might be problems of multicollinearity. The model suggests that low levels of education, high proportion of single people and being in the south each tend to increase the homicide rate. Neither the urban percent or the percent living in poverty seem to have any influence. However, the auxiliary regressions suggest that the effects of low educational levels and poverty may be influenced by multicollinearity. It is not surprising that these two variables are correlated with each other. In a larger research project we might try to create an index by combining the two into a measure that assesses the number of people with limited resources. But for simplicity, we will rerun the regression without the education variable. It may be that education matters but because it is so strongly related to the other variables it does not have much unique explanatory power

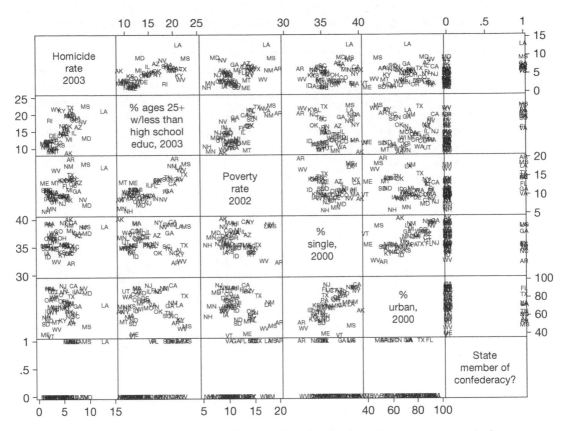

Figure 14A.1 Scatterplot of state homicide rate, education level, poverty rate, percent single, percent urban and region, N = 50
Data source: US Census Bureau, 2000, 2002, 2003, analyzed with SPSS.

Table 14A.1 Regression of homicide rate on full set of independent variables, N = 50

	Full model	R^2 from auxiliary regression
Percent with less than a high school education	0.291**	0.639
Percent in poverty	−0.009	0.610
Percent single	0.348*	0.303
Percent urban population	0.015	0.400
South	2.003*	0.421
Intercept	−13.645*	
R^2	0.551	

*p<0.05, **p<0.01, ***p<0.001
Data source: US Census Bureau, 2000, 2002, 2003, analyzed with SPSS.

Table 14A.2   Regression of homicide rate on reduced set of independent variables, N = 50

	Reduced model	R^2 from auxiliary regression
Percent with less than a high school education	–	–
Percent in poverty	0.223*	0.340
Percent single	0.334*	0.303
Percent urban population	0.033	0.347
South	2.921**	0.302
Intercept	−12.988*	
R^2	0.473	

*p<0.05, **p<0.01, ***p<0.001
Data source: US Census Bureau, 2000, 2002, 2003, analyzed with SPSS.

attributed to it. By dropping the most collinear variable, we may reduce this problem.

Table 14A.2 shows the new model. The levels of multicollinearity are all now below the rule of thumb of 0.4. Now we find that the percent in poverty has a significant effect. This bolsters our suspicion that in the larger regression it was not significant because of collinearity. From this we would conclude that the larger the population with a lack of resources (income or education) the larger the homicide rate, and that education seems more important than poverty but that because of multicollinearity we can't draw very strong conclusions about the relative importance of education and poverty from this data set. However, the results are quite clear that the more single people in a population the higher the homicide rate. Urbanization does not seem to have an effect when we control for other factors. And the effect of being in South that we have seen in other chapters is somewhat reduced by controlling for other factors but still remains statistically significant.

If we were to do further research on this topic, we might try to find variables that capture what is really going on with the effect of being a southern state. There have been some efforts to quantify differences in the "culture of violence" across states (Baron and Straus, 1988) and this might "unpack" what being in the south means. However, to go much further with the analysis of state level homicide rates it would be useful to expand the data set to include more observations. Obviously, we can't get data on more states, but we could get data for more years, converting our cross-sectional data set into a panel with observations for every state for a number of years. This would be a very powerful tool to apply to understanding why states differ in their homicide rates.

Example 2:   Animal concern

We have examined various factors that might influence concern for animals. Here we propose that animal concern varies as a result of gender, rural residence, age, and education. We have coded males as 1 and females as 2 to measure the effects of gender. We have created five binary variables, one for each ten-year age groups from age 30 on. We have used the 18–29 year olds as the category not included so all the regression coefficients for the other categories are comparing the effect of being in that age group to the views of the 18–29 year olds. We have coded rural residents as 1 and others as 0. For simplicity, we have used education as years of education. Table 14A.3 displays the results.

Women appear to endorse animal concern more strongly than men. There is a steady progression across the age categories, with the youngest group most pro-animal and the oldest group least pro-animal, although the very oldest group is slightly less pro-animal than the 60–69 year olds. Rural residents are less pro-animal than those living in the suburbs or cities, and education seems to decrease the endorsement of medical research on animals.

To keep our analyses relatively simple, we have so far not looked at differences across countries. For our final analysis of this issue, we will examine how being a resident of a country influences animal concern, net of the demographic variables. To keep things simple, we will use age as a single continuous measure rather than using the dummy variables for each age group. This may miss some subtleties, but since overall the age pattern seems to be one of steady decline in concern across age groups, this simplification won't do much harm. To capture country effects we have created a dummy variable for each country. We need to leave one country out

Table 14A.3   Regression of animal concern on full set of independent variables

Variable	Regression coefficient
Female	0.212***
Age 18–29	0
Age 30–39	−0.107***
Age 40–49	−0.105***
Age 50–59	−0.156***
Age 60–69	−0.270***
Age 70 and over	−0.228***
Rural	−0.042*
Education	−0.007***
Intercept	2.586***
R^2	0.013***
N	23,202

*$p<0.10$, **$p<0.05$, ***$p<0.001$
Data source: 2000 ISSP data set, analyzed with SPSS.

Table 14A.4　Regression of animal concern on demographics and country of respondent, N = 23,141

Variable	Sex	Age	Rural	Education	Intercept
Regression coefficient	0.212***	−0.006***	−0.031	−0.017***	3.030***
Country	Austria	Bulgaria	Canada	Chile	Czech Republic
Regression coefficient (deviation from US)	−0.070	−0.790***	−0.268***	0.374**	−0.365***
Country	Finland	East Germany	West Germany	Great Britain	Japan
Regression coefficient (deviation from US)	−0.132**	−0.407***	−0.001	0.194***	0.540***
Country	Latvia	Mexico	Netherlands	New Zealand	Norway
Regression coefficient (deviation from US)	−0.371***	−0.408***	0.076	0.075	−0.156**
Country	Philippines	Portugal	Russia	Slovenia	Spain
Regression coefficient (deviation from US)	−0.384***	−0.495***	−0.906***	−0.344***	−0.564***
Country	Sweden	Switzerland	United States		
Regression coefficient (deviation from US)	−0.174***	−0.037	0		

*p<0.05, **p<0.01, ***p<0.001; R^2 = 0.100
Data source: 2000 ISSP data set, analyzed with SPSS.

of the regression so that it can serve as the basis for comparison. Since the sample size is the US is one of the larger in the survey, we will leave it out, so all the coefficients reported show how residence in that particular country, controlling for gender, age, rural residence, and education, make the respondent have different views from a respondent in the US. We have to drop Denmark, Ireland and Israel from the sample because the surveys for those countries didn't include rural residence and Northern Ireland because that survey did not include education. These results are presented in Table 14A.4.

Both the demographic variables and the country variables have significant effects on the animal concern score, so both the respondent's demographic characteristics and her or his country seem to influence their views on the use of

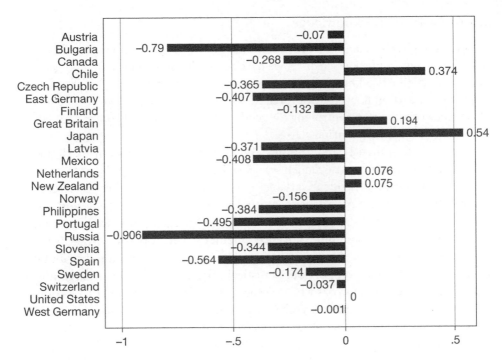

Figure 14A.2   Effect of nation on animal concern

animals in medical research. Figure 14.A.2 plots the country specific effects to make them easier to interpret. Note that this plot would be the same as a plot of the deviation of the mean for each country from the US mean *if and only if* the demographic variables had no effect. This plot shows how much the countries differ from the US *controlling for* sex, age, education, and rural residence.

Respondents from Japan, Chile and Great Britain are likely to offer strong animal concern compared to those in the US, while those in Russia, Bulgaria, Spain, and Portugal are likely to have the least endorsement of animal concern. The highest and lowest groups of countries appear to be quite diverse. It is interesting to speculate about what drives these patterns. We might hypothesize that religion plays a role because Spain and Portugal are mostly Catholic, but so is Chile, which is at the other end of the distribution of effects, while Mexico is relatively low. Bulgaria and Russia are both part of the former Soviet Union, and the Czech Republic, East Germany, Lativa and Slovenia also have relatively low scores, so that seems a more consistent pattern than we found with religion. In the normal course of research on this issue we would consider what makes theoretical sense and add such variables to the model to see if they clarify the pattern.

Example 3:   Environmental treaty participation

Treaty participation examples were developed throughout the text of this chapter, so we will not use this illustration in the Applications.

Example 4:   AIDS knowledge

In this example, we examine a more complex model of AIDS knowledge than we have up to now. In previous chapters we have found that both education and gender had an effect on AIDS knowledge. We now want to see if controlling for the effects of gender impact on how education relates to AIDS knowledge (and how gender's effects may change when controlling for gender). Then we will add controls for rural residence, religion, and marital status.

As we noted in the last applications section, regression analysis in its simplest form assumes a continuous dependent variable. Regression-like techniques that work well with categorical dependent variables are beyond what we will cover here. But as we noted, treating a binary dependent variable as a continuous variable and using continuous and categorical independent variables to predict it usually gives results quite similar to those from more advanced methods. This is particularly true with a large sample where we don't need to make assumptions about the distribution of the residual terms to support our statistical inferences – the law of large numbers applied to regression works whatever the distribution of the residuals. Of course, if we were to conduct further research on the subject we would use one of the forms of regression appropriate for categorical dependent variables. These are called logit (or logistic) and probit regression. We have checked the results we report here against logit and probit regressions and we get the same conclusions no matter how we analyze the data.

We will use education as a continuous variable – the number of years of education. Gender will be a binary variable, coded 0 for male and 1 for female. Table 14A.5 shows the results of the regression using both gender and education as independent variables. In this and the next table we include the t values for each variable. Remember that there are many alternative ways of reporting the results of a regression equation.

Table 14A.5   Regression of AIDS knowledge on education and gender, N = 8,306

	Regression coefficient (t value)
Female	−0.04
	(−3.92)***
Education	0.02
	(19.58)***
Intercept	0.72
	(33.61)***
R^2	0.049
	($F_{2,8303} = 214.78$, <.0001)

*p<0.05, ** p<0.01, ***p<0.001
Data source: 2000 Uganda DHS data set, analyzed with SPSS.

Table 14A.6   Regression of AIDS knowledge on full set of independent variables, N = 8,300

Variable	Regression coefficient (t value)
Female	−0.064 (−5.79)***
Education	0.019 (15.48)***
Age	−0.005 (−7.98)***
Rural residence	−0.059 (−5.72)***
Catholic	0.014 (1.45)
Protestant	0
Muslim	0.064 (4.58)***
Other religion	−0.045 (−2.13)*
Never married	−0.050 (−3.97)***
Currently married	0
Formerly married	0.023 (1.60)
Constant	0.880
R^2	0.065 ($F_{9,8290}$ = 64.22, p < .0001)

*p<0.05, ** p<0.01, ***p<0.001
Data source: 2000 Uganda DHS data set, analyzed with SPSS.

Here we see that both gender and education have a statistically significant effect on AIDS knowledge. The R^2 for the entire equation is also statistically significant. The intercept tells us the value for a man with no formal education (0 years), is 0.72, that is, the model predicts that 72 percent of men with no formal education understood the relationship between AIDS transmission and condom use.

Now we turn to the more elaborate model which we report in Table 14A.6 Note that since we are using more variables the sample size has dropped from the previous regression because data on some of these new variables is not available for some respondents. Gender and education continue to have significant effects, with women 6 percent less likely than men to know that condom use can prevent AIDS and each additional year of education increasing the probability of a correct answer by nearly 2 percent. Each additional year of age decreases the probability of a correct answer by about half a percent. Catholics and Protestants are equally likely to answer the question correctly, while Muslims are 6 percent more likely to get the answer correct and those of other religions 4.5 percent less likely to know the relationship between condom use and AIDS transmission. Currently married and formerly married people are equally likely to have answered correctly (the effect of being formerly married is not significantly different from the reference category of currently married), while those who have never been married are 5 percent less likely to get the correct answer. Because we have included age as an independent variable, the constant has no direct interpretation because we can't have someone with zero years of age in the study population (recall that the constant is the value we predict on the dependent variable when all the independent variables are equal to zero.)

Exercises

1. In the last chapter, we examined the relationships between six variables and job satisfaction using bivariate regression. Now let's examine these relationships using multiple regression and include gender also. Be sure to refer back to Chapter 13 to remember how the variables were measured.

a) What is the major difference between the R^2 and Adjusted R^2?
b) What does the F statistic tell us?
c) What does the beta value of $-.152$ for difficulty fulfilling family responsibilities indicate?
d) Write the regression equation line.
e) What does the constant value indicate in this equation?

f) What does the t value for number of young children indicate?
g) Summarize how well this model predicts the variance in job satisfaction. What are the statistically significant predictors in this model? How did you determine this?
h) Refer back to the bivariate analyses conducted in the exercise in the prior chapter. Have any of the statistically significant relationships found at the bivariate level changed in the multivariate model? Why might a relationship that was statistically significant at the bivariate level become non-significant when examined using multivariate regression?

Table 14E.1 Regression of job satisfaction on seven predictor variables

Statistics

R^2	0.037	F	8.595
Adjusted R^2	0.033	P value	0.000

Coefficients

	Unstandardized coefficients		Standardized coefficients		Sig.
	B	Std. Error	Beta	t	
Constant	5.241	0.238	–	21.981	<0.001
Age of respondent	−0.002	0.004	−0.012	−0.486	0.627
Number of hours worked weekly	0.004	0.002	0.059	2.161	0.031
Number of hours doing housework	−0.002	0.002	−0.028	−1.028	0.304
Job is rarely stressful	0.090	0.024	0.094	3.708	<0.001
Difficult to fulfill family responsibility	−0.171	0.029	−0.152	−5.858	<0.001
Number of young children	0.008	0.024	0.009	0.340	0.734
Respondent's sex	0.023	0.064	0.010	0.361	0.718

Dependent variable: Job satisfaction; *Data source*: 2000 ISSP.

2. Now let's reexamine the same multiple regression model, adding views about whether "life at home is rarely stressful" (1 = strongly agree to 5 = strongly disagree) and whether one is "too tired from work to do duties at home" (1 = several times a week to 4 = never) to the model.

a) How much explained variance does the inclusion of the two variables add to the model?
b) Are views about life at home being stressful and work being too tiring statistically significant predictors of job satisfaction?

c) Did the addition of the two variables affect any of the other relationships? If yes, what changes occurred, and why might these changes have taken place?
d) Now take this final model with nine predictor variables and draw the most logical causal model, indicating which variables had statistically significant direct relationships with job satisfaction. In drawing this diagram, think through whether some of the variables may have indirect relationships with job satisfaction.

Table 14E.2 Regression of job satisfaction on nine predictor variables

Statistics

R^2	0.041	F	7.193
Adjusted R^2	0.035	P value	<0.001

Coefficients

	Unstandardized Coefficients		Standardized Coefficients		Sig.
	B	Std. Error	Beta	t	
Constant	4.840	0.288	–	0.055	<0.001
Age of respondent	−0.001	0.004	−0.008	−0.335	0.738
Number of hours worked weekly	0.005	0.002	0.062	2.259	0.024
Number of hours doing housework	−0.003	0.002	−0.028	−1.040	0.298
Job is rarely stressful	0.083	0.025	0.087	3.257	0.001
Difficult to fulfill family responsibility	−0.119	0.035	−0.106	−3.443	0.001
Number of young children	0.005	0.025	0.005	0.197	0.844
Respondent's sex	0.054	0.066	0.024	0.821	0.412
Life at home rarely stressful	0.001	0.026	0.001	0.055	0.956
Work too tiring for home duties	0.096	0.034	0.084	2.794	0.005

Dependent variable: Job satisfaction; *Data source*: 2000 ISSP.

3. A researcher is interested in understanding what predicts the amount of monetary contributions people give to various charities and organizations. The researcher selected five potential predictors from the US General Social Survey: age; sex; income; number of children in the household; and perceptions about income. The multivariate regression output is presented in Table 14E.3.

a) The researcher hypothesized that older adults, those with more income, females, those with fewer children in the household, and those who assess their income more positively would contribute more money. Is this researcher correct in her hypotheses?

Table 14E.3 Regression of monetary contributions to charities and organizations on five predictor variables

Statistics

R^2	0.030	F	14.952
Adjusted R^2	0.028	P value	<0.001

Coefficients

	Unstandardized Coefficients		Standardized Coefficients		Sig.
	B	Std. Error	Beta	t	
Constant	−888.890	277.625		−3.202	0.001
Age of respondent	4.929	2.779	0.039	1.774	0.076
Total family income	25.072	18.665	0.029	1.343	0.179
Respondent's sex	−152.879	82.565	−0.037	−1.852	0.064
Number of children	55.158	26.954	0.045	2.046	0.041
Opinion of family income	319.785	50.430	0.138	6.341	<0.001

Dependent variable: Total amount of money donated to various charities
Data source: 1996 General Social Survey.

References

Baron, L. and Straus, M. A. 1988. Cultural and economic sources of homicide in the United States. *The Sociological Quarterly* 29, 371–390.

Cohen, J. and Cohen, P. 1975. *Applied Multiple Regression/Correlation Analysis for Behavioral Sciences*. Hillside, NJ: Lawrence Erlbaum Associates.

International Social Survey Programme (ISSP). 2000. 2000 Environment II data set. www.issp.org. Catalog no. ZA 3440. Cologne, Germany: GESIS-ZA Central Archive for Empirical Research.

Ip, E. H. S. 2001. Visualizing multiple regression. *Journal of Statistics Education* 9(1).

Kaufmann, D., Kraay, A., and Zoido-Lobaton, P. (Jan. 2002). Governance Matters II: Updated Indicators for 2000/01. Policy Research Working Paper no. 2772. The World Bank Research Development Group and World Bank Institute; Governance, Regulation and Finance Division (http://hdr.undp.org/reports/global/2002/en/).

Kennedy, P. E. 2002. More on Venn diagrams for regression. *Journal of Statistics Education* 10(1).

Roberts, J. T., Parks, B. C., and Vasquez, A. A. 2004. Who ratifies environmental treaties and why?

Institutionalism, structuralism and participation of 192 nations in 22 treaties. *Global Environmental Politics* 4(3), 22–64.

United Nations Development Programme (UNDP). 2002. Human Development Report. Deepening Democracy in a Fragmented World. New York: Oxford University Press (http://hdr.undp.org/en/reports/global/hdr2002/).

US Census Bureau 2000. Table 33. Urban and rural population, and by state: 1990 and 2000. ("http://www.census.gov/prod/cen2000/index.html"\t"_blank"www.census.gov/prod/cen2000/index.html).

US Census Bureau 2002. Historical poverty tables: Table 21. Number of poor and poverty rate, by state: 1980 to 2006. Year 2002 ("http://www.census.gov/hhes/www/poverty/histpov/"\t"_blank"www.census.gov/hhes/www/poverty/histpov/hstpov21.html).

US Census Bureau 2003. Table 295. Crime rates by state, 2002 and 2003, and by type, 2003 ("http://www.census.gov/prod/2005pubs/06stata"www.census.gov/prod/2005pubs/06statab/law.pdf).

APPENDIX A: SUMMARY OF VARIABLES USED IN EXAMPLES

This appendix summarizes the variables we are using throughout the book for our four examples. The information has been organized by dataset and includes the variable name as it is listed in the dataset, a description of the variable, and how the variable was coded (including category labels).

Regardless of whether or not you are using the datasets and working through the examples with the original data, you will likely find this appendix helpful. When interpreting tables, figures, and graphs, it is important to remember how the variables are measured. This appendix provides a quick guide on all the variables.

Example 1: Homicide Rate of US states

These data were taken from various government sources, as indicated throughout the text.

State – State name
Abbreviation – 2-letter abbreviation for each state
Region – Region of the country
   1   South
   2   East
   3   Midwest
   4   West
Homicide03 – Homicides per 100,000 population per state (data for 2003)
Lesshs03 – Percent of the state population ages 25+ with less than a high school education (data for 2003)
Poverty02 – Percent of the state population in poverty (data from 2002)
Divorced00 – Percent of the state population that is divorced (data from 2000)

Poverty10 – Poverty data (poverty02) divided into ten categories (1 = States with lowest percent of population in poverty; 10 = States with highest percent of population in poverty)

Income02 – Per capita income (data from 2002)

Single00 – Percent of state population that are single (includes divorced and never married persons) (data from 2000)

Confederate – Is the state a former "Confederate"/Southern state? (0 = no, not a Confederate state; 1 = yes, a Confederate state)

Povertyhalf – Poverty data dichotomized into two categories: lower and higher poverty states (1 = Percent in poverty is in lower half of states (under the median) and 2 = Percent in poverty is in higher half of states (above the median))

Povertyconf – Combination of poverty rate and being Confederacy state:
1   Northern (non-Confederate) state, lower poverty state
2   Northern state, higher poverty state
3   Southern (Confederate) state, lower poverty state
4   Southern state, higher poverty state

Urban00 – Percent of state population living in urban area (data from 2000)

Deathpen – State has the death penalty? (1 = Not a death penalty state; and 2 = Death penalty state)

Homicidedi – Homicide rate dichotomized into lower and higher homicide rates (1 = Homicide rate is in lower half of states (under the median); 2 = Homicide rate is in higher half of states (above the median))

Zhomicide03 – Standardized scores of homicide03 data

Example 2: Animal Concern

These data are taken from the 2000 International Social Survey Programme (ISSP) Environment II survey module. www.issp.org (data catalog #ZA 3440).

Country – Country*
[*note: the coding of this variable was done by the ISSP and has not been changed]
2   West Germany
3   East Germany
4   Great Britain
5   Northern Ireland
6   United States
7   Austria
10   Ireland
11   Netherlands
12   Norway
13   Sweden
14   Czech Republic

15 Slovenia

17 Bulgaria

18 Russia

19 New Zealand

20 Canada

21 Philippines

22 Israel

24 Japan

25 Spain

26 Latvia

30 Portugal

31 Chile

32 Denmark

33 Switzerland

37 Finland

38 Mexico

Animalx – Belief that "It is right to use animals for medical testing if it might save human lives"

1 Strongly agree

2 Agree

3 Neither agree nor disagree

4 Disagree

5 Strongly disagree

Humanim – Belief that humans developed from animals (belief in evolution)

1 Definitely true

2 Probably true

3 Probably untrue

4 Definitely untrue

Sex – Respondent's sex (1 = male; 2 = female)

Age – Respondent's age (measured in years)

Marital – Respondent's marital status

1 Married/living as married

2 Widowed

3 Divorced

4 Separated

5 Single/never married

Urbrural – Type of community respondent lives in (self-reported data)

1 Urban area

2 Suburb, city, town, county seat

3 Rural area

Age6cat – Age recoded into six categories

1 ages 18–29

2 ages 30–39

3 ages 40–49

4   ages 50–59
5   ages 60–69
6   ages 70+

Agedi – Age dichotomized (1 = Age 45 and under; 2 = Ages 46 and older)

Humanimdi – Belief that humans derived from animals (humanism) dichotomized (1 = Yes, believe humans derived from animals (definitely and probably true); 2 = No, do not believe humans derived from animals (definitely and probably untrue))

Educ – Number of years of education completed

Agesex – Age and sex of respondent combined into four categories
1   Male, ages 45 and under
2   Male, ages 46 and older
3   Female, ages 45 and under
4   Female, ages 46 and older

Agegrp1 – Is respondent age 18–29?
0   not in age group
1   in age group

Agegrp2 – Is respondent age 30–39?
0   not in age group
1   in age group

Agegrp3 – Is respondent age 40–49?
0   not in age group
1   in age group

Agegrp4 – Is respondent age 50–59?
0   not in age group
1   in age group

Agegrp5 – Is respondent age 60–69?
0   not in age group
1   in age group

Agegrp6 – Is respondent age 70+?
0   not in age group
1   in age group

Rmarital – Martial status recoded (combine maried and separated)
1   Married/living as married/separated
2   Widowed
3   Divorced
5   Single/never married

Age3cat – Age collapsed into three categories
1   Young (ages 39 and under)
2   Middle aged (ages 40 to 59)
3   Older (ages 60 and older)

Animalx3 – Animalx collapsed into three categories
1   Low animal support (codes 1 and 2)
2   Moderate animal support (code 3)
3   High animal support (codes 4 and 5)

Agegender6 – Age and gender combined into six categories
1   Young men
2   Middle aged men
3   Older men
4   Young women
5   Middle aged women
6   Older women

Animalxdi – Animalx dichotomized (1 = Lower animal support (codes 1 and 2); 2 = Higher animal support (codes 3, 4, and 5))

Evolagedi – Age and belief in evolution (humanim) combined into four categories
1   Younger age, believes in evolution
2   Younger age, doesn't believe in evolution
3   Older age, believes in evolution
4   Older age, doesn't believe in evolution

Example 3: Country Environmental Treaty Participation

These country data were gathered and computed by Roberts et al. (2004) and Kaufmann et al. (2002).

Country – Country name

Popmill – Population of country is at least one million
0   Population of country is under one million
1   Population of country is one million or more

Envtreat – Number of environmental treaties each country participated in. The possible range = 0 to 16

Voice – Index measuring the extent citizens in a country have voice (freedom of expression) and governments are accountable to their citizens. The possible range = −2.5 to +2.5

Abbrev – Three-letter country abbreviation

Voicedi – Voice and accountability scale dichotomized into low and high voice (data from 2000) (0 = Lower level of voice and accountability in country (below median); 1 = Higher level of voice and accountability in country (above median))

Ngo – Number of NGOs in a country (data from 2000)

Ngodi – Number of NGOs dichotomized into low and high presence of NGOs (0 = Lower presence of NGOs in country (below median); 1 = Greater presence of NGOs in country (above median))

Ngovoice4 – NGO and voice data collapsed into four categories
1   Few NGOs, low voice
2   Few NGOs, high voice
3   Many NGOs, low voice
4   Many NGOs, high voice

Continent – Continent the country is in
1   North America
2   South America
3   Europe
4   Asia
5   Africa
6   Australia

Envtreatdi – Environmental treaty participation dichotomized (0 = Lower level of treaty participation (<6 treaties); 1 = Higher level of treaty participation (6+ treaties))

Goveff – Index measuring the extent a country's government is effective (data from 2000). The possible range = −2.5 to +2.5

Example 4: AIDS knowledge

This data set is taken from the 2000 Ugandan-Demographic and Health Survey (DHS) (www.measuredhs.com).

ID – Respondent's ID number (as designated by DHS)
Sex – Respondent's sex
1   Male
2   Female
Age – Respondent's age
Residence – Type of community respondent lives in (1 = Urban; 2 = Rural)
Religion – Respondent's religious affiliation
1   Catholic
2   Protestant
3   Muslim
4   Other religious affiliation
Educ – Respondent's highest educational attainment
0   No education
1   Incomplete primary
2   Completed primary
3   Incomplete secondary
4   Completed secondary
5   Higher attainment than secondary school
Marital – Respondent's marital status
0   Never married
1   Married
2   Living together
3   Widowed
4   Divorced
5   Not living together

Rmarital – Marital status collapsed into three categories
- 0 Never married
- 1 Currently married
- 2 Formerly married

Sexres – Sex and type of residence combined into four categories
- 1 Male urban dweller
- 2 Female urban dweller
- 3 Male rural dweller
- 4 Female rural dweller

AIDScon – Knowledgeable about AIDS transmission (know condoms can help prevent the transmission of AIDS)?
- 0 No
- 1 Yes

Catholic – Respondent is Catholic?
- 0 No
- 1 Yes

Muslim – Respondent is Muslim?
- 0 No
- 1 Yes

Othrelig – Respondent has other religious affiliation (not Catholic, Protestant or Muslim)?
- 0 No
- 1 Yes

Nevmarried – Is respondent never married?
- 0 No
- 1 Yes

Formmarr – Is respondent formerly married?
- 0 No
- 1 Yes

APPENDIX B: MATHEMATICS REVIEW

Many students enrolled in a course in social statistics have not exercised their mathematics skills in some time. There are a few basic concepts from earlier mathematics courses that you will need for your statistics course. We review them here.

Arithmetic with Signed Numbers

In statistical calculations we often have to use numbers that have signs, that is, numbers that are positive or negative. If we follow a few simple rules, we can do arithmetic with signed numbers just as easily as with unsigned numbers. Here are the rules.

Adding signed numbers

Step 1 Group together all numbers with the same sign
Example: Suppose we have to add together $(+7) + (-2) + (-5) + (+5) + (-4)$. Step 1 says group all the numbers with like signs together, so that would be:

$$((+7) + (+5)) + ((-2) + (-5) + (-4)) \tag{AB.1}$$

Step 2 Add together all the positive numbers and add together all the negative numbers, ignoring the signs within each group
Example: So we would add $7 + 5 = 12$ for the positive numbers and $2 + 5 + 4 = 11$ for the negative numbers. Thus our example becomes:

$$(12) + (-11) \tag{AB.2}$$

Step 3 Subtract the smaller number from the larger number not worrying about the signs of the two numbers
Example: 12 is larger than 11 so 12 − 11 = 1.

Step 4 Give the result the sign of the larger number
Example: 12 is larger than 11 so the result is +1.

Subtracting signed numbers

Remember that we only define subtraction in terms of two numbers, the number we start with (sometimes called the minuend) and the number we subtract from it (sometimes called the subtrahend). This yields the result (the difference).

Step 1 Change the sign of the number being subtracted (the subtrahend)
Example: To find 6 − (−8), change the sign of the number being subtracted. So −8 becomes +8

Step 2 Add the two numbers following the rules for adding above
There are only two numbers. Both are now positive. 6 + 8 = 14. Since both are positive we assign a positive sign, so the result is +14.

  Another example: −9 − (−5). Step one changes −5 to +5. Then the rules say to subtract 5 from 9 to get 4. The negative 9 is larger than the positive 5 so the sign of the result is negative. That is:

$$-9 - (-5) = -4 \qquad\qquad\qquad (\text{AB.3})$$

Multiplying signed numbers

Step 1 Multiply the numbers together, ignoring the signs
Example: (+5) * (−2) * (−3) * (−5). Multiplying 5 * 2 * 3 * 5 = 150.

Step 2 If there is an even number of negative signs, the result is positive.
If there are an odd number of negative signs, the result is negative
There are three negative signs here, which is an odd number, so the result is negative. So:

$$(+5) * (-2) * (-3) * (-5) = -150 \qquad\qquad (\text{AB.4})$$

Dividing signed numbers

Remember that as with subtraction, there are two numbers we are using at a time. The denominator is the number we are dividing into the numerator.

Step 1 Divide the numbers
Example: 9/−3. The division yields 3.

Step 2 If both are of the same sign, the result is positive. If the numbers are of opposite sign, the result is negative
The numbers are of opposite signs, so the result is negative. The result of this division then is −3.

Summations

In statistics, we often need to add together long strings of numbers. In explaining statistics we have to have a way to find a way to discuss summations. The standard way to do this is with the summation operation, which is a Greek capital letter Σ ("sigma") with little subscripts and superscripts added. At first this may seem a bit daunting, but you will quickly get used to it.

Let's suppose we have 10 numbers we want to work with: 1, 3, 6, 4, 2, 7, 2, 6, 9, 3. When we refer to these numbers, without specifying which one we mean, we will call the numbers X. So X takes on 10 values: 1, 3, 6, 4, 2, 7, 2, 6, 9, 3. It is often useful to use the slightly fancier notation X_i to indicate one of the Xs without being specific about which one. X_i means "any X." We can also substitute a specific number of i. So X_1 is 1, X_2 is 3, X_{10} is 3 and so on.

Suppose we want to add them up. In Σ notation this would be written as:

$$\sum_{i=1}^{10} (X_i) \tag{AB.5}$$

We read this as "the sum from i equals 1 to 10 of X sub i." Let's look at it piece by piece. The Σ tells us that we are going to add things together. The little number (subscript) below the Σ says that we will use the index i to keep track of what we are adding. We are using "i" as an arbitrary symbol to "index" the things to be added. The subscript tells us where to start the additions, in this case with the first of the Xs. The superscript on top of the Σ tells us where to stop adding, in this case at the tenth X. Then the thing to be added is to the right of the Σ. The subscript on X reminds us to check back with the information above and below the Σ about where to start and where to end the addition.

So, for the little set of numbers we are using as an example:

$$\sum_{i=1}^{10} (X_i) = 1 + 3 + 6 + 4 + 2 + 7 + 2 + 6 + 9 + 3 = 43 \tag{AB.6}$$

Suppose we wanted to find the average of these numbers. To find the average (called the mean in statistics), we add up all the numbers and divide by how many we have. In statistics we usually call the number of data points we have N. To indicate division by N, we often multiply by 1/N which is the same as dividing by N. Or sometimes we just put the N as the divisor. It depends on what seems to make the equation easier to read. So, to find the average of the Xs, the equation could be written two ways:

$$Average = \left(\frac{1}{N}\right)\sum_{i=1}^{N}(X_i)$$ (AB.7)

or

$$Average = \frac{\sum_{i=1}^{N}(X_i)}{N}$$ (AB.8)

So, using the first equation, we would have

$$Average = \frac{1}{N}\sum_{i=1}^{N}(X_i) = \frac{1}{10}*(1 + 3 + 6 + 4 + 2 + 7 + 2 + 6 + 9 + 3)$$ (AB.9)

or

$$Average = \frac{1}{10}*(43) = \frac{43}{10} = 4.3$$ (AB.10)

Sometimes we have to do something a bit complicated to the numbers before we add them up. For example, we might subtract something from each number before we add. We will often subtract the average of all the numbers from each number before adding. The summation sign for that would be:

$$\sum_{i=1}^{N}(X_i - Average)$$ (AB.11)

For this data set, we've just calculated the average to be 4.3, so then the sum becomes:

$$\sum_{i=1}^{N}(X_i - Average) = \sum_{i=1}^{10}(X_i - 4.3) = (1 - 4.3) + (3 - 4.3) + (6 - 4.3)$$
$$+ (4 - 4.3) + (2 - 4.3)$$
$$+ (7 - 4.3) + (2 - 4.3) + (6 - 4.3)$$
$$+ (9 - 4.3) + (3 - 4.3)$$ (AB.12)

Then doing the subtractions, being careful to keep track of positive and negative signs (see the rules for signed numbers above) we have:

$$\sum_{i=1}^{N}(X_i - Average) = \sum_{i=1}^{10}(X_i - 4.3) = (-3.3) + (-1.3) + (+1.7) + (-0.3)$$
$$+ (-2.3) + (+2.7) + (-2.3)$$
$$+ (+1.7) + (+4.7) + (-1.3) \qquad \text{(AB.13)}$$

Using our rules for adding signed numbers above we clump together the positives and the negatives:

$$\sum_{i=1}^{N}(X_i - Average) = (+1.7) + (+2.7) + (+1.7) + (+4.7) + (-3.3)$$
$$+ (-1.3) + (-0.3) + (-2.3) + (-2.3) + (-1.3) \qquad \text{(AB.14)}$$

Then adding within each group of same sign numbers:

$$\sum_{i=1}^{N}(X_i - Average) = (+10.8) + (-10.8) \qquad \text{(AB.15)}$$

The final step is to subtract. Since both numbers are equal, the rule of subtracting the smaller from the larger works either way as the result is still zero. Since zero is neither positive nor negative, we don't have to assign a sign to it. So:

$$\sum_{i=1}^{N}(X_i - Average) = 0 \qquad \text{(AB.16)}$$

Usually we will let the computer do any calculation involving more than a few numbers but it is important to walk through these steps to understand what we mean when we use the Σ process.

Finally, when it is very clear that we are summing up all the data, from 1 to N, to simplify things we drop the subscript and the superscript on the Σ. Thus we could write:

$$Average = \left(\frac{1}{N}\right)\sum_{i=1}^{N}(X_i) \qquad \text{(AB.17)}$$

or

$$Average = \left(\frac{1}{N}\right)\sum(X_i) \qquad \text{(AB.18)}$$

Both mean the same thing.

Because we are always adding up all the data points, in the text we drop the subscripts and superscripts on Σ and the subscript "i" to make things simpler.

The Equation for a Straight Line

In the text we will explain how the equation translates into a straight line. You may have seen the equation for a straight line expressed as:

$$Y = mX + b. \tag{AB.19}$$

In statistics, we usually use slightly different symbols. For now, we just want to remind you of the basic terms. The equation for a straight line is:

$$Y = A + (B{*}X) \tag{AB.20}$$

Y is often called the dependent variable. X is often called the independent variable. In the text we explain why we use all these terms, so this is just to give you a quick overview. B is usually called the slope. It tells us how much the dependent variable Y changes as X changes by 1. A is called the intercept. It tells us the value of Y when $X = 0$ (If $X = 0$ then $B{*}X = 0$, so $Y = A$.) Sometimes we will use subscripts to indicate that we have more than one value for X and Y – that's why we call them variables. Then the equation for a straight line can be written as:

$$Y_i = A + (B{*}X_i) \tag{AB.21}$$

Because there is only one value for A and one value for B in the equation, we sometimes call them constants.

Order of Operations and Symbols for Operations

Students sometimes get confused about what order they should do the arithmetic operations when they see the equation for a straight line or other equations that involve both multiplication or division and addition or subtraction. The formal rule is to multiple or divide first and then add or subtract. But we will try to make this clear by using parentheses. Always do what is inside the innermost parentheses first. So, in the equation for a straight line:

$$Y_i = A + (B{*}X_i) \tag{AB.22}$$

our first step is to multiply B times the value we have for X_i and then add the result to A. Here we have used the asterisk (*) to indicate multiplication. Some texts will use a small dot. Generally statistics texts don't use "x" to indicate multiplication because we use x and other letters of the alphabet to indicate variables and constants. Sometimes when what we intend is very clear, we leave the * out. So we could also write the equation for a straight line as:

$$Y_i = A + (BX_i) \tag{AB.23}$$

APPENDIX C: STATISTICAL TABLES

Table of t and Z Values

How to use this table

The columns give you the α value for testing hypotheses or building confidence intervals. Remember that a confidence interval is based on $(1 - \alpha)$, so to build a 95 percent confidence interval, you need a t or Z value corresponding to $\alpha = 0.05$.

The rows give you the value for the number of degrees of freedom (df). To keep the size of the table manageable, we don't give every possible number of degrees of freedom. For working exercises, you can approximate for values that aren't in the table. If you use a value of t based on fewer degrees of freedom than you actually have, your results will be conservative in the sense that your confidence intervals will be a bit bigger than they need to be and you will not reject some hypotheses you might with the correct t value for the actual number of degrees of freedom. For research, t values are readily available from spreadsheets, statistical packages and on the web.

The last row in the table provides Z values.

We built this table using the procedures described at http://www.coventry.ac.uk/ec/~nhunt/tables.htm (Hunt, 1997).

Table A.1   t and Z values for varying α and df

df	α				
	0.1	0.05	0.025	0.01	0.001
1	6.314	12.706	25.452	63.657	636.619
2	2.920	4.303	6.205	9.925	31.599
3	2.353	3.182	4.177	5.841	12.924
4	2.132	2.776	3.495	4.604	8.610
5	2.015	2.571	3.163	4.032	6.869
6	1.943	2.447	2.969	3.707	5.959
7	1.895	2.365	2.841	3.499	5.408
8	1.860	2.306	2.752	3.355	5.041
9	1.833	2.262	2.685	3.250	4.781
10	1.812	2.228	2.634	3.169	4.587
15	1.753	2.131	2.490	2.947	4.073
20	1.725	2.086	2.423	2.845	3.850
25	1.708	2.060	2.385	2.787	3.725
30	1.697	2.042	2.360	2.750	3.646
40	1.684	2.021	2.329	2.704	3.551
50	1.676	2.009	2.311	2.678	3.496
60	1.671	2.000	2.299	2.660	3.460
70	1.667	1.994	2.291	2.648	3.435
80	1.664	1.990	2.284	2.639	3.416
90	1.662	1.987	2.280	2.632	3.402
100	1.660	1.984	2.276	2.626	3.390
120	1.658	1.980	2.270	2.617	3.373
150	1.655	1.976	2.264	2.609	3.357
200	1.653	1.972	2.258	2.601	3.340
500	1.648	1.965	2.248	2.586	3.310
1000	1.646	1.962	2.245	2.581	3.300
Z	1.645	1.960	2.241	2.576	3.291

Table of Chi Square (χ^2) Values

How to use this table

The columns give you the α value for testing hypotheses.

The rows give you the value for the number of degrees of freedom (df). To keep the size of the table manageable, we don't give every possible number of degrees of freedom. For working exercises, you can approximate for values that aren't in the table. If you use a value of χ^2 based on fewer degrees of freedom than you actually have, your results will be conservative in the sense that you will not reject some hypotheses you might with the correct χ^2 for the actual number of degrees of freedom. For research, χ^2 values are readily available from spreadsheets, statistical packages and on the web.

We built this table using the procedures described at http://www.coventry.ac.uk/ec/~nhunt/tables.htm (Hunt, 1997).

Table A.2   Critical values of chi-square (χ^2) by degrees of freedom for varying α

df	α				
	0.100	0.050	0.025	0.010	0.001
1	2.706	3.841	5.024	6.635	10.828
2	4.605	5.991	7.378	9.210	13.816
3	6.251	7.815	9.348	11.345	16.266
4	7.779	9.488	11.143	13.277	18.467
5	9.236	11.070	12.833	15.086	20.515
6	10.645	12.592	14.449	16.812	22.458
7	12.017	14.067	16.013	18.475	24.322
8	13.362	15.507	17.535	20.090	26.124
9	14.684	16.919	19.023	21.666	27.877
10	15.987	18.307	20.483	23.209	29.588
15	22.307	24.996	27.488	30.578	37.697
20	28.412	31.410	34.170	37.566	45.315
25	34.382	37.652	40.646	44.314	52.620
30	40.256	43.773	46.979	50.892	59.703
40	51.805	55.758	59.342	63.691	73.402
50	63.167	67.505	71.420	76.154	86.661
60	74.397	79.082	83.298	88.379	99.607
70	85.527	90.531	95.023	100.425	112.317
80	96.578	101.879	106.629	112.329	124.839
90	107.565	113.145	118.136	124.116	137.208
100	118.498	124.342	129.561	135.807	149.449
120	140.233	146.567	152.211	158.950	173.617
150	172.581	179.581	185.800	193.208	209.265
200	226.021	233.994	241.058	249.445	267.541
500	540.930	553.127	563.852	576.493	603.446
1000	1057.724	1074.679	1089.531	1106.969	1143.917

Table of F Values

How to use this table

F values depend on three numbers: the α value for testing hypotheses, the numerator degrees of freedom and the denominator degrees of freedom.

There is one table for each of four commonly used α values: 0.10, 0.05, 0.01 and 0.001.

The columns give you the numerator degrees of freedom. The rows give you the denominator degrees of freedom. To keep the size of the table manageable, we don't give every possible number of degrees of freedom. For working exercises, you can approximate for values that aren't in the table. If you use a value of F based on fewer degrees of freedom than you actually have, your results will be conservative in the sense that you will not reject some hypotheses you might with the correct F for the actual number of degrees of freedom. For research, F values are readily available from spreadsheets, statistical packages and on the web.

We built these tables using the procedures described at http://www.coventry.ac.uk/ec/~nhunt/tables.htm (Hunt, 1997).

Reference

Hunt, N. 1997. What price statistical tables now? *Teaching Statistics* 19, pp. 49–51.

Table A.3 F values ($\alpha = 0.10$)

Denominator df	Numerator df															
	1	2	3	4	5	6	7	8	9	10	15	20	30	50	100	200
1	39.863	49.500	53.593	55.833	57.240	58.204	58.906	59.439	59.858	60.195	61.220	61.740	62.265	62.688	63.007	63.167
2	8.526	9.000	9.162	9.243	9.293	9.326	9.349	9.367	9.381	9.392	9.425	9.441	9.458	9.471	9.481	9.486
3	5.538	5.462	5.391	5.343	5.309	5.285	5.266	5.252	5.240	5.230	5.200	5.184	5.168	5.155	5.144	5.139
4	4.545	4.325	4.191	4.107	4.051	4.010	3.979	3.955	3.936	3.920	3.870	3.844	3.817	3.795	3.778	3.769
5	4.060	3.780	3.619	3.520	3.453	3.405	3.368	3.339	3.316	3.297	3.238	3.207	3.174	3.147	3.126	3.116
6	3.776	3.463	3.289	3.181	3.108	3.055	3.014	2.983	2.958	2.937	2.871	2.836	2.800	2.770	2.746	2.734
7	3.589	3.257	3.074	2.961	2.883	2.827	2.785	2.752	2.725	2.703	2.632	2.595	2.555	2.523	2.497	2.484
8	3.458	3.113	2.924	2.806	2.726	2.668	2.624	2.589	2.561	2.538	2.464	2.425	2.383	2.348	2.321	2.307
9	3.360	3.006	2.813	2.693	2.611	2.551	2.505	2.469	2.440	2.416	2.340	2.298	2.255	2.218	2.189	2.174
10	3.285	2.924	2.728	2.605	2.522	2.461	2.414	2.377	2.347	2.323	2.244	2.201	2.155	2.117	2.087	2.071
15	3.073	2.695	2.490	2.361	2.273	2.208	2.158	2.119	2.086	2.059	1.972	1.924	1.873	1.828	1.793	1.774
20	2.975	2.589	2.380	2.249	2.158	2.091	2.040	1.999	1.965	1.937	1.845	1.794	1.738	1.690	1.650	1.629
30	2.881	2.489	2.276	2.142	2.049	1.980	1.927	1.884	1.849	1.819	1.722	1.667	1.606	1.552	1.507	1.482
50	2.809	2.412	2.197	2.061	1.966	1.895	1.840	1.796	1.760	1.729	1.627	1.568	1.502	1.441	1.388	1.359
100	2.756	2.356	2.139	2.002	1.906	1.834	1.778	1.732	1.695	1.663	1.557	1.494	1.423	1.355	1.293	1.257
200	2.731	2.329	2.111	1.973	1.876	1.804	1.747	1.701	1.663	1.631	1.522	1.458	1.383	1.310	1.242	1.199

Table A.4 F values ($\alpha = 0.05$)

Denominator df	Numerator df															
	1	2	3	4	5	6	7	8	9	10	15	20	30	50	100	200
1	161.448	199.500	215.707	224.583	230.162	233.986	236.768	238.883	240.543	241.882	245.950	248.013	250.095	251.774	253.041	253.677
2	18.513	19.000	19.164	19.247	19.296	19.330	19.353	19.371	19.385	19.396	19.429	19.446	19.462	19.476	19.486	19.491
3	10.128	9.552	9.277	9.117	9.013	8.941	8.887	8.845	8.812	8.786	8.703	8.660	8.617	8.581	8.554	8.540
4	7.709	6.944	6.591	6.388	6.256	6.163	6.094	6.041	5.999	5.964	5.858	5.803	5.746	5.699	5.664	5.646
5	6.608	5.786	5.409	5.192	5.050	4.950	4.876	4.818	4.772	4.735	4.619	4.558	4.496	4.444	4.405	4.385
6	5.987	5.143	4.757	4.534	4.387	4.284	4.207	4.147	4.099	4.060	3.938	3.874	3.808	3.754	3.712	3.690
7	5.591	4.737	4.347	4.120	3.972	3.866	3.787	3.726	3.677	3.637	3.511	3.445	3.376	3.319	3.275	3.252
8	5.318	4.459	4.066	3.838	3.687	3.581	3.500	3.438	3.388	3.347	3.218	3.150	3.079	3.020	2.975	2.951
9	5.117	4.256	3.863	3.633	3.482	3.374	3.293	3.230	3.179	3.137	3.006	2.936	2.864	2.803	2.756	2.731
10	4.965	4.103	3.708	3.478	3.326	3.217	3.135	3.072	3.020	2.978	2.845	2.774	2.700	2.637	2.588	2.563
15	4.543	3.682	3.287	3.056	2.901	2.790	2.707	2.641	2.588	2.544	2.403	2.328	2.247	2.178	2.123	2.095
20	4.351	3.493	3.098	2.866	2.711	2.599	2.514	2.447	2.393	2.348	2.203	2.124	2.039	1.966	1.907	1.875
30	4.171	3.316	2.922	2.690	2.534	2.421	2.334	2.266	2.211	2.165	2.015	1.932	1.841	1.761	1.695	1.660
50	4.034	3.183	2.790	2.557	2.400	2.286	2.199	2.130	2.073	2.026	1.871	1.784	1.687	1.599	1.525	1.484
100	3.936	3.087	2.696	2.463	2.305	2.191	2.103	2.032	1.975	1.927	1.768	1.676	1.573	1.477	1.392	1.342
200	3.888	3.041	2.650	2.417	2.259	2.144	2.056	1.985	1.927	1.878	1.717	1.623	1.516	1.415	1.321	1.263

Table A.5 F values (α = 0.01)

Denominator df	Numerator df															
	1	2	3	4	5	6	7	8	9	10	15	20	30	50	100	200
1	4052.181	4999.500	5403.352	5624.583	5763.650	5858.986	5928.356	5981.070	6022.473	6055.847	6157.285	6208.730	6260.649	6302.517	6334.110	6349.967
2	98.503	99.000	99.166	99.249	99.299	99.333	99.356	99.374	99.388	99.399	99.433	99.449	99.466	99.479	99.489	99.494
3	34.116	30.817	29.457	28.710	28.237	27.911	27.672	27.489	27.345	27.229	26.872	26.690	26.505	26.354	26.240	26.183
4	21.198	18.000	16.694	15.977	15.522	15.207	14.976	14.799	14.659	14.546	14.198	14.020	13.838	13.690	13.577	13.520
5	16.258	13.274	12.060	11.392	10.967	10.672	10.456	10.289	10.158	10.051	9.722	9.553	9.379	9.238	9.130	9.075
6	13.745	10.925	9.780	9.148	8.746	8.466	8.260	8.102	7.976	7.874	7.559	7.396	7.229	7.091	6.987	6.934
7	12.246	9.547	8.451	7.847	7.460	7.191	6.993	6.840	6.719	6.620	6.314	6.155	5.992	5.858	5.755	5.702
8	11.259	8.649	7.591	7.006	6.632	6.371	6.178	6.029	5.911	5.814	5.515	5.359	5.198	5.065	4.963	4.911
9	10.561	8.022	6.992	6.422	6.057	5.802	5.613	5.467	5.351	5.257	4.962	4.808	4.649	4.517	4.415	4.363
10	10.044	7.559	6.552	5.994	5.636	5.386	5.200	5.057	4.942	4.849	4.558	4.405	4.247	4.115	4.014	3.962
15	8.683	6.359	5.417	4.893	4.556	4.318	4.142	4.004	3.895	3.805	3.522	3.372	3.214	3.081	2.977	2.923
20	8.096	5.849	4.938	4.431	4.103	3.871	3.699	3.564	3.457	3.368	3.088	2.938	2.778	2.643	2.535	2.479
30	7.562	5.390	4.510	4.018	3.699	3.473	3.304	3.173	3.067	2.979	2.700	2.549	2.386	2.245	2.131	2.070
50	7.171	5.057	4.199	3.720	3.408	3.186	3.020	2.890	2.785	2.698	2.419	2.265	2.098	1.949	1.825	1.757
100	6.895	4.824	3.984	3.513	3.206	2.988	2.823	2.694	2.590	2.503	2.223	2.067	1.893	1.735	1.598	1.518
200	6.763	4.713	3.881	3.414	3.110	2.893	2.730	2.601	2.497	2.411	2.129	1.971	1.794	1.629	1.481	1.391

Table A.6 F values ($\alpha = 0.001$)

Denominator df	Numerator df															
	1	2	3	4	5	6	7	8	9	10	15	20	30	50	100	200
1	405284.068	499999.500	540379.202	562499.583	576404.556	585937.111	592873.288	598144.156	602283.992	605620.971	615763.662	620907.673	626098.958	630285.380	633444.344	635029.88
2	998.500	999.000	999.167	999.250	999.300	999.333	999.357	999.375	999.389	999.400	999.433	999.450	999.467	999.480	999.490	999.495
3	167.029	148.500	141.108	137.100	134.580	132.847	131.583	130.619	129.860	129.247	127.374	126.418	125.449	124.664	124.069	123.770
4	74.137	61.246	56.177	53.436	51.712	50.525	49.658	48.996	48.475	48.053	46.761	46.100	45.429	44.883	44.469	44.261
5	47.181	37.122	33.202	31.085	29.752	28.834	28.163	27.649	27.244	26.917	25.911	25.395	24.869	24.441	24.115	23.951
6	35.507	27.000	23.703	21.924	20.803	20.030	19.463	19.030	18.688	18.411	17.559	17.120	16.672	16.307	16.028	15.887
7	29.245	21.689	18.772	17.198	16.206	15.521	15.019	14.634	14.330	14.083	13.324	12.932	12.530	12.202	11.951	11.824
8	25.415	18.494	15.829	14.392	13.485	12.858	12.398	12.046	11.767	11.540	10.841	10.480	10.109	9.804	9.571	9.453
9	22.857	16.387	13.902	12.560	11.714	11.128	10.698	10.368	10.107	9.894	9.238	8.898	8.548	8.260	8.039	7.926
10	21.040	14.905	12.553	11.283	10.481	9.926	9.517	9.204	8.956	8.754	8.129	7.804	7.469	7.193	6.980	6.872
15	16.587	11.339	9.335	8.253	7.567	7.092	6.741	6.471	6.256	6.081	5.535	5.248	4.950	4.702	4.508	4.408
20	14.819	9.953	8.098	7.096	6.461	6.019	5.692	5.440	5.239	5.075	4.562	4.290	4.005	3.765	3.576	3.478
30	13.293	8.773	7.054	6.125	5.534	5.122	4.817	4.581	4.393	4.239	3.753	3.493	3.217	2.981	2.792	2.693
50	12.222	7.956	6.336	5.459	4.901	4.512	4.222	3.998	3.818	3.671	3.204	2.951	2.679	2.441	2.246	2.140
100	11.495	7.408	5.857	5.017	4.482	4.107	3.829	3.612	3.439	3.296	2.840	2.591	2.319	2.076	1.867	1.749
200	11.155	7.152	5.634	4.812	4.287	3.920	3.647	3.434	3.264	3.123	2.672	2.424	2.151	1.902	1.682	1.552

GLOSSARY OF KEY TERMS

This glossary is intended as an aid to learning from the text. More formal and more general definitions of these and other terms can be found in Vogt's (1993) statistical dictionary.

Adjacent values – statistic displayed in a boxplot; high and low values are selected so that anything beyond them can be considered an outlier.

Alpha error – see type I error.

Alpha level – chances one is willing to take in making a Type I error when testing a hypothesis; also known as the p value, probability level or critical value. Most commonly the alpha value is set to 0.05, 0.01, or 0.10. If 0.05 is selected, this means the researcher will be wrong in rejecting the null hypothesis (make a Type I error) 5 percent of the time.

Alternative hypothesis – when testing a hypothesis, the alternative hypothesis is the alternative to the hypothesis being tested, which is the null hypothesis; also known as the research hypothesis.

Analysis of variance – a statistical procedure to determine whether there is a significant difference among three or more sample means.

Asymptotic properties – properties of, and results from, large samples.

Auxiliary regression – regression analyses in which the independent variables are used to predict each other to determine the level of collinearity.

Bar chart – graph used to depict the distribution of a nominal or ordinal variable; bars are used to represent the count, percent, or proportion of each category of the variable.

Bayesian – Bayesian approaches view probabilities as something in our minds, it is a more subjectivist approach to probability than the frequentist approach. If we say a coin has a probability of coming up heads 50 percent of the time we are really saying that the best guess is that we don't know whether it will come up tails or heads, and there is no reason to predict either tails rather than heads or heads rather than tails.

Bell-shaped curve – see Normal distribution.

Best-fit line – the line on a scatterplot that lies closer to all data points than any other line; also called the regression line. Usually "best fit" is defined in terms of the sum of the squared errors, resulting in the "least squares" line.

Beta – standardized regression coefficient that measures the size of the influence of multiple independent variables on a dependent variable when the independent and dependent variables have all been standardized (converted to Z scores). This is sometimes seen as useful for comparing the effects of independent variables that are measured on very different scales.

Beta error – see type II error.

Bimodal – a distribution with two modes (peaks).

Binary variable – a variable with only two categories (for example, gender); also known as binary nominal variable. It is sometimes called a "dummy variable."

Binomial distribution – plot of the outcomes of a binomial process.

Binomial process – random process with only two outcomes (for example, heads or tails when flipping a coin).

Bivariate analysis – analysis based on two variables.

Bivariate regression – analysis of the relationship between one independent and one dependent variable.

Box and whisker diagram – see boxplot.

Boxplot – graph used to summarize a variable's distribution; the graph displays a variable's median, upper and lower quartile, adjacent values, and outlying data points; also known as box and whisker diagram.

Breakdown point – for a descriptive statistic, this is the proportion of the data set that can take on extreme values without seriously affecting the summary statistic. The mean has a breakdown point of zero (i.e., the mean cannot tolerate any extreme values), while the median has a breakdown point of 50 percent.

Causal order – the time ordering of a series of variables; identification of which variable occurred first in time and therefore may influence/predict another variable.

Causality – see causal order.

Causation – an association/correlation between two variables in which one variable – the independent variable – is believed to influence the other variable – the dependent variable.

Central Limit Theorem – theorem stating that when the sum or mean of large samples of random samples are plotted, the shape will have a special bell-shape, called a Normal distribution. Also called the Law of Large Numbers.

Central tendency measures – statistics used to describe what is typical in the data and the extent observations differ from each other (for example, mean, median, mode).

Chi-square test – a number used to determine the statistical significance of a hypothesis in a contingency table. The test is based on comparing the frequencies expected if the two variables are not related with the observed frequencies. If the statistic is less than the critical value, differences between observed and expected frequencies are not statistically significant, and we conclude the two variables are not related.

Cleveland dotplot – graph plotting both the distribution of a variable and the values of all cases on that variable.

Cluster sample – method of sampling based on selecting groups from a population and sampling from the groups rather than individual cases in the population (for example, if we want to study students, we could select a sample based on first selecting classrooms then students rather than sampling from a list of all students).

Conditional effect – see interaction effect.

Conditional value – the values of one variable as predicted based on the values of other variables.

Confidence interval – range of values that is likely to contain the true value of the population parameter; commonly we use a 90 percent, 95 percent or 99 percent confidence interval.

Constant – a number that is the same for every case in a data set.

Contingency table – table depicting the relationship between two or more variables. In a contingency table for two variables one variable is represented in rows and the other variable is represented in columns. For more than two variables there is one two variable table (also called a two-way table) combination of the values of the other variables. Each cell presents the frequency or percent of cases that fall into a particular category of each variable.

Continuous variable – variable that can take on any possible value; measured at an interval/ratio level.

Control group – in experimental research, this is the set of participants who do not receive the experimental "manipulation."

Control variable – a variable whose effect on the dependent variable is taken into account (controlled for) in a study.

Convenience sample – selecting sample members who are readily available; a non-representative sample.

Correlation – an association between two variables. Correlations can be positive – when an increase in one variable is linked to an increase in the other variable – or negative – when an increase in one variable is linked to a decrease in the other variable.

Correlation coefficient – measure of the strength and direction of the relationship between two variables. Possible range is between −1 and +1 with scores closer to +1 or −1 indicating a stronger relationship. A correlation of 0 indicates that the two variables are not related. The most common correlation coefficient is Pearson's r.

Critical value – a number that indicates a point at which some result from the sample is so improbable that we don't believe the assumptions that led us to calculate that value. The alternative is to believe that a very unusual sample has occurred. Also known as the alpha level.

Cross-sectional data – data that are collected at one point in time.

Cumulative distribution graph – graph with a line that indicates the percentage of cases that fall below each value on the X-axis. The cumulative distribution must always increase or stay flat as we move across values of the variable being plotted on the X axis.

Cumulative frequency – in a frequency table, this column shows the number of data points falling into a category and all preceding categories.

Cumulative percent – in a frequency table, this column shows the percent of data points falling into a category and all preceding categories.

Degrees of freedom – an indication of how much information is available in an analysis. When we examine the mean of one sample the number of degrees of freedom is N − 1 (the sample size minus 1); abbreviated as (df).

Dependent variable – the variable that we are trying to predict/explain; often labeled as Y in statistical equations.

Descriptive statistics – tools used to summarize and describe data.

Deviation from the mean – how far each data point is from the average value; see also squared deviations about the mean.

Direct relationship/effect – an independent variable predicts/influences a dependent variable.

Discrete variable – another name for a nominal or qualitative variable.

Distribution – how the values of a variable are spread over all the possible values.

Dotplot histogram – a graph that provides the same type of information as a bar chart but uses stacks of plot symbols rather than bars to indicate how many cases are in each category.

Ecological fallacy – studying one kind of thing (i.e., unit of analysis) and making conclusions about another unit of analysis.

Endogenous variable – a variable that is explained by other variables in the model (i.e., variable that does not come first in a causal order sequence).

Equiprobability – every case in a population has an equal probability of being included in the sample.

Error term – the term in a model that indicates random error. Most of this is amount by which the independent variable(s) misses predicting the dependent variable.

Event history analysis – analytic techniques used to study the occurrence of events (for example, wars, divorces).

Exogenous variable – a variable whose causes are not included in a model (i.e., the variable that comes first in a causal sequence).

Expected cell frequency – number of cases in a cell of a contingency table predicted by the null hypothesis model.

Expected value – the most likely value of a sample statistic (for example, the mean) when we look at the values that arise across many samples. The expected value is the mean of the sampling distribution for a sample statistic.

Experiment – a research design in which subjects are divided (typically randomly) into two or more groups: experimental/treatment groups and control groups. All groups receive the same experience except for the factors being studied, which are varied systematically across groups.

Experimental group – in experimental research, this is the set of participants who receive the experimental "manipulation."

Explanation – the ability to predict one variable based on another variable(s).

Explanatory variable – a variable for which data are available and which is the focus of the research.

External validity – degree to which the results of a study done in one situation generalizes to other situations.

Extraneous variable – a variable that may have an influence on the dependent variable but which is not being studied.

Extreme case – a data point that is far from the other cases on a particular variable; also known as an outlier.

F ratio – statistic computed in analysis of variance to determine whether there are differences between group means or in regression to determine if one or more independent variables explain variance in the dependent variable.

Frequency table – a table showing the values of a variable and how many data points fall into each value.

Frequentist – approach to probability where the probability of an outcome is thought of as the proportion of times that the outcome occurs when the event is repeated many times.

Function – equation that links the independent variable(s) to the dependent variable.

Gaussian curve – see Normal curve; a theoretical bell-shaped curve. It sometimes arises in data and often describes the results of sampling experiments.

Goodness of fit – how well a model or hypothesis matches the observed data.

Goodness of fit test – statistic to compare sample data to what we would expect if the null hypothesis is true (for example, chi-square statistic).

Heteroscedastic – population variances are not equal.

Histogram – a bar graph showing the distribution of a continuous variable.

Homoscedastic – population variances are equal.

Hypothesis – a statement about the relationship between variables that can be tested using data.

Hypothesis of no difference – expectation that there is no difference across groups in the variable of interest.

Hypothesis test – way to assess the degree to which some assertion about the population is reasonable to believe given the information about the population we have in our sample.

Independence – knowing one situation does not allow you to predict the other; selecting one member of a sample does not affect the probability of other members being selected in the sample.

Independent variable – variable that is used to try to explain/predict a dependent variable; usually known as X in statistical equations.

Indirect relationship/effect – when there is no direct relationship between an independent variable (X1) and a dependent variable (Y); rather the relationship between X1 and Y is mediated by a third variable. For example X1 has an effect on another variable X2, and X2 in turn has an effect on Y. Variables can have both direct and indirect effects.

Inferential statistics – tools used to make statements about a population using sample data in the face of error.

Interaction effect – the effect of one variable (X2) on a dependent variable (Y) depends on the value of another variable (X1); in other words, the relationship between X2 and Y differs across level/values of X1. For example, education might have a different effect on attitudes for men than for women.

Intercept – in the regression equation, it is the value of the dependent variable that is expected when the independent variable or variables equal zero.

Interquartile range – distance that one must go to span the middle 50 percent of the data; calculated as the difference between the upper and lower quartiles of the data set.

Interval – level of measurement in which observations are assigned numbers and numbers indicate how far apart observations are from each other.

Interval estimate – range that we can be quite certain includes the real population statistic of interest, such as the mean.

Intervening variable – a variable (X2) that is affected by an independent variable (X1) and has an effect on the dependent variable (Y). Therefore, the relationship between X1 and Y is indirect via X2 (in other words, X1 affects X2 and X2 affects Y).

Law of Large Numbers – theorem stating that when the sum or mean of large samples of random samples are plotted, the shape will have a special bell-shape, called a Normal distribution. Also called the Central Limit Theorem.

Least squares – a calculation that best summarizes a relationship by making the sum of the squared errors as small as possible.

Left skew – see skew.

Level of measurement – see nominal, ordinal, interval, and ratio.

Linear relationship – a relationship between two variables that can be best depicted with a straight line; the direction and degree of change in one variable matches the other variable.

Listwise deletion – cases are omitted from analysis when there is missing data on any one of the variables being examined.

Longitudinal – a study of data that are collected on the same cases at multiple points in time. When there are many observations on one or a few units the term time series is often used. When there are multiple observations over time on many units the terms panel or pooled time series cross section are often used.

Lower quartile – The value below which 25 percent of cases fall.

Margin of error – see confidence interval.

Marginal – summary of frequencies of an entire column or row in a contingency table.

Mean – average score on a variable; calculated as the sum of all values divided by the total number of data points; used with continuous and binary variables.

Measurement error – extent to which variables being studied are mis-measured.

Median – the 50th percentile of a variable; half the data points fall above the median value and half the data points fall below the median value. When there is an even number of data points, the median is the average of the middle two values. Used with continuous and ordinal variables.

Median absolute deviation from the median (MAD) – measure of variability around the median; calculated as the median value of the absolute values of the deviations from the median.

Median depth – how many cases have to be counted into the data to find the place where half the data points will be above and half below; calculated as (number of data points + 1)/2.

Mediating variable – see intervening variable.

Missing data – information on a variable that is not available for a case (for example, a person does not answer a survey question or a country does not report a statistic).

Mode – most frequently occurring value in the data; can be used with nominal, ordinal, and interval/ratio data.

Model – a description designed to understand relationships between variables. A model is often a proposal of relationships that we test using statistics.

Moderating effect – see interaction effect.

Multicollinearity – high correlations among independent variables in a regression. If independent variables are too highly correlated, then the ability of multiple regression to estimate the effects of each variable is weakened, and it is difficult to make conclusions about the effects of the independent variables.

Multinomial process – a process with a countable number of possible outcomes, but more than two. See binomial process.

Multiple regression – analysis in which two or more independent variables are used to predict the dependent variable and thus explain its variability. Multiple regression produces estimates of the influence of each independent variable on the dependent variable taking into account the impact of other independent variables.

Multivariate analysis – analysis based on three or more variables.

Negative correlation – see correlation.

Negative skew – see skew.

Nominal – level of measurement in which observations are assigned to categories with no ordering among the categories (for example, ethnicity, religious affiliation).

Non-linear relationship – the relationship between two variables that is not best represented with a straight (linear) line (for example, a relationship may resemble a curve).

Non-representative sample – a non-probability, non-random sample (for example, convenience sample). There is no way to know the probability that a case will be selected for the sample nor how representative the sample is of the population.

Normal curve – the sum or average of uniform random numbers take on a unimodal bell-shape distribution; also referred to as the Gaussian or bell-shaped curve.

Null hypothesis – a hypothesis that two or more variables, groups or means are not related or different; if the null hypothesis is rejected, the alternative hypothesis is accepted.

Observed cell frequency – cell frequencies in a contingency table that are based on the actual data.

Observed variable – a variable that is measured.

One-sided test – a hypothesis test in which only values of the sample statistic that are too high (or only those that are too low) are considered evidence to reject the hypothesis. In two-sided tests, any values of the sample statistics that deviate substantially from the null hypothesis lead to rejection of the hypothesis.

One-way analysis of variance – see analysis of variance.

One-way scatterplot – a univariate graph that uses vertical lines to show the value of each case and consequently where in the range of the variable being graphed most cases fall.

Ordinal – level of measurement in which observations are assigned to categories; the categories have a rank ordering but the distance between the categories is unknown.

Outlier – a case whose value on a variable falls outside the typical pattern (either much higher or lower than other values).

p value – see alpha level.

Pairwise deletion – all cases are used that have complete data for pairs of variables; if more than two variables are being examined, the sample size statistics based on each pair of variables may differ.

Panel study – a longitudinal study of the same set of cases. See longitudinal study.

Parameter – a statistic based on data from a population.

Part correlation – correlation between an independent and dependent variable, controlling for other variables (the square root of the amount of total variance in Y accounted for by X controlling for all other independent variables). See Part R^2.

Part R^2 – the unique explained variance of an independent variable on a dependent variable controlling for the effect of other independent variables. See Part correlation.

Partial correlation – the amount of variance an independent variable accounts for out of the variance left over to be explained in the dependent variable.

Partial relationship – the relationship between an independent variable and a dependent variable when the effects of one or more other independent variables are held constant.

Path diagram – visual diagram to depict a proposition about how variables are related to one another.

Pearson correlation coefficient – coefficient indicating the strength of the relationship between two variables and the direction of the relationship between two variables; usually labeled r; see also correlation coefficient.

Perceptual error – error occurring because it can be challenging to see or recognize patterns in the data accurately.

Pie chart – graph used to depict the distribution of a nominal or ordinal variable; a circle is divided into wedges, with each wedge representing the frequency, proportion, or percent of each category of the variable.

Point estimate – use of the sample statistic such as the mean to estimate the population parameter of interest. See interval estimate.

Population – all the possible units being studied (for example, all countries, all residents of a designated city in a given year). See sample.

Positive correlation – see correlation.

Positive skew – see skew.

Power – chances of not making a type II error.

Probability – the likelihood that something will occur.

Probability distribution – frequency distribution of a random variable. The probability distribution shows the likelihood that various values of the variable will occur.

Probability level – see alpha level.

Probability sample – a randomly selected sample; with this sampling process the probability that each unit in a population can be selected is known.

Qualitative variable – a variable measured at the nominal level; also known as discrete variable.

Quantitative variable – a variable measured at an ordinal, interval, or ratio level.

Quartile depth – calculated as (median depth + 1)/2.

Quartiles – values that divide a variable into four parts of equal size.

R^2 – the proportion of variation in the dependent variable explained by the independent variable or variables.

Random events – events that are not predictable.

Random assignment – occurs in experimental research when participants are randomly assigned to the control and experimental groups (for example, assignment by toss of a coin).

Random error – errors that occur due to random and unpredictable events occurring during the research process.

Random selection – process by which everyone in a population has an equal chance of being included in the sample; results in a representative sample. Also known as a probability sample.

Randomization error – when doing an experiment, the random assignment of subjects to experimental and control groups may randomly produce differences between the two groups (control and experimental), resulting in error.

Range – the difference between the lowest and highest values in a variable's distribution.

Ratio – level of measurement in which observations are assigned numbers and numbers indicate how far apart observations are from each other. Distinguished from interval level of measurement in that ratio variables have a natural zero point.

Regression line – see best-fit line.

Representative sample – sample in which every member of the population has an equal chance of being elected; also see probability sample.

Residual term – another name for error term.

Right skew – see skew.

Sample – a subset of a population.

Sampling distribution – a probability distribution that shows what happens when we repeat the sampling process many times.

Sampling error – extent to which the sample does not reflect the population due to differences between the population and the sample generated by random selection of cases.

Sampling experiment – a process of drawing many samples from an imaginary population to see what happens when samples are drawn.

Sampling with replacement – sampling strategy in which a member of the population could be selected more than once. Selection of each member of the sample is truly independent of selection of every other member.

Sampling without replacement – sampling strategy in which a member of the population could be selected only once. This violates the strictest sense of independence of selection since knowing a member of the population has been drawn into the sample means that they cannot be drawn into it again.

Scatterplot – graph of the relationship between two continuous variables; also called a X–Y plot or scattergram.

Scatterplot matrix – display of a series of scatterplots in which every variable is shown as both an independent and a dependent variable.

Simple random sample – samples based on drawing observations from the population with equal probability and independence.

Skew – extent the distribution of scores has some values that are extremely high or extremely low. Left/negative skew is when a variable has a distribution with more cases on the low end than in a normal distribution. Right/positive skew is when a variable has a distribution with more cases on the high end than in a normal distribution.

Slope – the change in the dependent variable that is expected when the independent variable changes by one unit. If there are multiple independent variables, the value when there is no change in the other independent variables.

Spurious relationship – two variables are correlated but they are associated only because both are causally influenced by a third variable. When the effects of the third variable are taken into account, the relationship between the two variables disappears.

Squared deviations about the mean – indicator of the variability around the mean. Calculated by squaring the distance between each case's value on a variable from the variable's mean. The sum of these squared values is an indicator of variability.

Standard deviation – number commonly used to describe the variation in a variable's distribution; the square root of the variance.

Standard error – the standard deviation of the sampling distribution.

Standardized coefficient – see beta.

Standardized score – see Z score.

Statistical significance – a result that is too improbable to be considered to have been generated by random error.

Statistics – 1) numbers that represent some aspect of life. 2) Numbers that are calculated from a sample to represent parameters in the population. 3) The study of using numbers to represent various aspects of life and especially of drawing conclusions in the face of error.

Stem and leaf diagram – a graph used to display the distribution of a continuous variable; it shows the numeric values in a distribution using a single digit to represent it.

Stratified sample – taking random samples from designated strata, which are population groupings (for example, gender, age categories). In some instances the samples of each strata are selected in sizes equal to their proportion in the population, but in other instances some strata are oversampled.

Subject variable – personal characteristics of individuals that are often included in research, such as gender, ethnicity, and age.

Subjectivist approach to probability – see Bayesian approach to probability.

Sum of squares – see squared deviations about the mean.

Superpopulation – A hypothetical infinite population of all possible members. It is invoked to understand the role of statistical inference when data is available from an "apparent" population, such as all nation states at a particular point in time. The superpopulation would be the set of all hypothetically possible nation states.

Symmetrical distribution – a variable whose distribution has no skew; from the central point in the data set, the reflection of the two sides from the center will be identical. A Normal distribution is symmetrical.

t distribution – distribution used to find the critical region for tests of sample means when the sample size is small and the population is normally distributed.

Time series – analysis of data over time.

Time series graph – plot of a variable over time (time being the independent variable).

Treatment group – see control group.

Trimmed mean – calculating the mean value by taking out a designated percent of data points with the highest and lowest values; used to reduce the influence of outliers on the mean value. A 5 percent trimmed mean, for instance, removes the highest 5 percent of data points and lowest 5 percent of data points and calculates the mean based on the values for the remaining 90 percent of data points.

Two-sided test – see two-tailed test.

Two-tailed test – hypothesis test of whether a population parameter falls to either side of the hypothesized value; also known as two-tailed test.

Two-way table – see contingency table.

Type I error – error made when the null hypothesis is rejected when it is true; also known as alpha error.

Type II error – error made when the null hypothesis is accepted even though it is not true; also known as beta error.

Uncontrolled variable – a variable whose effect on the dependent variable is not taken into account in a study.

Uniform distribution – when every number in a given range is equally likely to show up.

Unimodal – a distribution with one mode (peak).

Unit of analysis – the thing on which data were collected (for example, countries, individuals).

Univariate analysis – analysis using one variable.

Unstandardized coefficient – in a regression analysis, the unstandardized coefficient is the slope of each independent variable with the dependent and independent variables expressed in the units in which they were measured. The unstandardized coefficient indicates how many units of change in the dependent variable is expected if the independent variable changes by one unit. This is in contrast to the standardized coefficient or beta value where change is in standard deviation units.

Upper quartile – 25 percent of cases fall above this value.

Variable – properties or characteristics of something (for example, countries, people). The properties of a variable differ (vary) across observations.

Variance – a measure of a variable's dispersion; the average of the squared deviations from the mean.

Variance explained – the amount of variance in a dependent variable that can be predicted by one or more independent variables.

Variation – the extent to which values differ across observations.

Variation explained – the amount of variance in a dependent variable that can be predicted by one or more independent variables.

Weighted average – when each case is given a different weight when calculating the mean. Weighted averages are used with stratified samples to take account of the fact that there are more cases sampled from some groups than from others than would be the case with a simple random sample.

Z score – a standardized score; when values are transformed in a way that allows for comparisons between variables measured on different scales; distance is measured from the mean and relative to the standard deviation of the variables. Thus a z score is calculated by subtracting the mean from the individual score and dividing by the standard deviation.

Reference

Vogt, W. P. 1993. *Dictionary of Statistics and Methodology. A Nontechnical Guide for the Social Sciences*. Newbury Park: Sage Publications.

NOTES

Preface

1 The history of probability and statistics has long been of interest to statisticians, and a number of excellent studies review that history. Perhaps the most accessible are Bennett (1998) and Salsburg (2001).

2 Important papers from this literature are collected in Kahneman, Slovic and Tversky (1982).

3 This and many other quotations we use are from Gaither and Cavazos-Gaither (1996).

4 The one most focused on sociology is Freedman (1991b), with commentary by Berk (1991), Blalock (1991) and Mason (1991), three very distinguished quantitative sociologists and rejoined by Freedman (1991a).

5 Braverman (1974) provides an excellent discussion of deskilling, scientific management and the evolution of labor in the twentieth century.

6 The term journey*man* is of course gendered. This is particularly unfortunate because traditional women's work in most Western societies has always been craftwork, learned and honed by continuous interaction with elders and peers and ever creative and adaptive to changing circumstances. It has more often been men's work that has lost craft through mechanization. If we view statistical analysis as craftwork, then perhaps a truly creative use of quantitative methods would be consistent with, and benefit from, feminist analysis, feminist critiques of common statistical practice not withstanding.

Chapter 1

1 Of course, models can be deceptive, as the example of the fashion model indicates. The most famous fashion models have highly unusual physiques that are not at all typical of most people.

2 To be as general as possible, the model should allow for E to be complicated, or show multiple components representing multiple factors acting on Y. But for now, a simple E term will suffice.

3   In many experiments there may be several experimental groups each of which receives a different treatment and even several kinds of control groups. But the simple example with two groups is sufficient to explain the role of randomization error in models of experimental effects.

4   This example is based on Kalof (1999).

5   John Tukey's (1977) *Exploratory Data Analysis* was a turning point in thinking about statistical graphics. Cleveland (1993; 1994) examined many key issues in statistical graphics. Tufte (1982; 1990; 1997) has produced three beautiful books showing some of the best graphical displays of information every produced. Browsing any of these books shows how creative and useful graphics can be.

6   We owe the material in the section to Dr Scott W. Williams, Professor of Mathematics at SUNY, Buffalo, and his comprehensive web site on "Mathematicians of the African Diaspora" (www.math.buffalo.edu/mad).

7   This biographical material is from The Society for Advancement of Chicanos and Native Americans in Science (SACNAS) and their Biography Project which is designed to teach students about the accomplishments of Chicano/Latino and Native American scientists (www.sacnas.org/bio/index.html).

8   This biographical material is from Agnes Scott College's "Biographies of Women Mathematicians" (www.scottlan.edu/lriddle/women/alpha.htm). Salsburg's (2001) Chapter 15 is a short biography of F.N. David and Chapter 19 is a short biography of Gertrude Cox.

Chapter 2

1   Not every person in a survey answers every question. In working with data, a missing data code is needed for every variable that indicates that there is no data for that observation. Thus if someone answered this question with the response "I do not know" we might code them a 9 or a −9 to indicate that the individual didn't have an answer. The missing data code is some number or character that cannot be confused with a valid response. We then tell our computer software which codes go with the missing data, and it automatically keeps track of who answered and who didn't and handles the data appropriately. But it is not uncommon for beginning researchers to forget to tell the computer that, for example, 9 is missing data. Then the non-respondents might appear to be people who *really* disagree with the statement (a score of 9). This mistake can lead to some very strange results.

2   The discussion between Bollen and Barb on the one hand and O'Brien on the other is a good place to start reading about this issue, though it requires an understanding of Pearson's correlation coefficient that we won't cover until much later in the book. See (Bollen and Barb, 1981) and (Bollen and Barb, 1983; O'Brien 1983). Binder (1984) discusses the implications of level of measurement in the context of criminal justice. Ferrando (1999) and Krieg (1999) are recent discussions of the effect of level of measurement on statistical analysis that require a bit of statistical sophistication.

3   Calling nominal qualitative and both ordinal and interval quantitative is the most common way to use these terms. But some researchers call *both* nominal and ordinal variables qualitative while reserving the term quantitative for interval level data.

4   Or at least any value that makes sense for that variable. Think about the age of a person in a survey like the ISSP. Countries in the ISSP only interview people 18 and

over (in a few countries the minimum age is 16), so age cannot take on values below 16, and the upper limit will be the oldest person in each country who still lives in a household. But in that range, any value is possible. Practically speaking, we only measure age in whole years in most surveys. But we could ask people for their dates of birth and calculate their ages to the day, if we had some reason to do that.

5 King (1997) has shown how to make inferences across levels of aggregation. Using his methods it is sometimes possible to learn about individual voting patterns from data on the voting patterns of cities or other units that aggregate together many people.

6 Of course, it is important not to be naïve in using time ordering as an argument for causal order. People, organizations, and nations may anticipate things to come and act accordingly.

7 But if we were studying political attitudes and a major scandal broke out while we were doing the survey, or if we were studying investments and savings and the stock market crashed while we doing the survey, we'd have to take into account that those events might influence people's responses. Then what started as a cross-sectional study might be considered a pooled time-series cross-section study.

8 But some of the complexities of time series data analysis can creep into the analysis of panel data as well.

9 Researchers conducting ethnographic studies and in-depth interviews tend to work with small sample sizes because they collect a lot of data per person interviewed or a lot of information on each situation observed. This is an understandable constraint on their approach, and it can limit the amount of variation in their samples. They also tend to rely on convenience samples rather than probability samples. Sometimes this is necessary because there is no way to draw a probability sample of the units to be studied, but other times it seems a flaw in the research design – a probability design would have ensured a more representative sample.

10 Since gender is a social or cultural category rather than a biological one, we would rather call this variable "gender." But the ISSP, the data set we are using, is widely used and has always labeled this variable "sex" so to avoid confusion we will stay with that convention. But be careful to remember that when we do an analysis using the variable "sex" we are actually analyzing gender.

11 While we will not focus on differences between countries in this book, it is important to recognize that some questions and topics may be influenced by cultural differences. For instance, a popular measure of depressive symptoms, known as the CES-D instrument, includes an item asking how frequently a person "feels blue." When translated into Japanese, the question was meaningless since that phrase has no comparable meaning in Japan – the Japanese don't associate the term "blue" with feelings of depression.

12 See note 10.

Chapter 3

1 Friendly and Denis (http://www.math.yorku.ca/SCS/Gallery/milestone/) provide a brief history of statistical graphics along with extensive links to sites providing more detail.

2 A timeline of important developments in statistical graphics is maintained by the Mathematics Department at York University: (http://www.math.yorku.ca/SCS/Gallery/milestone/).

3   This work is summarized in Cleveland (1985), especially pp. 262–9.

4   Nightingale was a major figure in the development of modern nursing, epidemiology and public health, despite the considerable obstacles she encountered as a woman in Victorian England. For more information about her, visit the website of the Florence Nightingale Museum (http://www.florence-nightingale.co.uk/).

5   It is sometimes said that the term "rule of thumb" refers to an English common law tradition that it was permissible for husbands to beat their wives as long as the stick used was smaller in diameter than the man's thumb. If the phrase actually had this misogynist meaning we wouldn't use it. But apparently the phrase "rule of thumb" does not come from the travesty of wife beating but from the idea that a skilled carpenter can successfully use her or his thumb as a measuring device. It is in that sense, with a reverence for craftwork, that we use the phrase – a useful approximation.

6   Even here we have to be careful. Sometimes odd data points tell us a great deal. For example, when readings of the level of ozone above Antarctica first came back very low, scientists thought that there was some error in measurement. It took some time to realize that this was not measurement error that needed to be corrected but rather a sign that human use of the chemicals chlorofluorocarbons was depleting the ozone in the stratosphere.

Chapter 4

1   There are several areas where this growth is taking place. One is the huge amount of information that comes from analyzing the gene. Another is the information coming from satellite observations of the earth. A third is the increasing consolidation of private databases on individuals – linking of credit card records, bank accounts, and other information. A fourth is the Internet, which is not only a source of information but is a subject of research in itself.

2   The word average comes from the Latin "havaria." Havaria meant damage to a cargo being shipped by sea. At times part of the cargo had to be thrown overboard to lighten the ship for safety. The cost of the lost goods was divided among all the people who had goods on the ship – sharing the risk. Havaria came to be the amount each shipper paid towards the lost goods. This would be the amount of the loss divided by the number of people shipping goods on the ship – an average loss.

3   Baseball fans will know that the number of "official at bats" is a bit complicated. The number of "official at bats" is defined as: Number of appearances at the plate minus the sum of (walks + hit by the ball + sacrifices).

4   Later in the text we will have to make clear the difference between numbers that describe the sample and numbers that describe the population from which the sample was drawn. We can calculate the sample mean. But we may be interested in the population mean and if we only have data on the sample we can't directly calculate the population mean. By convention, $\bar{X}$ refers to the mean of a sample, μ to the mean of a population. Before we begin discussing statistical inference – how to use sample information to learn about the population – it is not clear which symbol is more appropriate because the difference between samples and populations (or, more generally, observation with and without sampling error) is not meaningful except in the context of inference. We will use $\bar{X}$ because it is less intimidating than the Greek symbol.

5   Of course, the result won't be exactly zero if you don't carry enough decimal places in the calculation of the mean, the deviations, and the sum of the deviations. In statistics, getting answers that are different from the correct answer because of not carrying enough decimal places in the calculation is called rounding error. As a general rule, it is a good idea to carry through the calculations to two more decimal places than there are in the variable you are using. So if the data are whole numbers, carry two decimal places. In computer programming, great care must be taken to prevent rounding error from building up in complex calculations.

6   The mean is in fact the expected value of a sample in a very technical sense that will be discussed later. But here we intend the common sense view that the mean is a typical number for a batch of data.

7   Statistics texts often say that in skewed distributions this relationship between the mean and the median is *always* true. But while it is usually true, there are exceptions (von Hippel, 2005).

8   We always take the positive square root in calculating the standard deviation.

9   These depth formulas may seem rather complicated. It may be easier to think of counting in half way from the top or bottom to find the median, and a quarter of the way from the top (three quarters of the way from the bottom) to find the upper quartile and a quarter of the way from the bottom (three quarters of the way form the top) to find the lower quartile. While this is logical, it turns out that the depth formulas are a better guide to finding the quartiles.

10   Many applications differentiate between outliers and extreme outliers that are more than 3 IQR above the upper quartile or more than 3 IQR below the lower quartile. A different plotting system is used for the "regular" outliers and the "extreme" outliers. We will not make this distinction.

11   This calculation is based on the properties of the Normal distribution we will discuss later in the book. Hoaglin, Iglewicz and Tukey (1986) found that in random samples from Normal populations, outliers are even more common than in the Normal distribution itself. It is their simulation work that provides the basis for using the 1.5 IQR definition for a fence that is standard in applications of the boxplot.

12   There are other versions of the boxplot, including one where the width of the box is proportional to the square root of the sample size. The square root of the sample size is directly related to using samples to draw inferences about populations, as we will see in later chapters. Frigee, Hoaglin and Iglewicz (1989) identify a variety of other variations.

13   Quetelet and the other researchers we discuss here were very prolific and their thinking evolved over time. So our quick synopsis of their ideas oversimplifies the subtlety in their thinking, but we hope it captures the essence of their main concerns.

14   Galton did not focus on these other factors, which he once referred to as a "host of petty disturbing influences" (quoted in Bennett, 1998, p. 104). But his conceptualization opens the door for thinking about variation and its causes in the way we do throughout the book.

15   Charles Darwin's work inspired many eugenecists. But Darwin himself was not an advocate of eugenics nor of "Social Darwinism," which justifies poverty because the poor are somehow "less fit." Indeed, the phrase "survival of the fittest" was not coined by Darwin, but by Herbert Spencer, a sociologist much influenced by Darwin but who developed a very unusual version of Darwin's ideas.

16   Looking back on the analyses that Galton and other eugenicists conducted, it is easy for anyone who has had an introduction to research methods to see the flaws of logic in their analysis. But we must also remember that these researchers were conducting work that was "state of the art" in methods at the time it was done. If we can see the flaws in their thinking, it is in large part because science is a social activity in which the insights of individuals pass on into common understanding.

17   Note that most statistical packages report an estimated population variance and standard deviation. That is, they divide the sum of squares for the variable by $N - 1$ in calculating the variance. We divide the sum of squares by N so that our variance for a batch of data is the average of the squared deviations from the mean. To convert from the estimated population variance calculated with the $N - 1$ formula to the variance based on N, multiply by $(N - 1)/N$. For 50 states the correction factor for the variance is 0.98.

Chapter 6

1   There is a popular genre of science fiction or fantasy writing, called "alternative history" that conducts thought experiments of this sort: "What would have happened if . . ." Some historians have also written in this genre.

2   The problem of causation has complexities beyond what is appropriate to discuss here. Philosophers have examined what a causal statement means, and recently there has been some debate about the proper use of causal statements to describe the results of non-experimental research. But here we will focus on conventional ideas of causation.

3   These are well reviewed in Campbell and Stanley (1963) and Cook and Campbell (1979). These texts also address threats to the validity of conclusions drawn from experimental data. Achen (1986) offers some cautionary notes about the limits of drawing causal inference from non-experimental data and Sobel (1995; 1996) offers strong criticism of standard practice. A careful approach to evaluating scientific information can be found in Stern and Kalof (1996).

4   Of course, good experiments can be very difficult to design. There is a craft involved here too.

5   It is always possible to think of counter-examples and hypothetical situations that contradict any causal assumption. The craft of good research involves making reasonable assumptions that are generally true, and being attentive to when the assumptions don't hold and the consequences of those situations. Confusing the philosophy and sociology of science with the practice of science is rather like confusing the philosophy and sociology of art with painting, dance, or music.

6   Unfortunately, different texts use somewhat different terms and definitions for concepts such as explanatory and extraneous variables. If you have taken, or take later, a course that uses different terms or different definitions, remember that the concepts are the same across all the disciplines but that the language can be a bit different from field to field, course to course and textbook to textbook.

7   Blalock (1985) reprints many of the classic papers on path and causal analysis.

8   You may notice that the sample sizes in this table are higher than the sample sizes listed in Table 6.2. The reason for this is that all cases had valid data for gender and type of residence, which means we know the gender and residence type of all those

who participated in the survey. However, not all survey participants answered the AIDS knowledge question. The sample size is reduced when we examine this variable since those participants who did not answer the question are excluded from analyses.

9   The language and logic of tabular analysis was a product of World War II, when sociologists and many other scientists were pressed into military service. These social scientists examined a number of issues critical for managing a large army of draftees and volunteers from across the country. A famous series of studies called "The American Solider" laid the foundations for many modern data analysis strategies. This was a sort of "Manhattan Project" for sociologists. See Stouffer (1962) for some fine examples of quantitative analysis by a master of the craft who was a key figure in the World War II studies.

10   While the argument for causal ordering for membership in the Confederacy is fairly strong, the argument that poverty is causally prior to homicide is less certain. But it does seem sufficiently plausible that it is a reasonable first approximation.

Chapter 7

1   Technically, chaos and complexity theory show that for many non-linear systems (those in which the equations contain squares of variables or other complications beyond linear terms) outcomes are tremendously sensitive to even trivial differences in initial conditions. The systems are predictable in principle, but are so complex and sensitive to minor changes that they are very hard to predict in detail. Chris Langton (1992; 1995) of the Santa Fe Institute (www.santafe.edu) has suggested that life, and other complex adaptive systems, sits on the edge between deterministic and fully random systems.

2   Isaacson (2007, p. 609) says the quote was "ob der liebe Gott würfelt." Niels Bohr, one of the creators of quantum theory, said he replied: "But still, it cannot be for us to tell God how he is to run the world."

3   The leading technical definition of random is parallel to this approach (Bennett, 1998). A truly random sequence of numbers is one that cannot be described by an algorithm less complex than the sequence itself – there are no regularities or predictability that allow a simpler description. An algorithm is just a procedure for describing the sequence. If the sequence is just a string of "1"s it is completely predictable and can be described as "Write 1 over and over." If the sequence was 01010101 . . . and so on, then the description would be "Write 01 over and over," a slightly longer description. The more complex the algorithm needed to describe the sequence, the more random the sequence.

4   Many of the key papers on this subject are collected in Kahneman, Slovic, and Tversky (1982). Kahneman won the Noble Prize in Economics in 1992 for his work on the problems of human decision making under uncertainty.

5   Some of the key studies making this theoretical argument are Gigerenzer (1998) and Cosmides and Tooby (1996). The implications for decision making are discussed in Gigerenzer and Hoffrage (1995) and Hoffrage et al. (2000). The example is taken from Kurzenhauser and Hoffrage (nd).

6   The human tendency to find pattern in random data is very strong. When we do these kinds of simulations, people looking at the data can usually find some pattern – per-

haps the chances of tails is usually above 0.5 once there are 6 or more heads in a row. That is the case for this experiment, but it is pure chance. When we run the experiment again, we may see fewer tails after the long strings of heads. Statistics is helpful precisely because we try to see a pattern where there is none, and over-generalize from very limited evidence.

7   Generating independent random events on a computer is a difficult task but an increasingly important one as we find more and more uses for random numbers. Dietz inherited from one of his mentors a book entitled *One Million Random Digits and 100,000 Standard Deviations*. In the days before desktop computers, researchers needing random numbers often consulted this book. Even now, building the random number generator for a computer is a difficult task (Bennett, 1998, ch. 8).

8   Blackjack is an exception. Card counters try to watch which cards are shown as they are dealt from a blackjack deck. By knowing what has been dealt, the card counter can mentally adjust the probability of what is not showing and bet accordingly. They are taking account of the fact that the probability of a particular card coming up shifts as other cards are dealt and there are fewer cards remaining in the deck.

9   We know there were dice games with three dice at least since the Roman Empire. Three dice leads to $6 \times 6 \times 6 = 216$ outcomes but only 56 sums, some of which will be more probable than others, as we have seen in the two dice example. Yet it apparently took a millennium before it became clear to most gamblers that some of the sums were more likely than others. The first known description of those outcomes dates to the middle of the 13th century (Bennett, 1998, ch. 4).

10 Complex role-playing games have created a demand for many-sided dice. This has been nice for statisticians who can buy a dice with as many sides as they might need, up to a dozen or so. Of course, we have to worry about whether or not the dice are manufactured well enough to be equiprobability. Mostly we use these dice for the classroom and rely on random number generators in the computer or a random number table in designing experiments.

11 Designing good survey questions that are easy to understand, not leading and that get at the subject being studied is not an easy task. Indeed in many ways the hardest part of survey research is developing good questions.

12 There are more sophisticated methods that allow us to see if it is reasonable to assume that two (or more) questions are measuring the same thing and that allow for one (or several) of the questions to have less measurement error than the other(s). However, they are beyond the scope of this text.

13 Averaging across many individual measurements only works if the average of all the measurement errors is zero. Suppose someone in an interview doesn't want the interviewer to think they are too concerned about the welfare of animals. They might answer with less pro-animal responses than might otherwise be the case. This kind of problem occurs especially when there are responses that are "socially desirable." When questions systematically over- or underestimate what they are trying to measure, we call the problem "bias." Statistical theory can handle it too, but it's a bit more complex than random error that averages out to zero.

14 Technically, it is $P(x) = \dfrac{1}{\sigma\sqrt{2\pi}} e^{-(x-\mu)^2/(2\sigma^2)}$ where $P(x)$ is the probability of a value equal to x, σ is the standard deviation of the distribution, μ is the mean of the distribution, e is the base of the natural logarithms and π is the constant from geometry.

15   Western (1996) and Berk, Western, and Weiss (1995) use Bayesian methods to address the problem of using statistical tools when sampling error, measurement error, and random assignment do not seem to describe the data. This is an alternative to the hypothetical populations approach we describe in thinking about error in models where we have data from all units, such as all nations.

Chapter 8

1   To be precise, the General Social Survey sample has some restrictions, so we are generalizing to a population slightly different from that of the whole US population. For example, interviews were conducted in English, so those who don't speak English or were unwilling to be interviewed in English will not be represented in the sample. If we were studying, for example, attitudes towards bilingual education or if we believed that recent immigrants have different levels of education than others, these restrictions on the sample would be a more serious limitation to our study conclusions. We should also note that the statistical procedures we use in this and subsequent chapters assume a simple random sample, but the GSS is a cluster sample so all the formulas we use will give answers that are a little off. Again, for our purposes the differences are small and don't matter. But in reading or designing research, it is important to keep in mind who was sampled, how the sampling was done and how those things might influence the results. Sometimes such details matter a great deal, though often they don't.
2   To be perfectly proper, the expected value of the sampling distribution is defined in terms of probabilities for a random variable, but the idea of the mean of the sampling distribution will be adequate for this discussion.
3   In the late nineteenth and early twentieth century it was thought by many scientists that physical features could differentiate criminals from non-criminals. This is part of a long history in science of conflating the biological with the social, cultural, and psychological.
4   Sociological applications of the bootstrap are described in Dietz, Frey, and Kalof (1987) and in Dietz, Kalof, and Frey (1991). Efron and Tibshirani (1993) is the standard text on the technique.

Chapter 9

1   There is a subfield of statistics (beyond the scope of this book) called decision theory that suggests how sure we should be by calculating the costs of being wrong and the costs of collecting data.

Chapter 10

1   The standards are a bit more complicated than this, trying to allow some flexibility for small water supply systems and to take extra steps to protect school and daycare centers. See http://www.epa.gov/safewater/lead/index.html
2   In addition to fundamental contributions to statistics, Fisher also founded much of quantitative genetics. In some sense he put the theories of Darwin and Mendel on a solid

mathematical basis, and posed problems that are still worthy of investigation today. As we have mentioned, Fisher was a sharp rival of Karl Pearson, whose work we will encounter several times in this book.

3   Egon Pearson was the son of Karl Pearson. Karl Pearson closely followed the work of Galton, who was a cousin of Charles Darwin. Science can be a small world.

4   R.A. Fisher suggested the 0.05 level (1 chance in 20 of a Type I error) and it has stuck. You may have noticed that we use α in testing hypotheses and $1 - \alpha$ for confidence intervals. We will explore the connection between confidence intervals and hypothesis tests shortly.

5   Perhaps we are suspicious of a con game in which we are being "set up" by a string of good luck to be cheated later.

6   We are assuming here a causal model in which the death penalty influences the homicide rate. But one might argue the other way – that the homicide rate generates social concern that leads to the death penalty. Ironically, implicit in that causal argument is an assumption that the public accepts the hypothesis that the death penalty deters homicide.

7   Finifter (1972) provides a classic overview of this approach.

Chapter 11

1   When we have data from a single point in time, age groups are the same as cohorts. Cohorts are groups of people who are the same age at the same time and thus go through important historical events at the same age. In the US, we commonly refer to the "baby boom" cohort (usually defined as those born between 1946 and 1960) or to Gen X (usually defined as those born between 1961 and 1981). Cohorts are a very useful concept in the social sciences but with data from a single point in time, a birth cohort (defined by the year in which people were born) can't be differentiated from age, so we will just refer to our groups as age groups even though they also represent birth cohorts.

2   A bit of algebra is involved in going from Equation 11.7 to Equation 11.8. If you like to do algebra, first multiply out the left hand side

$$(Y - \bar{Y})(Y - \bar{Y}) = Y^2 - Y\bar{Y} - \bar{Y}Y + \bar{Y}^2 = Y^2 - 2Y\bar{Y} + \bar{Y}^2$$

Then with a good bit more work you can multiply out the right hand side as well:

$$((Y - \hat{Y}) + (\hat{Y} - \bar{Y}))^2 = (Y - \hat{Y} + \hat{Y} - \bar{Y})(Y - \hat{Y} + \hat{Y} - \bar{Y}) =$$
$$YY - Y\hat{Y} + Y\hat{Y} - Y\bar{Y}$$
$$-\hat{Y}Y + \hat{Y}\hat{Y} - \hat{Y}\hat{Y} + \hat{Y}\bar{Y}$$
$$+\hat{Y}Y - \hat{Y}\hat{Y} + \hat{Y}\hat{Y} - \hat{Y}\bar{Y}$$
$$-\bar{Y}Y + \bar{Y}\hat{Y} - \bar{Y}\hat{Y} + \bar{Y}\bar{Y}$$

Lots of terms in this long expression cancel and get us to the same thing as when we squared the left hand side.

3   Some texts prefer the spelling homoskedastic and heteroskedastic.

4   The ISSP originally coded "separated but married" separately. For this analysis, this small subgroup – 2.6 percent of the sample – was coded as married.

Chapter 12

1   It won't surprise students of the social sciences to learn that in fact the idea was the average man, with women considered a deviant group.
2   For a thoughtful exposition of the modern view, see Stephen Jay Gould's fascinating book, *Full House*.
3   We could state Pearson's results as a sampling theorem just as we did Gosset's Theorem and the Law of Large Numbers. But the presentation would be too complex to aid your understanding.
4   It is possible to have more complex hypotheses for a frequency table and sometimes a different number of degrees of freedom.
5   There are more complex methods that can deal with tables with more than two dimensions (that is, more than two variables). The logic is the same as chi-square – state a hypothesis about relationships between variables and see how well the data fit it.
6   If you get confused about rows and columns remember that columns are vertical, just as they are in buildings.
7   Some readers may ask why, if we split around the median, we don't have 50 percent of people expressing high animal concern and 50 percent low. There are two reasons. First, in the table we are using both age and the animal concern variable. There are 29,486 people for whom we have data on animal concern but 29,193 for whom we have data on both animal concern and age. Second, unless the median falls between categories of a variable, we have to include the value onto which the median fell into either the high or low group, thus throwing the split off by a small amount. But splitting at the median is just a convenient way to turn a continuous variable into a dichotomous one. It doesn't matter if the split is exactly even as long as we get enough people in the marginals of the table so that each expected cell is greater than 5.

Chapter 13

1   This history is based on Stanton (2001).
2   In the start of our discussions, we will use capital letters for both variables (e.g. X2, X3) and for coefficients (e.g. A, B). However, when we move to inference we will need to make a distinction between the value of coefficients in the sample and the value in the population from which the sample was drawn. Then we will often use capital letters for the population values (e.g. A, B) and lower case letters for the sample values (e.g. a, b). We will always use capital N to refer to the sample size. As we move through the chapter we will begin to leave out the * used for multiplication (e.g. B*X) to make the equations less cluttered (e.g. BX).
3   You may remember from geometry that B is the slope of the line linking X to Y. If you've had calculus you will recognize B as the first derivative of Y with respect to X. When we get to multiple regression, the regression coefficients that link each independent variable to the dependent variable are the partial derivatives of Y with respect to each independent variable, controlling for the other independent variables.

4   Note that the vertical distance is not the same as the Euclidean distance, which is the shortest path between the line and the point. Note also that the vertical distance is not the same as the horizontal distance. We mention these two distinctions because the distance between the point and the line depends on which variable is plotted on the horizontal axis and which on the vertical axis.

5   We use several decimal places in doing the regression calculations to avoid errors that come from rounding. Even with four decimal places our answers will likely differ in the fourth decimal place from those we would get using statistical software on computers. Computer software routinely carries 16 digits in the calculations, which for the number we are using would be 14 decimal places.

6   It might seem that if you switched dependent and independent variables the new regression coefficient would be (1/B) the inverse of the other coefficient. This would be true in geometry where we have only the straight line. But in statistics we are dealing with data. Recall that we are taking the vertical difference between the predicted value and the actual value as our measure of error. This means that if we calculate the regression to get B, then switch the variables the new regression coefficient won't be exactly 1/B unless all the points fall exactly on the line – something we would never see with real data.

7   While fixed X is the simplest way to think about the issue of inference in regression, there are other approaches. See Press (1972, pp. 187–98) for a discussion of the differences at a more sophisticated level.

8   The most common of these methods are logit and probit analysis. They are described in detail in Long (1997), Menard (2002), and Pampel (2000) though the treatment is rather technical.

9   We also ran a logit analysis, which is the preferred method for analyzing the effects of a continuous independent variable on a categorical dependent variable. We found that the p values for the effects of education was <0.001, which is identical to the results from the OLS regression. It is often the case that with large samples OLS regression performs very well even when there are better methods to deploy.

Chapter 14

1   Some have argued that using crosstabulations may prove a better approach to causation than regression. There is some merit to this argument, but the use of tables has a great weakness. It is common to think that several independent variables may influence the dependent variable, and some or all of those variables have many categories. Suppose we want to examine the effects of 4 independent variables on a dependent variable, and in the simplest situation every variable has just two categories. Then we need a $2 \times 2 \times 2 \times 2 \times 2$ table, which will have 32 cells. If we want to have at least five cases per cell (in the hope that this will give us at least five cases in every expected cell in a chi-square or related test), we need a sample size of at least 160. That's not too bad. But it works *only* if there are the same number of cases in every cell. If that's the case, then the variables are uncorrelated with one another, and the data aren't very interesting. Suppose the smallest cell has half the cases of the average cell. Then we would need something like 320 cases. Very quickly, contingency tables become too large to be useful with most data sets.

2   The term *multivariate regression* is used for analyses with multiple dependent as well as independent variables. Not all texts are consistent on this point, some using the term multivariate regression for multiple independent as well as multiple dependent variables. Multiple regression is one of a family of multivariate statistical techniques, as is multivariate regression. Other members of the family include factor analysis, discriminant analysis and canonical correlation analysis. All of these methods can be subsumed under the term structural equation models (which also includes path analysis or simultaneous equations). Structural equation models allow for complex combination of multiple independent and dependent variables, including both observed and unobserved variables.

3   Note that there are nearly as many notations as authors, so in reading outside this text you must keep track of what is meant by the notation, not simply memorize the symbols.

4   Recent discussions of the use of Venn diagrams can be found in Ip (2001) and Kennedy (2002). The first use of Venn diagrams for regression was in the textbook by Cohen and Cohen (1975).

5   The part correlations are not the same as the partial correlations found in the literature years ago that are much less popular today. A **partial correlation** is the amount of variance an independent variable accounts for out of the variance left over to be explained in the dependent variable. The **part R** is essentially the correlation between an independent and dependent variable, controlling for other variables (the amount of total variance in y accounted for by x). The **part R^2** is the unique explained variance of an independent variable on a dependent variable. The partial is more likely to overestimate explained variance and thus is not as often used.

6   The Government Effectiveness Index, like the Voice and Accountability index, takes many indicators from multiple sources into account when determining a nation's score. Kaufmann and colleagues (Kaufmann et al., 2002) said the index "combine(s) perceptions of the quality of public service provision, the quality of the bureaucracy, the competence of civil servants, the independence of the civil service from political pressures, and the credibility of the government's commitment to policies into a single grouping" (p. 5). Many of these indicators are complicated to measure, so scores are based on multiple subjective assessments from various organizations and institutions, like the World Bank.

7   In the days of computation by hand and even with early computer software for regression, variables were in fact added "one step at a time" in a process called stepwise regression. But now we simply run a regression with the variables that we are interested in, at least up to the point where the number of variables is so large compared to the number of observations that we can't believe the results.

INDEX

NB Page locators in *italics* refer to a figure; page locators in **bold** to a table.

Printed and bound by CPI Group (UK) Ltd, Croydon, CR0 4YY